Lecture Notes in Computer Science   **10491**

Commenced Publication in 1973
Founding and Former Series Editors:
Gerhard Goos, Juris Hartmanis, and Jan van Leeuwen

More information about this series at http://www.springer.com/series/7407

Serge Gaspers · Toby Walsh (Eds.)

Theory and Applications of Satisfiability Testing – SAT 2017

20th International Conference
Melbourne, VIC, Australia, August 28 – September 1, 2017
Proceedings

Editors
Serge Gaspers (iD)
UNSW Sydney and Data61, CSIRO
Sydney, NSW
Australia

Toby Walsh (iD)
UNSW Sydney and Data61, CSIRO
Sydney, NSW
Australia

ISSN 0302-9743   ISSN 1611-3349 (electronic)
Lecture Notes in Computer Science
ISBN 978-3-319-66262-6   ISBN 978-3-319-66263-3 (eBook)
DOI 10.1007/978-3-319-66263-3

Library of Congress Control Number: 2017949510

LNCS Sublibrary: SL1 – Theoretical Computer Science and General Issues

Printed on acid-free paper

This Springer imprint is published by Springer Nature
The registered company is Springer International Publishing AG
The registered company address is: Gewerbestrasse 11, 6330 Cham, Switzerland

Preface

This volume contains the papers presented at SAT 2017, the 20th International Conference on Theory and Applications of Satisfiability Testing, held from 28 August until 1 September 2017 in Melbourne, Australia. This 20th edition of SAT was co-located with CP 2017 (23rd International Conference on Principles and Practice of Constraint Programming) and ICLP 2017 (33rd International Conference on Logic Programming), featuring joint workshops, tutorials, invited talks, social events, and a doctoral consortium. It was held at the Melbourne Convention and Exhibition Centre, starting with a workshop day on 28 August 2017.

The International Conference on Theory and Applications of Satisfiability Testing (SAT) is the premier annual meeting for researchers focusing on the theory and applications of the satisfiability problem, broadly construed. In addition to plain propositional satisfiability, it also includes Boolean optimization (such as MaxSAT and Pseudo-Boolean (PB) constraints), Quantified Boolean Formulas (QBF), Satisfiability Modulo Theories (SMT), and Constraint Programming (CP) for problems with clear connections to Boolean-level reasoning.

Many hard combinatorial problems can be tackled using SAT-based techniques including problems that arise in Formal Verification, Artificial Intelligence, Operations Research, Computational Biology, Cryptography, Data Mining, Machine Learning, Mathematics, etc. Indeed, the theoretical and practical advances in SAT research over the past 25 years have contributed to making SAT technology an indispensable tool in a variety of domains.

SAT 2017 aimed to further advance the field by welcoming original theoretical and practical contributions in these areas with a clear connection to satisfiability. Specifically, SAT 2017 welcomed scientific contributions addressing different aspects of SAT interpreted in a broad sense, including (but not restricted to) theoretical advances (such as algorithms, proof complexity, parameterized complexity, and other complexity issues), practical search algorithms, knowledge compilation, implementation-level details of SAT solvers and SAT-based systems, problem encodings and reformulations, applications (including both novel application domains and improvements to existing approaches), as well as case studies and reports on findings based on rigorous experimentation.

A total of 65 papers were submitted to SAT, including 47 long papers, 13 short papers, and 5 tool papers. The Program Committee chairs decided on a summary reject for one short paper. Each of the remaining 64 submissions was reviewed by at least 3 Program Committee members, sometimes with the help of expert external reviewers. The committee decided to accept 30 papers, consisting of 22 long papers, 5 short papers, and 3 tool papers. This included two conditional accepts that were accepted after a revision. There was no reclassification of papers.

The Program Committee decided to single out the following three papers:

- Joshua Blinkhorn and Olaf Beyersdorff, Shortening QBF Proofs with Dependency Schemes
  for the Best Paper Award,
- Miguel Terra-Neves, Inês Lynce, and Vasco Manquinho, Introducing Pareto Minimal Correction Subsets
  for the Best Student Paper Award, and
- Jia Liang, Vijay Ganesh, Krzysztof Czarnecki, Pascal Poupart, and Hari Govind V K, An Empirical Study of Branching Heuristics through the Lens of Global Learning Rate for a Best Student Paper Honourable Mention.

In addition to the contributed talks, the program featured five invited talks, which were joint invited talks with CP 2017 and ICLP 2017:

- The Role of SAT, CP, and Logic Programming in Computational Biology
  by Agostino Dovier (University of Udine, Italy),
- The Best of Both Worlds: Machine Learning Meets Logical Reasoning
  by Holger H. Hoos (Leiden Institute of Advanced Computer Science, Netherlands and University of British Columbia, Canada),
- Recent Advances in Maximum Satisfiability
  by Nina Narodytska (VMware Research, USA),
- Back to the Future - Parallelism and Logic Programming
  by Enrico Pontelli (New Mexico State University, USA), and
- Constraints and the 4th Industrial Revolution
  by Mark Wallace (Monash University and Opturion, Australia).

The joint program also featured tutorials organized to enable participants to engage in the program of the other co-located conferences:

- Introduction to Constraint Programming - If You Already Know SAT or Logic Programming
  by Guido Tack (Monash University, Australia),
- An Introduction to Satisfiability
  by Armin Biere (Johannes Kepler University Linz, Austria),
- Introduction to Machine Learning and Data Science
  by Tias Guns (Vrije Universiteit Brussel, Belgium), and
- Mixed Integer Nonlinear Programming: An Introduction
  by Pietro Belotti (FICO, UK).

All 8 workshops were also affiliated with SAT 2017, CP 2017, and ICLP 2017:

- Workshop on Parallel Methods for Constraint Solving (PaSeO 2017)
  organized by Philippe Codognet, Salvador Abreu, and Daniel Diaz,
- Pragmatics of Constraint Reasoning (PoCR 2017)
  organized by Daniel Le Berre and Pierre Schaus,
- Workshop on Answer Set Programming and Its Applications (IWASP/ASPIA 2017)
  organized by Kewen Wang and Yan Zhang,

- Workshop on Constraint Solvers in Testing, Verification and Analysis (CSTVA 2017)
  organized by Zakaria Chihani, Sébastien Bardin, Nikolai Kosmatov, and Arnaud Gotlieb,
- Workshop on Logic and Search (LaSh 2017)
  organized by David Mitchell,
- Progress Towards the Holy Grail (PTHG 2017)
  organized by Eugene Freuder, Christian Bessiere, Narendra Jussien, Lars Kotthoff, and Mark Wallace,
- International Workshop on Constraint Modeling and Reformulation (ModRef 2017)
  organized by Özgür Akgün, and
- Colloquium on Implementation of Constraint Logic Programming Systems (CICLOPS 2017)
  organized by Jose F. Morales and Nataliia Stulova.

As in previous years, the results of several competitive events were announced at SAT 2017:

- the 2017 MaxSAT Evaluation (MSE 2017)
  organized by Carlos Ansótegui, Fahiem Bacchus, Matti Järvisalo, and Ruben Martins,
- the 2017 Competitive Evaluation of QBF Solvers (QBFEVAL'17)
  organized by Luca Pulina and Martina Seidl, and
- the 2017 SAT Competition
  organized by Marijn Heule, Matti Järvisalo, and Tomas Balyo.

SAT 2017 also hosted a Doctoral Program, jointly organized by the three co-located conferences and chaired by Christopher Mears and Neda Saeedloei.

We thank everyone who contributed to making SAT 2017 a success. We are indebted to the Program Committee members and the external reviewers, who dedicated their time to review and evaluate the submissions to the conference. We thank the authors of all submitted papers for their contributions, the SAT association for their guidance and support in organizing the conference, the EasyChair conference management system for facilitating the submission and selection of papers, as well as the assembly of these proceedings. We also thank the SAT 2017 organizing committee for handling the practical aspects of the organization of the conference, and the workshop chair, Stefan Rümmele, for organizing the workshop program in collaboration with the workshop chairs of CP 2017 and ICLP 2017. We also thank IJCAI 2017, held the preceding week at the same location, and anyone else who helped promote the conference. Finally, we thank CP 2017 and ICLP 2017 for a smooth collaboration in the co-organization of the three conferences.

The SAT Association greatly helped with financial support for students attending the conference and Springer sponsored the Best Paper Awards for SAT 2017. SAT, CP, and ICLP 2017 were also sponsored by the Association for Constraint Programming, the Association for Logic Programming, CSIRO Data61, Monash University, the

VIII   Preface

University of Melbourne, COSYTEC, Satalia, Google, CompSustNet, Cosling, the
Artificial Intelligence journal, and EurAI. Thank you!

July 2017   Serge Gaspers
                                                             Toby Walsh

Organization

Program Committee

Gilles Audemard	CRIL, Artois University, France
Fahiem Bacchus	University of Toronto, Canada
Valeriy Balabanov	Mentor Graphics, USA
Bernd Becker	Albert-Ludwigs-University Freiburg, Germany
Olaf Beyersdorff	University of Leeds, UK
Armin Biere	Johannes Kepler University Linz, Austria
Nadia Creignou	Aix-Marseille Université, France
Uwe Egly	TU Wien, Austria
Serge Gaspers	UNSW Sydney and Data61, CSIRO, Australia
Carla Gomes	Cornell University, USA
Marijn Heule	The University of Texas at Austin, USA
Alexey Ignatiev	University of Lisbon, Portugal
Mikolas Janota	University of Lisbon, Portugal
Michael Kaufmann	Wilhelm-Schickard-Institute for Informatics, University of Tuebingen, Germany
Chu-Min Li	Université de Picardie Jules Verne, France
Florian Lonsing	TU Wien, Austria
Ines Lynce	University of Lisbon, Portugal
Vasco Manquinho	University of Lisbon, Portugal
Carlos Mencía	University of Oviedo, Spain
Nina Narodytska	VMware Research, USA
Alessandro Previti	University of Lisbon, Portugal
Enric Rodríguez Carbonell	Technical University of Catalonia, Spain
Vadim Ryvchin	Technion, Israel
Bart Selman	Cornell University, USA
Laurent Simon	Labri, Bordeaux Institute of Technology, France
Carsten Sinz	Karlsruhe Institute of Technology (KIT), Germany
Friedrich Slivovsky	TU Wien, Austria
Ofer Strichman	Technion, Israel
Stefan Szeider	TU Wien, Austria
Allen Van Gelder	University of California, Santa Cruz, USA
Toby Walsh	Data61, CSIRO and UNSW Sydney, Australia

Additional Reviewers

Angelini, Patrizio
Argelich, Josep
Bayless, Sam
Bekos, Michael
Bjorck, Johan
Björck, Johan
Blinkhorn, Joshua
Burchard, Jan
Capelli, Florent
Chew, Leroy
Clymo, Judith
de Haan, Ronald
De Rougemont, Michel
Galesi, Nicola
Ganian, Robert
Gurfinkel, Arie
Guthmann, Ofer
Hinde, Luke
Ivrii, Alexander
Katz, Emilia
Kiesl, Benjamin
Klieber, William
Marin, Paolo

Martin, Barnaby
Meel, Kuldeep S.
Meiners, Chad
Morgado, Antonio
Nadel, Alexander
Nieuwenhuis, Robert
Oetsch, Johannes
Oliveras, Albert
Ordyniak, Sebastian
Ravelomanana, Vlady
Reimer, Sven
Scheibler, Karsten
Schmidt, Johannes
Schutt, Andreas
Sigley, Sarah
Strozecki, Yann
Terra-Neves, Miguel
Truchet, Charlotte
Wimmer, Ralf
Winterer, Leonore
Wu, Xiaojian
Xue, Yexiang

Contents

Parallel SAT Solving

Quantified Boolean Formulas

Satisfiability Modulo Theories

SAT Encodings

Tool Papers

Algorithms, Complexity, and Lower Bounds

Probabilistic Model Counting with Short XORs

Dimitris Achlioptas[1] and Panos Theodoropoulos[2]([✉])

[1] Department of Computer Science, University of California Santa Cruz,
Santa Cruz, USA
optas@cs.ucsc.edu
[2] Department of Informatics and Telecommunications,
University of Athens, Athens, Greece
ptheodor@di.uoa.gr

Abstract. The idea of counting the number of satisfying truth assignments (models) of a formula by adding random parity constraints can be traced back to the seminal work of Valiant and Vazirani, showing that NP is as easy as detecting unique solutions. While theoretically sound, the random parity constraints in that construction have the following drawback: each constraint, on average, involves half of all variables. As a result, the branching factor associated with searching for models that also satisfy the parity constraints quickly gets out of hand. In this work we prove that one can work with much shorter parity constraints and still get rigorous mathematical guarantees, especially when the number of models is large so that many constraints need to be added. Our work is based on the realization that the essential feature for random systems of parity constraints to be useful in probabilistic model counting is that the geometry of their set of solutions resembles an error-correcting code.

1 Introduction

Imagine a blind speaker entering an amphitheater, wishing to estimate the number of people present. She starts by asking "Is anyone here?" and hears several voices saying "Yes." She then says "Flip a coin inside your head; if it comes up heads, please never answer again." She then asks again "Is anyone here?" and roughly half of the people present say "Yes." She then asks them to flip another coin, and so on. When silence occurs after i rounds, she estimates that approximately 2^i people are present.

Given a CNF formula F with n variables we would like to approximate the size of its set of satisfying assignments (models), $S = S(F)$, using a similar approach. Following the path pioneered by [13, 16, 18], in order to check if $|S(F)| \approx 2^i$ we form a random set $R_i \subseteq \{0, 1\}^n$ such that $\Pr[\sigma \in R_i] = 2^{-i}$ for every $\sigma \in \{0, 1\}^n$ and determine if $|S(F) \cap R_i| \approx 1$. The key point is that we will represent R_i *implicitly* as the set of solutions to a system of i random

D. Achlioptas—Research supported by NSF grant CCF-1514128 and grants from Adobe and Yahoo!

P. Theodoropoulos—Research supported by the Greek State Scholarships Foundation (IKY).

S. Gaspers and T. Walsh (Eds.): SAT 2017, LNCS 10491, pp. 3–19, 2017.
DOI: 10.1007/978-3-319-66263-3_1

linear equations modulo 2 (parity constraints). Thus, to determine $|S(F) \cap R_i|$ we simply add the parity constraints to F and invoke CryptoMiniSAT [15] to determine the number of solutions of the combined set of constraints (clauses and parity equations).

There has already been a long line of practical work along the lines described above for model counting, including [6–10] and [1–5,11]. In most of these works, each parity constraint includes each variable independently and with probability $1/2$, so that each parity constraint includes, on average, $n/2$ variables. While such long parity constraints have the benefit that membership in their set of solutions enjoys pairwise independence, making the probabilistic analysis very simple, the length of the constraints can be a severe limitation. This fact was recognized at least as early as [8], and efforts have been made in some of the aforementioned works to remedy it by considering parity equations where each constraint still includes each variable independently, but with probability $p < 1/2$. While such sparsity helps with computing $|S(F) \cap R|$, the statistical properties of the resulting random sets, in particular the variance of $|S(F) \cap R|$, deteriorate rapidly as p decreases [7].

In this work we make two contributions. First, we show that bounding $|S(F)|$ from below and from above should be thought of as two separate problems, the former being much easier. Secondly, we propose the use of random systems of parity equations corresponding to the parity-check matrices of low density binary *error-correcting codes*. These matrices are also sparse, but their entries are *not indepedent*, causing them to have statistical properties dramatically better than those of similarly sparse i.i.d. matrices. As a result, they can be used to derive *upper* bounds on $|S(F)|$, especially when $\log_2 |S(F)| = \Omega(n)$.

2   Lower Bounds are Easy

For a distribution $\mathcal{D}$, let $R \sim \mathcal{D}$ denote that random variable R is distributed according to $\mathcal{D}$.

Definition 1. *Let $\mathcal{D}$ be a distribution on subsets of a set U and let $R \sim \mathcal{D}$. We say that $\mathcal{D}$ is i-uniform if $\Pr[\sigma \in R] = 2^{-i}$ for every $\sigma \in U$.*

When $U = \{0,1\}^n$, some examples of i-uniform distributions are:

(i) R contains each $\sigma \in \{0,1\}^n$ independently with probability 2^{-i}.
(ii) R is a uniformly random subcube of dimension $n - i$.
(iii) $R = \{\sigma : A\sigma = b\}$, where $A \in \{0,1\}^{i \times n}$ is *arbitrary* and $b \in \{0,1\}^i$ is *uniformly* random.

We can use *any* i-uniform distribution, $\mathcal{D}_i$, to compute a *rigorous lower bound* on the number of satisfying truth assignments of a formula (models), as follows. We note that CryptoMiniSAT has an optional cutoff parameter $s \geq 1$ such that if s solutions are found, the search stops without completing. We use this capacity in line 6 in Algorithm 1 and lines 6, 23 of Algorithm 3.

The only tool we use to analyze Algorithm 1 is Hoeffding's Inequality.

Algorithm 1. Decides if $|S| \geq 2^i$ with 1-sided error probability $\theta > 0$

1: $t \leftarrow \lceil 8\ln(1/\theta) \rceil$ ▷ θ is the desired error probability bound
2: $Z \leftarrow 0$
3: $j \leftarrow 0$
4: **while** $j < t$ and $Z < 2t$ **do**   ▷ The condition $Z < 2t$ is an optimization
5:   Sample $R_j \sim \mathcal{D}_i$ ▷ Where $\mathcal{D}_i$ is any i-uniform distribution
6:   $Y_j \leftarrow \min\{4, |S(F) \cap R_j|\}$ ▷ Search for up to 4 elements of $S(F) \cap R_j$
7:   $Z \leftarrow Z + Y_j$
8:   $j \leftarrow j + 1$
9: **if** $Z/t \geq 2$ **then**
10:   **return** "Yes, I believe that $|S(F)| \geq 2^i$"
11: **else**
12:   **return** "Don't know"

Lemma 1 (Hoeffding's Inequality). *If* $Z = Y_1 + \cdots + Y_t$, *where* $0 \leq Y_i \leq b$ *are independent random variables, then for any* $w \geq 0$,

$$\Pr[Z/t \geq \mathbb{E}Z/t + w] \leq e^{-2t(w/b)^2} \quad and \quad \Pr[Z/t \leq \mathbb{E}Z/t - w] \leq e^{-2t(w/b)^2} .$$

Theorem 1. *The output of Algorithm 1 is correct with probability at least* $1 - \theta$.

Proof. Let $S = S(F)$. For the algorithm's answer to be incorrect it must be that $|S| < 2^i$ and $Z/t \geq 2$. If $|S| < 2^i$, then $\mathbb{E}Y_j \leq |S|2^{-i} < 1$, implying $\mathbb{E}Z/t < 1$. Since Z is the sum of t i.i.d. random variables $0 \leq Y_j \leq 4$, Hoeffding's inequality implies that $\Pr[Z/t \geq 2] \leq \Pr[Z/t \geq \mathbb{E}Z/t + 1] \leq \exp(-t/8)$.

Notably, Theorem 1 does not address the *efficacy* of Algorithm 1, i.e., the probability of "Yes" when $|S| \geq 2^i$. As we will see, bounding this probability from below requires much more than mere i-uniformity.

2.1 Searching for a Lower Bound

We can derive a lower bound for $\log_2 |S(F)|$ by invoking Algorithm 1 with $i = 1, 2, 3, \ldots, n$ sequentially. Let ℓ be the greatest integer for which the algorithm returns "Yes" (if any). By Theorem 1, $\log_2 |S(F)| \geq \ell$ with probability at least $1 - \theta n$, as the probability that at least one "Yes" answer is wrong is at most θn.

There is no reason, though, to increase i sequentially. We can be more aggressive and invoke Algorithm 1 with $i = 1, 2, 4, 8, \ldots$ until we encounter our first "Don't know", say at $i = 2^u$. At that point we can perform binary search in $\{2^{u-1}, \ldots, 2^u - 1\}$, treating every "Don't know" answer as a (conservative) imperative to reduce the interval's upper bound to the midpoint and every "Yes" answer as an allowance to increase the interval's lower bound to the midpoint. We call this scheme "doubling-binary search". In fact, even this scheme can be accelerated by running Algorithm 1 with the full number of iterations only for values of i for which we have good evidence of being lower bounds for $\log_2 |S(F)|$.

Let $A1(i, z)$ denote the output of Algorithm 1 if line 1 is replaced by $t \leftarrow z$.

Algorithm 2. Returns i such that $|S| \geq 2^i$ with probability at least $1 - \theta$

1: $j \leftarrow 0$ ▷ Doubling search until first "Don't Know"
2: **while** $\{A1(2^j, 1) =$ "Yes"$\}$ **do**
3:      $j \leftarrow j + 1$
4: **if** $j = 0$ **then return** 0
5:
6: $h \leftarrow 2^j - 1$ ▷ First "Don't Know" occurred at $h + 1$
7: $i \leftarrow 2^{j-1}$ ▷ Last "Yes" occurred at i
8:
9: **while** $i < h$ **do**   ▷ Binary search for "Yes" in $[i, h]$
10:      $m \leftarrow i + \lceil (h - i)/2 \rceil$ ▷ where i holds the greatest seen "Yes" and
11:      **if** $A1(m, 1) =$ "Yes" **then**   ▷ $h + 1$ holds the smallest seen "Don't know"
12:          $i \leftarrow m$
13:      **else**
14:          $h \leftarrow m - 1$
15:          ▷ i holds the greatest seen "Yes"
16: $j \leftarrow 1$
17: **repeat**   ▷ Doubling search backwards starting
18:      $i \leftarrow i - 2^j$ ▷ from $i - 2$ for a lower bound that
19:      $j \leftarrow j + 1$ ▷ holds with probability $1 - \theta/\lceil \log_2 n \rceil$
20: **until** $i \leq 0$ **or** $A1(i, \lceil 8 \ln(\lceil \log_2 n \rceil/\theta) \rceil) =$ "Yes"
21: **return** $\max\{0, i\}$

Theorem 2. *If $S \neq \emptyset$, the probability that the output of Algorithm 2 exceeds $\log_2 |S|$ is at most θ.*

Proof. For the answer to be wrong it must be that some invocation of Algorithm 1 in line 20 with $i > \log_2 |S|$ returned "Yes". Since Algorithm 2 invokes Algorithm 1 in line 20 at most $\lceil \log_2 n \rceil$ times, and in each such invocation we set the failure probability to $\theta/\lceil \log_2 n \rceil$, the claim follows.

2.2 Choice of Constants

We can make Algorithm 1 more likely to return "Yes" instead of "Don't know" by increasing the number 4 in line 6 and/or decreasing the number 2 in lines 4, 9. Each such change, though, increases the number of iterations, t, needed to achieve the same probability of an erroneous "Yes". The numbers 4, 2 appear to be a good balance in practice between the algorithm being fast and being useful.

2.3 Dealing with Timeouts

Line 6 of Algorithm 1 requires determining $\min\{4, |S(F) \cap R_j|\}$. Timeouts may prevent this from happening, since in the allotted time the search may only find $s < 4$ elements of $S(F) \cap R_j$ but not conclude that no other such elements exist. Nevertheless, if Y_j is always set to a number *no greater* than $\min\{4, |S(F) \cap R_j|\}$, then both Theorem 1 and its proof remain valid. So, for example, whenever timeout occurs, we can set Y_j to the number $s < 4$ of elements of $S(F) \cap R_j$ found so far. Naturally, the modification may increase the probability of "Don't know", e.g., if we trivially always set $Y_j \leftarrow 0$.

3   What Does it Take to Get a *Good* Lower Bound?

The greatest $i \in [n]$ for which Algorithm 1 will return "Yes" may be arbitrarily smaller than $\log_2 |S|$. The reason for this, at a high level, is that even though i-uniformity is enough for the *expected* size of $S \cap R$ to be $2^{-i}|S|$, the actual size of $S \cap R$ may behave like the winnings of a lottery: typically zero, but huge with very small probability. So, if we add $j = \log_2 |S| - \Theta(1)$ constraints, if a lottery phenomenon is present, then even though the expected size of $S \cap R$ is greater than 1, in any realistic number of trials we will always see $S \cap R = \emptyset$, in exactly the same manner that anyone playing the lottery a realistic number of times will, most likely, never experience winning. Yet concluding $|S| < 2^j$ is obviously wrong.

The above discussion makes it clear that the heart of the matter is controlling the *variance* of $|S \cap R|$. One way to do this is by bounding the "lumpiness" of the sets in the support of the distribution $\mathcal{D}_i$, as measured by the quantity defined below, measuring lumpiness at a scale of M (the smaller the quantity in (1), the less lumpy the distribution, the smaller the variance).

Definition 2. *Let $\mathcal{D}$ be any distribution on subsets of $\{0,1\}^n$ and let $R \sim \mathcal{D}$. For any fixed $M \geq 1$,*

$$\text{Boost}(\mathcal{D}, M) = \max_{\substack{S \subseteq \{0,1\}^n \\ |S| \geq M}} \frac{1}{|S|(|S| - 1)} \sum_{\substack{\sigma, \tau \in S \\ \sigma \neq \tau}} \frac{\Pr[\sigma, \tau \in R]}{\Pr[\sigma \in R]\Pr[\tau \in R]} . \qquad (1)$$

To get some intuition for (1) observe that the ratio inside the sum equals the factor by which the a priori probability that a truth assignment belongs in R is modified by conditioning on some other truth assignment belonging in R. For example, if membership in R is pairwise independent, then $\text{Boost}(\mathcal{D}, \cdot) = 1$, i.e., there is no modification. Another thing to note is that since we require $|S| \geq M$ instead of $|S| = M$ in (1), the function $\text{Boost}(\mathcal{D}, \cdot)$ is non-increasing.

While pairwise independence is excellent from a statistical point of view, the corresponding geometry of the random sets R tend to be make determining $|S \cap R|$ difficult, especially when $|R|$ is far from the extreme values 2^n and 1. And while there are even better distributions statistically, i.e., distributions for which $\text{Boost}(\mathcal{D}, \cdot) < 1$, those are even harder to work with. In the rest of the paper we restrict to the case $\text{Boost}(\mathcal{D}, \cdot) \geq 1$ (hence the name Boost). As we will see, the crucial requirement for an i-uniform distribution $\mathcal{D}_i$ to be useful is that $\text{Boost}(\mathcal{D}_i, \Theta(2^i))$ is relatively small, i.e., an i-uniform distribution can be useful even if $\text{Boost}(\mathcal{D}_i)$ is large for sets of size much less than 2^i. The three examples of i-uniform distributions discussed earlier differ dramatically in terms of Boost.

4   Accurate Counting Without Pairwise Independence

For each $i \in \{0, 1, \ldots, n\}$ let $\mathcal{D}_i$ be some (arbitrary) i-uniform distribution on subsets of $\{0, 1\}^n$. (As mentioned, we consider F, and thus n, fixed, so that we can write $\mathcal{D}_i$ instead of $\mathcal{D}_{i,n}$ to simplify notation.) We will show that given any $0 \leq L \leq |S|$, we can get a *guaranteed* probabilistic approximation of $|S|$ in time proportional to the square of $B = \max_{\ell \leq i \leq n} \mathrm{Boost}(\mathcal{D}_i, 2^i)$. We discuss specific distributions and approaches to bounding B in Sects. 7 and 8.

Algorithm 3. Returns a number in $(1 \pm \delta)|S|$ with probability at least $1 - \theta$

1: Choose the approximation factor $\delta \in (0, 1/3]$ ▷ See Remark 1
2: Choose the failure probability $\theta > 0$
3:
4: Receive as input any number $0 \leq L \leq |S|$ ▷ We can set $L = 0$ by default
5:
6: **if** $|S| < 4/\delta$ **then return** $|S|$ ▷ If $L < 4/\delta$ then search for up to $4/\delta$ models of F
7:
8: Let $\ell = \max\{0, \lfloor \log_2(\delta L/4) \rfloor\}$
9:
10: **if** all distributions $\{\mathcal{D}_i\}_{i=\ell}^n$ enjoy pairwise independence **then**
11:     $B \leftarrow 1$
12: **else**
13:     Find $B \geq \max_{\ell \leq i \leq n} \mathrm{Boost}(\mathcal{D}_i, 2^i)$ ▷ See Sects. 7, 8
14:
15: $\xi \leftarrow 8/\delta$
16: $b \leftarrow \lceil \xi + 2(\xi + \xi^2(B - 1)) \rceil$ ▷ If $B = 1$, then $b = \lceil 24/\delta \rceil$
17: $t \leftarrow \lceil (2b^2/9) \ln(2n/\theta) \rceil$
18:
19: **for** i from ℓ to n **do**
20:     $Z_i \leftarrow 0$
21:     **for** j from 1 to t **do**
22:         Sample $R_{i,j} \sim \mathcal{D}_i$
23:         Determine $Y_{i,j} = \min\{b, |S \cap R_{i,j}|\}$ ▷ Seek up to b elements of $S(F) \cap R_{i,j}$
24:         $Z_i \leftarrow Z_i + Y_{i,j}$
25:     $A_i \leftarrow Z_i/t$
26: $j \leftarrow \max\{i \geq \ell : A_i \geq (1 - \delta)(4/\delta)\}$
27: **return** $\max\{L, A_j 2^j\}$

Theorem 3. *The probability that the output of Algorithm 3 does not lie in the range* $(1 \pm \delta)|S|$ *is at most* θ.

Remark 1. Algorithm 3 can be modified to have accuracy parameters $\beta \in (0, 1)$ and $\gamma \in (1, \infty)$ so that its output is between $\beta|S|$ and $\gamma|S|$ with probability at least $1 - \theta$. At that level of generality, both the choice of ξ and the criterion for choosing j in line 26 must be adapted to β, γ, θ. Here we focus on the high-accuracy case, choosing ξ, b, t with simple form.

Let $q = \lfloor \log_2(\delta|S|/4) \rfloor$. We can assume that $q \geq 0$, since otherwise the algorithm reports $|S|$ and exits. The proof of Theorem 3, presented in Sect. 6, boils down to establishing the following four propositions:

(a) The probability that $A_q 2^q$ is not in $(1 \pm \delta)|S|$ is at most $e^{-9t/(2b^2)}$.
(b) The probability that $A_{q+1} 2^{q+1}$ is not in $(1 \pm \delta)|S|$ is at most $e^{-9t/(2b^2)}$.
(c) If $A_q 2^q$ is in the range $(1 \pm \delta)|S|$, then the maximum in line 26 is at least q (deterministically).
(d) For each $i \geq q + 2$, the probability that the maximum in line 26 equals i is at most $\exp\left(-8t/b^2\right)$.

These propositions imply that the probability of failure is at most the sum of the probability of the bad event in (a), the bad event in (b), and the (at most) $n - 2$ bad events in (d). The fact that each bad event concerns only one random variable A_j allows a significant acceleration of Algorithm 3, discussed next.

5 Nested Sample Sets

In Algorithm 3, for each $i \in [n]$ and $j \in [t]$, we sample each set $R_{i,j}$ from an i-uniform distribution on subsets of $\{0,1\}^n$ independently. As was first pointed out in [5], this is both unnecessary and inefficient. Specifically, as a thought experiment, imagine that we select all random subsets of $\{0,1\}^n$ that we may need *before* Algorithm 3 starts, in the following manner (in reality, we only generate the sets as needed).

Algorithm 4. Generates t monotone decreasing sequences of sample sets

$R_{0,j} \leftarrow \{0,1\}^n$ for $j \in [t]$
for i from 1 to n **do**
    for j from 1 to t **do**
        Select $R_{i,j} \subseteq R_{i-1,j}$ from a 1-uniform distribution on $R_{i-1,j}$

Organize now these sets in a matrix whose rows correspond to values of $0 \leq i \leq n$ and whose columns correspond to $j \in [t]$. It is easy to see that:

1. For each (row) $i \in [n]$:
   (a) Every set $R_{i,j}$ comes from an i-uniform distribution on $\{0,1\}^n$.
   (b) The sets $R_{i,1}, \ldots, R_{i,t}$ are mutually independent.
2. For each column $j \in [t]$:
   (a) $R_{0,j} \supseteq R_{1,j} \supseteq \cdots \supseteq R_{n-1,j} \supseteq R_{n,j}$.

With these new random sets Propositions (a)–(d) above hold exactly as in the fully independent sets case, since for each fixed $i \in [n]$ the only relevant sets are the sets in row i and their distribution, per (1a)–(1b), did not change. At the same time, (2a) ensures that $Y_{1,j} \geq Y_{2,j} \geq \cdots \geq Y_{n,j}$ for every $j \in [t]$ and, as a result, $Z_1 \geq Z_2 \geq \cdots \geq Z_n$. Therefore, the characteristic function of $A_i = Z_i/t \geq (1 - \delta)(4/\delta)$ is now *monotone decreasing*. This means that in order to compute j in line 26, instead of computing Z_i for i from ℓ to n, we can compute $A_\ell, A_{\ell+1}, A_{\ell+2}, A_{\ell+4}, A_{\ell+8}, \ldots$ until we encounter our first k such that $A_k < (1 - \delta)(4/\delta)$, say at $k = \ell + 2^c$, for some $c \geq 0$. At that point, if $c \geq 1$, we can perform binary search for $j \in \{A_{\ell+2^{c-1}}, \ldots, A_{\ell+2^c-1}\}$ etc., so that the number of times the loop that begins in line 21 is executed is *logarithmic* instead of linear in $n - \ell$. Moreover, as we will see, the number of iterations t of this loop can now be reduced from $O(\ln(n/\theta))$ to $O(\ln(1/\theta))$, shaving off another $\log n$ factor from the running time.

6   Proof of Theorem 3

To prove Theorem 3 we will need the following tools.

Lemma 2. *Let $X \geq 0$ be an arbitrary integer-valued random variable. Write $\mathbb{E}X = \mu$ and $\mathrm{Var}(X) = \sigma^2$. Let $Y = \min\{X, b\}$, for some integer $b \geq 0$. For any $\lambda > 0$, if $b \geq \mu + \lambda\sigma^2$, then $\mathbb{E}Y \geq \mathbb{E}X - 1/\lambda$.*

Proof. Omitted due to space limitations. The main idea is to use that if Z is a non-negative, integer-valued random variable, then $\mathbb{E}Z = \sum_{j>0} \Pr[Z \geq j]$, and then apply Chebychev's inequality.

Lemma 3. *Let $\mathcal{D}$ be any i-uniform distribution on subsets of $\{0,1\}^n$. For any fixed set $S \subseteq \{0,1\}^n$, if $R \sim \mathcal{D}$ and $X = |S \cap R|$, then $\mathrm{Var}(X) \leq \mathbb{E}X + (\mathrm{Boost}(\mathcal{D}, |S|) - 1)(\mathbb{E}X)^2$.*

Proof. Omitted due to space limitations. The main idea is to write X as a sum of $|S|$ indicator random variables and use linearity of expectation.

Proof (Proof of Theorem 3). Let $q = \lfloor \log_2(\delta|S|/4) \rfloor$. Recall that if $|S| < 4/\delta$, the algorithm returns $|S|$ and exits. Therefore, we can assume without loss of generality that $q \geq 0$. Fix any $i = q + k$, where $k \geq 0$. Let $X_{i,j} = |S \cap R_{i,j}|$ and write $\mathbb{E}X_{i,j} = \mu_i$, $\mathrm{Var}(X_{i,j}) = \sigma_i^2$.

To establish propositions (a), (b) observe that the value ℓ defined in line 8 is at most q, since $L \leq |S|$, and that $|S| \geq 2^{q+1}$, since $\delta \leq 2$. Thus, since $\mathrm{Boost}(\mathcal{D}, M)$ is non-increasing in M, and $\max_{\ell \leq i \leq n} \mathrm{Boost}(\mathcal{D}_i, 2^i) \leq B$,

$$\max_{k \in \{0,1\}} \mathrm{Boost}(\mathcal{D}_{q+k}, |S|) \leq \max\{\mathrm{Boost}(\mathcal{D}_q, 2^q), \mathrm{Boost}(\mathcal{D}_{q+1}, 2^{q+1})\} \leq B.$$

Therefore, we can apply Lemma 3 for $i \in \{q, q + 1\}$ and conclude that $\sigma_i^2 \leq \mu_i + (B - 1)\mu_i^2$ for such i. Since $\mu_i < 8/\delta$ for all $i \geq q$ while $\xi = 8/\delta$, we see that $b = \lceil \xi + 2(\xi + \xi^2(B - 1)) \rceil \geq \mu_i + 2\sigma_i^2$. Thus, we can conclude that for

$i \in \{q, q+1\}$ the random variables $X_{i,j}, Y_{i,j}$ satisfy the conditions of Lemma 2 with $\lambda = 2$, implying $\mathbb{E}Y_{i,j} \geq \mathbb{E}X_{i,j} - 1/2$. Since Z_i is the sum of t independent random variables $0 \leq Y_{i,j} \leq b$ and $\mathbb{E}Z_i/t \geq \mu_i - 1/2$, we see that for $i \in \{q, q+1\}$ Hoeffding's inequality implies

$$\Pr[Z_i/t \leq (1-\delta)\mu_i] \leq \exp\left(-2t\left(\frac{\delta\mu_i - 1/2}{b}\right)^2\right). \qquad (2)$$

At the same time, since Z_i is the sum of t independent random variables $0 \leq Y_{i,j} \leq b$ and $\mathbb{E}Z_i/t \leq \mu_i$, we see that for all $i \geq q$, Hoeffding's inequality implies

$$\Pr[Z_i/t \geq (1+\delta)\mu_i] \leq \exp\left(-2t\left(\frac{\delta\mu_i}{b}\right)^2\right). \qquad (3)$$

To conclude the proof of propositions (a) and (b) observe that $\mu_{q+k} \geq 2^{2-k}/\delta$. Therefore, (2) and (3) imply that for $k \in \{0, 1\}$, the probability that $A_{q+k}2^{q+k}$ is outside $(1 \pm \delta)|S|$ is at most

$$2\exp\left(-2t\left(\frac{2^{2-k} - 1/2}{b}\right)^2\right) < 2\exp(-9t/(2b^2)).$$

To establish proposition (c) observe that if $A_q \geq (1-\delta)\mu_q$, then $A_q \geq (1-\delta)(4/\delta)$ and, thus, $j \geq q$. Finally, to establish proposition (d) observe that $\mu_i < 2/\delta$ for all $i \geq q+2$. Thus, for any such i, requiring $\mu_i + w \geq (1-\delta)(4/\delta)$, implies $w > 2(1 - 2\delta)/\delta$, which, since $\delta \leq 1/3$, implies $w > 2$. Therefore, for every $k \geq 2$, the probability that $j = q + k$ is at most $\exp(-8t/b^2)$.

Having established propositions (a)–(d) we argue as follows. If $A_{q+k}2^{q+k}$ is in the range $(1 \pm \delta)|S|$ for $k \in \{0, 1\}$ and smaller than $(1-\delta)(4/\delta)$ for $k \geq 2$, then the algorithm will report either A_q2^q or $A_{q+1}2^{q+1}$, both of which are in $(1 \pm \delta)|S|$. Therefore, the probability that the algorithm's answer is incorrect is at most $2 \cdot 2\exp(-9t/(2b^2)) + n \cdot \exp(-8t/b^2) < \theta$, for $n > 2$.

6.1 Proof for Monotone Sequences of Sample Sets

Theorem 4. *For any $s > 0$, if the sets $R_{i,j}$ are generated by Algorithm 4 and $t \geq (2b^2/9)\ln(5s)$ in line 17, then the output of Algorithm 3 lies in the range $(1 \pm \delta)|S|$ with probability at least $1 - \exp(-s) > 0$.*

Proof. Observe that for any fixed i, since the sets $R_{i,1}, \ldots, R_{i,t}$ are mutually independent, Eqs. (2) and (3) remain valid and, thus, propositions (a)–(c) hold. For proposition (d) we note that if the inequality $A_{q+k}2^{q+k} < (1-\delta)(4/\delta)$ holds for $k = 2$, then, by monotonicity, it holds for all $k \geq 2$. Thus, all in all, when monotone sequences of sample sets are used, the probability that the algorithm fails is at most $4\exp(-9t/(2b^2)) + \exp(-8t/b^2)$, a quantity smaller than $\exp(-s)$ for all $t \geq (2b^2/9)\ln(5s)$.

7 Low Density Parity Check Codes

In our earlier discussion of i-uniform distributions we saw that if both $A \in \{0,1\}^{i \times n}$ and $b \in \{0,1\}^i$ are uniformly random, then membership in the random set $R = \{\sigma : A\sigma = b\}$ enjoys the (highly desirable) property of pairwise independence. Unfortunately, this also means that each of the i parity constraints involves, on average, $n/2$ variables, making it difficult to work with when i and n are large (we are typically interested in the regime $\log_2 |S(F)| = \Omega(n)$, thus requiring $i = \Omega(n)$ constraints to be added).

The desire to sparsify the matrix A has long been in the purview of the model counting community. Unfortunately, achieving sparsity by letting each entry of A take the value 1 independently with probability $p < 1/2$ is not ideal [7]: the resulting random sets become dramatically "lumpy" as $p \to 0$.

A motivation for our work is the realization that if we write $|S| = 2^{\alpha n}$, then as α grows it is possible to choose A so that it is both very sparse, i.e., with each row having a *constant* non-zero elements, and so that the sets R have relatively low lumpiness *at the $2^{\alpha n}$ scale*. The key new ingredient comes from the seminal work of Sipser and Spielman on expander codes [14] and is this:

Require each *column* of A to have at least 3 elements.

Explaining why this very simple modification has profound implications is beyond the scope of this paper. Suffice it to say, that it is precisely this requirement of minimum variable degree that dramatically reduces the correlation between elements of R and thus Boost($\mathcal{D}$).

For simplicity of exposition, we only discuss matrices $A \in \{0,1\}^{i \times n}$ where:

- Every column (variable) has exactly $1 \geq 3$ non-zero elements.
- Every row (parity constraint) has exactly $r = 1n/i \in$ non-zero elements.

Naturally, the requirement $1n/i \in \mathbb{N}$ does not always hold, in which case some rows have $\lfloor 1n/i \rfloor$ variables, while the rest have $\lceil 1n/i \rceil$ variables, so that the average is $1n/i$. To simplify discussion we ignore this point in the following.

Given n, i, and 1 a (bi-regular) Low Density Parity Check (LDPC) code is generated by selecting a uniformly random matrix as above[1] and taking the set of codewords to be the set of solutions of the linear system $A\sigma = \mathbf{0}$. (While, for model counting we must also take the right hand side of the equation to be a uniformly random vector, due to symmetry, we can assume without loss of generality that $b = \mathbf{0}$.) In particular, note that $\sigma = \mathbf{0}$ is always a solution of the system and, therefore, to discuss the remaining solutions (codewords) instead of referring to them by their distance from our reference solution $\sigma = \mathbf{0}$ we can refer to them by their weight, i.e., their number of ones.

---

[1] Generating such a matrix can be done by selecting a random permutation of $\lceil 1n \rceil$ and using it to map each of the $1n$ non-zeros to equations, r non-zeros at a time; when $1, r \in O(1)$, the variables in each equation will be distinct with probability $\Omega(1)$, so that a handful of trials suffice to generate a matrix as desired.

It is well-known [12] that the expected number of codewords of weight w in a bi-regular LDPC code is given by the following (rather complicated) expression.

Lemma 4 (Average weight-distribution of regular LDPC ensembles).
The expected number of codewords of weight w in a bi-regular LDPC code with n variables and i parity equations, where each variable appears in l equations and each equation includes r variables equals the coefficient of x^{wl} in the polynomial

$$\binom{n}{w} \frac{\left(\sum_i \binom{r}{2i} x^{2i}\right)^{n\frac{l}{r}}}{\binom{nl}{wl}} . \tag{4}$$

We will denote the quantity described in Lemma 4 by codewords(w).

8   Tractable Distributions

Let $\mathcal{D}_i$ be any i-uniform distribution on subsets of $\{0,1\}^n$.

Definition 3. *Say that $\mathcal{D}_i$ is* tractable *if there exists a function f, called the* density *of $\mathcal{D}_i$, such that for all $\sigma, \tau \in \{0,1\}^n$, if $R \sim \mathcal{D}_i$, then $\Pr[\tau \in R \mid \sigma \in R] = f(\text{Hamming}(\sigma, \tau))$, where*

- $f(j) \geq f(j+1)$ for all $j < n/2$, and,
- either $f(j) \geq f(j+1)$ for all $j \geq n/2$, or $f(j) = f(n-j)$ for all $j \geq n/2$.

For any $S \subset \{0,1\}^n$ and $\sigma \in S$, let $H_\sigma(d)$ denote the number of elements of S at Hamming distance d from σ. Recalling the definition of Boost in (5), we get (6) by i-uniformity and (7) by tractability,

$$\text{Boost}(\mathcal{D}_i, M) = \max_{\substack{S \subseteq \{0,1\}^n \\ |S| \geq M}} \frac{1}{|S|(|S|-1)} \sum_{\substack{\sigma,\tau \in S \\ \sigma \neq \tau}} \frac{\Pr[\sigma, \tau \in R]}{\Pr[\sigma \in R]\Pr[\tau \in R]} \tag{5}$$

$$= \max_{\substack{S \subseteq \{0,1\}^n \\ |S| \geq M}} \frac{2^i}{|S|(|S|-1)} \sum_{\sigma \in S} \sum_{\tau \in S-\sigma} \Pr[\tau \in S \mid \sigma \in S] \tag{6}$$

$$= \max_{\substack{S \subseteq \{0,1\}^n \\ |S| \geq M}} \frac{2^i}{|S|(|S|-1)} \sum_{\sigma \in S} \sum_{d=1}^{n} H_\sigma(d) f(d) \tag{7}$$

$$\leq \max_{\substack{S \subseteq \{0,1\}^n \\ |S| \geq M \\ \sigma \in S}} \frac{2^i}{|S|-1} \sum_{d=1}^{n} H_\sigma(d) f(d). \tag{8}$$

Let z be the unique integer such that $|S|/2 = \binom{n}{0} + \binom{n}{1} + \cdots + \binom{n}{z-1} + \alpha\binom{n}{z}$, for some $\alpha \in [0,1)$. Since $z \leq n/2$, tractability implies that $f(j) \geq f(j+1)$ for all $0 \leq d < z$, and therefore that

$$\frac{\sum_{d=1}^{n} H_\sigma(d) f(d)}{|S| - 1} \leq \frac{\sum_{d=0}^{n} H_\sigma(d) f(d)}{|S|} \tag{9}$$

$$\leq \frac{\sum_{d=0}^{n/2} H_\sigma(d) f(d) + \sum_{d>n/2} H_\sigma(d) f(n-d)}{|S|} \tag{10}$$

$$\leq \frac{2 \left(\sum_{d=0}^{z-1} \binom{n}{d} f(d) + \alpha \binom{n}{z} f(z) \right)}{|S|} \tag{11}$$

$$= \frac{\sum_{d=0}^{z-1} \binom{n}{d} f(d) + \alpha \binom{n}{z} f(z)}{\sum_{d=0}^{z-1} \binom{n}{d} + \alpha \binom{n}{z}} \tag{12}$$

$$\leq \frac{\sum_{d=0}^{z-1} \binom{n}{d} f(d)}{\sum_{d=0}^{z-1} \binom{n}{d}} \tag{13}$$

$$:= B(z). \tag{14}$$

To bound $B(z)$ observe that $B(j) \geq B(j+1)$ for $j < n/2$, inherited by the same property of f. Thus, to bound $B(z)$ from above it suffices to bound z for below. Let $h : x \mapsto -x \log_2 x - (1-x) \log_2 x$ be the binary entropy function and let $h^{-1} : [0,1] \mapsto [0,1]$ map y to the smallest number x such that $h(x) = y$. It is well-known that $\sum_{d=0}^{z} \binom{n}{d} \leq 2^{nh(z/n)}$, for every integer $0 \leq z \leq n/2$. Therefore, $z \geq \lceil nh^{-1}(\log_2(|S|/2)/n) \rceil$, which combined with (8) implies the following.

Theorem 5. *If $\mathcal{D}_i$ is a tractable i-uniform distribution with density f, then*

$$\text{Boost}(\mathcal{D}_i, M) \leq 2^i B \left(\left\lceil nh^{-1} \left(\frac{\log_2 M - 1}{n} \right) \right\rceil \right), \tag{15}$$

where $B(z) = \sum_{d=0}^{z-1} \binom{n}{d} f(d) / \sum_{d=0}^{z-1} \binom{n}{d}$ and $h^{-1} : [0,1] \mapsto [0,1]$ maps y to the smallest number x such that $h(x) = y$, where h is the binary entropy function.

Before proceeding to discuss the tractability of LDPC codes, let us observe that the bound in (14) is essentially tight, as demonstrated when S comprises a Hamming ball of radius z centered at σ and a Hamming ball of radius z centered at σ's complement (in which case the only (and miniscule) compromise is (9)). On the other hand, the passage from (7) to (8), mimicking the analysis of [7], allows the aforementioned worst case scenario to occur *simultaneously* for every $\sigma \in S$, an impossibility. As $|S|/2^n$ grows, this is increasingly pessimistic.

8.1   The Lumpiness of LDPC Codes

Let $\mathcal{D}_i$ be the i-uniform distribution on subsets of $\{0,1\}^n$ that results when $R = \{\sigma : A\sigma = b\}$, where A corresponds to a biregular LDPC code with i parity equations. The row- and column-symmetry in the distribution of A implies that the function $f(d) = \text{codewords}(d)/\binom{n}{d}$ is the density of $\mathcal{D}_i$. Regarding tractability, it is easy to see that if 1 is odd, then codewords$(2j+1) = 0$ for all j and that if r is even, then codewords(d) = codewords$(n-d)$ for all d. Thus, for simplicity of exposition, we will restrict to the case where both l and r are even, noting that this is not a substantial restriction.

With l, r even, we are left to establish that $f(j) \geq f(j+1)$ for all $0 \leq j < n/2$. Unfortunately, this is not true for a trivial reason: f is non-monotone in the vicinity of $n/2$, exhibiting minisucle finite-scale-effect fluctuations (around its globally minimum value). While this renders Theorem 5 inapplicable, it is easy to overcome. Morally, because f is *asymptotically* monotone, i.e., for any fixed $\beta \in [0, 1/2)$, the inequality $f(\beta n) \geq f(\beta n+1)$ holds for all $n \geq n_0(\beta)$. Practically, because for the proof of Theorem 5 to go through it is enough that $f(j) \geq f(j+1)$ for all $0 \leq j < z$ (instead of all $0 \leq j < n/2$), something which for most sets of interest holds, as $z \ll n/2$. Thus, in order to provide a rigorous upper bound on Boost, as required in Algorithm 3, it is enough to verify the monotonicity of f up to z in the course of evaluating $B(z)$. This is precisely what we did with $l = 8$, $\log_2 M = 2n/5$, and $n \in \{100, 110, \ldots, 200\}$, resulting in $r = 20$, i.e., equations of length 20. The resulting bounds for B are in Table 1 below.

Table 1. Upper bounds for Boost for equations of length 20

n	100	110	120	130	140	150	160	170	180	190	200
Boost	75	50	35	26	134	89	60	44	34	154	105

Several comments are due here. First, the non-monotonicity of the bound is due to the interaction of several factors in (15), most anomalous of which is the ceiling. Second, recall that the running time of Algorithm 3 is proportional to the square of our upper bound for Boost. While the bounds in Table 1 are not ideal, they do allow us to derive rigorous results for systems with $40 - 80$ equations and $n \in [100, 200]$ after $\sim 10^4$ (parallelizable) solver invocations. Results for such settings are completely outside the reach of CryptoMiniSAT (and, thus, ApproxMC2) when equations of length $n/2$ are used. Finally, as we will see in Sect. 9, these bounds on Boost appear to be *extremely* pessimistic in practice.

9 Experiments

The goal of our this section is to demonstrate the promise of using systems of parity equations corresponding to LDPC codes *empirically*. That is, we will use such systems, but make *far fewer* solver invocations than what is mandated by our theoretical bounds for a high probability approximation. In other words, **our results are not guaranteed**, unlike those of ApproxMC2. The reason we do this is because we believe that while the error-probability analysis of Theorems 3 and 4 is not too far off the mark, the same can no be said for Theorem 5 providing our rigorous upper bounds on Boost.

Formula name	#SAT	LDPC	AMC2	1/4	#SAT	LDPC	AMC2	1/4
jburnim_morton.sk_13_530	NA	248.49	NA	NA	NA	27826.4	NA	NA
blasted_case37	NA	151.02	NA	NA	NA	4149.9	NA	NA
blasted_case_0_b12_even1	NA	147.02	NA	NA	NA	1378.8	NA	NA
blasted_case_2_b12_even1	NA	147.02	NA	NA	NA	1157.5	NA	NA
blasted_case42	NA	147.02	NA	NA	NA	1008.0	NA	NA
blasted_case_1_b12_even1	NA	147.02	NA	NA	NA	1102.0	NA	NA
blasted_case_0_b12_even2	NA	144.02	NA	NA	NA	881.6	NA	NA
blasted_case_1_b12_even2	NA	144.02	NA	NA	NA	1156.3	NA	NA
blasted_case_2_b12_even2	NA	144.02	NA	NA	NA	1050.5	NA	NA
blasted_case_3_4_b14_even	NA	138.02	NA	NA	NA	293.4	NA	NA
blasted_case_1_4_b14_even	NA	138.02	NA	NA	NA	472.6	NA	NA
log2.sk_72_391	—	136.00	NA	NA	—	12811.1	NA	NA
blasted_case1_b14_even3	NA	122.02	NA	NA	NA	169.6	NA	NA
blasted_case_2_b14_even	NA	118.02	NA	NA	NA	89.2	NA	NA
blasted_case3_b14_even3	NA	118.02	NA	NA	NA	107.7	NA	NA
blasted_case_1_b14_even	NA	118.02	NA	NA	NA	94.7	NA	NA
partition.sk_22_155	NA	107.17	NA	NA	NA	5282.3	NA	NA
scenarios_tree_delete4.sb.pl.sk_4_114	—	105.09	NA	NA	—	708.4	NA	NA
blasted_case140	NA	103.02	NA	NA	NA	1869.0	NA	NA
scenarios_tree_search.sb.pl.sk_11_136	NA	96.46	NA	NA	NA	3314.2	NA	NA
s1423a_7_4	90.59	90.58	NA	NA	6.2	32.4	NA	NA
s1423a_3_2	90.16	90.17	NA	NA	5.7	28.3	NA	NA
s1423a_15_7	89.84	89.83	NA	NA	13.6	44.8	NA	NA
scenarios_tree_delete1.sb.pl.sk_3_114	—	89.15	NA	NA	—	431.3	NA	NA
blasted_case_0_ptb_2	NA	88.02	NA	NA	NA	463.6	NA	NA
blasted_case_0_ptb_1	NA	87.98	NA	NA	NA	632.0	NA	NA
scenarios_tree_delete2.sb.pl.sk_8_114	—	86.46	NA	NA	—	210.3	NA	NA
scenarios_aig_traverse.sb.pl.sk_5_102	NA	86.39	NA	NA	NA	3230.0	NA	NA
54.sk_12_97	82.50	81.55	NA	NA	20.4	235.8	NA	NA
blasted_case_0_b14_1	79.00	79.09	NA	NA	28.8	33.5	NA	NA
blasted_case_2_ptb_1	NA	77.02	NA	NA	NA	10.1	NA	NA
blasted_case_1_ptb_1	NA	77.02	NA	NA	NA	9.5	NA	NA
blasted_case_1_ptb_2	NA	77.02	NA	NA	NA	17.8	NA	NA
blasted_case_2_ptb_2	NA	77.00	NA	NA	NA	25.0	NA	NA
blasted_squaring70	66.00	66.04	NA	NA	5822.7	87.7	NA	NA
blasted_case19	66.00	66.02	NA	NA	25.1	6.9	NA	NA
blasted_case20	66.00	66.02	NA	NA	2.0	4.4	NA	NA
blasted_case15	65.00	65.02	NA	NA	172.3	12.4	NA	NA
blasted_case10	65.00	65.02	NA	NA	209.8	8.8	NA	NA
blasted_TR_b12_2_linear	NA	63.93	NA	NA	NA	1867.1	NA	NA
blasted_case12	NA	62.02	NA	NA	NA	21.5	NA	NA
blasted_case49	61.00	61.02	NA	NA	8.9	15.6	NA	NA
blasted_TR_b12_1_linear	NA	59.95	NA	NA	NA	767.9	NA	NA
scenarios_tree_insert_insert.sb.pl.sk_3_68	—	51.86	NA	NA	12.1	54.3	NA	NA
blasted_case18	NA	51.00	NA	NA	NA	16.7	NA	NA
blasted_case14	49.00	49.07	NA	NA	117.2	7.6	NA	NA
blasted_case9	49.00	49.02	NA	NA	123.6	7.1	NA	NA
blasted_case61	48.00	48.02	NA	NA	154.2	6.7	NA	NA
ProjectService3.sk_12_55	—	46.55	46.58	46.55	—	12.9	273.4	267.1

(continued)

(continued)

Formula name	#SAT	LDPC	AMC2	1/4	#SAT	LDPC	AMC2	1/4
blasted_case145	46.00	46.02	NA	46.02	29.2	8.4	NA	5570.4
blasted_case146	46.00	46.02	46.02	NA	29.3	4.8	9528.6	NA
ProcessBean.sk_8_64	—	42.83	42.91	42.83	—	17.0	323.2	207.3
blasted_case106	42.00	42.02	42.02	42.02	10.2	3.3	325.0	14728.3
blasted_case105	41.00	41.00	41.04	NA	7.5	4.0	368.5	NA
blasted_squaring16	40.76	40.83	NA	41.07	99.4	50.3	NA	1633.3
blasted_squaring14	40.76	40.70	NA	41.00	102.1	34.3	NA	2926.5
blasted_squaring12	40.76	40.61	NA	41.00	117.3	39.6	NA	1315.6
blasted_squaring7	38.00	38.29	38.00	38.11	45.4	34.9	432.4	263.2
blasted_squaring9	38.00	38.04	37.98	38.15	36.3	24.2	489.8	238.6
blasted_case_2_b12_2	38.00	38.02	38.02	38.00	29.3	4.4	186.8	87.2
blasted_case_0_b11_1	38.00	38.02	38.02	38.04	45.5	2.5	190.4	180.7
blasted_case_0_b12_2	38.00	38.02	38.02	38.02	29.2	3.8	181.1	69.9
blasted_case_1_b11_1	38.00	38.02	38.02	37.81	45.2	3.5	159.5	119.2
blasted_case_1_b12_2	38.00	38.02	38.02	38.02	30.6	2.9	185.3	80.0
blasted_squaring10	38.00	38.02	37.91	38.04	17.6	32.0	415.1	221.7
blasted_squaring11	38.00	37.95	38.02	38.09	19.8	19.7	470.1	207.3
blasted_squaring8	38.00	37.93	38.09	39.00	18.6	28.0	431.5	727.8
sort.sk_8_52	—	36.43	36.43	36.36	—	92.0	339.2	156.8
blasted_squaring1	36.00	36.07	36.07	36.00	6.6	20.0	367.8	156.9
blasted_squaring6	36.00	36.04	36.00	35.93	8.5	17.1	429.1	170.5
blasted_squaring3	36.00	36.02	36.02	36.02	7.7	18.7	397.3	198.5
blasted_squaring5	36.00	35.98	36.02	36.04	8.5	28.8	384.0	228.2
blasted_squaring2	36.00	35.98	36.00	36.07	7.5	30.6	411.5	195.8
blasted_squaring4	36.00	35.95	36.04	35.98	7.9	23.2	469.8	180.0
compress.sk_17_291	NA	34.00	NA	NA	NA	1898.2	NA	NA
listReverse.sk_11_43	NA	32.00	32.00	32.00	NA	2995.3	2995.3	2995.7
enqueueSeqSK.sk_10_42	NA	31.49	31.39	31.43	NA	67.6	252.0	124.6
blasted_squaring29	26.25	26.36	26.29	26.39	1.3	42.7	218.7	75.2
blasted_squaring28	26.25	26.32	26.36	26.36	1.9	57.6	185.1	59.0
blasted_squaring30	26.25	26.25	26.29	26.17	1.6	40.9	179.8	60.8
tutorial3.sk_4_31	NA	25.29	25.32	25.25	NA	3480.5	19658.2	2414.7
blasted_squaring51	24.00	24.11	24.15	24.07	1.6	4.8	49.3	5.3
blasted_squaring50	24.00	23.86	24.00	24.02	1.3	4.7	54.2	5.1
NotificationServiceImpl2.sk_10_36	—	22.64	22.49	22.55	—	13.7	29.6	9.6
karatsuba.sk_7_41	—	20.36	NA	20.52	—	24963.0	NA	19899.0
LoginService.sk_20_34	—	19.49	19.39	19.43	—	28.1	33.0	20.7
LoginService2.sk_23_36	—	17.55	17.43	17.43	—	72.9	40.8	32.6

To illuminate the bigger picture, besides ApproxMC2 we also included in the comparison the exact sharpSAT model counter of Thurley [17], and the modification of ApproxMC2 in which each equation involves each variable independently with probability $p = 1/2^j$, for $j = 2, \ldots, 5$. (ApproxMC2 corresponds to $j = 1$).

To make the demonstration as transparent as possible, we only made two modifications to ApproxMC2 and recorded their impact on performance.

- We incorporated Algorithm 2 for quickly computing a lower bound.
- We use systems of equations corresponding to LDPC codes instead of systems where each equation involves $n/2$ variables on average (as ApproxMC2 does).

Algorithm 2 is invoked at most once, while the change in the systems of equations is entirely encapsulated in the part of the code generating the random systems. No other changes to ApproxMC2 (AMC2) were made.

We consider the same 387 formulas as [5]. Among these are 2 unsatisfiable formulas which we removed. We also removed 9 formulas that were only solved by sharpSAT and 10 formulas whose number of solutions (and, thus, equations) is so small that the LDPC equations devolve into long XOR equations. Of the remaining 366 formulas, sharpSAT solves 233 in less than 1 s, in every case significantly faster than all sampling based methods. At the other extreme, 46 formulas are not solved by any method within the given time limits, namely 8 h per method-formula pair (and 50 min for each solver invocation for the sampling based algorithms). We report on our experiments with the remaining 87 formulas. All experiments were run on a cluster of 13 nodes, each with 16 cores and 128 GB RAM.

Our findings can be summarized as follows:

1. The LDPC-modified version of AMC2 has similar accuracy to AMC2, even though the number of solver invocations is much smaller than what theory mandates for an approximation guarantee. Specifically, the counts are very close to the counts returned by AMC2 and sharpSAT *in every single formula*.
2. The counts with $p = 1/4$ are as accurate as with $p = 1/2$. But for $p \leq 1/8$, the counts were often significantly wrong and we don't report results.
3. The LDPC-modified version of AMC2 is faster than AMC2 in all but one formulas, the speedup typically exceeding 10x and often exceeding 50x.
4. In formulas where both sharpSAT and the LDPC-modified version of AMC2 terminate, sharpSAT is faster more often than not. That said, victories by a speed difference of 50x occur for both algorithms.
5. The LDPC-modified version of AMC2 did not time out on *any* formula. In contrast, sharpSAT timed out on 38% of the formulas, AMC2 with $p = 1/4$ on 59% of the formulas, and AMC2 on 62%.

In the following table, the first four columns report the binary logarithm of the estimate of $|S|$ returned by each algorithm. The next four columns report the time taken to produce the estimate, in seconds. We note that several of the 87 formulas come with a desired *sampling set*, i.e., a subset of variables V such that the goal is to count the size of the projection of the set of all models on V. Since, unlike AMC2, sharpSAT does not provide such constrained counting functionality, to avoid confusion, we do not report a count for sharpSAT for these formulas, writing "—" instead. Timeouts are reported as "NA".

References

1. Chakraborty, S., Fremont, D.J., Meel, K.S., Seshia, S.A., Vardi, M.Y.: Distribution-aware sampling and weighted model counting for SAT. In: Brodley, C.E., Stone, P. (eds.) Proceedings of the Twenty-Eighth AAAI Conference on Artificial Intelligence, Québec City, Québec, Canada, 27–31 July 2014, pp. 1722–1730. AAAI Press (2014)
2. Chakraborty, S., Meel, K.S., Vardi, M.Y.: A scalable and nearly uniform generator of SAT witnesses. In: Sharygina, N., Veith, H. (eds.) CAV 2013. LNCS, vol. 8044, pp. 608–623. Springer, Heidelberg (2013). doi:10.1007/978-3-642-39799-8_40
3. Chakraborty, S., Meel, K.S., Vardi, M.Y.: A scalable approximate model counter. In: Schulte, C. (ed.) CP 2013. LNCS, vol. 8124, pp. 200–216. Springer, Heidelberg (2013). doi:10.1007/978-3-642-40627-0_18

4. Chakraborty, S., Meel, K.S., Vardi, M.Y.: Balancing scalability and uniformity in SAT witness generator. In: The 51st Annual Design Automation Conference 2014, DAC 2014, San Francisco, CA, USA, 1–5 June 2014, pp. 60:1–60:6. ACM (2014)
5. Chakraborty, S., Meel, K.S., Vardi, M.Y.: Algorithmic improvements in approximate counting for probabilistic inference: from linear to logarithmic SAT calls. In: Kambhampati, S. (ed.) Proceedings of the Twenty-Fifth International Joint Conference on Artificial Intelligence, IJCAI 2016, New York, NY, USA, 9–15 July 2016, pp. 3569–3576. IJCAI/AAAI Press (2016)
6. Ermon, S., Gomes, C.P., Sabharwal, A., Selman, B.: Taming the curse of dimensionality: discrete integration by hashing and optimization. In: Proceedings of the 30th International Conference on Machine Learning (ICML) (2013)
7. Ermon, S., Gomes, C.P., Sabharwal, A., Selman, B.: Low-density parity constraints for hashing-based discrete integration. In: Proceedings of the 31st International Conference on Machine Learning (ICML), pp. 271–279 (2014)
8. Gomes, C.P., Hoffmann, J., Sabharwal, A., Selman, B.: Short XORs for model counting: from theory to practice. In: Marques-Silva, J., Sakallah, K.A. (eds.) SAT 2007. LNCS, vol. 4501, pp. 100–106. Springer, Heidelberg (2007). doi:10.1007/978-3-540-72788-0_13
9. Gomes, C.P., Sabharwal, A., Selman, B.: Model counting: a new strategy for obtaining good bounds. In: Proceedings of the 21st National Conference on Artificial Intelligence (AAAI), pp. 54–61 (2006)
10. Gomes, C.P., Sabharwal, A., Selman, B.: Near-uniform sampling of combinatorial spaces using XOR constraints. In: Advances in Neural Information Processing Systems (NIPS) (2006)
11. Ivrii, A., Malik, S., Meel, K.S., Vardi, M.Y.: On computing minimal independent support and its applications to sampling and counting. Constraints 21(1), 41–58 (2016)
12. Richardson, T., Urbanke, R.: Modern Coding Theory. Cambridge University Press, New York (2008)
13. Sipser, M.: A complexity theoretic approach to randomness. In: Proceedings of the 15th ACM Symposium on Theory of Computing (STOC), pp. 330–335 (1983)
14. Sipser, M., Spielman, D.A.: Expander codes. IEEE Trans. Inf. Theory 42(6), 1710–1722 (1996)
15. Soos, M.: Cryptominisat-a sat solver for cryptographic problems (2009). http://www.msoos.org/cryptominisat4
16. Stockmeyer, L.: On approximation algorithms for #P. SIAM J. Comput. 14(4), 849–861 (1985)
17. Thurley, M.: sharpSAT – counting models with advanced component caching and implicit BCP. In: Biere, A., Gomes, C.P. (eds.) SAT 2006. LNCS, vol. 4121, pp. 424–429. Springer, Heidelberg (2006). doi:10.1007/11814948_38
18. Valiant, L.G., Vazirani, V.V.: NP is as easy as detecting unique solutions. Theoret. Comput. Sci. 47, 85–93 (1986)

Backdoor Treewidth for SAT

Robert Ganian[✉], M.S. Ramanujan, and Stefan Szeider

Algorithms and Complexity Group, TU Wien, Vienna, Austria
{ganian,ramanujan,sz}@ac.tuwien.ac.at

Abstract. A strong backdoor in a CNF formula is a set of variables such that each possible instantiation of these variables moves the formula into a tractable class. The algorithmic problem of finding a strong backdoor has been the subject of intensive study, mostly within the parameterized complexity framework. Results to date focused primarily on backdoors of small size. In this paper we propose a new approach for algorithmically exploiting strong backdoors for SAT: instead of focusing on small backdoors, we focus on backdoors with certain structural properties. In particular, we consider backdoors that have a certain tree-like structure, formally captured by the notion of *backdoor treewidth*.

First, we provide a fixed-parameter algorithm for SAT parameterized by the backdoor treewidth w.r.t. the fundamental tractable classes Horn, Anti-Horn, and 2CNF. Second, we consider the more general setting where the backdoor decomposes the instance into components belonging to different tractable classes, albeit focusing on backdoors of treewidth 1 (i.e., acyclic backdoors). We give polynomial-time algorithms for SAT and #SAT for instances that admit such an acyclic backdoor.

1 Introduction

SAT is the problem of determining whether a propositional formula in conjunctive normal form (CNF) is satisfiable. Since SAT was identified as the first NP-complete problem, a significant amount of research has been devoted to the identification of "islands of tractability" or "tractable classes," which are sets of CNF formulas on which SAT is solvable in polynomial time. The notion of a strong backdoor, introduced by Williams et al. [28], allows one to extend these polynomial-time results to CNF formulas that do not belong to an island of tractability but are close to one. Namely, a *strong backdoor* is a set of variables of the given CNF formula, such that for all possible truth assignments to the variables in the set, applying the assignment moves the CNF formula into the island of tractability under consideration. In other words, using a strong backdoor consisting of k variables transforms the satisfiability decision for one general CNF formula into the satisfiability decision for 2^k tractable CNF formulas. A natural way of exploiting strong backdoors algorithmically is to search for *small* strong backdoors. For standard islands of tractability, such as the class of Horn

Supported by the Austrian Science Fund (FWF), project P26696. Robert Ganian is also affiliated with FI MU, Brno, Czech Republic.

© Springer International Publishing AG 2017
S. Gaspers and T. Walsh (Eds.): SAT 2017, LNCS 10491, pp. 20–37, 2017.
DOI: 10.1007/978-3-319-66263-3_2

formulas and the class of 2CNF formulas, one can find a strong backdoor of size k (if it exists) in time $f(k)L^c$ (where f is a computable function, c is a constant, and L denotes the length of the input formula) [22]; in other words, the detection of a strong backdoor of size k to Horn or 2CNF is *fixed-parameter tractable* [10]. The parameterized complexity of backdoor detection has been the subject of intensive study. We refer the reader to a survey article [16] for a comprehensive overview of this topic.

In this paper we propose a new approach for algorithmically exploiting strong backdoors for SAT. Instead of focusing on small backdoors, we focus on backdoors with certain structural properties. This includes backdoors of unbounded size and thus applies in cases that were not accessible by previous backdoor approaches. In particular, we consider backdoors that, roughly speaking, can be arbitrarily large but have a certain "tree-like" structure. Formally, this structure is captured in terms of the *treewidth* of a graph modeling the interactions between the backdoor and the remainder of the CNF formula (this is called the *backdoor treewidth* [14]). *Treewidth* itself is a well-established structural parameter that can be used to obtain fixed-parameter tractability of SAT [26]. The combination of strong backdoors and treewidth, as considered in this paper, gives rise to new tractability results for SAT, not achievable by backdoors or treewidth alone.

The notion of backdoor treewidth outlined above was recently introduced in the context of the *constraint satisfaction problem* (CSP) [14]. However, a direct translation of those results to SAT seems unlikely. In particular, while the results for CSP can be used "out-of-the-box" for CNF formulas of bounded clause width, additional machinery is required in order to handle CNF formulas of *unbounded* clause width.

The first main contribution of our paper is hence the following.

(1) SAT is fixed-parameter tractable when parameterized by the backdoor treewidth w.r.t. any of the following islands of tractability: Horn, Anti-Horn, and 2CNF (Theorem 1).

For our second main contribution, we consider a much wider range of islands of tractability, namely every island of tractability that is closed under partial assignments (a property shared by most islands of tractability studied in the literature). Moreover, we consider backdoors that split the input CNF formula into components where each of them may belong to a different island of tractability (we can therefore speak of an "archipelago of tractability" [15]). This is a very general setting, and finding such a backdoor of small treewidth is a challenging algorithmic task. It is not at all clear how to handle even the special case of backdoor treewidth 1, i.e., acyclic backdoors.

In this work, we take the first steps in this direction and settle this special case of acyclic backdoors. We show that if a given CNF formula has an acyclic backdoor into an archipelago of tractability, we can test its satisfiability in polynomial time. We also obtain an equivalent result for the model counting problem #SAT (which asks for the number of satisfying assignments of the given formula).

(2) SAT and #SAT are solvable in polynomial time for CNF formulas with an acyclic backdoor into any archipelago of tractability whose islands are closed under partial assignments (Theorems 2 and 3).

We note that the machinery developed for backdoor treewidth in the CSP setting cannot be used in conjunction with islands of tractability in the general context outlined above; in fact, we leave open even the existence of a polynomial-time algorithm for backdoor treewidth 2. An interesting feature of the algorithms establishing Theorems 2 and 3 is that they do not explicitly *detect* an acyclic backdoor. Instead, we only require the *existence* of such a backdoor in order to guarantee that our algorithm is correct on such inputs. In this respect, our results add to the rather small set of backdoor based algorithms for SAT (see also [13]) which only rely on the existence of a specific kind of backdoor rather than computing it, in order to solve the instance.

We now briefly mention the techniques used to establish our results. The general idea behind the proof of Theorem 1 is the translation of the given CNF formula F into a "backdoor-equivalent" CNF formula F' which has bounded clause width. Following this, we can safely perform a direct translation of F' into a CSP instance $\mathcal{I}$ which satisfies all the conditions for invoking the previous results on CSP [14]. For Theorems 2 and 3, we consider the biconnected components of the incidence graph of the CNF formula and prove that the existence of an acyclic backdoor imposes useful structure on them. We then design a dynamic programming algorithm which runs on the block decomposition of the incidence graph of the CNF formula and show that it correctly determines in polynomial time whether the input CNF formula is satisfiable (or counts the number of satisfying assignments, respectively) *as long as* there is an acyclic backdoor of the required kind.

2   Preliminaries

Parameterized Complexity. We begin with a brief review of the most important concepts of parameterized complexity. For an in-depth treatment of the subject we refer the reader to textbooks [8,10].

The instances of a parameterized problem can be considered as pairs (I, k) where I is the *main part* of the instance and k is the *parameter* of the instance; the latter is usually a non-negative integer. A parameterized problem is *fixed-parameter tractable* (FPT) if instances (I, k) of size n (with respect to some reasonable encoding) can be solved in time $O(f(k)n^c)$ where f is a computable function and c is a constant independent of k. The function f is called the *parameter dependence.*

We say that parameter X *dominates* parameter Y if there exists a computable function f such that for each CNF formula F we have $X(F) \leq f(Y(F))$ [25]. In particular, if X dominates Y and SAT is FPT parameterized by X, then SAT is FPT parameterized by Y [25]. We say that two parameters are *incomparable* if neither dominates the other.

Satisfiability. We consider propositional formulas in conjunctive normal form (CNF), represented as sets of clauses. That is, a *literal* is a (propositional) variable x or a negated variable $\overline{x}$; a *clause* is a finite set of literals not containing a complementary pair x and $\overline{x}$; a *formula* is a finite set of clauses. For a literal $l = \overline{x}$ we write $\overline{l} = x$; for a clause c we set $\overline{c} = \{\overline{l} \mid l \in c\}$. For a clause c, $var(c)$ denotes the set of variables x with $x \in c$ or $\overline{x} \in c$, and the *clause width* of c is $|var(c)|$. Similarly, for a CNF formula F we write $var(F) = \bigcup_{c \in F} var(c)$. The *length* (or *size*) of a CNF formula F is defined as $\sum_{c \in F} |c|$. We will sometimes use the a graph representation of a CNF formula F called the *incidence graph* of F and denoted $\mathsf{Inc}(F)$. The vertices of $\mathsf{Inc}(F)$ are variables and clauses of F and two vertices a, b are adjacent if and only if a is a clause and $b \in var(a)$.

A *truth assignment* (or *assignment*, for short) is a mapping $\tau : X \to \{0, 1\}$ defined on some set X of variables. We extend τ to literals by setting $\tau(\overline{x}) = 1 - \tau(x)$ for $x \in X$. $F[\tau]$ denotes the CNF formula obtained from F by removing all clauses that contain a literal x with $\tau(x) = 1$ and by removing from the remaining clauses all literals y with $\tau(y) = 0$; $F[\tau]$ is the *restriction* of F to τ. Note that $var(F[\tau]) \cap X = \emptyset$ holds for every assignment $\tau : X \to \{0, 1\}$ and every CNF formula F. An assignment $\tau : X \to \{0, 1\}$ *satisfies* a CNF formula F if $F[\tau] = \emptyset$, and a CNF formula F is satisfiable if there exists an assignment which satisfies F. In the SAT problem, we are given a CNF formula F and the task is to determine whether F is satisfiable.

Let $X \subseteq var(F)$. Two clauses c, c' are *X-adjacent* if $(var(c) \cap var(c')) \setminus X \neq \emptyset$. We say that two clauses c, d are *X-connected* if there exists a sequence $c = c_1, \ldots, c_r = d$ such that each consecutive c_i, c_{i+1} are X-adjacent. An *X-component* Z of a CNF formula F is then an inclusion-maximal subset of X-connected clauses, and its boundary is the set $var(Z) \cap X$. An $\emptyset$-component of a CNF formula F is also called a *connected component* of F.

A class $\mathcal{F}$ of CNF formulas is *closed under partial assignments* if, for each $F \in \mathcal{F}$ and each assignment τ of a subset of $var(F)$, it holds that $F[\tau] \in \mathcal{F}$. Examples of classes that are closed under partial assignment include 2CNF, Q-Horn, hitting CNF formulas and acyclic CNF formulas (see, e.g., the Handbook of Satisfiability [2]).

Backdoors and Tractable Classes for SAT. Backdoors are defined relative to some fixed class $\mathcal{C}$ of instances of the problem under consideration (i.e., SAT); such a class $\mathcal{C}$ is then often called the *base class*. One usually assumes that the problem is tractable for instances from $\mathcal{C}$, as well as that the recognition of $\mathcal{C}$ is tractable; here, tractable means solvable by a polynomial-time algorithm.

In the context of SAT, we define a *strong backdoor set* into $\mathcal{F}$ of a CNF formula F to be a set B of variables such that $F[\tau] \in \mathcal{F}$ for each assignment $\tau : B \to \{0, 1\}$. If we know a strong backdoor of F into $\mathcal{F}$, we can decide the satisfiability of F by looping over all assignments τ of the backdoor variables and checking the satisfiability of the resulting CNF formulas $F[\tau]$ (which belong to $\mathcal{F}$). Thus SAT decision is fixed-parameter tractable in the size k of the strong backdoor, assuming we are given such a backdoor as part of the input and $\mathcal{F}$ is tractable. We note that every natural base class $\mathcal{F}$ has the property that if

$B \subseteq var(F)$ is a backdoor of F into $\mathcal{F}$, then B is also a backdoor of every B-component of F into $\mathcal{F}$; indeed, each such B-component can be treated separately for individual assignments of B. We will hence assume that all our base classes have this property.

Here we will be concerned with three of the arguably most prominent polynomially tractable classes of CNF formulas: *Horn*, *Anti-Horn* and *2CNF*, defined in terms of syntactic properties of clauses. A clause is (a) *Horn* if it contains at most one positive literal, (b) *Anti-Horn* if it contains at most one negative literal, (c) *2CNF* if it contains at most two literals,[1]

A CNF formula belongs to the class *Horn*, *Anti-Horn*, or *2CNF* if it contains only Horn, Anti-Horn, or 2CNF clauses, respectively; each of these classes is known to be tractable and closed under partial assignments. It is known that backdoor detection for each of the classes listed above is FPT, which together with the tractability of these classes yields the following.

Proposition 1 ([16]). *SAT is fixed-parameter tractable when parameterized by the size of a minimum backdoor into $\mathcal{F}$, for each $\mathcal{F} \in \{Horn, Anti-Horn, 2CNF\}$.*

We note that in the literature also other types of backdoors (weak backdoors and deletion backdoors) have been considered; we refer to a survey article for examples [16]. In the sequel our focus lies on strong backdoors, and for the sake of brevity will refer to them merely as *backdoors*.

The Constraint Satisfaction Problem. Let $\mathcal{V}$ be a set of variables and $\mathcal{D}$ a finite set of values. A *constraint of arity ρ over* $\mathcal{D}$ is a pair (S, R) where $S = (x_1, \ldots, x_\rho)$ is a sequence of variables from $\mathcal{V}$ and $R \subseteq \mathcal{D}^\rho$ is a ρ-ary relation. The set $var(C) = \{x_1, \ldots, x_\rho\}$ is called the *scope* of C. An *assignment* $\alpha : X \to \mathcal{D}$ is a mapping of a set $X \subseteq \mathcal{V}$ of variables. An assignment $\alpha : X \to \mathcal{D}$ satisfies a constraint $C = ((x_1, \ldots, x_\rho), R)$ if $var(C) \subseteq X$ and $(\alpha(x_1), \ldots, \alpha(x_\rho)) \in R$. For a set $\mathcal{I}$ of constraints we write $var(\mathcal{I}) = \bigcup_{C \in \mathcal{I}} var(C)$ and $rel(\mathcal{I}) = \{R \mid (S, R) \in C, C \in \mathcal{I}\}$. A finite set $\mathcal{I}$ of constraints is *satisfiable* if there exists an assignment that simultaneously satisfies all the constraints in $\mathcal{I}$. The *Constraint Satisfaction Problem* (CSP, for short) asks, given a finite set $\mathcal{I}$ of constraints, whether $\mathcal{I}$ is satisfiable.

Next, we will describe how a partial assignment alters a given CSP instance. Let $\alpha : X \to \mathcal{D}$ be an assignment. For a ρ-ary constraint $C = (S, R)$ with $S = (x_1, \ldots, x_\rho)$ and $R \in \mathcal{D}^\rho$, we denote by $C|_\alpha$ the constraint (S', R') obtained from C as follows. R' is obtained from R by (i) deleting all tuples $(d_1, \ldots, d_\rho)$ from R for which there is some $1 \leq i \leq \rho$ such that $x_i \in X$ and $\alpha(x_i) \neq d_i$, and (ii) removing from all remaining tuples all coordinates d_i with $x_i \in X$. S' is obtained from S by deleting all variables x_i with $x_i \in X$. For a set $\mathcal{I}$ of constraints we define $\mathcal{I}|_\alpha$ as $\{C|_\alpha \mid C \in \mathcal{I}\}$. It is important to note that there is a crucial distinction between assignments in SAT and assignments in CSP: while in SAT one deletes all clauses which are already satisfied by a given assignment,

---

[1] A clause containing exactly two literals is also known as a *Krom clause*.

in CSP this is not the case – instead, constraints are restricted to the tuples which match the assignment (but never deleted).

A *constraint language* (or *language*, for short) Γ over a domain $\mathcal{D}$ is a set of relations (of possibly various arities) over $\mathcal{D}$. By $\mathrm{CSP}(\Gamma)$ we denote CSP restricted to instances $\mathcal{I}$ with $\mathrm{rel}(\mathcal{I}) \subseteq \Gamma$. A constraint language is *tractable* if for every finite subset $\Gamma' \subseteq \Gamma$, the problem $\mathrm{CSP}(\Gamma')$ can be solved in polynomial time. Let Γ be a constraint language and $\mathcal{I}$ be an instance of CSP. A variable set X is a *(strong) backdoor* to $\mathrm{CSP}(\Gamma)$ if for each assignment $\alpha : X \to \mathcal{D}$ it holds that $\mathcal{I}|_\alpha \in \mathrm{CSP}(\Gamma)$. A language is *closed under partial assignments* if for each $\mathcal{I} \in \mathrm{CSP}(\Gamma)$ and for each assignment α, it holds that $\mathcal{I}|_\alpha \in \mathrm{CSP}(\Gamma)$. Finally, observe that the notions of being *X-adjacent*, *X-connected* and being a *X-component* can be straightforwardly translated to CSP. For example, a CSP instance containing three constraints $((a, b, d), R_1)$, $((c, d), R_2)$, and $((e, b), R_3)$ would contain two $\{b\}$-components: one containing $((a, b, d), R_1)$, $((c, d), R_2)$ and the other containing $((e, b), R_3)$.

Each SAT instance (i.e., CNF formula) admits a direct encoding into a CSP instance (over the same variable set and with domain $\{0, 1\}$), which transforms each clause into a relation containing all tuples which do not invalidate that clause. Note that this will exponentially increase the size of the instance if the CNF formula contains clauses of unbounded clause width; however, for any fixed bound on the clause width of the original CNF formula, the direct encoding into CSP will only increase the bit size of the instance by a constant factor.

Treewidth and Block Decompositions. The set of *internal vertices* of a path P in a graph is denoted by $V_{\mathrm{int}}(P)$ and is defined as the set of vertices in P which are not the endpoints of P. We say that two paths P_1 and P_2 in an undirected graph are *internally vertex-disjoint* if $V_{\mathrm{int}}(P_1) \cap V_{\mathrm{int}}(P_2) = \emptyset$. Note that under this definition, a path consisting of a single vertex is internally vertex-disjoint to every other path in the graph.

The graph parameter *treewidth* will be of particular interest in the context of this paper. Let G be a simple, undirected, finite graph with vertex set $V = V(G)$ and edge set $E = E(G)$. A *tree decomposition* of G is a pair $(\{B_i : i \in I\}, T)$ where $B_i \subseteq V$, $i \in I$, and T is a tree with elements of I as nodes such that (a) for each edge $uv \in E$, there is an $i \in I$ such that $\{u, v\} \subseteq X_i$, and (b) for each vertex $v \in V$, $T[\{i \in I \mid v \in B_i\}]$ is a (connected) tree with at least one node. The *width* of a tree decomposition is $\max_{i \in I} |B_i| - 1$. The *treewidth* [21, 24] of G is the minimum width taken over all tree decompositions of G and it is denoted by $\mathrm{tw}(G)$. We call the elements of I *nodes* and B_i *bags*. It is well known that, for every clique over $Z \subseteq V(G)$ in G, it holds that every tree decomposition of G contains an element B_i such that $Z \subseteq B_i$ [21].

We now recall the definitions of *blocks* and *block decompositions* in a graph. A *cut-vertex* in an undirected graph H is a vertex whose removal disconnects the connected component the vertex belongs to.

A maximal connected subgraph without a cut-vertex is called a *block*. Every block of a graph G is either a maximal 2-connected subgraph, or an isolated vertex or a path on 2 vertices (see, e.g., [9]).

By maximality, different blocks of a graph H overlap in at most one vertex, which is then easily seen to be a cut-vertex of H. Therefore, every edge of H lies in a unique block and H is the union of its blocks.

Let A denote the set of cut-vertices of H and B the set of its blocks. The *block-graph* of H is the bipartite graph on $A \cup B$ where $a \in A$ and $b \in B$ are adjacent if and only if $a \in b$. The set of vertices in B are called *block-vertices*.

Proposition 2 ([9]). *The block-graph of a connected undirected graph is a tree.*

Due to the above proposition, we will henceforth refer to block-graphs of connected graphs as block-trees. Furthermore, this proposition implies that the block-graph of a disconnected graph is precisely the disjoint union of the block-trees of its connected components. As a result, we refer to block-graphs in general as block-forests. Finally, it is straightforward to see that the leaves of the block-tree are all block-vertices.

The block decomposition of a connected graph G is a pair $(T, \eta : V(T) \to 2^{V(G)})$ where T is the block-tree, and (a) for every $t \in V(T)$ such that t is a block-vertex, $G[\eta(t)]$ is the block of G corresponding to this block-vertex and (b) for every $t \in V(T)$ such that t is a cut-vertex, $\eta(t) = \{t\}$. For a fixed *root* of the tree T and a vertex $t \in V(T)$, we denote by $\mathsf{child}(t)$ the set of children of t with respect to this root vertex.

Proposition 3 ([7, 18]). *There is an algorithm that, given a graph G, runs in time $\mathcal{O}(m + n)$ and outputs the block decomposition of G.*

3 Backdoor Treewidth

The core idea of *backdoor treewidth* is to fundamentally alter how the *quality* of a backdoor is measured. Traditionally, the aim has always been to seek for backdoors of small *size*, since these can easily be used to obtain fixed-parameter algorithms for SAT. Instead, one can define the treewidth of a backdoor—obtained by measuring the treewidth of a graph obtained by "collapsing" the CNF formula into the backdoor—and show that this is more beneficial than focusing merely on its size. In line with existing literature in graph algorithms and theory, we call the resulting "collapsed graph" the *torso*.

Definition 1. *Let F be a CNF formula and X be a backdoor in F to a class $\mathcal{F}$. Then the X-torso graph G_F^X of F is the graph whose vertex set is X and where two variables x, y are adjacent if and only if there exists an X-component A such that $x, y \in var(A)$. The treewidth of X is then the treewidth of G_F^X, and the backdoor treewidth of F w.r.t. $\mathcal{F}$ is the minimum treewidth of a backdoor into $\mathcal{F}$.*

Given the above definition, it is not difficult to see that backdoor treewidth w.r.t. $\mathcal{F}$ is upper-bounded by the size of a minimum backdoor to $\mathcal{F}$, but can be arbitrarily smaller than the latter. As a simple example of this behavior, consider the CNF formula $\{\{x_1, x_2\}, \{x_2, x_3\}, \ldots, \{x_{i-1}, x_i\}\}$, which does not contain a constant-size backdoor to Horn but has backdoor treewidth 1 w.r.t. Horn (one can use, e.g., a backdoor containing every variable with an even index). This motivates the design of algorithms for SAT which are fixed-parameter tractable when parameterized not by the *size* of a smallest backdoor into a particular base class but by the value of the backdoor treewidth with respect to the base class. Our first main contribution is the following theorem.

Theorem 1. *SAT is fixed-parameter tractable when parameterized by the backdoor treewidth w.r.t. any of the following classes: Horn, Anti-Horn, 2CNF.*

In order to prove Theorem 1, we first show that a backdoor of small treewidth can be used to design a fixed-parameter algorithm for solving SAT *if* such a backdoor is provided in the input; we note that the proof of this claim proceeds along the same lines as the proof for the analogous lemma in the case of constraint satisfaction problems [14].

Lemma 1. *Let F be a CNF formula and let X be a strong backdoor of F into a tractable class $\mathcal{F}$. There is an algorithm that, given F and X, runs in time $2^{tw(G_F^X)}|F|^{\mathcal{O}(1)}$ and correctly decides whether F is satisfiable or not.*

Proof. We prove the lemma by designing an algorithm which constructs an equivalent CSP instance $\mathcal{I}$ (over domain $\{0,1\}$) and then solves the instance in the specified time bound. The variables of $\mathcal{I}$ are precisely the set X. For each X-component Z of F with boundary B, we add a constraint c_B into $\mathcal{I}$ over $var(Z)$, where c_B contains one tuple for each assignment of B which can be extended to a satisfying assignment of Z. For instance, if $B = \{a, b, c\}$ and the only assignment which can be extended to a satisfying assignment for Z is $a \mapsto 0, b \mapsto 1, c \mapsto 1$, then c_B will have the scope (a, b, c) and the relation with a single tuple $(0, 1, 1)$.

We note that since B is a backdoor of Z to $\mathcal{F}$ of size at most $k = G_F^X$ (as follows from the fact that B is a clique in G_F^X and hence must fully lie in some bag of T, and from our discussion following the introduction of backdoors), we can loop through all assignments of B and test whether each such assignment can be extended to satisfy Z or not in time at most $2^k \cdot |B|^{\mathcal{O}(1)}$. Consequently, we can construct $\mathcal{I}$ from F and X in time $2^k|F|^{\mathcal{O}(1)}$. As an immediate consequence of this construction, we see that any assignment satisfying $\mathcal{I}$ can be extended to a satisfying assignment for F, and vice-versa the restriction of any satisfying assignment for F onto X is a satisfying assignment in $\mathcal{I}$.

Next, in order to solve $\mathcal{I}$, we recall the notion of the *primal graph* of a CSP instance. The primal graph of $\mathcal{I}$ is the graph whose vertex set is X and where two vertices a, b are adjacent iff there exists a constraint whose scope contains both a and b. Observe that in the construction of $\mathcal{I}$, two variables a, b will be adjacent if and only if they occur in some X-component of F, i.e., a, b are adjacent in $\mathcal{I}$ iff they are adjacent in G_F^X. Hence the primal graph of $\mathcal{I}$ is isomorphic to G_F^X, and in particular the primal graph of $\mathcal{I}$ must have treewidth at most k. To conclude the proof, we use the well-known fact that boolean CSP can be solved in time $2^t \cdot n^{\mathcal{O}(1)}$, where n is the number of variables and t the treewidth of the primal graph [27]. $\qquad\square$

With Lemma 1 in hand, it remains to show that a backdoor of small treewidth can be found efficiently (if such a backdoor exists). The main tool we will use in this respect is the following result, which solves the problem of finding backdoors of small treewidth in the context of CSP for classes defined via finite languages; we note that backdoor treewidth on backdoors for CSP is defined analogously as for SAT.

Proposition 4 ([14]). *Let Γ be a finite language. There exists a fixed-parameter algorithm which takes as input a CSP instance $\mathcal{I}$ and a parameter k, and either finds a backdoor of treewidth at most k into $CSP(\Gamma)$ or correctly determines that no such backdoor exists.*

In order to prove Proposition 4, Ganian et al. [14] introduced a subroutine that is capable of replacing large parts of the input CSP by a strictly smaller CSP in such a way that the existence of a backdoor of treewidth at most k into $CSP(\Gamma)$ is not affected. Their approach was inspired by the graph replacement tools dating back to the results of Fellows and Langston [11] and further developed by Arnborg, Bodlaender, and other authors [1,3–5,12]. Subsequently, they utilized the *recursive-understanding* technique, introduced by Grohe et al. [17] to show that as long as the instance has a size exceeding some function of k, then it is possible to find a large enough part of the input instance which can then be strictly reduced. Their theorem then follows by a repeated application of this subroutine, followed by a brute-force step at the end when the instance has size bounded by a function of k.

There are a few obstacles which prevent us from directly applying Proposition 4 to our SAT instances and backdoors to our classes of interest (Horn, Anti-Horn and 2CNF). First of all, while SAT instances admit a direct encoding into CSP, in case of unbounded clause width this can lead to an exponential blowup in the size of the instance. Second, the languages corresponding to Horn, Anti-Horn and 2CNF in such a direct encoding are not finite. Hence, instead of immediately using the direct encoding, we will proceed as follows: given a CNF formula F, we will first construct an auxiliary CNF formula F' of clause width at most 3 which is equivalent as far as the existence of backdoors is concerned. We note that the CNF formula F' constructed in this way will *not* be satisfiability-equivalent to F. But since F' has bounded clause width, we can then use a direct encoding of F' into CSP and apply Proposition 4 while circumventing the above obstacles.

Lemma 2. *Let $\mathcal{F} \in \{Horn, Anti\text{-}Horn, 2CNF\}$. There exists a polynomial-time algorithm which takes as input a CNF formula F and outputs a 3-CNF formula F' such that $var(F) = var(F')$ with the following property: for each $X \subseteq var(F)$, X is a backdoor of F into $\mathcal{F}$ if and only if X is a backdoor of F' into $\mathcal{F}$.*

Proof. We first describe the construction and then complete the proof by arguing that the desired property holds. We construct the CNF formula F' as follows: for each clause $c \in F$ of width at least 4, we loop over all sets of size 3 of literals from c and add each such set into F; afterwards, we remove c. Observe that

$|F'| \leq (2 \cdot |var(F)|)^3$ and each clause in F' is either equal to or "originated from" at least one clause in F.

Now consider a set $X \subseteq var(F)$ which is a backdoor of F into $\mathcal{F}$ (for an arbitrary choice of $\mathcal{F} \in \mathcal{H}$). We claim that X must also be a backdoor of F' into $\mathcal{F}$. Indeed, if $\mathcal{F} =$ Horn then for each assignment τ of X, each clause c in $F[\tau]$ must be a horn clause, i.e., cannot contain more than one positive literal. But then each c' in $F'[\tau]$ that originated from taking a subset of literals from c must also be a horn clause. If $\mathcal{F} = $ 2CNF then for each assignment τ of X, each clause c in $F[\tau]$ must contain at most two literals; once again, each c' in $F'[\tau]$ that originated from taking a subset of literals from c must also contain at most 2 literals. Finally, the same argument shows that the claim also holds if $\mathcal{F} = $ Anti-Horn.

On the other hand, consider a set $X \subseteq var(F)$ which is a backdoor of F' into some $\mathcal{F} \in \mathcal{H}$. Once again, we claim that X must also be a backdoor of F into $\mathcal{F}$. Indeed, let $\mathcal{F} = $ Horn and assume for a contradiction that there exists an assignment τ of X such that $F[\tau]$ contains a clause c with at least two positive literals, say a and b. Then $F'[\tau]$ must either also contain c, or it must contain a subset c' of c such that $a, b \in c'$; in either case, we arrive at a contradiction with X being a backdoor of F' into Horn. The argument for $\mathcal{F} = $ Anti-Horn is, naturally, fully symmetric. Finally, let $\mathcal{F} = $ 2CNF and assume for a contradiction that there exists an assignment τ of X such that $F[\tau]$ contains a clause c or width at least 3; let us pick three arbitrary literals in c, say a_1, a_2, a_3, and observe that these cannot be contained in X. Then, by construction, $F'[\tau]$ contains the clause $\{a_1, a_2, a_3\}$—contradicting the fact that X is a backdoor of F' into 2CNF. □

Our final task in this section is to use Lemma 2 to obtain an algorithm to detect a backdoor of small treewidth, which along with Lemma 1 implies Theorem 1.

Lemma 3. *Let $\mathcal{F} \in \{Horn, Anti\text{-}Horn, 2CNF\}$. There exists a fixed-parameter algorithm which takes as input a CNF formula F and a parameter k, and either finds a backdoor of treewidth at most k into $\mathcal{F}$ or correctly determines that no such backdoor exists.*

Proof. We begin by invoking Lemma 2 to obtain a 3-CNF formula F' which preserves the existence of backdoors. Next, we construct the direct encoding of F' into a CSP instance $\mathcal{I}$; since F' has bounded clause width, it holds that $|\mathcal{I}| \in \mathcal{O}(|F'|)$. For each choice of class $\mathcal{F}$, we construct the language Γ corresponding to the class in the setting of CSPs with arity at most 3. Specifically, if $\mathcal{F}$ is 2CNF, then Γ will contain all relations of arity at most 2; if $\mathcal{F}$ is Horn, then for each Horn clause of width at most 3, Γ will contain a relation with all tuples that satisfy the clause; and analogously for Anti-Horn. Finally, since Γ is a finite language, we invoke Proposition 4 to compute a backdoor of width at most k or correctly determine that no such backdoor exists in $\mathcal{I}$. Correctness follows from Lemma 2, the equivalence of $\mathcal{F}$ and CSP(Γ), and the natural correspondence of backdoors of F' into $\mathcal{F}$ to backdoors of $\mathcal{I}$ into CSP(Γ). □

4   Acyclic Backdoors to Scattered Classes for SAT

In this section, we build upon the notion of so-called *scattered classes* [15] and backdoor treewidth to introduce a new polynomial time tractable class of CNF formulas.

Definition 2. *Let $\mathcal{U} = \{\mathcal{F}_1, \dots, \mathcal{F}_r\}$ be a set of classes of CNF formulas. For a CNF formula F, we say that $X \subseteq var(F)$ is a backdoor of F into $\mathcal{U}$ if for every X-component F' and every partial assignment τ to $var(F') \cap X$, the CNF formula $F'[\tau]$ belongs to a class in $\mathcal{U}$.*

We let $\mathsf{btw}_\mathcal{U}(F) = \min\{\mathsf{tw}(G_F^X) \mid X \text{ is a strong backdoor of } F \text{ into } \mathcal{U}\}$, and observe that if $\mathcal{U} = \{\mathcal{F}\}$ then backdoors into $\mathcal{U}$ coincide with backdoors into $\mathcal{F}$. Next, we define the general property we will require for our base classes. We note that these precisely coincide with the notion of *permissive classes* [23], and that being *permissively tractable* (see Definition 3) is clearly a necessary condition for being able to use any sort of backdoor into $\mathcal{F}$.

Definition 3. *Let $\mathcal{F}$ be a class of CNF formulas closed under partial assignments. Then $\mathcal{F}$ is called* permissively tractable *if there is a polynomial time algorithm that, given a CNF formula F, either correctly concludes that $F \notin \mathcal{F}$ or correctly decides whether F is satisfiable.*

Similarly, the class $\mathcal{F}$ is called #-permissively tractable if there is a polynomial time algorithm that, given a CNF formula F either correctly concludes that $F \notin \mathcal{F}$ or correctly returns the number of satisfying assignments of F.

We now state the two main results of this section.

Theorem 2. *Let $\mathcal{U} = \{\mathcal{F}_1, \dots, \mathcal{F}_r\}$ be a set of permissively tractable classes. There is a polynomial time algorithm that decides whether any given CNF formula F with $\mathsf{btw}_\mathcal{U}(F) = 1$ is satisfiable.*

Theorem 3. *Let $\mathcal{U} = \{\mathcal{F}_1, \dots, \mathcal{F}_r\}$ be a set of #-permissively tractable classes. There is a polynomial time algorithm that counts the number of satisfying assignments for any given CNF formula F with $\mathsf{btw}_\mathcal{U}(F) = 1$.*

We call S an *acyclic* (strong) backdoor of F into $\mathcal{U}$ if it is a backdoor of F into $\mathcal{U}$ and the treewidth of the S-torso graph G_F^S is 1. Let us now illustrate a general high-level example of CNF formulas that can now be considered as polynomial time tractable due to Theorem 2. Let $\mathcal{U}$ contain the tractable class *Q-Horn* [6] and the tractable class of *hitting CNF formulas* [19,20]. The class *Q-Horn* is a proper superset of 2CNF, Horn and Anti-Horn, and *hitting CNF formulas* are those CNF formulas where, for each pair of clauses, there exists a variable x which occurs positively in one clause and negatively in the other. We can solve any CNF formula F which is iteratively built from "building blocks", each containing at most 2 backdoor variables such that any assignment of these variables turns the block into a hitting CNF formula or Q-Horn formula. Next we provide one example of such a building block (with backdoor variables x_1, x_2).

$$F' = \big\{ \{x_1, x_2, a, b, c\}, \{\overline{x_1}, x_2, \overline{a}, \overline{c}\}, \{\overline{x_1}, x_2, a, c\}, \{x_1, \overline{x_2}, \overline{a}, \overline{b}, c\},$$
$$\{\overline{x_1}, \overline{x_2}, a, b, c\}, \{\overline{x_1}, \overline{x_2}, a, \overline{b}, c\}, \{\overline{x_1}, \overline{x_2}, \overline{a}, b, \overline{c}\} \big\}$$

Observe that, for each assignment to x_1, x_2, we are either left with a Q-Horn formula (in case of all assignments except for $x_1, x_2 \mapsto 1$) or a hitting CNF formula (if $x_1, x_2 \mapsto 1$). Now we can use F' as well as any other building blocks with this general property (including blocks of arbitrary size) to construct a CNF formula F of arbitrary size by identifying individual backdoor variables inside the blocks in a tree-like pattern. For instance, consider the CNF formula F defined as follows. Let n be an arbitrary positive integer and define $var(F) = \{y_1, \ldots, y_n\} \bigcup \{a_i, b_i, c_i \mid 1 \le i \le n - 1\}$. We define

$$F = \bigcup_{i=1}^{n-1} \big\{ \{y_i, y_{i+1}, a_i, b_i, c_i\}, \{\overline{y_i}, y_{i+1}, \overline{a_i}, \overline{c_i}\}, \{\overline{y_i}, y_{i+1}, a_i, c_i\},$$
$$\{y_i, \overline{y_{i+1}}, \overline{a_i}, \overline{b_i}, c_i\}, \{\overline{y_i}, \overline{y_{i+1}}, a_i, b_i, c_i\},$$
$$\{\overline{y_i}, \overline{y_{i+1}}, a_i, \overline{b_i}, c_i\}, \{\overline{y_i}, \overline{y_{i+1}}, \overline{a_i}, b_i, \overline{c_i}\} \big\}.$$

Observe that $Y = \{y_1, \ldots, y_n\}$ is a strong backdoor of F into $\{$ *Q-Horn, Hitting CNF formulas* $\}$ and the Y-torso graph is a path.

We now proceed to proving Theorems 2 and 3. For technical reasons, we will assume that every clause in the given CNF formula occurs exactly twice. Observe that this does not affect the satisfiability of the CNF formula or the fact that a set of variables is an acyclic strong backdoor into $\mathcal{U}$. However, it does impose certain useful structure on the incidence graph of F, as is formalized in the following observation (recall that a *cut-vertex* is a vertex whose deletion disconnects at least 2 of its neighbors). For the following, recall that $\mathsf{Inc}(F)$ denotes the incidence graph of F.

Observation 1. *Let F be a CNF formula where every clause has a duplicate. Then, every cut-vertex of the graph $\mathsf{Inc}(F)$ corresponds to a variable-vertex.*

Proof. Let c and c' be the clause-vertices in $\mathsf{Inc}(F)$ corresponding to a clause and its duplicate. Observe that if c were a cut-vertex, then deleting it must disconnect at least two variables contained in c. But this leads to a contradiction due to the presence of the clause-vertex c'. □

The following lemma is a consequence of Proposition 3 and the fact that the size of the incidence graph of F is linear in the size of F.

Lemma 4. *There is an algorithm that takes as input a CNF formula F of size n and outputs the block decomposition of $\mathsf{Inc}(F)$ in time $\mathcal{O}(n)$.*

Henceforth, we will drop the explicit mention of F having duplicated clauses. We will also assume without loss of generality that $\mathsf{Inc}(F)$ is connected. Let (T, η)

be the block decomposition of $\mathsf{Inc}(F)$. For every $t \in T$, we define the tree β_t as the subtree of T rooted at t and we denote by γ_t the subformula of F induced by the clauses whose corresponding vertices are contained in the set $\bigcup_{t' \in V(\beta_t)} \eta(t')$.

We provide a useful observation which allows us to move freely between speaking about the CNF formula and its incidence graph. For a graph G and $X \subseteq V(G)$, we denote by $\mathsf{Torso}_G(X)$ the graph defined over the vertex set X as follows: for every pair of vertices $x_1, x_2 \in X$, we add the edge (x_1, x_2) if (a) $(x_1, x_2) \in E(G)$ or (b) x_1 and x_2 both have a neighbor in the same connected component of $G - X$.

Observation 2. *Let F be a CNF formula, $G = \mathsf{Inc}(F)$ and $X \subseteq var(X)$. Then, there is an edge between x_1 and x_2 in the X-torso graph G_F^X if and only if there is an edge between x_1 and x_2 in the graph $\mathsf{Torso}_G(X)$.*

The following lemma formalizes the crucial structural observation on which our algorithm is based.

Lemma 5. *Let F be a CNF formula and let $S \subseteq var(F)$ be such that the S-torso graph is a forest. Let $G = \mathsf{Inc}(F)$. Then, the following statements hold.*

1. *No block of G contains more than two variables of S.*
2. *In the rooted block decomposition $(T, \eta(T))$ of G, no block vertex has three distinct children x_0, x_1, x_2 such that $var(\gamma_{x_i})$ intersects S for every $i \in \{0, 1, 2\}$.*

Proof. Due to Observation 2, it follows that the graph $\mathsf{Torso}_G(S)$ is acyclic. Now, suppose that the first statement does not hold and let v_0, v_1, v_2 be variables of S contained in the same block of G. We prove that for every $i \in \{0, 1, 2\}$, there is a pair of internally vertex-disjoint v_i-$v_{i+1 \ (mod 3)}$ and v_i-$v_{i+2 \ (mod 3)}$ paths in $\mathsf{Torso}_G(S)$. This immediately implies that the graph $\mathsf{Torso}_G(S)$ contains a cycle; indeed, the existence of vertex-disjoint v_0-v_1 and v_0-v_2 paths would mean that any v_1-v_2 path disjoint from v_0 guarantees a cycle, and the existence of vertex-disjoint v_1-v_2 and v_1-v_0 paths means that such a v_1-v_2 path must exist. We only present the argument for $i = 0$ because the other cases are analogous.

Observe that since v_0, v_1, v_2 are in the same block of G, there are paths P and Q in G where P is a v_0-v_1 path, Q is a v_0-v_2 path and P and Q are internally vertex-disjoint. Let $p_1, \ldots, p_s$ be the vertices of S which appear (in this order) when traversing P from v_0. If no such vertex appears as an internal vertex of P then we set $p_1 = p_s = v_1$. Similarly, let $q_1, \ldots, q_r$ be the vertices of S which appear when traversing Q from v_0 and if no such vertex appears as an internal vertex of Q, then we set $q_1 = q_r = v_2$.

It follows from the definition of $\mathsf{Torso}_G(S)$ that if $s = 1$ ($r = 1$), then there are edges (v_0, v_1) and (v_0, v_2) in this graph. Otherwise, there are edges $(v_0, p_1), (p_1, p_2), \ldots, (p_s, v_1)$ and edges $(v_0, q_1), (q_1, q_2), \ldots, (q_r, v_2)$. In either case, we have obtained a pair of internally vertex-disjoint paths in $\mathsf{Torso}_G(S)$; one from v_0 to v_1 and the other from v_0 to v_2. This completes the argument for the first statement.

The proof for the second statement proceeds along similar lines. Suppose to the contrary that there are three cut-vertices x_0, x_1, x_2 which are children of a block-vertex b such that $var(\gamma_{x_i})$ intersects S for every $i \in \{0, 1, 2\}$ and let v_i be a variable of S chosen arbitrarily from γ_{x_i} (see Fig. 1). Observe that there are paths P_0, P_1, P_2 in G such that for every $i \in \{0, 1, 2\}$, the path P_i is a v_i-x_i path which is vertex-disjoint from any vertex (variable or clause) of G in $\gamma_{x_{i+1 \pmod 3}}$ or $\gamma_{x_{i+2 \pmod 3}}$. Without loss of generality, we assume that v_i is the *only* vertex of S on the path P_i.

Now, following the same argument as that for the first statement, the paths P_0, P_1, P_2 and the fact that x_0, x_1, x_2 are contained in the same block of G together imply that $\mathsf{Torso}_G(S)$ has a pair of internally vertex-disjoint v_i-$v_{i+1 \pmod 3}$ and v_i-$v_{i+2 \pmod 3}$ paths for every $i \in \{0, 1, 2\}$. This in turn implies that v_0, v_1, v_2 are in the same block of $\mathsf{Torso}_G(S)$. Hence, we conclude that $\mathsf{Torso}_G(S)$ is not acyclic, a contradiction. $\square$

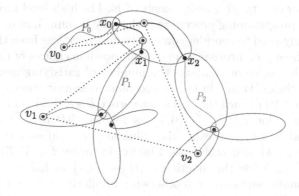

Fig. 1. An illustration of the configuration in the second statement of Lemma 5. The vertices denoted as concentric circles correspond to S. The dotted edges are edges of $\mathsf{Torso}_G(S)$.

We are now ready to present the proofs of Theorems 2 and 3. Since the proofs of both theorems are similar, we present them together. In what follows, we let $\mathcal{U} = \{\mathcal{F}_1, \ldots, \mathcal{F}_r\}$ be a set of permissively tractable classes and let $\mathcal{U}^\star = \{\mathcal{F}_1^\star, \ldots, \mathcal{F}_q^\star\}$ be a set of #-permissively tractable classes. For each $i \in \{1, \ldots, r\}$, we let $\mathcal{A}_i$ denote a polynomial time algorithm that certifies that $\mathcal{F}_i$ is permissively tractable and for each $i \in \{1, \ldots, q\}$, we let $\mathcal{A}_i^\star$ denote a polynomial time algorithm that certifies that $\mathcal{F}_i^\star$ is permissively tractable. Finally, let F be a CNF formula such that $\mathsf{btw}_{\mathcal{U}}(F) = \mathsf{btw}_{\mathcal{U}^\star}(F) = 1$.

Proof (of Theorems 2 and 3). Let S be a hypothetical strong backdoor of F into $\mathcal{U}$ ($\mathcal{U}^\star$) such that the S-torso graph is acyclic. We first handle the case when $|S| \leq 2$. In this case, we simply go over all possible pairs of variables and by assuming that they form a strong backdoor of F into $\mathcal{U}$ (respectively $\mathcal{U}^\star$), go over all instantiations of this pair of variables and independently solve (count the satisfying assignments of) each distinct connected component of the resulting CNF formula. For this, we make use of the polynomial time algorithms $\mathcal{A}_1, \ldots, \mathcal{A}_r$ and $\mathcal{A}_1^\star, \ldots, \mathcal{A}_q^\star$ respectively.

This step takes polynomial time and if $|S| \leq 2$, then for *some* guess of a pair of variables, we will be able to correctly determine whether F is satisfiable (correctly compute the number of satisfying assignments of F). In the case when for every pair $x, y \in var(F)$ there is an assignment $\tau : \{x, y\} \to \{0, 1\}$ and a connected component of $F[\tau]$ which is found to be not in any class of $\mathcal{U}$ (respectively $\mathcal{U}^\star$), it must be the case that $|S| > 2$.

Since S must have at least 3 variables, it follows from Lemma 5 (1) that S cannot be contained in a single block of $\mathsf{Inc}(F)$, which implies that $\mathsf{Inc}(F)$ has at least 2 distinct blocks. We now invoke Lemma 4 to compute (T, η), the block decomposition of $\mathsf{Inc}(F)$ (recall that by our assumption, $\mathsf{Inc}(F)$ is connected) and pick an arbitrary cut-vertex as the root for T. We will now execute a bottom up parse of T and use dynamic programming to test whether F is satisfiable and count the number of satisfying assignments of F. The high-level idea at the core of this dynamic programming procedure is that, by Lemma 5, at each block and cut-vertex we only need to consider constantly many vertices from the backdoor; by "guessing" these (i.e., brute-forcing over all possible choices of these), we can dynamically either solve or compute the number of satisfying assignments for the subformula "below them" in the rooted block decomposition.

For every $t, b \in V(T)$ such that t is a cut-vertex, b is a child of t and $i \in \{0, 1\}$, we define the function $\delta_i(t, b) \to \{0, 1\}$ as follows: $\delta_0(t, b) = 1$ if $\gamma_b[t \mapsto 0]$ is satisfiable and $\delta_0(t, b) = 0$ otherwise. Similarly, $\delta_1(t, b) = 1$ if $\gamma_b[t \mapsto 1]$ is satisfiable and $\delta_1(t, b) = 0$ otherwise. Finally, for every $t \in V(T)$ such that t is a cut-vertex, we define the function $\alpha_i(t) \to \{0, 1\}$ as follows. $\alpha_0(t) = 1$ if $\gamma_t[t = 0]$ is satisfiable and $\alpha_0(t) = 0$ otherwise. Similarly, $\alpha_1(t) = 1$ if $\gamma_t[t = 1]$ is satisfiable and $\alpha_1(t) = 0$ otherwise. Clearly, the formula is satisfiable if and only if $\alpha_1(\hat{t}) = 1$ or $\alpha_0(\hat{t}) = 1$, where $\hat{t}$ is the root of T.

Similarly, for every $t, b \in V(T)$ such that t is a cut-vertex, b is a child of t and $i \in \{0, 1\}$, we define $\delta_i^\star(t, b)$ to be the number of satisfying assignments of the CNF formula $\gamma_b[t = i]$. For every $t \in V(T)$ such that t is a cut-vertex and $i \in \{0, 1\}$, we define $\alpha_i^\star(t)$ to be the number of satisfying assignments of $\gamma_t[t = i]$. Clearly, the number of satisfying assignments of F is $\alpha_0^\star(\hat{t}) + \alpha_1^\star(\hat{t})$, where $\hat{t}$ denotes the root of T.

Due to Observation 1, every cut-vertex corresponds to a variable and hence the functions δ and $\delta^\star$ are well-defined. We now proceed to describe how we compute the functions δ, $\delta^\star$, α, and $\alpha^\star$ at each vertex of T assuming that the corresponding functions have been correctly computed at each child of the vertex.

We begin with the leaf vertices of T. Let b be a leaf in T. We know that b corresponds to a block of $\mathsf{Inc}(F)$ and it follows from Lemma 5 (1) that S contains at most 2 variables of γ_b. Let $Z_b = S \cap var(\eta(b))$. We guess (i.e., branch over all possible choices for) the set Z_b and for every $\tau : Z_b \cup \{t\} \to \{0, 1\}$, we run the algorithms $\mathcal{A}_1, \ldots, \mathcal{A}_r$ (respectively $\mathcal{A}_1^\star, \ldots, \mathcal{A}_q^\star$) on the CNF formula $\gamma_b[\tau]$ to decide whether $\gamma_b[\tau]$ is satisfiable or unsatisfiable (respectively count the satisfying assignments of $\gamma_b[\tau]$) or it is not in $\mathcal{F}$ for any $\mathcal{F} \in \mathcal{U}$. By going over all possible partial assignments $\tau : Z_b \cup \{t\} \to \{0,1\}$ in this way, we can compute $\delta_i(t,b)$ and $\delta_i^*(t,b)$ for $i \in \{0,1\}$. Hence, we may assume that we have computed the functions $\delta_i(t,b)$ and $\delta_i^*(t,b)$ for $i \in \{0,1\}$ for every leaf b. We now proceed to describe the computation of α, α^*, δ and δ^* for the internal nodes of the tree.

Let t be a cut-vertex in T such that $\delta_i(t,b)$ $(\delta_i^*(t,b))$ has been computed for every $i \in \{0,1\}$ and $b \in \mathsf{child}(t)$. Then, $\alpha_i(t)$ is simply defined as $\bigwedge_{b \in \mathsf{child}(t)} \delta_i(t,b)$ for each $i \in \{0,1\}$. On the other hand, for each $i \in \{0,1\}$ $\alpha_i^*(t)$ is defined as $\prod_{b \in \mathsf{child}(t)} \delta_i(t,b)$.

Finally, let b be a block-vertex in T such that for every $i \in \{0,1\}$, the value $\alpha_i(t)$ $(\alpha_i^*(t))$ has been computed for every child t of b. Let t^* be the parent of b in T. It follows from Lemma 5 (2) that for at most 2 children $t_1, t_2 \in \mathsf{child}(b)$, the CNF formulas γ_{t_1} and γ_{t_2} contain a variable of S. Furthermore, it follows from Lemma 5 (1) that at most 2 variables of S are contained in $\eta(b)$. This implies that the CNF formula $\gamma_b \backslash (\gamma_{t_1} \cup \gamma_{t_2})$ has a strong backdoor of size at most 2 into $\mathcal{U}$ (respectively $\mathcal{U}^*$). Hence, we can simply guess the set $Z = \{t_1, t_2\} \cup (S \cap \eta(b))$ which has size at most 4. We can then use the polynomial time algorithms $\mathcal{A}_1, \ldots, \mathcal{A}_r$ $(\mathcal{A}_1^*, \ldots, \mathcal{A}_q^*)$ to solve (count the satisfying assignments of) the CNF formula $(\gamma_b \backslash (\gamma_{t_1} \cup \gamma_{t_2}))[\tau]$ for every partial assignment $\tau : Z \cup \{t^*\} \to \{0,1\}$ and along with the pre-computed functions $\alpha_i(t_j)$, $\alpha_i^*(t_j)$ for $i \in \{0,1\}, j \in \{1,2\}$, compute $\delta_p(t^*,b)$ and $\delta_p^*(t^*,b)$ for each $p \in \{0,1\}$. While computing $\delta_p(t^*,b)$ is straightforward, note that $\delta_p^*(t^*,b)$ is defined as $\sum_{\tau:\tau(t^*)=p} (\alpha_{\tau(t_1)}(t_1) \cdot \alpha_{\tau(t_2)}(t_2) \cdot \ell_\tau)$, where ℓ_τ is the number of satisfying assignments of $(\gamma_b \backslash (\gamma_{t_1} \cup \gamma_{t_2}))[\tau]$.

The only remaining technical subtlety in the case of counting satisfying assignments is the following. If b has exactly one child, then t_2 is left undefined and we call the unique child t_1 and work only with it in the definition of $\delta_p(b,t^*)$. In other words, we remove the term $\alpha_{\tau(t_2)}(t_2)$ from the definition of $\delta_p^*(t^*,b)$. On the other hand, if b has at least two children but there is exactly one $t \in \mathsf{child}(b)$ such that γ_t contains a variable of S, then we set $t_1 = t$ and t_2 to be an arbitrary child of b distinct from t. Finally, if b has at least two children and there is no $t \in \mathsf{child}(b)$ such that γ_t contains a variable of S, then we define t_1 and t_2 to be an arbitrary pair of children of b. Since the set of possibilities has constant size, we can simply iterate over all of them.

Since we go over a constant number (at most 2^5) of partial assignments of $Z \cup \{t^*\}$, we will execute the algorithms $\mathcal{A}_1, \ldots, \mathcal{A}_r$ $(\mathcal{A}_1^*, \ldots, \mathcal{A}_q^*)$ only a constant number of times each. Therefore, this step also takes polynomial time, and the algorithm as a whole runs in polynomial time. This completes the proof of both theorems.   □

5   Conclusions

We have introduced the notion of backdoor treewidth in the context of SAT and developed algorithms for deciding the satisfiability of formulas of small backdoor treewidth: (1) a fixed-parameter tractability result for backdoor treewidth with respect to Horn, Anti-Horn, and 2CNF, and (2) a polynomial-time result for backdoor treewidth 1 with respect to a wide range or archipelagos of tractability. Both results significantly extend the borders of tractability for SAT. Our work also points to several avenues for interesting future research. In particular, our first result raises the question of whether there are further tractable classes w.r.t. which backdoor treewidth allows fixed-parameter tractability of SAT. Our second result provides a promising starting point towards the goal of obtaining a polynomial time algorithm for SAT (and #-SAT) for *every fixed* value of the backdoor treewidth with respect to a set of permissively tractable classes.

References

1. Arnborg, S., Courcelle, B., Proskurowski, A., Seese, D.: An algebraic theory of graph reduction. In: Ehrig, H., Kreowski, H.-J., Rozenberg, G. (eds.) Graph Grammars 1990. LNCS, vol. 532, pp. 70–83. Springer, Heidelberg (1991). doi:10.1007/BFb0017382
2. Biere, A., Heule, M., van Maaren, H., Walsh, T. (eds.): Handbook of Satisfiability. Frontiers in Artificial Intelligence and Applications, vol. 185. IOS Press, Amsterdam (2009)
3. Bodlaender, H.L., Fluiter, B.: Reduction algorithms for constructing solutions in graphs with small treewidth. In: Cai, J.-Y., Wong, C.K. (eds.) COCOON 1996. LNCS, vol. 1090, pp. 199–208. Springer, Heidelberg (1996). doi:10.1007/3-540-61332-3_153
4. Bodlaender, H.L., van Antwerpen-de Fluiter, B.: Reduction algorithms for graphs of small treewidth. Inf. Comput. **167**(2), 86–119 (2001)
5. Bodlaender, H.L., Hagerup, T.: Parallel algorithms with optimal speedup for bounded treewidth. In: Fülöp, Z., Gécseg, F. (eds.) ICALP 1995. LNCS, vol. 944, pp. 268–279. Springer, Heidelberg (1995). doi:10.1007/3-540-60084-1_80
6. Boros, E., Hammer, P.L., Sun, X.: Recognition of q-Horn formulae in linear time. Discr. Appl. Math. **55**(1), 1–13 (1994)
7. Cormen, T.H., Leiserson, C.E., Rivest, R.L., Stein, C.: Introduction to Algorithms, 3rd edn. MIT Press, Cambridge (2009). http://mitpress.mit.edu/books/introduction-algorithms
8. Cygan, M., Fomin, F.V., Kowalik, L., Lokshtanov, D., Marx, D., Pilipczuk, M., Pilipczuk, M., Saurabh, S.: Parameterized Algorithms. Springer, Cham (2015). doi:10.1007/978-3-319-21275-3
9. Diestel, R.: Graph Theory. Graduate Texts in Mathematics, vol. 173, 4th edn. Springer, Heidelberg (2012). doi:10.1007/978-3-662-53622-3
10. Downey, R.G., Fellows, M.R.: Fundamentals of Parameterized Complexity. Texts in Computer Science. Springer, London (2013). doi:10.1007/978-1-4471-5559-1
11. Fellows, M.R., Langston, M.A.: An analogue of the Myhill-Nerode theorem and its use in computing finite-basis characterizations (extended abstract). In: FOCS, pp. 520–525 (1989)
12. de Fluiter, B.: Algorithms for graphs of small treewidth. Ph.D. thesis, Utrecht University (1997)
13. Fomin, F.V., Lokshtanov, D., Misra, N., Ramanujan, M.S., Saurabh, S.: Solving d-SAT via backdoors to small treewidth. In: Proceedings of the Twenty-Sixth Annual ACM-SIAM Symposium on Discrete Algorithms, SODA 2015, San Diego, CA, USA, pp. 630–641, 4–6 January 2015 (2015)

14. Ganian, R., Ramanujan, M.S., Szeider, S.: Combining treewidth and backdoors for CSP. In: 34th Symposium on Theoretical Aspects of Computer Science (STACS 2017). Leibniz International Proceedings in Informatics (LIPIcs), vol. 66, pp. 36:1–36:17. Schloss Dagstuhl-Leibniz-Zentrum fuer Informatik (2017)
15. Ganian, R., Ramanujan, M.S., Szeider, S.: Discovering archipelagos of tractability for constraint satisfaction and counting. ACM Trans. Algorithms **13**(2), 29:1–29:32 (2017). http://doi.acm.org/10.1145/3014587
16. Gaspers, S., Szeider, S.: Backdoors to satisfaction. In: Bodlaender, H.L., Downey, R., Fomin, F.V., Marx, D. (eds.) The Multivariate Algorithmic Revolution and Beyond. LNCS, vol. 7370, pp. 287–317. Springer, Heidelberg (2012). doi:10.1007/978-3-642-30891-8_15
17. Grohe, M., Kawarabayashi, K., Marx, D., Wollan, P.: Finding topological subgraphs is fixed-parameter tractable. In: Proceedings of the 43rd ACM Symposium on Theory of Computing, STOC 2011, San Jose, CA, USA, pp. 479–488, 6–8 June 2011
18. Hopcroft, J.E., Tarjan, R.E.: Efficient algorithms for graph manipulation [H] (algorithm 447). Commun. ACM **16**(6), 372–378 (1973)
19. Kleine Büning, H., Kullmann, O.: Minimal unsatisfiability and autarkies, Chap. 11. In: Biere, A., Heule, M.J.H., van Maaren, H., Walsh, T. (eds.) Handbook of Satisfiability. Frontiers in Artificial Intelligence and Applications, vol. 185, pp. 339–401. IOS Press, Amsterdam (2009)
20. Kleine Büning, H., Zhao, X.: Satisfiable formulas closed under replacement. In: Kautz, H., Selman, B. (eds.) Proceedings for the Workshop on Theory and Applications of Satisfiability. Electronic Notes in Discrete Mathematics, vol. 9. Elsevier Science Publishers, North-Holland (2001)
21. Kloks, T. (ed.): Treewidth: Computations and Approximations. LNCS, vol. 842. Springer, Heidelberg (1994). doi:10.1007/BFb0045375
22. Nishimura, N., Ragde, P., Szeider, S.: Detecting backdoor sets with respect to Horn and binary clauses. In: Proceedings of Seventh International Conference on Theory and Applications of Satisfiability Testing (SAT 2004), Vancouver, BC, Canada, pp. 96–103, 10–13 May 2004
23. Ordyniak, S., Paulusma, D., Szeider, S.: Satisfiability of acyclic and almost acyclic CNF formulas. Theor. Comput. Sci. **481**, 85–99 (2013)
24. Robertson, N., Seymour, P.D.: Graph minors. II. Algorithmic aspects of tree-width. J. Algorithms **7**(3), 309–322 (1986)
25. Samer, M., Szeider, S.: Fixed-parameter tractability, Chap. 13. In: Biere, A., Heule, M., van Maaren, H., Walsh, T. (eds.) Handbook of Satisfiability, pp. 425–454. IOS Press, Amsterdam (2009)
26. Samer, M., Szeider, S.: Algorithms for propositional model counting. J. Discrete Algorithms **8**(1), 50–64 (2010)
27. Samer, M., Szeider, S.: Constraint satisfaction with bounded treewidth revisited. J. Comput. Syst. Sci. **76**(2), 103–114 (2010)
28. Williams, R., Gomes, C., Selman, B.: On the connections between backdoors, restarts, and heavy-tailedness in combinatorial search. In: Informal Proceedings of the Sixth International Conference on Theory and Applications of Satisfiability Testing (SAT 2003), S. Margherita Ligure - Portofino, Italy, pp. 222–230, 5–8 May 2003

New Width Parameters for Model Counting

Robert Ganian[✉] and Stefan Szeider

Algorithms and Complexity Group, TU Wien, Vienna, Austria
`{ganian,sz}@ac.tuwien.ac.at`

Abstract. We study the parameterized complexity of the propositional model counting problem #SAT for CNF formulas. As the parameter we consider the treewidth of the following two graphs associated with CNF formulas: the consensus graph and the conflict graph. Both graphs have as vertices the clauses of the formula; in the consensus graph two clauses are adjacent if they do not contain a complementary pair of literals, while in the conflict graph two clauses are adjacent if they do contain a complementary pair of literals. We show that #SAT is fixed-parameter tractable for the treewidth of the consensus graph but W[1]-hard for the treewidth of the conflict graph. We also compare the new parameters with known parameters under which #SAT is fixed-parameter tractable.

1 Introduction

Propositional model counting (#SAT) is the problem of determining the number of models (satisfying truth assignments) of a given propositional formula in conjunctive normal form (CNF). This problem arises in several areas of artificial intelligence, in particular in the context of probabilistic reasoning [1, 23]. The problem is well-known to be #P-complete [29], and remains #P-hard even for monotone 2CNF formulas and Horn 2CNF formulas. Thus, in contrast to the decision problem SAT, restricting the syntax of instances does not lead to tractability.

An alternative to restricting the syntax is to impose *structural restrictions* on the input formulas. Structural restrictions can be applied in terms of certain parameters (invariants) of *graphical models*, i.e., of certain graphs associated with CNF formulas. Among the most frequently used graphical models are *primal graphs* (sometimes called *variable interaction graphs* or *VIGs*), *dual graphs*, and *incidence graphs* (see Fig. 1 for definitions and examples).

The most widely studied and prominent graph parameter is *treewidth*, which was introduced by Robertson and Seymour in their Graph Minors Project. Small treewidth indicates that a graph resembles a tree in a certain sense (e.g., trees have treewidth 1, cyles have treewidth 2, cliques on $k+1$ vertices have treewidth k). Many otherwise NP-hard graph problems are solvable in polynomial time for graphs of bounded treewidth. It is generally believed that many practically

---

Supported by the Austrian Science Fund (FWF), project P26696. Robert Ganian is also affiliated with FI MU, Brno, Czech Republic.

S. Gaspers and T. Walsh (Eds.): SAT 2017, LNCS 10491, pp. 38–52, 2017.
DOI: 10.1007/978-3-319-66263-3_3

relevant problems actually do have low treewidth [2]. Treewidth is based on certain decompositions of graphs, called tree decompositions, where sets of vertices ("bags") of a graph are arranged at the nodes of a tree such that certain conditions are satisfied (see Sect. 2.3). If a graph has treewidth k then it admits a tree decomposition of *width* k, i.e., a tree decomposition where all bags have size at most $k + 1$.

Depending on whether we consider the treewidth of the primal, dual, or incidence graph of a given CNF formula, we speak of the primal, dual, or incidence treewidth of the formula, respectively. It is known that the number of models of a CNF formula of size L with primal, dual, or incidence treewidth k can be computed in time $f(k)L^c$ for a computable function f and a constant c which is independent of k; in other words, #SAT is *fixed-parameter tractable* parameterized by primal, dual, or incidence treewidth (see, e.g., [26]).

1.1   Contribution

In this paper we consider the treewidth of two further graphical models: the *consensus graph* and the *conflict graph* (see, e.g., [10,18,27]), giving rise to the parameters *consensus treewidth* and *conflict treewidth*, respectively. Both graphs have as their vertices the clauses of the formula. In the consensus graph two clauses are adjacent if they do not contain a complementary pair of literals; in the conflict graph, two clauses are adjacent if they do contain a complementary pair of literals (see Fig. 1 for examples). Here, we study the parameterized complexity of #SAT with respect to the new parameters and provide a comparison to known parameters under which #SAT is fixed-parameter tractable.

Our main result regarding consensus treewidth is a novel fixed-parameter algorithm for model counting (Theorem 1). The algorithm is based on dynamic programming along a tree decomposition of the consensus graph. This result is particularly interesting as none of the known parameters under which #SAT is fixed-parameter tractable dominates consensus treewidth, in the sense that there are instances of small consensus treewidth where all the other parameters can be arbitrarily large (Proposition 1). Hence consensus treewidth pushes the state-of-the-art for fixed-parameter tractability of #SAT further, and moreover does so via a parameter that forms a natural counterpart to the already established primal, dual and incidence treewidth parameters. We also note that the presented fixed-parameter algorithm generalizes the polynomial-time algorithm on hitting formulas (see Fact 1 below).

This positive result is complemented by our results on conflict treewidth. First we observe that when considering the conflict treewidth one needs to restrict the scope to formulas without pure literals: recall that #SAT remains #P-complete for monotone 2-CNF formulas, and the conflict graph of such formulas is edge-less and therefore of treewidth 0. We show that conflict treewidth in its general form does not provide a parameter under which #SAT is fixed-parameter tractable, even for formulas without pure literals (subject to the well-established complexity theoretic assumption $W[1] \neq FPT$ [8]). In fact, we show that already the decision problem SAT for formulas without pure literals is W[1]-hard when parameterized by conflict treewidth, or even by a weaker parameter, the size of a smallest vertex cover of the conflict graph (Proposition 2). However, if we bound in addition also the width of clauses (i.e., the number of literals in clauses), then #SAT becomes fixed-parameter tractable for formulas without pure literals. This result, however, does not add anything new to the complexity landscape, as we show that the incidence treewidth of a formula without pure literals is upper bounded by a function of conflict treewidth and clause width (Proposition 3).

2 Preliminaries

The set of natural numbers (that is, positive integers) will be denoted by $\mathbb{N}$. For $i \in \mathbb{N}$ we write $[i]$ to denote the set $\{1, \ldots, i\}$.

2.1 SAT and #SAT

We consider propositional formulas in conjunctive normal form (CNF), represented as sets of clauses. That is, a *literal* is a (propositional) variable x or a negated variable $\overline{x}$; a *clause* is a finite set of literals not containing a complementary pair x and $\overline{x}$; a *formula* is a finite set of clauses.

For a literal $l = \overline{x}$ we write $\overline{l} = x$; for a clause C we set $\overline{C} = \{\overline{l} \mid l \in C\}$. For a clause C, $var(C)$ denotes the set of variables x with $x \in C$ or $\overline{x} \in C$, and the *width* of C is $|var(C)|$. Similarly, for a formula F we write $var(F) = \bigcup_{C \in F} var(C)$. The *length* of a formula F is the total number of literals it contains, i.e., $\sum_{C \in F} |var(C)|$. We say that two clauses C, D *overlap* if $C \cap D \neq \emptyset$; we say that C and D *clash* if C and $\overline{D}$ overlap. Note that two clauses can clash and overlap at the same time. Two clauses C, D are *adjacent* if $var(C) \cap var(D) \neq \emptyset$. A variable is *pure* if it only occurs as either a positive literal or as a negative literal; the literals of a pure variable are then called pure literals.

The *dual graph* of a formula F is the graph whose vertices are clauses of F and whose edges are defined by the adjacency relation of clauses. We will also make references to the *primal graph* and the *incidence graph* of a formula F. The former is the graph whose vertices are the variables of F and where two variables a, b are adjacent iff there exists a clause C such that $a, b \in var(C)$, while the latter is the graph whose vertices are the variables and clauses of F and where two vertices a, b are adjacent iff a is a clause and $b \in var(a)$ (see Fig. 1 for an illustration).

A *truth assignment* (or *assignment*, for short) is a mapping $\tau : X \to \{0, 1\}$ defined on some set X of variables. We extend τ to literals by setting $\tau(\overline{x}) = 1 - \tau(x)$ for $x \in X$. $F[\tau]$ denotes the formula obtained from F by removing all clauses that contain a literal x with $\tau(x) = 1$ and by removing from the remaining clauses all literals y with $\tau(y) = 0$; $F[\tau]$ is the *restriction* of F to τ. Note that $var(F[\tau]) \cap X = \emptyset$ holds for every assignment $\tau : X \to \{0, 1\}$ and every formula F. An assignment $\tau : X \to \{0, 1\}$ *satisfies* a formula F if $F[\tau] = \emptyset$.

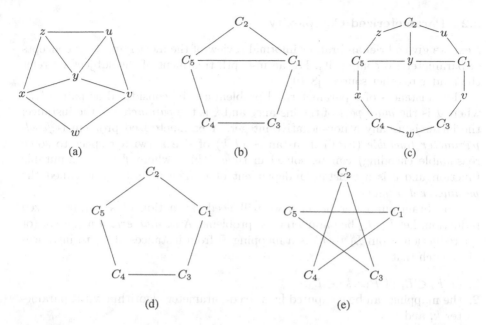

Fig. 1. The primal graph (a), dual graph (b), incidence graph (c), conflict graph (d) and consensus graph (e) of the formula $\{C_1, \ldots, C_5\}$ with $C_1 = \{u, \overline{v}, y\}$, $C_2 = \{\overline{u}, z, \overline{y}\}$, $C_3 = \{v, \overline{w}\}$, $C_4 = \{w, \overline{x}\}$, $C_5 = \{x, y, \overline{z}\}$. (a) The *primal graph* has as vertices the variables of the given formula, two variables are joined by an edge if they occur together in a clause. (b) The *dual graph* has as vertices the clauses, two clauses are joined by an edge if they share a variable. (c) The *incidence graph* is a bipartite graph where one vertex class consists of the clauses and the other consists of the variables; a clause and a variable are joined by an edge if the variable occurs in the clause. (d) The *conflict graph* has as vertices the clauses of the formula, two clauses are joined by an edge if they do contain a complementary pair of literals. (e) The *consensus graph* has as vertices the clauses of the formula, two clauses are joined by an edge if they do not contain a complementary pair of literals.

A truth assignment $\tau : \mathit{var}(F) \rightarrow \{0, 1\}$ that satisfies F is a *model* of F. We denote by $\#(F)$ the number of models of F. A formula F is *satisfiable* if $\#(F) > 0$. In the SAT problem, we are given a formula F and the task is to determine whether F is satisfiable. In the #SAT problem, we are also given a formula F and the task is to compute $\#(F)$.

A *hitting formula* is a CNF formula with the property that any two of its clauses clash (see [14,15,20]). The following result makes SAT and #SAT easy for hitting formulas.

Fact 1 ([13]). *A hitting formula F with n variables has exactly $2^n - \sum_{C \in F} 2^{n-|C|}$ models.*

2.2 Parameterized Complexity

Next we give a brief and rather informal review of the most important concepts of parameterized complexity. For an in-depth treatment of the subject we refer the reader to other sources [8,19].

The instances of a parameterized problem can be considered as pairs (I, k) where I is the *main part* of the instance and k is the *parameter* of the instance; the latter is usually a non-negative integer. A parameterized problem is *fixed-parameter tractable* (FPT) if instances (I, k) of size n (with respect to some reasonable encoding) can be solved in time $f(k)n^c$ where f is a computable function and c is a constant independent of k. The function f is called the *parameter dependence*.

To obtain our lower bounds, we will need the notion of a parameterized reduction. Let L_1, L_2 be parameterized problems. A *parameterized reduction* (or fpt-reduction) from L_1 to L_2 is a mapping P from instances of L_1 to instances of L_2 such that

1. $(x, k) \in L_1$ iff $P(x, k) \in L_2$,
2. the mapping can be computed by a fixed-parameter algorithm w.r.t. parameter k, and
3. there is a computable function g such that $k' \leq g(k)$, where $(x', k') = P(x, k)$.

The class W[1] captures parameterized intractability and contains all parameterized decision problems that are fpt-reducible to MULTICOLORED CLIQUE (defined below) [8]. Showing W[1]-hardness for a problem rules out the existence of a fixed-parameter algorithm unless the Exponential Time Hypothesis fails.

MULTICOLORED CLIQUE
Instance: A k-partite graph $G = (V, E)$ with a partition $V_1, \ldots, V_k$ of V.
Parameter: The integer k.
Question: Are there vertices $v_1, \ldots, v_k$ such that $v_i \in V_i$ and $\{v_i, v_j\} \in E$ for all i and j with $1 \leq i < j \leq k$ (i.e. the subgraph of G induced by $\{v_1, \ldots, v_k\}$ is a clique of size k)?

2.3 Treewidth

Let G be a simple, undirected, finite graph with vertex set $V = V(G)$ and edge set $E = E(G)$. A *tree decomposition* of G is a pair $(T, \{B_i : i \in I\})$ where $B_i \subseteq V$, T is a tree, and $I = V(T)$ such that:

1. for each edge $uv \in E$, there is an $i \in I$ such that $\{u, v\} \subseteq B_i$, and
2. for each vertex $v \in V$, $T[\{i \in I \mid v \in B_i\}]$ is a (connected) tree with at least one node.

The *width* of a tree decomposition is $\max_{i \in I} |B_i| - 1$. The *treewidth* [16,22] of G is the minimum width taken over all tree decompositions of G and it is denoted by $\mathbf{tw}(G)$. We call the elements of I *nodes* and B_i *bags*. As an example, consider the graphs depicted in Fig. 1: graphs (b), (d), (e) have treewidth 2, while graphs (a) and (c) have treewidth 3.

While it is possible to compute the treewidth exactly using a fixed-parameter algorithm [3], the asymptotically best running time is achieved by using the recent state-of-the-art 5-approximation algorithm of Bodlaender et al. [4].

Fact 2 ([4]). *There exists an algorithm which, given an n-vertex graph G and an integer k, in time $2^{\mathcal{O}(k)} \cdot n$ either outputs a tree decomposition of G of width at most $5k + 4$ and $\mathcal{O}(n)$ nodes, or correctly determines that $\mathrm{tw}(G) > k$.*

For other standard graph-theoretic notions not defined here, we refer to [7]. It is well known that, for every clique over $Z \subseteq V(G)$ in G, it holds that every tree decomposition of G contains an element B_i such that $Z \subseteq B_i$ [16]. Furthermore, if i separates a node j from another node l in T, then B_i separates $B_j \setminus B_i$ from $B_l \setminus B_i$ in G [16]; this *inseparability* property will be useful in some of our later proofs..

A tree decomposition $(T, \mathcal{B})$ of a graph G is *nice* if the following conditions hold:

1. T is rooted at a node r.
2. Every node of T has at most two children.
3. If a node t of T has two children t_1 and t_2, then $B_t = B_{t_1} = B_{t_2}$; in that case we call t a *join node*.
4. If a node t of T has exactly one child t', then exactly one of the following holds:
   (a) $|B_t| = |B_{t'}| + 1$ and $B_{t'} \subset B_t$; in that case we call t an *introduce node*.
   (b) $|B_t| = |B_{t'}| - 1$ and $B_t \subset B_{t'}$; in that case we call t a *forget node*.
5. If a node t of T is a leaf, then $|B_t| = 1$; we call these *leaf nodes*.

The main advantage of nice tree decompositions is that they allow the design of much more transparent dynamic programming algorithms, since one only needs to deal with four very specific types of nodes. It is well known (and easy to see) that for every fixed k, given a tree decomposition of a graph $G = (V, E)$ of width at most k and with $\mathcal{O}(|V|)$ nodes, one can construct in linear time a nice tree decomposition of G with $\mathcal{O}(|V|)$ nodes and width at most k [5]. We say that a vertex v was *forgotten* below a node $t \in V(T)$ if the subtree rooted at t contains a (forget) node s with a child s' such that $B_{s'} \setminus B_s = \{v\}$.

Finally, we summarize known algorithms for SAT and #SAT when parameterized by the treewidth of the three natural graph representations discussed in previous Subsect. 2.1; we note that the original results assumed that a tree decomposition is supplied as part of the input, and we can obtain one using Fact 2 (even while retaining the running time bounds).

Fact 3 ([26]). *#SAT is FPT when parameterized by the treewidth of any of the following graphical models of the formula: the incidence graph, the primal graph, or the dual graph.*

3 Consensus Treewidth

Recall that the consensus graph of a CNF formula F is the graph G whose vertices are the clauses of F and which contains an edge ab iff clauses a and b do not clash. Observe that the consensus graph of a hitting formula is edgeless. The consensus treewidth of F, denoted $\mathbf{contw}(F)$, is then the treewidth of its consensus graph.

Before proceeding to the algorithm, we make a short digression comparing the new notion of consensus treewidth to established parameters for SAT. We say that parameter X *dominates* parameter Y if there exists a computable function f such that for each formula F we have $X(F) \leq f(Y(F))$ [25]. In particular, if X dominates Y and SAT is FPT parameterized by X, then SAT is FPT parameterized by Y [25]. We say that two parameters are *incomparable* if neither dominates the other. We note that in our comparison, we only consider parameters which are known to give rise to fixed-parameter algorithms for SAT (i.e., not *incidence clique-width* [21]) and can be used without requiring additional information from an oracle (i.e., not *PS-width* [24]).

In the following, we show that consensus treewidth is incomparable with the *signed clique-width* [6,28] (the clique-width of the signed incidence graph; we note that a decomposition for signed clique-width can be approximated by using signed rank-width [11]), with *clustering-width* [20] (the smallest number of variables whose deletion results in a variable-disjoint union of hitting formulas) and with *h-modularity* [12] (a structural parameter inspired by the community structure of SAT instances). We remark that the former claim implies that consensus treewidth is not dominated by the treewidth of neither the incidence nor the primal graph, since these parameters are dominated by signed clique-width [28]. Furthermore, consensus treewidth is also not dominated by signed rank-width [11], which both dominates and is dominated by signed clique-width.

Proposition 1. *The following claims hold.*

1. *Signed clique-width and consensus treewidth are incomparable.*
2. *Clustering-width and consensus treewidth are incomparable.*
3. *H-modularity and consensus treewidth are incomparable.*

Proof. We prove these claims by showing that there exist classes of formulas such that each formula in the class has one parameter bounded while the other parameter can grow arbitrarily. For a formula F, let $\mathbf{scw}(F)$ and $\mathbf{clw}(F)$ denote its signed clique-width and clustering width, respectively.

Let us choose an arbitrary positive integer $i \in \mathbb{N}$. For the first claim, it is known that already the class of all hitting formulas has unbounded $\mathbf{scw}$ [20]. In particular, this means that there exists a hitting formula F_1 such that $\mathbf{scw}(F_1) \geq i$. Observe that the consensus graph of F_1 is edgeless, and hence $\mathbf{contw}(F_1) = 0$.

Conversely, consider the following formula $F_2 = \{c_1, \ldots, c_i\}$. The formula contains variables $x_1, \ldots x_i$, and each variable x_ℓ occurs only in clause c_ℓ. Since the incidence graph of F_2 is just a matching, its signed clique-width is bounded by a constant (in particular, it will be 2). However, the consensus graph of F_2 is a complete graph on i vertices, and it is known that such graphs have treewidth precisely $i - 1$, hence $\mathbf{contw}(F_2) = i - 1$.

We proceed similarly for the second and third claims; in fact, we can use a single construction to deal with both h-modularity and clustering width. Let us once again fix some $i \in \mathbb{N}$, let F_1' be the union of two variable-disjoint hitting formulas each containing i clauses. Both h-modularity and clustering width have a value of 0 for variable-disjoint hitting formulas. However, the consensus graph of F_1' is a complete bipartite graph with each side containing precisely i vertices, and it is well-known that such graphs have treewidth i; hence, $\mathbf{contw}(F_1') = i$.

Conversely, consider the formula F_2' over variable sets $Y = \{y_1, \ldots, y_i\}$ and $X = \{x_1, \ldots, x_i\}$. For each subset α of X, we will add two clauses to F_2':

– c_α contains α as positive literals and $X \setminus \alpha$ as negative literals;
– c_α^y contains α as positive literals, $X \setminus \alpha$ as negative literals, and all variables in Y as positive literals.

We observe that for each α, clause c_α^y clashes with all other clauses except for c_α (and vice-versa for c_α). This implies that the consensus graph of F_2' is a matching, and hence $\mathbf{contw}(F_2') = 1$. On the other hand, note that for each distinct pair of subsets $\alpha, \beta \subseteq X$, the clauses c_α, c_β, c_α^y, c_β^y form a formula which is not a variable-disjoint union of hitting formulas. However, deleting a subset of X from F_2' will only resolve this obstruction for choices of α and β which differed in X; for instance, even if we deleted all of X except for a single variable (w.l.o.g. say x_1), the resulting formula would still not be a disjoint union of hitting formulas (it would contain clauses $\{x_1\} \cup Y, \{x_1\}, \{\overline{x_1}\} \cup Y, \{\overline{x_1}\}$). Similarly, deleting any proper subset $Y' \subset Y$ will also clearly not result in a disjoint union of hitting formulas (it would, in fact, not change the consensus graph at all), and the same goes for any combination of deleting Y' along with a proper subset of X. Hence we conclude that $\mathbf{clw}(F_2') \geq i$.

Finally, we argue that F_2' has h-modularity at least i. We note that we will not need the definition of h-modularity to do so, as it will suffice to follow the proof of Lemma 1 in the paper [12] which provides a suitable lower-bound for h-modularity. In particular, closely following that proof, let us fix $q = i$ and a clause $c \in F_2'$. Then:

1. the set Z_0 defined in the proof will be equal to F_2';
2. the set Z_1 defined in the proof will be empty;
3. the set Z defined in the proof will be equal to F_2';
4. the set W defined in the proof will be equal to F_2';
5. since W is not a hitting formula, by point 3 of the proof it holds that F_2' has h-modularity greater than $q = i$.

The above general constructions show that for any choice of i, one can produce formulas with a gap of at least i between consensus treewidth and any of the three other measures under consideration. $\square$

Next, we proceed to our main algorithmic result. Our algorithm will in certain cases invoke the previously known algorithm [26] for #SAT parameterized by dual treewidth as a subroutine, and so we provide the full statement of its runtime below. We note that the runtime of that algorithm depends on the time required to multiply two n-bit integers, denoted δ.

Fact 4 ([26]). *Given a nice tree decomposition $(T, \mathcal{B})$ of the dual graph of a formula F, #SAT can be solved in time $2^k(k\ell + \delta)N$, where N is the number of nodes of T, k is its width, and ℓ is the maximum width of a clause in F.*

In the literature there exist several algorithms for multiplying two n-bit integers; we refer the interested reader to Knuth's in-depth overview [17]. One of the most prominent of these algorithms is due to Schönhage and Strassen [17] and runs in time $\mathcal{O}(n \log n \log \log n)$. Thus, we can assume that $\delta = \mathcal{O}(n \log n \log \log n)$, where n is the number of variables of the given CNF formula. Recently, Fürer [9] presented an even faster algorithm. If arithmetic operations are assumed to have constant runtime, that is, $\delta = \mathcal{O}(1)$, then we obtain an upper bound on the runtime of $2^{\mathcal{O}(k)} \cdot L^2$.

Theorem 1. *#SAT can be solved in time $2^{\mathcal{O}(k)} \cdot L(L + \delta)$, where L is the length of the formula and k is the consensus treewidth.*

Proof. Let F be an input formula over n variables, and let G be its consensus graph. Let $(T, \mathcal{B})$ be a nice tree decomposition of G of width at most $5k + 4$; recall that such $(T, \mathcal{B})$ can be computed in time $2^{\mathcal{O}(k)}$ by Fact 2. For brevity, we will use the following terminology: for a node t with bag B_t and a clause set $X \subseteq B_t$, we say that an assignment is X^t-validating if it satisfies all clauses in X but does not satisfy any clause in $B_t \setminus X$. For instance, if $X = \emptyset$ then a X^t-validating assignment cannot satisfy any clause in B_t, while if $X = B_t$ then a X^t-validating assignment must satisfy every clause in B_t.

Consider the following leaf-to-root dynamic programming algorithm $\mathcal{A}$ on T. At each bag B_t associated with a node t of T, $\mathcal{A}$ will compute two mappings ϕ_t^+, $\phi_t^\sim$, each of which maps each $X \subseteq B_t$ to an integer between 0 and 2^d. These mappings will be used to store the number of X^t-validating assignments of $var(F)$ under an additional restriction:

- in ϕ_t^+, we count only assignments which satisfy all clauses that were already forgotten below t, and
- in $\phi_t^\sim$, we count only assignments which invalidate at least one clause that was already forgotten below t.

Since we assume that the root r of a nice tree decomposition is an empty bag, the total number of satisfying assignments of F is equal to $\phi_r^+(\emptyset)$. The purpose of also keeping records for $\phi_t^\sim$ will become clear during the algorithm; in particular, they will be needed to correctly determine the records for ϕ_t^+ at certain stages.

At each node t, let σ_t be the set of clauses which were forgotten below t; for example, $\sigma_r = F$ and $\sigma_\ell = \emptyset$ for each leaf ℓ of T. We now proceed to explain how $\mathcal{A}$ computes the mappings ϕ_t^+, $\phi_t^\sim$ at each node t of T, starting from the leaves, along with arguing correctness of the performed operations.

1. *Leaf nodes.* Since σ_t is empty, $\phi_t^{\sim}$ will map each subset of B_t to 0. As for ϕ_t^+, we observe that there are precisely $2^{n-|c|}$ many assignments which invalidate a clause $c \in B_t$. Hence we correctly set $\phi_t^+(c) = 2^{n-|c|}$ and $\phi_t^+(\emptyset) = 2^n - 2^{n-|c|}$.

2. *Forget nodes.* Let t be a forget node with child p and let $B_p \setminus B_t = \{c\}$. We begin by observing that the number of X^t-validating assignments which satisfy all clauses in σ_t is precisely equal to the number of $(X\cup\{c\})^p$-validating assignments which satisfy all clauses in σ_p. In other words, for each $X \subseteq B_t$ we correctly set $\phi_t^+(X) = \phi_p^+(X \cup \{c\})$.

   On the other hand, X^t-validating assignments which do not satisfy at least one clause in σ_t are partitioned into the following mutually exclusive cases:
   (a) $(X\cup\{c\})^p$-validating assignments which do not satisfy at least one clause in σ_p;
   (b) X^p-validating assignments which do not satisfy at least one clause in σ_p;
   (c) X^p-validating assignments which satisfy all clauses in σ_p.
   Hence, we correctly set $\phi_t^{\sim}(X) = \phi_p^{\sim}(X \cup \{c\}) + \phi_p^{\sim}(X) + \phi_p^+(X)$.

3. *Join nodes.* Let t be a join node with children p, q. Recall that $\sigma_p \cap \sigma_q = \emptyset$ and $\sigma_t = \sigma_p \cup \sigma_q$ due to the properties of tree decompositions. Furthermore, an assignment satisfies all clauses in σ_t if and only if it satisfies all clauses in both σ_p and σ_q. In other words, X^t-validating assignments which do not satisfy at least one clause in σ_t are partitioned into the following mutually exclusive cases (recall that $B_p = B_q$ by the definition of join nodes):
   (a) X^p-validating assignments which do not satisfy at least one clause in σ_p but satisfy all clauses in σ_q;
   (b) X^p-validating assignments which do not satisfy at least one clause in σ_q but satisfy all clauses in σ_p;
   (c) X^p-validating assignments which invalidate at least one clause in σ_p and also at least one clause in σ_q.
   Recall that B_t is a separator between σ_p and σ_q, which means that every clause in σ_p clashes with ever clause in σ_q. That in turn implies that the number of assignments covered by point 3c must be equal to 0: every assignment that does not satisfy at least one clause in one of σ_p, σ_q must satisfy all clauses in the other set. Since we now know that every assignment which does not satisfy a clause in σ_p must satisfy all clauses in σ_q and vice-versa, we can correctly set $\phi_t^{\sim}(X) = \phi_q^{\sim}(X) + \phi_p^{\sim}(X)$. Finally, to compute $\phi_t^+(X)$ we can subtract $\phi_t^{\sim}(X)$ from the total number of X^t-validating assignments (which is equal to the sum of $\phi_p^+(X)$ and $\phi_p^{\sim}(X)$ and hence is known to us), i.e., we set $\phi_t^+(X) = \phi_p^+(X) + \phi_p^{\sim}(X) - \phi_t^{\sim}(X)$.

4. *Introduce nodes.* Let t be an introduce node with child p and let $B_t = B_p\cup\{c\}$. For each $X \subseteq B_p$, we consider two cases and proceed accordingly. On one hand, if $\phi_p^{\sim}(X) = 0$ (i.e., there exists no X^p-validating assignment invalidating at least one clause in σ_p), then clearly $\phi_p^{\sim}(X) = \phi_t^{\sim}(X) + \phi_t^{\sim}(X \cup \{c\}) = 0$ and in particular $\phi_t^{\sim}(X) = \phi_t^{\sim}(X \cup \{c\}) = 0$. On the other hand, assume $\phi_p^{\sim}(X) > 0$ and consider a X^p-validating assignment α which invalidates at least one clause in σ_p. Since c clashes with all clauses in σ_p, it follows that α must satisfy c. Consequently, we correctly set $\phi_t^{\sim}(X) = \phi_p^{\sim}(X \cup c)$ and

$\phi_t^{\sim}(X) = 0$. Since each subset of B_t is a subset of $B_p \cup \{c\}$, it follows that using the above rules $\mathcal{A}$ has computed the mapping $\phi_t^{\sim}$ for all $X' \subseteq B_t$.

The last remaining step is to compute $\phi_t^+(X')$ for each $X' \subseteq B_t$. In order to do so, we will first use Fact 4 to compute the number $s_{X'}$ of all X'^t-validating assignments of F. Since we are now interested in assignments which must invalidate all clauses in $B_t \setminus X'$, we can construct the subformula F' from F by

(a) removing all clauses except for those in X', i.e., $F' := X'$, and

(b) assigning all variables which occur in $B_t \setminus X'$ in order to invalidate clauses outside of X'. Formally, for each clause $c \in B_t \setminus X'$, we apply the partial assignment $x \mapsto 0$ whenever $x \in c$ and the partial assignment $x \mapsto 1$ whenever $\overline{x} \in c$. If a contradiction arises for some variable, then we know that there exists no X'-validating assignment and hence set $s_{X'} = 0$.

Clearly, F' can be constructed in time $\mathcal{O}(L)$ and satisfies $\#F' = s_{X'}$. Furthermore, since F' contains at most k clauses, we can construct a trivial nice tree decomposition of F' of width at most k containing at most $2k+1$ nodes in linear time by first consecutively introducing all of its nodes and then consecutively forgetting them. With this decomposition in hand, we invoke Fact 4 to compute $\#F'$ in time at most $2^k(kL + \delta)(2k+1)$, i.e., $2^{\mathcal{O}(k)} \cdot (L+\delta)$. Once we compute $s_{X'}$, we use the fact that $s_{X'} = \phi_t^{\sim}(X) + \phi_t^+(X)$ and correctly set $\phi_t^+(X) = s_{X'} - \phi_t^{\sim}(X)$.

Observe that the time requirements for performing the above-specified operations at individual nodes of T are dominated by the time requirements for processing introduce nodes, upper-bounded by $2^k \cdot (L + 2^{\mathcal{O}(k)} \cdot (L + \delta)) = 2^{\mathcal{O}(k)} \cdot (L+\delta)$. Furthermore, a nice tree decomposition with at most $\mathcal{O}(L)$ nodes and width at most $5k + 4$ can be obtained in time $2^{\mathcal{O}(k)} \cdot L$ by Fact 2. Hence we conclude that it is possible to compute $\phi_r^+(\emptyset) = \#(F)$ in time at most $2^{\mathcal{O}(k)} \cdot L(L + \delta)$. The correctness of the whole algorithm follows from the correctness of computing the mappings $\phi_t^{\sim}$ and ϕ_t^+ at each node t in T.   □

4   Conflict Treewidth

The algorithmic applications of the consensus graph, as detailed above, gives rise to a natural follow-up question: what can we say about its natural counterpart, the *conflict graph*? Recall that the conflict graph of a CNF formula F is the graph G whose vertices are the clauses of F and which contains an edge ab if and only if clauses a and b clash. Observe that the conflict graph of a hitting formula is a complete graph, and that the conflict graph is the complement graph of the consensus graph. The conflict treewidth of F is then the treewidth of its conflict graph.

Since the conflict graph is a subgraph of the dual graph, conflict treewidth can be much (and in fact arbitrarily) smaller than the dual treewidth. However, unlike the case of dual treewidth, we will show that SAT does not admit a fixed-parameter algorithm parameterized by conflict treewidth (unless $W[1] \neq FPT$).

Proposition 2. SAT *is W[1]-hard when parameterized by conflict treewidth. Furthermore, SAT remains W[1]-hard when parameterized by the size of a minimum vertex cover of the conflict graph, even for instances without pure literals.*

Proof. We provide a parameterized reduction from MULTICOLORED CLIQUE. Given an instance G of MULTICOLORED CLIQUE over vertex set $V = V_1 \cup \cdots \cup V_k$, we construct a formula F over the variable set V (i.e., each vertex in G is a variable in F). We add the following clauses to F (observe that F contains no pure literals):

1. for each $i \in [k]$, we add one clause containing one positive literal of each variable $x \in V_i$;
2. for each $i \in [k]$ and each distinct $x, y \in V_i$, we add one clause $\{\overline{x}, \overline{y}\}$;
3. for each non-edge between distinct vertices x, y in G, we add one clause $\{\overline{x}, \overline{y}\}$.

F can clearly be constructed from G in polynomial time. The intuition behind the construction is the following: variables set to true correspond to the vertices of a multicolored clique, clauses in groups 1 and 2 enforce the selection of a single vertex from each color class, and the remaining clauses ensure that the result is a clique.

To formally prove that the reduction is correct, consider a solution X to G, and consider the assignment α which sets variables in X to true and all other variables to false. Since X contains precisely one vertex from each color class V_i, α clearly satisfies all clauses in groups 1 and 2. Now consider any clause in group 3, and observe that it can only be invalidated if both of its variables are set to true. However, since X is a clique it must hold that for each pair of distinct variables $x, y \in C$ we'll never have a clause in group 3 between x and y, and hence in particular each such clause will always contain at least one variable that is set to false and that therefore satisfies it.

On the other hand, consider a satisfying assignments α for F. Then clauses in group 1 ensure that at least one variable is set to true in each color class, and clauses in group 2 ensure that at most one variable is set to true in each color class. Finally, clauses in group 3 prevent α from setting two variables to true if they are the endpoints of a non-edge in G. Consequently, the variables set to true by α must form a solution to the multicolored clique instance G.

Finally, we argue that the parameter values are bounded by k, as claimed by the hardness result. Observe that all literals in clause groups 2 and 3 are negative, which means that whenever two clauses clash, at least one of them must be in group 1. Furthermore, recall that there are precisely k clauses in group 1. Hence the clauses in group 1 form a vertex cover of size k in the conflict graph of F. It is well known (and easy to verify) that the vertex cover is an upper bound on the treewidth of a graph. □

Observe that Proposition 2 implies that there exist instances where the conflict treewidth is arbitrarily smaller than the incidence treewidth (since SAT is known to be FPT when parameterized by the latter). On the other hand, we can show that in the case of formulas of bounded clause width and without pure literals, conflict treewidth (denoted **conflict-tw**) is dominated by incidence treewidth.

Proposition 3. *For any formula F with clauses of width at most d and without pure literals, it holds that* $\mathbf{itw}(F) \leq (d+1) \cdot (\mathbf{conflict\text{-}tw}(F) + 1)$.

Proof. Let G be the conflict graph of F and $(T, \mathcal{B})$ be a tree decomposition of G of width k. Consider the structure $(T, \mathcal{B}')$ obtained as follows: for each $B_i \in \mathcal{B}$, we create a set B_i' in $\mathcal{B}'$ where $B_i' = B_i \cup \{ x \mid \exists c \in B_i : x \in var(c) \}$. Informally, the set $\mathcal{B}'$ is obtained by extending the bags in $(T, \mathcal{B})$ by the variables that occur in the clauses of that bag. We claim that $(T, \mathcal{B}')$ is a tree decomposition of the incidence graph G' of F.

Towards proving this claim, first observe that T is still a tree and each $B_i' \in \mathcal{B}'$ is a subset of $V(G')$. Furthermore, for any edge ab of G' between a clause a and variable b, it must hold that $a \in B_i$ for some $B_i \in \mathcal{B}$. By construction, B_i' must then contain both a and b and so condition 1 of the definition of tree decompositions is satisfied. As for condition 2, assume first for a contradiction that some vertex $v \in G'$ is not contained in any bag of $(T, \mathcal{B}')$. This clearly cannot happen if v is a clause, and so v must be a variable; furthermore, since F contains no pure literals, v must occur in at least two clauses.

It remains to show that all bags containing v induce a connected subtree of T. So, let us assume once more for a contradiction that this is not the case. By construction of $(T, \mathcal{B}')$ this implies that $(T, \mathcal{B})$ must contain a node t such that B_t separates some set of clauses containing v, say X_1, from all remaining clauses containing v, say X_2. Next, observe that $X_1 \cup X_2$ forms a complete bipartite graph in G: indeed, one side consists of all clauses containing v as a literal, while the other side consists of all clauses containing $\overline{v}$. But these two facts together contradict the inseparability property of tree decompositions: $X_1 \cup X_2$ induce a connected subgraph of G', and yet they are supposedly separated by B_t which does not intersect $X_1 \cup X_2$. Hence we conclude that no such node B_t exists and that the bags containing v indeed induce a connected subtree of T.

We conclude the proof by observing that the size of each bag $B_i' \in \mathcal{B}'$ is equal to $d+1$ times $|B_i|$, since we added at most d extra vertices for each vertex in B_i. $\qquad\square$

As a consequence of Proposition 3, restricted to formulas of bounded clause width, #SAT is FPT when parameterized by conflict treewidth, since in this case the parameter is dominated by incidence treewidth [26]. We note that the domination is strict: for each $i \in \mathbb{N}$ there exists a formula F_i of clause width 2 and without pure literals such that $\mathbf{itw}(F_i) = 1$ and $\mathbf{contw}(F_i) \geq i$. Indeed, one such example is the formula $F_i = \{\{y, x_1\}, \{\overline{x_1}\}, \{y, x_2\}, \{\overline{x_2}\}, \ldots, \{y, x_i\}, \{\overline{x_i}\}\} \cup \{\{\overline{y}, z_1\}, \{\overline{z_1}\}, \{\overline{y}, z_2\}, \{\overline{z_2}\}, \ldots, \{\overline{y}, z_i\}, \{\overline{z_i}\}\}$.

5   Concluding Remarks

We have considered two natural graphical models of CNF formulas and established whether #SAT is fixed-parameter tractable parameterized by their treewidth or not. The introduced notion of consensus treewidth generalizes and, in some sense, builds upon the classical #SAT algorithm on hitting formulas [13], and as such may be efficient in cases where other structural parameters fail. Our results show that it is worthwhile to consider further graphical models in addition to the already established ones such as primal, dual, and incidence graphs.

References

1. Bacchus, F., Dalmao, S., Pitassi, T.: Algorithms and complexity results for #SAT and Bayesian inference. In: 44th Annual IEEE Symposium on Foundations of Computer Science (FOCS 2003), pp. 340–351 (2003)
2. Bodlaender, H.L.: A tourist guide through treewidth. Acta Cybernetica **11**, 1–21 (1993)
3. Bodlaender, H.L.: A linear-time algorithm for finding tree-decompositions of small treewidth. SIAM J. Comput. **25**(6), 1305–1317 (1996)
4. Bodlaender, H.L., Drange, P.G., Dregi, M.S., Fomin, F.V., Lokshtanov, D., Pilipczuk, M.: A c^k n 5-approximation algorithm for treewidth. SIAM J. Comput. **45**(2), 317–378 (2016)
5. Bodlaender, H.L., Kloks, T.: Efficient and constructive algorithms for the path-width and treewidth of graphs. J. Algorithms **21**(2), 358–402 (1996)
6. Courcelle, B., Makowsky, J.A., Rotics, U.: On the fixed parameter complexity of graph enumeration problems definable in monadic second-order logic. Discr. Appl. Math. **108**(1–2), 23–52 (2001)
7. Diestel, R.: Graph Theory. Graduate Texts in Mathematics, vol. 173, 4th edn. Springer, New York (2010)
8. Downey, R.G., Fellows, M.R.: Fundamentals of Parameterized Complexity. Texts in Computer Science. Springer, London (2013)
9. Fürer, M.: Faster integer multiplication. SIAM J. Comput. **39**(3), 979–1005 (2009)
10. Galesi, N., Kullmann, O.: Polynomial time SAT decision, hypergraph transversals and the hermitian rank. In: Hoos, H.H., Mitchell, D.G. (eds.) SAT 2004. LNCS, vol. 3542, pp. 89–104. Springer, Heidelberg (2005). doi:10.1007/11527695_8
11. Ganian, R., Hlinený, P., Obdrzálek, J.: Better algorithms for satisfiability problems for formulas of bounded rank-width. Fund. Inform. **123**(1), 59–76 (2013)
12. Ganian, R., Szeider, S.: Community structure inspired algorithms for SAT and #SAT. In: Heule, M., Weaver, S. (eds.) SAT 2015. LNCS, vol. 9340, pp. 223–237. Springer, Cham (2015). doi:10.1007/978-3-319-24318-4_17
13. Iwama, K.: CNF-satisfiability test by counting and polynomial average time. SIAM J. Comput. **18**(2), 385–391 (1989)
14. Kleine Büning, H., Kullmann, O.: Minimal unsatisfiability and autarkies. In: Biere, A., Heule, M.J.H., van Maaren, H., Walsh, T. (eds.) Handbook of Satisfiability, Frontiers in Artificial Intelligence and Applications, vol. 185, chap. 11, pp. 339–401. IOS Press (2009)
15. Kleine Büning, H., Zhao, X.: Satisfiable formulas closed under replacement. In: Kautz, H., Selman, B. (eds.) Proceedings for the Workshop on Theory and Applications of Satisfiability. Electronic Notes in Discrete Mathematics, vol. 9. Elsevier Science Publishers, North-Holland (2001)
16. Kloks, T.: Treewidth: Computations and Approximations. Springer, Berlin (1994)
17. Knuth, D.E.: How fast can we multiply? In: The Art of Computer Programming. Seminumerical Algorithms, 3rd edn., vol. 2, chap. 4.3.3, pp. 294–318. Addison-Wesley (1998)

18. Kullmann, O.: The combinatorics of conflicts between clauses. In: Giunchiglia, E., Tacchella, A. (eds.) SAT 2003. LNCS, vol. 2919, pp. 426–440. Springer, Heidelberg (2004). doi:10.1007/978-3-540-24605-3_32

19. Niedermeier, R.: Invitation to Fixed-Parameter Algorithms. Oxford Lecture Series in Mathematics and its Applications. Oxford University Press, Oxford (2006)

20. Nishimura, N., Ragde, P., Szeider, S.: Solving #SAT using vertex covers. Acta Informatica 44(7–8), 509–523 (2007)

21. Ordyniak, S., Paulusma, D., Szeider, S.: Satisfiability of acyclic and almost acyclic CNF formulas. Theor. Comput. Sci. 481, 85–99 (2013)

22. Robertson, N., Seymour, P.D.: Graph minors. II. Algorithmic aspects of tree-width. J. Algorithms 7(3), 309–322 (1986)

23. Roth, D.: On the hardness of approximate reasoning. Artif. Intell. 82(1–2), 273–302 (1996)

24. Sæther, S.H., Telle, J.A., Vatshelle, M.: Solving #SAT and MAXSAT by dynamic programming. J. Artif. Intell. Res. 54, 59–82 (2015)

25. Samer, M., Szeider, S.: Fixed-parameter tractability. In: Biere, A., Heule, M., van Maaren, H., Walsh, T. (eds.) Handbook of Satisfiability, chap. 13, pp. 425–454. IOS Press (2009)

26. Samer, M., Szeider, S.: Algorithms for propositional model counting. J. Discrete Algorithms 8(1), 50–64 (2010)

27. Scheder, D., Zumstein, P.: How many conflicts does it need to be unsatisfiable? In: Kleine Büning, H., Zhao, X. (eds.) SAT 2008. LNCS, vol. 4996, pp. 246–256. Springer, Heidelberg (2008). doi:10.1007/978-3-540-79719-7_23

28. Szeider, S.: On fixed-parameter tractable parameterizations of SAT. In: Giunchiglia, E., Tacchella, A. (eds.) SAT 2003. LNCS, vol. 2919, pp. 188–202. Springer, Heidelberg (2004). doi:10.1007/978-3-540-24605-3_15

29. Valiant, L.G.: The complexity of computing the permanent. Theoret. Comput. Sci. 8(2), 189–201 (1979)

Hard Satisfiable Formulas for Splittings by Linear Combinations

Dmitry Itsykson and Alexander Knop[(⊠)]

Steklov Institute of Mathematics, Saint Petersburg, Russia
dmitrits@pdmi.ras.ru, aaknop@gmail.com

Abstract. Itsykson and Sokolov in 2014 introduced the class of DPLL($\oplus$) algorithms that solve Boolean satisfiability problem using the splitting by linear combinations of variables modulo 2. This class extends the class of DPLL algorithms that split by variables. DPLL($\oplus$) algorithms solve in polynomial time systems of linear equations modulo 2 that are hard for DPLL, PPSZ and CDCL algorithms. Itsykson and Sokolov have proved first exponential lower bounds for DPLL($\oplus$) algorithms on unsatisfiable formulas.

In this paper we consider a subclass of DPLL($\oplus$) algorithms that arbitrary choose a linear form for splitting and randomly (with equal probabilities) choose a value to investigate first; we call such algorithms drunken DPLL($\oplus$). We give a construction of a family of satisfiable CNF formulas Ψ_n of size poly(n) such that any drunken DPLL($\oplus$) algorithm with probability at least $1 - 2^{-\Omega(n)}$ runs at least $2^{\Omega(n)}$ steps on Ψ_n; thus we solve an open question stated in the paper [12]. This lower bound extends the result of Alekhnovich, Hirsch and Itsykson [1] from drunken DPLL to drunken DPLL($\oplus$).

1 Introduction

The Boolean satisfiability problem (SAT) is one of the most popular **NP**-complete problems. However, SAT solvers have different behaviors on satisfiable and unsatisfiable formulas. The protocol of a SAT solver on an unsatisfiable formula ϕ may be considered as a proof of unsatisfiability of ϕ. Therefore, the study of propositional proof systems is connected with the study of SAT solvers. It is well known that protocols of DPLL solvers on an unsatisfiable formula are equivalent to tree-like resolution proofs of this formula and protocols of CDCL solvers correspond to dag-like resolution proofs [3]. Thus lower bounds on the running time of DPLL and CDCL solvers on unsatisfiable instances follow from lower bounds on the size of tree-like and dag-like resolution proofs.

Satisfiable formulas are usually simpler for SAT solvers rather than unsatisfiable ones. Also, satisfiable instances are of much interest in the case where we reduce some **NP** search problem (for example, the factorization) to SAT.

The research is partially supported by the Government of the Russia (grant 14.Z50.31.0030).

S. Gaspers and T. Walsh (Eds.): SAT 2017, LNCS 10491, pp. 53–61, 2017.
DOI: 10.1007/978-3-319-66263-3_4

DPLL [7, 8] algorithm is a recursive algorithm. On each recursive call, it simplifies an input formula F (without affecting its satisfiability), chooses a variable v and makes two recursive calls on the formulas $F[v := 1]$ and $F[v := 0]$ in some order. Every DPLL algorithm is determined by a heuristic A that chooses a variable for splitting and by a heuristic B that chooses a value that should be investigated at first. If $\mathbf{P} = \mathbf{NP}$, then DPLL can solve all satisfiable instances in polynomial time: a heuristic B always chooses a correct value of a variable. Alekhnovich, Hirsch, and Itsykson [1] proved exponential lower bounds on satisfiable instances for two wide classes of DPLL algorithms: for myopic DPLL algorithms and drunken DPLL algorithms. In myopic DPLL the both heuristics have the following restrictions: they can see only the skeleton of the formula where all negation signs are erased, they also have access to the number of positive and negative occurrences of every variable and they are allowed to read $n^{1-\epsilon}$ of clauses precisely. Drunken algorithms have no restrictions on the heuristic A that chooses the variable, but the heuristic B chooses a value at random with equal probabilities. There are also known lower bounds on the complexity of the inversion of Goldreich's one-way function candidate by myopic and drunken DPLL algorithms [5, 10, 11].

Itsykson and Sokolov [12] introduced a generalization of DPLL algorithms that split by the value of linear combinations of variables modulo 2; we call them DPLL($\oplus$) algorithms. DPLL($\oplus$) algorithms quickly solves formulas that encode linear systems over GF(2) (unsatisfiable linear systems are hard for resolution [19] and even for bounded-depth Frege [2]; satisfiable linear systems are hard for drunken and myopic DPLL [1, 10] and PPSZ [16, 18]) and perfect matching principles for graphs with odd number of vertices (these formulas are hard for resolution [17] and bounded-depth Frege [4]).

It is well known that the tree-like resolution complexity (and hence the running time of DPLL) of the pigeonhole principle PHP_n^{n+1} is $2^{\Theta(n \log n)}$ [6]. Itsykson and Sokolov [12] proved a lower bound $2^{\Omega(n)}$ and Oparin recently proved an upper bound $2^{O(n)}$ [14] on the running time of DPLL($\oplus$) algorithms on PHP_n^{n+1}. There are three other families of formulas that are hard for DPLL($\oplus$) algorithms proposed by Itsykson and Sokolov [12], Krajíček [13], and Garlík and Kołodziejczyk [9].

Our results. Itsykson and Sokolov [12] formulated the following open question: to prove a lower bound on satisfiable formulas for drunken DPLL($\oplus$) algorithms that arbitrary choose a linear combination and randomly with equal probabilities chooses a value that would be investigated at first. In this paper we answer to the question and give a construction of a family of satisfiable formulas Ψ_n in CNF of size poly(n) such that any drunken DPLL($\oplus$) algorithm with probability $1 - 2^{-\Omega(n)}$ runs at least $2^{\Omega(n)}$ steps on Ψ_n.

In order to construct Ψ_n we take the pigeonhole principle PHP_n^{n+1} and manually add one satisfying assignment to it. We prove that with high probability a drunken DPLL($\oplus$) algorithm will make incorrect linear assumption and the algorithm will have to investigate a large subtree without satisfying assignments. To show that this subtree is indeed large we extend the technique that was used for PHP_n^{n+1} [12].

Further research. The constructed family Ψ_n has clauses with large width. It would be interesting to prove lower bounds for formulas in $O(1)$-CNF. It is also interesting to prove a lower bound for myopic DPLL($\oplus$) algorithms.

2   DPLL($\oplus$) and Parity Decision Trees

DPLL($\oplus$) algorithms are parameterized by two heuristics: A and B. The heuristic A takes a CNF formula and a system of linear equations and returns a linear combination (a DPLL($\oplus$) algorithm will use this linear combination for splitting). The heuristic B takes a CNF formula, a system of linear equations, and a linear combination and returns a value from $\{0,1\}$ (this value would be considered at first by a DPLL($\oplus$) algorithm). The set of DPLL($\oplus$) algorithms is the set of algorithms $\mathcal{D}_{A,B}$ for all heuristics A and B that are defined below. An algorithm $\mathcal{D}_{A,B}$ takes on the input a CNF formula Φ and a system of linear equations F (we may omit the second argument if the system F is empty) and works as follows:

1. If the system F does not have solutions (it can be verified in polynomial time), then return "Unsatisfiable".
2. If the system F contradicts to a clause C of the formula Φ (a system G contradicts a clause $\ell_1 \vee \ldots \vee \ell_k$ iff for all $i \in [k]$ the system $G \wedge (\ell_i = 1)$ is unsatisfiable, hence this condition may be verified in polynomial time), then return "Unsatisfiable".
3. If the system F has the unique solution τ (in variables of Φ) and this solution satisfies Φ (it can also be verified in polynomial time), then return τ.
4. $f := A(\Phi, F)$.
5. $\alpha := B(\Phi, F, f)$.
6. If $\mathcal{D}_{A,B}(\Phi, F \wedge (f = \alpha))$ returns an assignment, then return it.
7. Return the result of $\mathcal{D}_{A,B}(\Phi, F \wedge (f = 1 - \alpha))$.

The class of drunken DPLL($\oplus$) consists of all algorithms $\mathcal{D}_{A,rand}$, where A is an arbitrary heuristic and *rand* always returns a random element from $\{0,1\}$ with equal probabilities.

A *parity decision tree* for (Φ, F) is a rooted binary tree such that all its internal nodes are labeled with linear combinations of variables of Φ, for every internal node labeled with a linear form f one of its outgoing edge is labeled with $f = 0$ and the other one with $f = 1$. Let l be a leaf l of the tree, we denote by D_l the system of linear equations written on the edges of the path from the root to l. There are three kinds of leaves:

degenerate leaf: $F \wedge D_l$ does not have a solution;
satisfying leaf: $F \wedge D_l$ has the only solution in the variables of Φ and this solution satisfies Φ;
contradiction: $F \wedge D_l$ contradicts to a clause C of Φ.

If $\Phi \wedge F$ is unsatisfiable, then the recursion tree of $\mathcal{D}_{A,B}(\Phi, F)$ is a parity decision tree for (Φ, F) that does not contain satisfying leaves. Additionally the minimal size of a decision tree for (Φ, F) is a lower bound on the running time of $\mathcal{D}_{A,B}(\Phi, F)$. If $\Phi \wedge F$ is satisfiable then the recursion tree of $\mathcal{D}_{A,B}(\Phi, F)$ is a part of a parity decision tree since the execution stops when the algorithm $\mathcal{D}_{A,B}$ finds a satisfying assignment.

3   Lower Bound

In this section we construct a satisfiable formula that is hard for all drunken DPLL($\oplus$) algorithms. Our hard example is based on the pigeonhole principle. The pigeonhole principle (PHP_n^m) states that it is possible to put m pigeons into n holes such that every pigeon is in at least one hole and every hole contains at most one pigeon. For every pigeon $i \in [m]$ and hole $j \in [n]$ we introduce a variable $p_{i,j}$; $p_{i,j} = 1$ iff i-th pigeon is in the j-th hole. The formula PHP_n^m is the conjunction of the following clauses:

short clauses: $\neg p_{i,k} \vee \neg p_{j,k}$ for all $i \neq j \in [m]$ and $k \in [n]$;

long clauses: $\bigvee\limits_{k=1}^{n} p_{i,k}$ for all $i \in [m]$.

The formula PHP_n^m is unsatisfiable iff $m > n$. Let $\mathcal{P}_{m,n}$ denote the set of variables $\{p_{i,j} \mid i \in [m], j \in [n]\}$.

Let σ be a substitution to the variables $x_1, \ldots, x_n$ and Φ be a CNF formula on the variables $x_1, \ldots, x_n$. We denote by $\Phi + \sigma$ a CNF obtained from Φ in the following manner: for every clause C of Φ and variable x_i, the formula $\Phi + \sigma$ contains a clause $C \vee x_i^{\sigma(x_i)}$, where for every propositional variable x, x^0 denotes $\neg x$ and x^1 denotes x. Note that it is possible that $C \vee x_i^{\sigma(x_i)}$ is a trivial clause.

Proposition 1. *If Φ is unsatisfiable, then $\Phi + \sigma$ has the only satisfying assignment σ.*

Proof. It is straightforward that σ satisfies $\Phi + \sigma$. Assume that τ satisfies $\Phi + \sigma$ and $\tau \neq \sigma$. Consider $i \in [n]$ such that $\tau(x_i) \neq \sigma(x_i)$. Let for every CNF formula ϕ, variable x, and $\alpha \in \{0, 1\}$ $\phi[x := \alpha]$ denote the result of the substitution $x := \alpha$ applied to ϕ. Note that the formula $(\Phi + \sigma)[x_i := \tau(x_i)]$ contains all clauses of $\Phi[x_i := \tau(x_i)]$, but Φ is unsatisfiable, hence $(\Phi + \sigma)[x_i := \tau(x_i)]$ is unsatisfiable and τ can not satisfy $\Phi + \sigma$. $\qquad\square$

We call an assignment σ to the variables $\mathcal{P}_{m,n}$ *proper* if it satisfies all short clauses of PHP_n^m, that is there are no two pigeons in one hole in σ.

Let $f_1, f_2, \ldots, f_k$, and g be linear equations in variables $\mathcal{P}_{m,n}$. We say that $f_1, f_2, \ldots, f_k$ *properly implies* g iff every proper assignment that satisfies all f_1, $f_2, \ldots, f_k$ also satisfies g.

Let F be a linear system in variables $\mathcal{P}_{m,n}$. A *proper rank* of the system F is the size of the minimal set of equations from F such that linear equations from this set properly implies all other equations from F.

Notice that if F does not have a proper solutions, then its proper rank is the size of the minimal subsystem of F that has no proper solutions.

Proposition 2. *Let F and G be two linear systems in variables $\mathcal{P}_{m,n}$. Then the proper rank of $F \wedge G$ is at most the sum of the proper ranks of F and G.*

Proof. Let F' and G' be the minimal subsystems of F and G such that F' properly implies all equations from F and G' properly implies all equation from G. Hence $F' \cup G'$ properly implies all equations from $F \wedge G$.

Remark 1. In contrast to the case of the common rank it is possible that a linear system F does not properly implies linear equation f but the proper rank of $F \wedge f$ does not exceed the proper rank of F. For example, $p_{1,3} + p_{2,3} = 1$ does not properly implies $p_{2,3} = 1$ but the proper rank of $(p_{1,3} + p_{2,3} = 1) \wedge (p_{2,3} = 1)$ equals 1 since $p_{2,3} = 1$ properly implies $p_{1,3} + p_{2,3} = 1$.

Our goal is to prove the following theorem:

Theorem 1. *For every $m > n > 0$ and every proper assignment σ to the variables $\mathcal{P}_{m,n}$ the running time of any drunken $\mathrm{DPLL}(\oplus)$ algorithm $\mathcal{D}_{A,\mathrm{rand}}$ on the formula $\mathrm{PHP}_n^m + \sigma$ is at least $2^{\frac{n-1}{4}}$ with probability at least $1 - 2^{-\frac{n-1}{4}}$.*

In what follows we assume that $m > n$.

We use the following Lemma that was proposed in the paper of Itsykson and Sokolov [12]. We give its proof for the sake of completeness.

Lemma 1 ([12]). *Let us assume that a linear system $Ap = b$ in the variables $\mathcal{P}_{m,n}$ has at most $\frac{n-1}{2}$ equations and it has a proper solution. Then for every $i \in [m]$ this system has a proper solution that satisfies the long clause $p_{i,1} \vee \ldots \vee p_{i,n}$.*

Proof. Note that if we change 1 to 0 in a proper assignment, then it remains proper. Let the system have k equations; we know that $k \leq \frac{n-1}{2}$. We consider a proper solution π of the system $Ap = b$ with the minimum number of ones. We prove that the number of ones in π is at most k. Let the number of ones is greater than k. Consider $k + 1$ variables that take value 1 in π: $p_{r_1}, p_{r_2}, \ldots, p_{r_{k+1}}$. Since the matrix A has k rows, the columns that correspond to the variables $p_{r_1}, p_{r_2}, \ldots, p_{r_{k+1}}$ are linearly depended. Therefore, there exists a nontrivial solution π' of the homogeneous system $Ap = 0$ such that every variable with the value 1 in π' is from the set $\{p_{r_1}, p_{r_2}, \ldots, p_{r_{k+1}}\}$. The assignment $\pi' + \pi$ is also a solution of $Ap = b$ and is proper because $\pi' + \pi$ can be obtained from π by changing ones to zeros. Since π' is nontrivial, the number of ones in $\pi' + \pi$ is less than the number of ones in π and this statement contradicts the minimality of π.

The fact that π has at most k ones implies that π has at least $n - k$ empty holes. From the statement of the Lemma we know that $n - k \geq k + 1$; we choose $k + 1$ empty holes with numbers $l_1, l_2, \ldots, l_{k+1}$. We fix $i \in [m]$; the columns of A that correspond to the variables $p_{i,l_1}, \ldots, p_{i,l_{k+1}}$ are linearly depended, therefore, there exists a nontrivial solution τ of the system $Ap = 0$ such that every variable with value 1 in τ is from the set $\{p_{i,l_1}, \ldots, p_{i,l_{k+1}}\}$. The assignment $\pi + \tau$ is a solution of $Ap = b$; $\pi + \tau$ is proper since holes with numbers $l_1, l_2, \ldots, l_{k+1}$ are empty in π, and τ puts at most one pigeon to them (if τ puts a pigeon in a hole, then this is the i-th pigeon). The assignment $\pi + \tau$ satisfies $p_{i,1} \vee p_{i,2} \vee \cdots \vee p_{i,n}$ because τ is nontrivial. $\square$

Corollary 1. *If a linear system F in variables $\mathcal{P}_{m,n}$ has a proper solution and the proper rank of F is at most $\frac{n-1}{2}$ then for every $i \in [m]$ the system F has a proper solution that satisfies the long clause $p_{i,1} \vee \ldots \vee p_{i,n}$.*

Proof. Let F' be the minimal subsystem of F that properly implies all equations from F. The number of equations in F' is the proper rank of F that is at most $\frac{n-1}{2}$. F' also has a proper solution since it is a subsystem of F. Thus by Lemma 1 F' has a proper solution that satisfies $p_{i,1} \vee \ldots \vee p_{i,n}$. This solution should also be a solution of F by the choice of F'.   □

Corollary 2. *Assume that a linear system $Ap = b$ in the variables $\mathcal{P}_{m,n}$ has a proper solution and its proper rank is at most $\frac{n-1}{2}$, then this system has at least two proper solutions.*

Proof. Let σ be a solution of $Ap = b$. Since PHP_n^m is unsatisfiable there is a long clause C such that σ falsify this clause. Though, by Corollary 1 for any clause there is a proper solution τ of $Ap = b$ that satisfies the clause C. Hence $\tau \neq \sigma$, thus τ and σ are different proper solutions of $Ap = b$.   □

Lemma 2. *Let a system of linear equations $Ap = b$ in the variables $\mathcal{P}_{m,n}$ have a proper solution and let its proper rank be at most $\frac{n-1}{4}$. Then the size of any parity decision tree for $(\mathrm{PHP}_n^m, Ap = b)$ is at least $2^{\frac{n-1}{4}}$.*

Corollary 3. *Let a system $Ap = b$ of linear equations in the variables $\mathcal{P}_{m,n}$ have a proper solution and let its proper rank be at most $\frac{n-1}{4}$. Let σ be a proper assignment to variables $\mathcal{P}_{m,n}$ that does not satisfy $Ap = b$. Then the size of any parity decision tree for $(\mathrm{PHP}_n^m + \sigma, Ap = b)$ is at least $2^{\frac{n-1}{4}}$.*

Proof (Proof of Corollary 3). Consider a parity decision tree T for $(\mathrm{PHP}_n^m + \sigma, Ap = b)$. Since σ is the only satisfying assignment of $\mathrm{PHP}_n^m + \sigma$ and it does not satisfy $Ap = b$, there are no satisfying leaves in T. We claim that the tree T may be also considered as a parity decision tree for $(\mathrm{PHP}_n^m, Ap = b)$, and thus the size of T is at least $2^{\frac{n-1}{4}}$ by Lemma 2.

Consider any leaf of T. If this leaf corresponds to the situation where the system F contradicts to a clause C of $\mathrm{PHP}_n^m + \sigma$, then it also contradicts to some clause C' of PHP_n^m since every clause of $\mathrm{PHP}_n^m + \sigma$ is a superclause of a some clause of PHP_n^m Thus, tree T also may be considered as a parity decision tree for $(\mathrm{PHP}_n^m, Ap = b)$.   □

In order to prove Lemma 2 we generalize Prover–Delayer games introduced by Pudlak and Impagliazzo [15].

Consider the following game with two players: Prover and Delayer. They are given a CNF formula Φ and a system of linear equations F such that formula $\Phi \wedge F$ is unsatisfiable. On each step Prover chooses a linear form f that depends on variables of formula Φ, then Delayer may choose a value $\alpha \in \{0,1\}$ of f or return $*$. If Delayer returns $*$, then Prover chooses a value $\alpha \in \{0,1\}$ of f by himself. We add the equality $f = \alpha$ in the system F. The game ends if the current linear system F is inconsistent or refutes some clause of Φ. Delayer earns a coin for every $*$. The goal of Delayer is to earn the maximum number of coins and the goal of Prover is to minimize the number of coins earned by Delayer.

Lemma 3 (similar to [15]). *Consider some CNF formula Φ and linear system F such that $\Phi \wedge F$ is unsatisfiable. Assume that there is a strategy for Delayer that allows Delayer to earn t coins, then the size of any parity decision tree for (Φ, F) is at least 2^t.*

Proof. Consider some parity decision tree T for (Φ, F). We construct a probabilistic distribution on the leaves of T that corresponds to the strategy of Delayer and the following randomized strategy of Prover. Prover uses the tree T, initially he asks the question for the linear form in the root, if Delayer returns $*$, Prover chooses a value at random with equal probabilities and go to the next vertex along an edge labeled with the chosen value. By the statement of the Lemma the probability that the game will finish in every particular leaf is at most 2^{-t}. Since with probability 1 the game will finish in a leaf, the number of leaves of T is at least 2^t. □

Proof (Proof of Lemma 2). Let us construct a strategy for Delayer that will guarantee that Delayer earns at least $\frac{n-1}{4}$ coins. Let G be the current linear system that consists of all equations that are already made by Prover or Delayer in the game and equations from the system $Ap = b$. At the beginning G equals $Ap = b$. The strategy of the Delayer the following: assume that Prover chooses a linear form f, then if G properly implies $f = \alpha$ for some $\alpha \in \{0, 1\}$, then Delayer returns α, otherwise Delayer returns $*$.

We prove by induction on the number of steps that the following invariant holds: the system G always has a proper solution. Basis case is true since $Ap = b$ has a proper solution. Assume that Prover chooses a linear form f. If G has a proper solution, then either $G \wedge (f = 0)$ or $G \wedge (f = 1)$ has the same proper solution. Assume that for some $\alpha \in \{0, 1\}$, $G \wedge (f = \alpha)$ does not have proper solutions. In this case G properly implies $f = 1 - \alpha$ hence Delayer chooses the value $1 - \alpha$ and $F \wedge G \wedge (f = 1 - \alpha)$ has a proper solution. Consider three situations at the end of the game.

- The system G becomes unsatisfiable. This situation is impossible since G has a proper solution.
- The system G contradicts a short clause. This situation is also impossible since G has a proper solution.
- The system G contradicts a long clause $p_{i,1} \vee \ldots \vee p_{i,n}$. Let G' be a subsystem of G that corresponds to answers $*$ of Delayer. By the construction every equation from G is properly implied from $(Ap = b) \wedge G'$. Hence, $(Ap = b) \wedge G'$ does not have proper solutions that satisfy $p_{i,1} \vee \ldots \vee p_{i,n}$. Corollary 1 implies that the proper rank of the system $(Ap = b) \wedge G'$ is greater than $\frac{n-1}{2}$. By Proposition 2 the rank of G' is greater than $\frac{n-1}{4}$, hence G' contains more than $\frac{n-1}{4}$ equations. Note that Delayer earns a coin for every equation in G', hence Delayer earns more than $\frac{n-1}{4}$ coins. Hence by Lemma 3 the size of any parity decision tree for $(\mathrm{PHP}_n^m, Ap = b)$ is at least $2^{\frac{n-1}{4}}$. □

Now we are ready to prove Theorem 1.

Proof (Proof of Theorem 1). We may assume that a heuristic A does not use random bits. Indeed otherwise we may prove the lower bound for fixed random bits of heuristic A and then apply the averaging principle to handle the case of randomized A. If A is deterministic we may consider the whole parity decision tree T of the algorithm $\mathcal{D}_{A,B}$ on the input $\mathrm{PHP}_n^m + \sigma$ that corresponds to all possible answers of heuristic B. For every execution of the algorithm $\mathcal{D}_{A,B}$, this algorithm bypasses only a part of the tree T until it finds a satisfying assignment. Tree T contains exactly one branch that correspond to the satisfying assignment σ; we call this branch a satisfying path. We prove that with high probability the algorithm will deviate from the satisfying path and will fall in a hard unsatisfiable subtree.

Assume that algorithm is in the state in the accepting path, that is the current linear system F is satisfied by σ. There are two possibilities to deviate from the satisfying path:

1. The algorithm chooses an equation $f = \alpha$ such that the system $F \wedge (f = \alpha)$ has no proper solutions. In this case $f = 1 - \alpha$ is properly implied by F. Thus the adding of $f = 1 - \alpha$ to F does not increase the proper rank.
2. The algorithm chooses an equation $f = \alpha$ and the system $F \wedge (f = \alpha)$ has proper solutions but σ is not a solution of $F \wedge (f = \alpha)$. In this case the proper rank of $F \wedge (f = \alpha)$ may be larger by one then the proper rank of F (but it is also possible that the proper rank of $F \wedge (f = \alpha)$ does not exceed the proper rank of F, see Remark 1). If the proper rank of $F \wedge (f = \alpha)$ is at most $\frac{n-1}{4}$, then by Corollary 3 the algorithm falls in an unsatisfiable subtree of size at least $2^{(n-1)/4}$.

Consider the leaf of the satisfying path; the linear system F in this leaf has the only solution σ. Hence by Corollary 2 the proper rank of F is greater than $\frac{n-1}{2}$. The value of the proper rank of F in the root is zero, the rank of F increases along the satisfying path. Consider the first nodes of the path when the proper rank equals $1, 2, \ldots, \frac{n-1}{4}$. The algorithm should visit this nodes, hence it should visit the predecessors of these nodes. In every of these predecessors algorithm have chance $\frac{1}{2}$ to deviate from the acceptance path. And since the proper rank increases, all this deviations correspond to the case 2 of the above. Thus, with probability at least $1 - 2^{-\frac{n-1}{4}}$ the algorithm goes to an unsatisfiable subtree of size at least $2^{\frac{n-1}{4}}$. $\qquad\square$

Acknowledgements. The authors are grateful to Dmitry Sokolov for fruitful discussions.

References

1. Alekhnovich, M., Hirsch, E.A., Itsykson, D.: Exponential lower bounds for the running time of DPLL algorithms on satisfiable formulas. J. Autom. Reason. **35**(1–3), 51–72 (2005)
2. Ben-Sasson, E.: Hard examples for bounded depth frege. In: Proceedings of the Thiry-Fourth Annual ACM Symposium on Theory of Computing, pp. 563–572. ACM (2002)
3. Beame, P., Kautz, H.A., Sabharwal, A.: Towards understanding and harnessing the potential of clause learning. J. Artif. Intell. Res. (JAIR) **22**, 319–351 (2004)
4. Beame, P., Pitassi, T.: An exponential separation between the parity principle and the pigeonhole principle. Ann. Pure Appl. Logic **80**(3), 195–228 (1996)
5. Cook, J., Etesami, O., Miller, R., Trevisan, L.: Goldreich's one-way function candidate and myopic backtracking algorithms. In: Reingold, O. (ed.) TCC 2009. LNCS, vol. 5444, pp. 521–538. Springer, Heidelberg (2009). doi:10.1007/978-3-642-00457-5_31
6. Dantchev, S.S., Riis, S.: Tree resolution proofs of the weak pigeon-hole principle. In: Proceedings of the 16th Annual IEEE Conference on Computational Complexity, Chicago, Illinois, USA, pp. 69–75. IEEE Computer Society, 18–21 June 2001
7. Davis, M., Logemann, G., Loveland, D.W.: A machine program for theorem-proving. Commun. ACM **5**(7), 394–397 (1962)
8. Davis, M., Putnam, H.: A computing procedure for quantification theory. J. ACM **7**(3), 201–215 (1960)
9. Garlík, M., Kołodziejczyk, L.A.: Some subsystems of constant-depth Frege with parity (2017, Preprint)
10. Itsykson, D.: Lower bound on average-case complexity of inversion of Goldreich's function by drunken backtracking algorithms. Theor. Comput. Syst. **54**(2), 261–276 (2014)
11. Itsykson, D., Sokolov, D.: The complexity of inverting explicit Goldreich's function by DPLL algorithms. J. Math. Sci. **188**(1), 47–58 (2013)
12. Itsykson, D., Sokolov, D.: Lower bounds for splittings by linear combinations. In: Csuhaj-Varjú, E., Dietzfelbinger, M., Ésik, Z. (eds.) MFCS 2014. LNCS, vol. 8635, pp. 372–383. Springer, Heidelberg (2014). doi:10.1007/978-3-662-44465-8_32
13. Krajíček, J.: Randomized feasible interpolation and monotone circuits with a local oracle. CoRR, abs/1611.0 (2016)
14. Oparin, V.: Tight upper bound on splitting by linear combinations for pigeonhole principle. In: Creignou, N., Le Berre, D. (eds.) SAT 2016. LNCS, vol. 9710, pp. 77–84. Springer, Cham (2016). doi:10.1007/978-3-319-40970-2_6
15. Pudlák, P., Impagliazzo, R.: A lower bound for DLL algorithms for k-SAT (preliminary version). In: Proceedings of the Eleventh Annual ACM-SIAM Symposium on Discrete Algorithms, San Francisco, CA, USA, pp. 128–136, 9–11 January 2000
16. Pudlák, P., Scheder, D., Talebanfard, N.: Tighter Hard Instances for PPSZ. CoRR, abs/1611.0 (2016)
17. Razborov, A.A.: Resolution lower bounds for perfect matching principles. J. Comput. Syst. Sci. **69**(1), 3–27 (2004)
18. Scheder, D., Tang, B., Chen, S., Talebanfard, N.: Exponential Lower Bounds for the PPSZ k-SAT Algorithm. In: Khanna, S. (ed.) Proceedings of the Twenty-Fourth Annual ACM-SIAM Symposium on Discrete Algorithms, SODA 2013, New Orleans, Louisiana, USA, pp. 1253–1263. SIAM, 6–8 January 2013
19. Urquhart, A.: Hard examples for resolution. J. ACM **34**(1), 209–219 (1987)

Clause Learning and Symmetry Handling

On the Community Structure of Bounded Model Checking SAT Problems

Guillaume Baud-Berthier[1,3], Jesús Giráldez-Cru[2(✉)], and Laurent Simon[1]

[1] LaBRI, UMR 5800, University of Bordeaux, Bordeaux, France
`guillaume.baud-berthier@safe-river.com`, `lsimon@labri.fr`
[2] KTH, Royal Institute of Technology, Stockholm, Sweden
`giraldez@kth.se`
[3] SafeRiver, Paris, France

Abstract. Following the impressive progress made in the quest for efficient SAT solving in the last years, a number of researches has focused on explaining performances observed on typical application problems. However, until now, tentative explanations were only partial, essentially because the semantic of the original problem was lost in the translation to SAT.

In this work, we study the behavior of so called "modern" SAT solvers under the prism of the first successful application of CDCL solvers, i.e., Bounded Model Checking. We trace the origin of each variable w.r.t. its unrolling depth, and show a surprising relationship between these time steps and the communities found in the CNF encoding. We also show how the VSIDS heuristic, the resolution engine, and the learning mechanism interact with the unrolling steps. Additionally, we show that the Literal Block Distance (LBD), used to identify good learnt clauses, is related to this measure.

Our work shows that communities identify strong dependencies among the variables of different time steps, revealing a structure that arises when unrolling the problem, and which seems to be caught by the LBD measure.

1 Introduction

We observed in the last years an impressive explosion of A.I. and Formal Methods tools using SAT solvers as backbones. The practical interest in SAT solver technologies really took off in the early 2000's when Conflict-Driven Clause Learning (CDCL) algorithms were introduced [14,24]. This allowed huge SAT instances, encoding application problems, to be solved in practice, where previous *adhoc* methods had failed. One of the first successful application of SAT solvers, Bounded Model Checking (BMC), unrolls a transition system for a given

This work was partially supported by the European Research Council under the European Union's Seventh Framework Programme (FP7/2007–2013)/ERC grant agreement no. 279611, and by the French National Research Agency (ANR), with the ANR SATAS Project 2015 (ANR-15-CE40-0017-01), and SafeRiver.

S. Gaspers and T. Walsh (Eds.): SAT 2017, LNCS 10491, pp. 65–82, 2017.
DOI: 10.1007/978-3-319-66263-3_5

number of steps, quickly producing a huge number of clauses and variables. On this kind of applications, the laziness of data structures is crucial but such an efficiency comes with a price: SAT solvers are now some kind of complex systems for which performances can be hardly explained. We know how to build an *efficient*[1] SAT solver, but the reasons of its efficiency are not clearly known.

Moreover, it has been shown that SAT formulas, now arising from many different fields, are highly heterogeneous: some problems can have thousands of decisions levels whereas others just have a few dozens. Some problems with millions of variables can be solved in a few seconds whereas others still resist after a few hundreds. If this is due to the diversity of applications using SAT solvers, it makes harder to analyze and explain the reasons of why SAT solvers are good: possible explanations are probably not the same on all problems.

Thus, hidden behind the success story of SAT solvers, the fact that they are complex systems working on a wide range of different problems seems to have prevented any simple explanation for their efficiency to arise. This is a crucial issue for the future of SAT solving, because additional speed-ups will certainly come from a better understanding of all the components of SAT solvers (branching heuristic, learning mechanism, clause database cleaning). This is even more crucial for parallel SAT solving, where simply identifying a good clause to share or how to split the search space between cores is still unclear. Somehow paradoxically, the reasons of most of the improvements in sequential and parallel SAT solving are hardly understood yet. Of course, a number of works proposed some explanations. For instance it was shown that CDCL solvers are stronger than DPLL, unleashing the full power of propositional resolution [23]. More recently, it was shown that other branching heuristics were possible [15]. Closer to our current work, it was also shown that most industrial instances have a *community structure* [1], which could even be used as an estimator of solver success [21]. Moreover, it has been also shown that the measure used for learnt clause usefulness (the Literal Block Distance) is highly related to this notion of clusters, where the literals of the same decision level seem to belong to the same community. In this direction, some models of random SAT formulas with community structure have been presented to better understand SAT solvers components [11–13]. In fact, the community structure has been successfully used in some SAT and MaxSAT approaches [2,17,19].

However, none of these works link the structure of the SAT formulas with the original problem they encode. If communities are clearly identified as a key structure in real-world instances, their origin is unknown. The paper we propose aims at answering this crucial question. We propose an experimental paper, aiming at pointing out observed correlations between the high-level description of problems and the behavior of the main components of the SAT solver, thus tracing the origin of communities to the high-level description of the problem. We chose to focus on Bounded Model Checking, because of its historical importance in the rapid development of SAT solver technologies. More precisely, we study the

---

[1] CDCL is nowadays the dominant technique solving this kind of problems.

relationship between communities, clauses scoring, variable branching, conflict analysis w.r.t. the unrolling depth of variables.

Contributions of This Paper. We show that:

- communities are built on small unrolling of time steps, revealing a structure that is not present inside a single depth;
- LBD measures the proximity of literals in the clause, w.r.t. time steps;
- computation tends to produce clauses (proof) at larger and larger time steps;
- the learning mechanism implies touching literals when (1) deciding, (2) analyzing and (3) learning. We show that literals touched in (1), (2) and (3) show clearly distinct time step extents. Typically, resolution variables (phase 2) belong to more distant time steps than learnt clause literals (phase 3), which belong to more distant than decisions literals.

Let us emphasize here the first item in the above list. It shows that the general idea on the existence of communities in BMC is wrong, or at least only partial. It is indeed believed that communities are simply a side effect of the unrolling mechanism, each community (or set of communities) being simply inside a single time step. Our work shows that variables connections between communities are stronger than previously believed.

2 Preliminaries

We assume the reader familiar with SAT but let us just recall here the global schema of CDCL solvers [7,9,14,18,24]: a branch is a sequence of decisions (taken accordingly to the VSIDS heuristic), followed by unit propagations, repeated until a conflict is reached. Each decision literal is assigned at a distinct, increasing decision level, with all propagated literals assigned at the same level (we call "block of literals" to the set of literals assigned at the same decision level). Each time a conflict is reached, a series of resolution steps, performed during the conflict analysis, allows the solver to extract a new clause to learn. This clause is then added to the clause database and a *backjumping* is triggered, forcing the last learnt clause to be unit propagated. Solvers also incorporate other important components such as preprocessing [8], restarts and learnt clause database reduction policies. It was shown in [3] that the strategy based on Literal Block Distance (LBD) was a good way of scoring clauses. The LBD is computed during conflict analysis: it measures the number of distinct decisions levels occurring in the learnt clause. Restarts are commonly following the Luby series [16], but recent studies shown that LBD-based restarts is generally more efficient on real-world instances [4], especially on UNSAT instances [22].

Bounded Model Checking. In this paper, we focus on finite-state transition systems with Boolean variables only, which are commonly used to model sequential circuits. A transition system is a tuple $\mathcal{M} = \langle V, I, T, P \rangle$, where V is a set of variables, $I(V)$ and $P(V)$ are formulas over V representing the initial states and safe states, respectively. We also refer to $\neg P(V)$ or $Bad(V)$ to express the set of bad states. $T(V, V')$ is a transition relation over V, V' defining the accepted transitions. $V' = \{v' \mid v \in V\}$ is the primed version of the set V, used to represent next state variables. When multiple transition relations are required, we will use V_i to represent variables in V after i steps.

Bounded Model Checking [5,26] is an efficient bug-finding algorithm. BMC explores bounded paths of a transition system and checks if they can lead to a bad state. The main idea is to build a formula that is satisfiable if there exists a path of length k from the initial states to a bad state. To this end, BMC *unrolls* the transition relation k times, s.t.:

$$I(V_0) \wedge T(V_0, V_1) \wedge \ldots \wedge T(V_{k-1}, V_k) \wedge Bad(V_k)$$

This formula is usually translated into a CNF and then solved with a SAT solver. If the formula is unsatisfiable, P holds for all states reachable in k iterations. However, this definition does not ensure that Bad cannot be reached in less than k transitions. Hence, BMC_k is usually performed incrementally for $k = 0$ to n. Another approach consists in extending $Bad(V_k)$ to $\bigvee_{i=0}^{k} Bad(V_i)$. If the formula is satisfiable, the transition system has a *counterexample*, i.e. there exists a path in the transition system leading to a state that contradicts the property P.

Community Structure of SAT Instances in CNF. A graph is said to have community structure (or good *clustering*), if in a partition of its nodes into communities, most edges connect nodes of the same community. In order to analyze the quality of such partition, community structure is usually analyzed via scoring functions. The most popular one is *modularity* [10,20].

Recently, it has been shown that most industrial SAT instances used in SAT competitions have a clear community structure (or high modularity) [1], when the CNF formula is represented as a graph. In this paper, we use the same approach of [1] to analyze the community structure of CNF formulas. This is: creating the Variable-Incidence Graph[2] (VIG) of the CNF, and using the Louvain method [6] to compute a lower-bound of the maximal modularity Q, which also returns its associated partition P of the graph, i.e., a partition of the Boolean variables of the CNF. Since we analyze the relation between variables in the high-level BMC encoding and the low-level CNF formula, we consider VIG as the most suitable graph representation of the formula.

We also emphasize that, although we use an approximate method to compute the community structure, our conclusions do not seem to be a consequence of computing a *wrong* partition (much different to the optimal one), as we will show in the next section.

---

[2] In this model, the variables of the CNF are the nodes of the graph, and there is an edge between two variables if they appear together in a clause. In its weighted version –the one we use–, the clause size is also considered.

3 Communities and Unrolling Depth in SAT Encodings

The origins of community structure in industrial SAT formulas remain unknown. In previous works, the heterogeneous set of industrial benchmarks used in the competitions has been analyzed as a whole, regardless of where they come from or which problems they encode. Interestingly enough, it has been shown that (clear) community structure is a property in most of these instances. In this section, we provide an exhaustive analysis of the relations between the community structure and the high-level structure of the problem, on the case study of BMC problems.

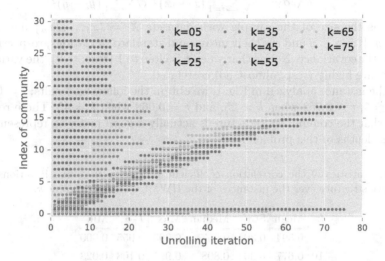

Fig. 1. Relation between unrolling iterations and community structure in the instance 6s7, for different number of unrolling timesteps k.

For a given problem, we can generate different CNF formulas representing BMC_k, for different values of k. For these formulas, each Boolean variable belongs to a certain unrolling depth x,[3] and it is also assigned to a certain community y by the clustering algorithm. Notice that it is possible that two (or more) variables are characterized by the same (x, y) coordinates, when they both belong to the same unrolling iteration and they both are assigned to the same community.

In Fig. 1, we represent the relation between the unrolling iterations and the community structure of the instances BMC_k encoding the problem 6s7, for different values of the unrolling depth $k = \{5, 15, 25, 35, 45, 55, 65, 75\}$. This is, the iteration x and the community y of the variables of each CNF. The y-value for some k are slightly shifted to improve the visualization.

We can observe that when the total number of unrolling iterations is small (see $k = 5$), the community structure is not related to the unrolling depth, since most communities have variables in all depths. The number of communities is

---

[3] For simplicity, we omit the very few variables which do not belong to any iteration.

greater than k. Interestingly enough, as we increase the total number of iterations k, there is a clear pattern, suggesting that the correlation may be very strong (see cases $k \geq 45$).

However, each point represented in the previous plot can be just due to a single variable. Therefore, it is hard to say if the unrolling iterations and the community structure are indeed correlated. In order to solve this problem, we can compute the Pearson's correlation coefficient r between these two variables X and Y over all these (x, y) points. The correlation coefficient is defined as:

$$r_{X,Y} = \frac{\sigma_{XY}}{\sigma_X \, \sigma_Y} = \frac{\sum_{i=1}^{n} (x_i - \bar{x})(y_i - \bar{y})}{\sqrt{\sum_{i=1}^{n}(x_i - \bar{x})^2}\sqrt{\sum_{i=1}^{n}(y_i - \bar{y})^2}}$$

where n is the size of the sample with datasets $X = \{x_1, \ldots, x_n\}$ and $Y = \{y_1, \ldots, y_n\}$, $\bar{x}$, σ_X, $\bar{y}$ and σ_Y their means and standard deviations, respectively, and σ_{XY} the covariance. Notice that when r is close to 1 (resp. to 0), the variables X and Y are highly (resp. almost no) correlated.

For the instance analyzed in Fig. 1, we obtain the following results: $r = 0.423$ when $k = 5$, $r = 0.769$ when $k = 25$, and $r = 0.968$ when $k = 45$. These results confirm that the community structure is actually a good proxy to represent the unrolling depths of this problem.

Table 1. Statistics of the correlation coefficient r between unrolling iterations and community structure over the instances of the HWMCC15, for different k.

k	Mean	Std	Median	Max	$P_{5\%}$	Min
5	0.594	0.304	0.667	0.995	0.055	0.000
10	0.677	0.304	0.808	0.997	0.108	0.023
20	0.856	0.170	0.918	0.999	0.492	0.108
40	0.892	0.155	0.956	0.999	0.580	0.109
60	0.904	0.152	0.973	0.999	0.643	0.131

Now the question is whether this occurs in all BMC problems. To answer this question, we analyze the 513 problems of the Hardware Model Checking 2015 competition (HWMCC15). For each of these benchmarks, we create different BMC_k, with $k = \{5, 10, 20, 40, 60\}$, compute the community structure of those, and finally calculate the correlation coefficient r as before. In this analysis, we have omitted those benchmarks for which the computation of the community structure requires more than 5000 s (notice that some problems become extremely large after a big number of unrolling iterations). Even though, the resulting sets of instances always contain more than 400 benchmarks; the size of the set varies for each value of k, from 400 instances when $k = 60$, to 509 instances when $k = 5$.

In Table 1, we represent the aggregated results of this analysis. In particular, for each depth k, we compute mean, standard deviation (std), median, max, min, and percentile 5% ($P_{5\%}$), over the r coefficients of all these instances.

It is clear that, in general, the correlation between the unrolling depth and the community structure is very strong (high mean and median, with small deviation). In fact, it becomes stronger for bigger values of k (as we expected from the previous experiment). We can also observe that such correlation is surprisingly small for some instances (very small min). However, it is only the case for a very reduced number of problems, as the percentile 5% indicates. *This analysis strongly suggests that the community structure is originated by the process of unrolling depth iterations to create the BMC formula.*

Additionally, an interesting observation depicted in Fig. 1 is that communities computed by the clustering algorithm contain variables from different unrolling depths, rather than just aggregating all variables of the same depth. Therefore, the community structure reveals a non-trivial structure existing in the CNF encoding: *Our analysis also suggests that communities are spread over successive time steps.*

3.1 Communities are Stronger When Spreading over Time Steps

Our last observation above is important enough to be checked in detail. It goes against the general belief about the existence of communities in BMC problems. We thus propose here to search for additional evidences validating it. We also need to check whether this is a property shared by all the BMC problems we gathered.

For this purpose, we consider the partition of the formula in which all the variables of the same unrolling depth conform a distinct community, and compute its associated modularity Q_d. In order to check that our results are not a consequence of computing a *wrong* partition, much different to the optimal one, we compare Q_d w.r.t. the modularity Q computed by the clustering algorithm. In Table 2, we represent the aggregated results of Q and Q_d over the set of benchmarks of the HWMCC15. Again, we represent some statistics for different values of k.

As expected, we observe that the modularities Q and Q_d increase for bigger values of the unrolling depth k. Interestingly, the actual modularity of the

Table 2. Statistics of the actual modularity Q and the modularity Q_d using as partition the variables of the same unrolling depth, over the instances of the HWMCC15, for different values of depth k. The highest values between $\bar{Q}$ and $\bar{Q}_d$ are highlighted, for each depth k.

	Q						Q_d					
k	Mean	Std	Median	Min	Max	$P_{5\%}$	Mean	Std	Median	Min	Max	$P_{5\%}$
5	**0.850**	0.053	0.857	0.681	0.970	0.747	0.616	0.083	0.607	0.480	0.826	0.387
10	**0.874**	0.047	0.877	0.701	0.977	0.786	0.687	0.085	0.676	0.549	0.901	0.450
20	**0.887**	0.043	0.894	0.731	0.981	0.813	0.728	0.086	0.717	0.588	0.945	0.488
40	**0.906**	0.038	0.907	0.759	0.986	0.837	0.751	0.086	0.739	0.609	0.968	0.509
60	**0.917**	0.036	0.917	0.776	0.988	0.856	0.761	0.094	0.767	0.614	0.976	0.516

formula Q is clearly greater than Q_d, independently of the number of iterations k the problem is unrolled. In fact, in many instances, the number of communities is smaller than the number of unrolling iterations k, when k is big enough (e.g., $k = 60$). Another interesting observation is that Q_d fluctuates less than the correlation coefficient r (see Table 1).

These observations support our hypothesis. Therefore, this analysis indicates that the existence of community structure in BMC problems encoded as SAT instances is due to the process of unrolling iterations in the high-level problem, and the larger the unrolling depth is, the more clear the community structure becomes. However, *this structure seems to identify strong dependencies among the Boolean variables in the CNF encoding, which is different from just the unrolling depths they belong to.*

4　Unrolling Depth of Decisions, Resolutions and Learnt Clauses

In the previous section, we have studied the static structure of SAT formulas w.r.t. the unrolling depths of propositional variables. We now propose to analyze where the solver performs the search in these formulas. In particular, we show that, although the solver is not aware about the semantics of each variable (i.e., the unrolling iteration it belongs to), it is able to exploit some features of the BMC encoding. This is probably due to the relation between the high-level BMC encoding and the CNF formula, captured by the community structure.

Our experimental investigations will be based on a set of 106 BMC instances. This set of formulas contains all the problems of the HWMCC15, unrolled until obtaining a satisfiable answer or not solved within a timeout of 6600 s (for the largest possible depth). We excluded the easy instances, solved in less than a minute. We base our experimental observations on an instrumented version of the solver `Glucose`.

For our analysis, we compute three sets of variables that characterize the solver at each conflict. First, the set *dec* contains all the decision variables, i.e., all the decisions stored in the trail of the solver. Some decisions may not be related to the current conflict, but we think that observing the evolution of *dec* may be a better trace of the solver search (for instance, top decision variables may be useful in conflicts). Second, the set *res* of variables used in all the resolution steps, performed during conflict analysis. And finally, the set *learnt* of literals in the learnt clause. One may notice here that resolution variables (*res*) are disjoint from decision and learnt variables (*dec* ∪ *learnt*). In some sense, *dec* will be the witness of the VSIDS heuristics (even if *dec* has a longer memory of variables decisions, because a variable with a low VSIDS score can still be in the *dec* if it was chosen at the top of the search tree and its score has been decreased). Moreover, one may notice that the VSIDS heuristic bumps all the variables in the set *res* ∪ *learnt*. It is thus very surprising to find different distributions of values in these 3 sets of variables, as we will show in Sect. 4.3.

For each of these sets, and for every conflict, we store the set size, and some statistics about the unrolling depths in the high-level problem. Namely, they are: min, max, mean, standard deviation (std), median, median absolute deviation (mad) and skewness. Clearly enough, on hard problems involving millions of conflicts, we cannot simply summarize all these data. Thus, we decided to take samples of 10,000 values and only report summaries of the above values for each sample. Every 10,000 conflicts, we aggregate the results by computing the same values (min, max, et cetera) for each measure (we thus recorded, for example, the median of min, the skewness of skewness, et cetera). After investigating all these data, we found that working on the mean of means for each set *dec*, *res* and *learnt* is sufficient to draw some conclusions; it is a good estimator of the measured values in the sample and it is easily understood. We thus report in this paper the mean values (computed over 10,000 conflicts) of the average number of unrolling depths found in each set *dec*, *res* and *learnt*. Note that we recorded this for SAT and UNSAT instances, and also for instances that timed-out, where final statistics were printed before exiting. We found 7 SAT instances, 61 UNSAT ones and 39 time-outs. We cumulated more than 1.2 billion conflicts, and 3.8 millions samples averaging each measure over 10,000 conflicts.

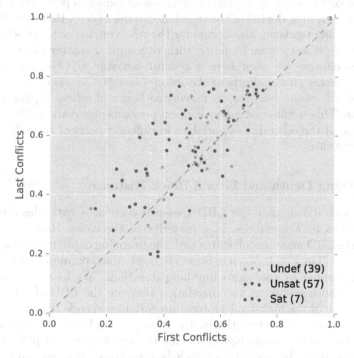

Fig. 2. Unrolling depth of the decision variables during the first conflict versus the last conflicts. Values are normalized w.r.t. the total depth of each BMC SAT formula.

4.1   Relation Between Solver Progress and Unrolling Depths

In our first experiment, we want to investigate whether the solver is first working on the variables of certain unrolling depths before moving to variables of other depths. For this, we compare the values of *dec*, *res* and *learnt* during the first conflicts w.r.t. to these values measured during the last conflicts. In Fig. 2, we represent this comparison for the set *dec*. The "first conflicts" are computed between the first 10,000 and the 20,000 conflicts (thus, avoiding any possible initialization phase in the very early conflicts, that could introduce a bias), and the value of the last conflicts is computed between the last $10,000 + X$ conflicts, where X is the total number of conflicts modulo 10,000. Results are normalized w.r.t. the total number of unrolling iterations of each SAT instances. A value close to 1 means that the set *dec* contains variables of the *last* unrolling iterations, whilst a value close to 0 means that it contains the ones of the first iterations.

As we can see, the unrolling iterations involved in the early conflicts are smaller than the ones at the latest stages of the search. This is the case in most of the formulas analyzed, regardless of the answer of the solver. Interestingly enough, this also happens in non-solved instances, indicating a common tendency which suggests that the solver tends to start the search by the first unrolling depths, and continues it by exploring variables of higher depths. We have only reported the unrolling depths of the decision variables (*dec*). However, the same kind of increasing tendency also occurs for the resolving variables (*res*), and the variables in the learnt clause (*learnt*), with very similar scatter plots.

As a conclusion, we show here a general behavior of CDCL solvers over BMC benchmarks: all the efforts of the solver (variable decisions, clause analysis / resolution, clause learning) is moving to larger unrolling dephts along the computation. This results cast a new light on previous observations [25] made on the evolution of the centrality of variables in different parts of the solver engine in the general case.

4.2   Unrolling Depth and Literal Block Distance

In [21], it was shown that the LBD measure correlates with the number of communities in SAT instances, in a majority of the cases. Here, we want to know how the LBD measure correlates with the unrolling depth of the variables of BMC formulas. The Fig. 1 shown in Sect. 3 suggests that communities are defined on a small number of successive unrolling iterations. We have tested a large number of hypotheses about the correlation between the LBD of a clause and the information of the high-level problem its variables encode (i.e., the unrolling depth they belong to). Interestingly, the highest correlation we found relates LBD with the $(max - min)$ measure. This is, the maximum depth minus the minimum depth of the variables of each learnt clause. We can represent this relation using a *heatmap* for every problem. In Fig. 3, we represent one of the most striking ones. There is clearly a relationship between these two values.

However, such a strong visual relationship does not occur in all formulas, even if we observed it in the immense majority of the cases. In order to summarize

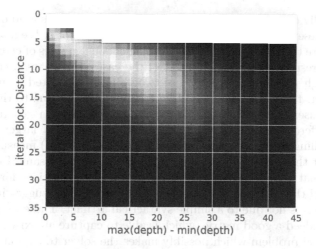

Fig. 3. Heatmap showing the correlation between LBD values of clauses and the $(max - min)$ value of the unrolling depths of its variables. The pearson correlation is only 0.315 on this example, showing that even small values are already good indicators for a good correlation on our set of problems.

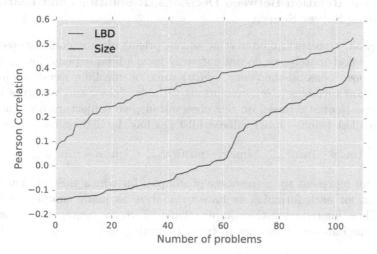

Fig. 4. CDF of pearson correlation coefficient of LBD against $(max - min)$ of unrolling depth. We also show the pearson coefficient for the sizes of clauses instead of LBD.

all the results, we decided to compute, for each problem, the Pearson correlation coefficient r between the LBD and the value of $(max - min)$ of all learnt clauses. Of course this correlation will not be perfect: the Pearson coefficient measures the correlation between two lines, and we immediately see in Fig. 3 that, even if we observe a clear tendency, the cloud of points are dispersed around a line. The Pearson coefficient here will just indicates the general tendency of the cloud, and we typically cannot expect values greater than 0.5 on this kind of clouds. In Fig. 4, we represent the cumulative distribution function of r. We also report on the same Figure the CDF of the Pearson coefficients using sizes of clauses instead of LBD. Our results show that, in most of the cases, the Pearson coefficient is sufficiently high to indicates that these two values are related in most of the cases. However, this correlation could be a simple artifact due to the length of the learnt clause. In particular, a larger clause can have, in general, a larger number of different decision levels, and as a consquence, a larger number of different unrolling iterations, thus increasing the $(max - min)$ measure as a side effect. We test this hypothesis, and show the results on the same Fig. 4. It can be observed that the CDF of the Pearson correlation coefficient for the clause size (instead of the LBD) shows a much weaker correlation, suggesting that the LBD correlation is not due to a simple syntactical artifact. These results suggest that LBD is indeed a good metric, which is able to capture an existing structure of the high-level problem which possibly makes the solver to exploit it.

As a conclusion, we show that, in the majority of the cases, the LBD measure is related to the max-min unrolling depth of clauses.

4.3   On the Relation Between Decisions, Resolutions and Learning During Solver Search

In this section, we report an interesting and surprising phenomenon we observed, for which, unfortunately, we do not currently have a final explanation.

Let us now focus on the $(max - min)$ value of unrolling iterations for the set of decision variables (dec), resolutions variables (res), and variables in the learnt clause ($learnt$). Based on our observations, we conjecture now that dec has the smallest $(max - min)$ value, whilst res has the highest one. This is:

$$(max - min)_{dec} \leq (max - min)_{learnt} \leq (max - min)_{res}$$

In Fig. 5, we represent the percentage of samples for which this relation holds. Notice that for each formula, we have to analyze as many samples as learnt clauses. Thus, for reducing the computational effort, we aggregated some data to reduce the number of samples to treat, by taking their average every 10,000 conflicts.

It can be observed that the previous relation occurs in an immense majority of the analyzed samples. *Decisions variables links fewer unrolling iterations than the variables in the learnt clauses, which relate less unrolling iterations than resolution variables.* The CDF plot is clearly showing that this hypothesis holds for almost all the samples we measured. This result is also surprising because variables from dec are chosen thanks to VSIDS, which bumps variables from res and $learnt$. Measuring a significant difference between these sets indicates that a strong mechanism is at work.

This phenomenon is possibly due to variable dependencies w.r.t unit propagations, which tends to imply variables of smaller/larger unrolling iterations as the solver goes deeper in the search tree. However, we have not been able

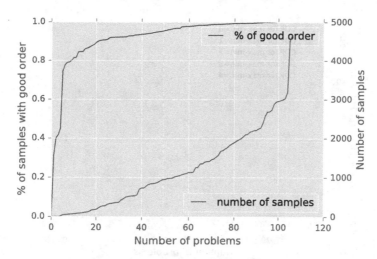

Fig. 5. CDF of the percentage of samples s.t. $(max - min)_{dec} \leq (max - min)_{learnt} \leq (max - min)_{res}$ (left Y-axis). We also report the CDF of the number of samples per problem (right Y-axis).

to simply link this to the high-level BMC problem. We also make the hypothesis that if the analysis is done on a greater number of unrolling depths, only a restricted set of iterations is bumped more often, allowing the solver to focus on a restricted set of iterations only. This means that, even if unit propagation tends to propagate variables in the next unrolling iterations (which is a reasonable hypothesis), only a few variables are more often bumped. These variables tends to be localized near the top decision variables of the current search tree.

4.4 On the Relation Between Variable Unrolling Depths and Variable Eliminations

The classical preprocessing used in Minisat, and hence in Glucose as well, is essentially built on top of the Variable Elimination (VE) principle [8]. Variable Elimination is crucial for many SAT problems, and particularly for BMC ones. On a typical BMC problem, hundreds of thousands variables can be eliminated. More precisely, the preprocessor orders the variables according to the (current) product of their positive and negative occurrences in the formula, thus trying to eliminate variables that will limit the combinatorial explosion first. Then, a variable is eliminated only if it does not increase the formula size too much (i.e., no more clause after the elimination, and no clause larger than a constant in the resolvents). We want to check here where the preprocessing is working. Our initial hypothesis is that in most of the cases, variables are eliminated inside a single unrolling depth. We thus build the following experimentation. We measure, for each eliminated variable, the maximal depth of the variables occurring in the set of all produced clauses (by the cross product) minus their minimal depth. Thus,

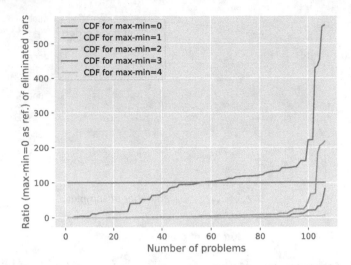

Fig. 6. CDF of the percentage of eliminated variables within the same depth, at distance 1, 2, ... The number of eliminated variable within the same depth is taken as a reference for scaling all the problems.

a $(max - min)$ value 0 means that eliminating a variable does not add any link between two unrolled depths.

In Fig. 6, we represent the results. This plot can be read as follows. On approximately 50 problems (where the blue curve representing $(max - min) = 1$ crosses the red curve representing $(max - min) = 0$) there are more eliminated variables inside a single unrolling iteration than connecting two of them. We can see that on almost all the problems, there are only very few variable eliminations that involve a large $(max - min)$ value of depths. It is however surprising to see that some problems involve the elimination of a larger set of variables not strictly within the same community (i.e., when the blue curve is above the red line). We think that an interesting mechanism may be at work here and further investigations on the role of these variables may be important. However, *in general, the elimination of the variable chosen by the classical preprocessing heuristic only involves clauses of very limited $(max - min)$ values*, as expected.

5   Discussions About Improving SAT Solvers

We strongly believe that a better understanding of what CDCL solvers are doing, and how they are doing that is needed by the community for improving SAT solving. Given the observations we made in this paper, we naturally tried a number of Glucose hacks in order to try to exploit the unrolling iterations of variables. However, until now, our attempts were at least partially unsuccessful.

5.1   Shifting Variables Scores

We tried, at each restart triggered by the solver, to force it to branch more deeper (or shallower) in the formula. For this, during the first descent right after a restart, we modified the variable picking heuristics as follows: each time a variable was picked, we took the same variable (1) one time step after, (2) on the last time step, (3) one time step before, (4) on the first time step. What we observed is that only the hack (2) was competitive with the original Glucose (it solved the same amount of problems). All the other versions were degrading the performances.

5.2   Scoring Clauses w.r.t. Max-Min of Unrolling Iterations

Of course, because of the clear relationship of LBD and the $(max - min)$ of unrolling iterations of variables, we tried to change Glucose clause scoring scheme. Instead of the LBD, we simply used this $(max - min)$ metric. We however observed a small degradation of its performance. Further experiments are needed here, essentially because the LBD mechanism is important in Glucose for scoring the clauses, but also for triggering restarts.

5.3   Forcing Variable Elimination According to Their Unrolling Iterations

We also tried to change the order of variable eliminations during the preprocessing step [8]. In our attempt, we first eliminate all the variables from within a single time step: when the cross product does not produce any clause with variable from different unrolling iterations. Then, variable elimination is allowed to produce clauses containing variables from two successive time steps. Then 3, 4, ... time steps. We observed, as expected, that almost all the variable eliminations are done during the first round (all clauses are generated within the same time step), even if other variables are eliminated in the other rounds. From a performance point of view, these versions show a slightly degradation of performance.

5.4   Further Possible Ways of Improvements

Despite the disappointing experimental improvements we reported above, we strongly believe that our work can lead to a significant improvement of SAT solvers over BMC problems. Our work advocates for a better specialization of SAT solvers. For instance, it may probably be possible to use the notion of variables time steps to estimate the progress of the solver along its computation. We are also thinking of using this knowledge for a better parallelization of SAT solvers, for instance, by splitting the search on deeper variables only.

6   Conclusions

Despite the empirical success of SAT solvers, they lack from a simple explanation supporting the observed performances. One intriguing question is why they are more efficient, at the CNF level, than *adhoc* approaches that can access to higher level semantic information. In our study, we focus on the historical success application of SAT solvers, i.e., Bounded Model Checking. We show that community structures, identified a few years ago in most of the SAT formulas encoding real-world problems, are not a trivial artifact due to the circuit unrolling when encoding BMC problems. Our work is an important effort for a SAT solver experimental study. We report evidences that the community structure identified in previous works, and well captured by current SAT solvers, is unveiling a structure that arise when unrolling the circuit only (on our set of BMC problems). This is an interesting finding, that answers open questions about the origin of communities. We also show that the proof built by the SAT solver, approximated in our study by the set of all learnt clauses is evolving along the computation, producing clauses of higher unrolling depths at the latest stages of the search, rather than at the beginning of the execution.

We plan to extend our work in many ways. First, we would like to study the semantic of SAT formulas in other domains, like CSP benchmarks encoding planning or scheduling problems, or cryptographic formulas. In all these cases, we think that working on benchmark structures could help SAT solver designers to specialize their solvers to some important applications. We could also expect that CSP solvers could benefit from an expertise of how SAT solvers are working on the underlying SAT formula representing such CSP problems. In particular, this may be useful to refine the heuristic used to solve such instances. Second, we would like to test whether SAT solvers could benefit from the unrolling depth information for BMC problems. We particularly think of studying the unsatisfiable proof in order to work on a better parallelization of SAT solvers *specialized* for BMC problems.

References

1. Ansótegui, C., Giráldez-Cru, J., Levy, J.: The community structure of SAT formulas. In: Cimatti, A., Sebastiani, R. (eds.) SAT 2012. LNCS, vol. 7317, pp. 410–423. Springer, Heidelberg (2012). doi:10.1007/978-3-642-31612-8_31
2. Ansótegui, C., Giráldez-Cru, J., Levy, J., Simon, L.: using community structure to detect relevant learnt clauses. In: Heule, M., Weaver, S. (eds.) SAT 2015. LNCS, vol. 9340, pp. 238–254. Springer, Cham (2015). doi:10.1007/978-3-319-24318-4_18
3. Audemard, G., Simon, L.: Predicting learnt clauses quality in modern sat solvers. In: Proceeding of IJCAI 2009, pp. 399–404 (2009)
4. Biere, A.: Splatz, Lingeling, Plingeling, Treengeling, YalSAT Entering the SAT Competition 2016. In: Proceeding of the SAT Competition 2016, Department of Computer Science Series of Publications B, vol. B-2016-1, pp. 44–45. University of Helsinki (2016)

5. Biere, A., Cimatti, A., Clarke, E., Zhu, Y.: Symbolic model checking without BDDs. In: Cleaveland, W.R. (ed.) TACAS 1999. LNCS, vol. 1579, pp. 193–207. Springer, Heidelberg (1999). doi:10.1007/3-540-49059-0_14
6. Blondel, V.D., Guillaume, J.L., Lambiotte, R., Lefebvre, E.: Fast unfolding of communities in large networks. J. Stat. Mech. Theor. Exper. **2008**(10), P10008 (2008)
7. Darwiche, A., Pipatsrisawat, K.: Complete Algorithms, chap. 3, pp. 99–130. IOS Press (2009)
8. Eén, N., Biere, A.: Effective preprocessing in SAT through variable and clause elimination. In: Bacchus, F., Walsh, T. (eds.) SAT 2005. LNCS, vol. 3569, pp. 61–75. Springer, Heidelberg (2005). doi:10.1007/11499107_5
9. Eén, N., Sörensson, N.: An extensible SAT-solver. In: Giunchiglia, E., Tacchella, A. (eds.) SAT 2003. LNCS, vol. 2919, pp. 502–518. Springer, Heidelberg (2004). doi:10.1007/978-3-540-24605-3_37
10. Fortunato, S.: Community detection in graphs. Phys. Rep. **486**(3–5), 75–174 (2010)
11. Giráldez-Cru, J., Levy, J.: A modularity-based random SAT instances generator. In: Proceeding of IJCAI 2015, pp. 1952–1958 (2015)
12. Giráldez-Cru, J., Levy, J.: Generating SAT instances with community structure. Artif. Intell. **238**, 119–134 (2016)
13. Giráldez-Cru, J., Levy, J.: Locality in random SAT instances. In: Proceeding of IJCAI 2017 (2017)
14. Gomes, C.P., Selman, B., Crato, N.: Heavy-tailed distributions in combinatorial search. In: Smolka, G. (ed.) CP 1997. LNCS, vol. 1330, pp. 121–135. Springer, Heidelberg (1997). doi:10.1007/BFb0017434
15. Liang, J.H., Ganesh, V., Poupart, P., Czarnecki, K.: Learning rate based branching heuristic for SAT solvers. In: Creignou, N., Le Berre, D. (eds.) SAT 2016. LNCS, vol. 9710, pp. 123–140. Springer, Cham (2016). doi:10.1007/978-3-319-40970-2_9
16. Luby, M., Sinclair, A., Zuckerman, D.: Optimal speedup of Las Vegas algorithms. Inf. Process. Lett. **47**(4), 173–180 (1993)
17. Martins, R., Manquinho, V., Lynce, I.: Community-based partitioning for MaxSAT solving. In: Järvisalo, M., Van Gelder, A. (eds.) SAT 2013. LNCS, vol. 7962, pp. 182–191. Springer, Heidelberg (2013). doi:10.1007/978-3-642-39071-5_14
18. Moskewicz, M., Madigan, C., Zhao, Y., Zhang, L., Malik, S.: Chaff: Engineering an efficient SAT solver. In: Proceeding of DAC 2001, pp. 530–535 (2001)
19. Neves, M., Martins, R., Janota, M., Lynce, I., Manquinho, V.: Exploiting resolution-based representations for MaxSAT solving. In: Heule, M., Weaver, S. (eds.) SAT 2015. LNCS, vol. 9340, pp. 272–286. Springer, Cham (2015). doi:10.1007/978-3-319-24318-4_20
20. Newman, M.E.J., Girvan, M.: Finding and evaluating community structure in networks. Phys. Rev. E **69**(2), 026113 (2004)
21. Newsham, Z., Ganesh, V., Fischmeister, S., Audemard, G., Simon, L.: Impact of community structure on SAT solver performance. In: Sinz, C., Egly, U. (eds.) SAT 2014. LNCS, vol. 8561, pp. 252–268. Springer, Cham (2014). doi:10.1007/978-3-319-09284-3_20
22. Oh, C.: Between SAT and UNSAT: the fundamental difference in CDCL SAT. In: Heule, M., Weaver, S. (eds.) SAT 2015. LNCS, vol. 9340, pp. 307–323. Springer, Cham (2015). doi:10.1007/978-3-319-24318-4_23
23. Pipatsrisawat, K., Darwiche, A.: On the power of clause-learning SAT solvers as resolution engines. Artif. Intell. **175**(2), 512–525 (2011)
24. Silva, J.P.M., Sakallah, K.A.: GRASP: A search algorithm for propositional satisfiability. IEEE Trans. Comput. **48**(5), 506–521 (1999)

25. Katsirelos, G., Simon, L.: Eigenvector centrality in industrial SAT instances. In: Milano, M. (ed.) CP 2012. LNCS, pp. 348–356. Springer, Heidelberg (2012). doi:10. 1007/978-3-642-33558-7_27
26. Strichman, O.: Accelerating bounded model checking of safety properties. Form. Methods Syst. Des. **24**(1), 5–24 (2004)

Symmetric Explanation Learning: Effective Dynamic Symmetry Handling for SAT

Jo Devriendt$^{(\boxtimes)}$, Bart Bogaerts$^{(\boxtimes)}$, and Maurice Bruynooghe

Department of Computer Science, KU Leuven,
Celestijnenlaan 200A, 3001 Heverlee, Belgium
{jo.devriendt,bart.bogaerts,maurice.bruynooghe}@cs.kuleuven.be

Abstract. The presence of symmetry in Boolean satisfiability (SAT) problem instances often poses challenges to solvers. Currently, the most effective approach to handle symmetry is by *static symmetry breaking*, which generates asymmetric constraints to add to the instance. An alternative way is to handle symmetry *dynamically* during solving. As modern SAT solvers can be viewed as propositional proof generators, adding a symmetry rule in a solver's proof system would be a straightforward technique to handle symmetry dynamically. However, none of these proposed *symmetrical learning* techniques are competitive to static symmetry breaking. In this paper, we present *symmetric explanation learning*, a form of symmetrical learning based on learning symmetric images of explanation clauses for unit propagations performed during search. A key idea is that these symmetric clauses are only learned when they would restrict the current search state, i.e., when they are unit or conflicting. We further provide a theoretical discussion on symmetric explanation learning and a working implementation in a state-of-the-art SAT solver. We also present extensive experimental results indicating that symmetric explanation learning is the first symmetrical learning scheme competitive with static symmetry breaking.

Keywords: Boolean satisfiability · Symmetry · Proof theory · Symmetric learning · Dynamic symmetry breaking

1 Introduction

Hard combinatorial problems often exhibit symmetry. When these symmetries are not taken into account, solvers are often needlessly exploring isomorphic parts of a search space. Hence, we need methods to handle symmetries that improve solver performance on symmetric instances.

One common method to eliminate symmetries is to add symmetry breaking formulas to the problem specification [1,10], which is called *static symmetry breaking*. For the Boolean satisfiability problem (SAT), the tools SHATTER [3] and BREAKID [14] implement this technique; they function as a preprocessor that can be used with any SAT solver.

© Springer International Publishing AG 2017
S. Gaspers and T. Walsh (Eds.): SAT 2017, LNCS 10491, pp. 83–100, 2017.
DOI: 10.1007/978-3-319-66263-3_6

Dynamic symmetry handling, on the other hand, interferes in the search process itself. For SAT, dynamic symmetry handling has taken on many forms. Early work on this topic dynamically detects symmetry after failing a search branch to avoid failing symmetrical search branches, using an incomplete symmetry detection strategy [7]. Next, *dynamic symmetry breaking* posts and retracts symmetry breaking formulas during search, dynamically detecting symmetry with graph automorphism techniques [5].

A more principled approach is implemented by SYMCHAFF, a *structure-aware* SAT solver [29]. Next to a conjunctive normal form (CNF) theory, SYMCHAFF assumes as input a special type of symmetry, structuring the Boolean variables from the theory in so-called *k-complete m-classes*. This structure is then used to branch over a subset of variables from the same class instead of over a single variable, allowing the solver to avoid assignments symmetric to these variables.

Arguably, the most studied dynamic symmetry handling approach is *symmetrical learning*, which allows a SAT solver to learn symmetrical clauses when constructing an unsatisfiability proof. The idea is that SAT solvers do not only *search* for a satisfying assignment, but simultaneously try to *prove* that none exists. For this, their theoretical underpinning is the propositional resolution proof system [28], which lets a SAT solver learn only those clauses that are resolvents of given or previously learned clauses. A SAT solver's proof system provides upper bounds on the effectiveness of SAT solvers when solving unsatisfiable instances. For instance, for encodings of the pigeonhole principle, no polynomial resolution proofs exist [19], and hence, a SAT solver cannot solve such encodings efficiently.

However, if one were to add a rule that under a symmetry argument, a symmetrical clause may be learned, then short proofs for problems such as the pigeonhole encoding exist [23]. As with the resolution rule, the central question for systems that allow the symmetry argument rule then becomes what selection of symmetrical clauses to learn, as learning all of them is infeasible [20] (nonetheless, some have experimented with learning all symmetrical clauses in a SAT solver [30]). The *symmetrical learning scheme* (SLS) only learns the symmetrical images of clauses learned by resolution, under some small set of generators of a given symmetry group [6]. Alternatively, *symmetry propagation* (SP) learns a symmetrical clause if it is guaranteed to propagate a new literal immediately [15]. Finally, for the graph coloring problem, symmetry-handling clauses can be learned based on *Zykov contraction* [20]. Unfortunately, none of these are competitive to state-of-the-art static symmetry breaking for SAT, as we will show with extensive experiments.

Symmetrical learning, as discussed in the previous paragraph differs significantly from the other methods discussed. These other methods all prune the search tree in a satisfiability-preserving way, but possibly also prune out models, for instance by adding symmetry breaking clauses or by not considering all possible choices at a given choice point. As such, they change the set of models of the theory, hence why we call them *symmetry breaking*. Symmetrical learning exploits symmetry in another way: if unsatisfiabilty of a certain branch of the search tree is concluded, it manages to learn that symmetrical parts of the search tree are also unsatisfiable; all clauses learned by symmetrical learning are consequences of the original specification. Hence, it never eliminates any models: symmetries are not broken, but merely exploited.

In this paper, we propose a new approach to symmetrical learning – *symmetric explanation learning* (SEL) – that improves upon our earlier work on symmetry propagation [15]. SEL's central idea is to learn a symmetric image of a clause only if (i) the clause is an explanation for a unit propagated literal and if (ii) the symmetric image itself is either unit or conflicting. In short, (i) limits the number of symmetric images under investigation to a manageable set, and (ii) guarantees that any learned symmetric clause is useful – it restricts the search state – at least once.

We experimentally validate this algorithm and conclude that SEL is the first dynamic symmetry exploitation approach to successfully implement a symmetric learning scheme. It performs on-par with the award winning static symmetry breaking tool BREAKID [14] and outperforms previous symmetrical learning algorithms such as SLS [6] and SP [15].

The rest of this paper is structured as follows. In Sect. 2 we recall some preliminaries on symmetry and satisfiability solving. Afterwards, we introduce our new algorithm in Sect. 3 and compare it to related work in Sect. 4. We present experimental results in Sect. 5 and conclude in Sect. 6.

2   Preliminaries

Satisfiability problem. Let Σ be a set of Boolean variables and $\mathbb{B} = \{\mathbf{t}, \mathbf{f}\}$ the set of Boolean values denoting true and false respectively. For each $x \in \Sigma$, there exist two *literals*; the *positive* literal denoted by x and the *negative* literal denoted by $\neg x$. The negation $\neg(\neg x)$ of a negative literal $\neg x$ is the positive literal x, and vice versa. The set of all literals over Σ is denoted $\overline{\Sigma}$. A *clause* is a finite disjunction of literals $(l_1 \vee \ldots \vee l_n)$ and a *formula* is a finite conjunction of clauses $(c_1 \wedge \ldots \wedge c_m)$. By this definition, we implicitly assume a formula is an expression in *conjunctive normal form* (CNF).

A *(partial) assignment* is a set of literals $(\alpha \subset \overline{\Sigma})$ such that α contains at most one literal over each variable in Σ. Under assignment α, a literal l is said to be *true* if $l \in \alpha$, *false* if $\neg l \in \alpha$, and *unknown* otherwise. An assignment α *satisfies* a clause c if it contains at least one true literal under α. An assignment α *satisfies* a formula φ, denoted $\alpha \models \varphi$, if α satisfies each clause in φ. If $\alpha \models \varphi$, we also say that φ *holds* in α. A formula is *satisfiable* if an assignment exists that satisfies it, and is *unsatisfiable* otherwise. The Boolean satisfiability (SAT) problem consists of deciding whether a formula is satisfiable. Two formulas are *equisatisfiable* if both are satisfiable or both are unsatisfiable.

An assignment α is *complete* if it contains exactly one literal over each variable in Σ. A formula ψ (resp. clause c) is a *logical consequence* of a formula φ, denoted $\varphi \models \psi$ (resp. $\varphi \models c$), if for all complete assignments α satisfying φ, α satisfies ψ (resp. α satisfies c). Two formulas are *logically equivalent* if each is a logical consequence of the other.

A clause c is a *unit clause* under assignment α if all but one literals in c are false. A clause c is a *conflict clause* (or *conflicting*) under α if all literals in c are false.

We often consider a formula φ in the context of some assignment α. For this, we introduce the notion of φ *under* α, denoted as $\varphi \downarrow \alpha$, which is the formula obtained by conjoining φ with a unit clause (l) for each literal $l \in \alpha$. Formally, $\varphi \downarrow \alpha$ is the formula

$$\varphi \wedge \bigwedge_{l \in \alpha} l$$

Symmetry in SAT. Let π be a permutation of a set of literals $\overline{\Sigma}$. We extend π to clauses: $\pi(l_1 \vee \ldots \vee l_n) = \pi(l_1) \vee \ldots \vee \pi(l_n)$, to formulas: $\pi(c_1 \wedge \ldots \wedge c_n) = \pi(c_1) \wedge \ldots \wedge \pi(c_n)$, and to assignments: $\pi(\alpha) = \{\pi(l) \mid l \in \alpha\}$. We write permutations in *cycle notation*. For example, $(a\ b\ c)(\neg a\ \neg b\ \neg c)(\neg d\ d)$ is the permutation that maps a to b, b to c, c to a, $\neg a$ to $\neg b$, $\neg b$ to $\neg c$, $\neg c$ to $\neg a$, swaps d with $\neg d$, and maps any other literals to themselves.

Permutations form algebraic groups under the composition relation ($\circ$). A set of permutations $\mathcal{P}$ is a set of *generators* for a permutation group $\mathbb{G}$ if each permutation in $\mathbb{G}$ is a composition of permutations from $\mathcal{P}$. The group $Grp(\mathcal{P})$ is the permutation group *generated* by all compositions of permutations in $\mathcal{P}$. The *orbit* $Orb_{\mathbb{G}}(x)$ of a literal or clause x under a permutation group $\mathbb{G}$ is the set $\{\pi(x) \mid \pi \in \mathbb{G}\}$.

A *symmetry* π of a propositional formula φ over Σ is a permutation over $\overline{\Sigma}$ that *preserves satisfaction to* φ; i.e., $\alpha \models \varphi$ iff $\pi(\alpha) \models \varphi$.

A permutation π of $\overline{\Sigma}$ is a symmetry of a propositional formula φ over Σ if the following sufficient syntactic condition is met:

– π commutes with negation: $\pi(\neg l) = \neg \pi(l)$ for all $l \in \overline{\Sigma}$, and
– π fixes the formula: $\pi(\varphi) = \varphi$.

It is easy to see that these two conditions guarantee that π maps assignments to assignments, preserving satisfaction to φ.

Typically, only this syntactical type of symmetry is exploited, since it can be detected with relative ease. One first converts a formula φ over variables Σ to a colored graph such that any automorphism – a permutation of a graph's nodes that maps the graph onto itself – of the graph corresponds to a permutation of $\overline{\Sigma}$ that commutes with negation and that fixes the formula. Next, the graph's automorphism group is detected by tools such as NAUTY [25], SAUCY [22] or BLISS [21], and is translated back to a symmetry group of φ.

The technique we present in this paper works for all kinds of symmetries, syntactical and others. However, our implementations use BREAKID for symmetry detection, which only detects a syntactical symmetry group by the method described above.

2.1   Conflict Driven Clause Learning SAT Solvers

We briefly recall some of the characteristics of modern conflict driven clause learning SAT (CDCL) solvers [24].

A CDCL solver takes as input a formula φ over a set of Boolean variables Σ. As output, it returns an (often complete) assignment satisfying φ, or reports that none exists.

Internally, a CDCL solver keeps track of a partial assignment α – called the *current assignment* – which initially is empty. At each search step, the solver chooses a variable x for which the current assignment α does not yet contain a literal, and adds either the positive literal x or the negative literal $\neg x$ to α. The added literal is now a *choice literal*, and may result in some clauses becoming unit clauses under the *refined* current assignment. This prompts a *unit propagation* phase, where for all unknown literals l occurring in a unit clause, the current assignment is extended with l. Such literals are *propagated literals*; we refer to the unit clause that initiated l's unit propagation as l's *explanation clause*. If no more unit clauses remain under the resulting assignment, the unit propagation phase ends, and a new search step starts by deciding on a next choice literal.

During unit propagation, a clause c can become conflicting when another clause propagates the last unknown literal l of c to false. At this moment, a CDCL solver will construct a *learned clause* by investigating the explanation clauses for the unit propagations leading to the conflict clause. This learned clause c is a logical consequence of the input formula, and using c in unit propagation prevents the conflict from occurring again after a *backjump*.[1] We refer to the set of learned clauses of a CDCL solver as the *learned clause store* Δ.

Formally, we characterize the *state* of a CDCL solver solving a formula φ by a quadruple $(\alpha, \gamma, \Delta, \mathcal{E})$, where

- α is the current assignment,
- $\gamma \subseteq \alpha$ is the set of choice literals – the set of literals $\alpha \setminus \gamma$ are known as *propagated* literals,
- Δ is the learned clause store,
- $\mathcal{E}$ is a function mapping the propagated literals $l \in \alpha \setminus \gamma$ to their explanation clause $\mathcal{E}(l)$, which can be either a clause from the input formula φ or from the learned clause store Δ.

During the search process, the invariant holds that the current assignment is a logical consequence of the decision literals, given the input formula. Formally:

$$\varphi \downarrow \gamma \models \varphi \downarrow \alpha.$$

Secondly, the learned clauses are logical consequences of the input formula:

$$\varphi \models c \text{ for each } c \in \Delta.$$

3   Symmetric Explanation Learning

From the definition of symmetry, the following proposition easily follows:

---

[1] *Backjumping* is a generalization of the more classical *backtracking* over choices in combinatorial solvers.

Proposition 1. *Let φ be a propositional formula, π a symmetry of φ, and c a clause. If $\varphi \models c$, then also $\varphi \models \pi(c)$.*

Proof. If $\varphi \models c$ then $\pi(\varphi) \models \pi(c)$, as π renames the literals in formulas and clauses. Symmetries preserve models, hence $\pi(\varphi)$ is logically equivalent to φ, hence $\varphi \models \pi(c)$.

Since learned clauses are always logical consequences of the input formula, every time a CDCL solver learns a clause c, one may apply Proposition 1 and add $\pi(c)$ as a learned clause for every symmetry π of some symmetry group of $\mathbb{G}$. This is called *symmetrical learning*, which extends the resolution proof system underpinning a SAT solver's learning technology with a symmetry rule.

Symmetrical learning can be used as a symmetry handling tool for SAT: because every learned clause prevents the solver from encountering a certain conflict, the orbit of this clause under the symmetry group will prevent the encounter of all symmetrical conflicts, resulting in a solver never visiting two symmetrical parts of the search space.

However, since the size of permutation groups can grow exponentially in the number of permuted elements, learning all possible symmetrical clauses will in most cases add too many symmetrical clauses to the formula to be of practical use. Symmetrical learning approaches need to limit the amount of symmetrical learned clauses [20].

Given a set of input symmetries $\mathcal{P}$, the idea behind symmetric explanation learning (SEL) is to aim at learning symmetrical variants of learned clauses on the moment these variants propagate. A naive way to obtain this behaviour would be to check at each propagation phase for each clause $c \in \Delta$ and each symmetry $\pi \in \mathcal{P}$ whether $\pi(c)$ is a unit clause. Such an approach would have an unsurmountable overhead. Therefore, we implemented SEL using two optimizations.

The first is that we make a selection of "interesting" clauses: the symmetrical variants of clauses in Δ that are explanation clauses of some propagation in the current search state. The intuition is that an explanation clause c contains mostly false literals, so, assuming that the number of literals permuted by some symmetry π is much smaller than the total number of literals in the formula, $\pi(c)$ has a good chance of containing the same mostly false literals.

Secondly, we store those promising symmetrical variants in a separate *symmetrical learned clause store* Θ. Clauses in this store are handled similar to clauses in Δ, with the following differences:

1. propagation with Δ is always prioritized over propagation with Θ,
2. whenever a clause in Θ propagates, it is added to Δ,
3. whenever the solver backjumps over a propagation of a literal l, all symmetrical clauses $\pi(\mathcal{E}(l))$ are removed from Θ.

The first two points ensure that no duplicate clauses will ever be added to Δ without the need for checking for duplicates. Indeed, by prioritizing propagation with Δ, a clause in Θ can only propagate if it is not a part of Δ yet. The third point guarantees that Θ contains only symmetrical variants of clauses that have shown to be relevant in the current branch of the search tree.

On a technical note, Θ contains clauses $\pi(\mathcal{E}(l))$ from the moment l is propagated until a backjump unassigns l. As a result, it is useless to add $\pi(\mathcal{E}(l))$ to Θ if it is satisfied at the moment l is propagated, as it will never become an unsatisfied unit clause before backjumping over l. Similarly, it is not necessary to store any literals of $\pi(\mathcal{E}(l))$ that are false at the moment l is propagated, as these will not change status before backjumping over l. To combat this, Θ contains an approximation $\pi(\mathcal{E}(l))^*$ of $\pi(\mathcal{E}(l))$, which excludes any literals that are false at the moment l is propagated. If $\pi(\mathcal{E}(l))^*$ ever becomes unit, so does $\pi(\mathcal{E}(l))$. At this point, we recover the original clause $\pi(\mathcal{E}(l))$ from some stored reference to π and l, by simply applying π to $\mathcal{E}(l)$ again. Additionally, before adding a unit $\pi(\mathcal{E}(l))$ as a learned clause to Δ, our implementation performs a self-subsumption clause simplification step, as this is a simple optimization leading to stronger learned clauses [31].

Finally, keeping track of unit clauses in Θ during refinement of the current assignment is efficiently done by the well-known two-watched literal scheme [27].

We give pseudocode for SEL's behavior during a CDCL solver's propagation phase in Algorithm 1.

---

data: a formula φ, a set of symmetries $\mathcal{P}$ of φ, a partial assignment α, a set of learned clauses Δ, an explanation function $\mathcal{E}$, a set of symmetrical explanation clauses Θ

```
1  repeat
2      foreach unsatisfied unit clause c in φ or Δ do
3          let l be the unassigned literal in c;
4          add l to α;
5          set c as E(l);
6          foreach symmetry π in P do
7              if π(E(l)) is not yet satisfied by α then
8                  add the approximation π(E(l))* to Θ;
9              end
10         end
11     end
12     if an unsatisfied unit clause π(E(l))* in Θ exists then
13         add the self-subsumed simplification of π(E(l)) to Δ;
14     end
15 until no new literals have been propagated or a conflict has occurred;
```

Algorithm 1. propagation phase of a CDCL solver using SEL

Example 1 presents a unit propagation phase with the SEL technique.

Example 1. Let a CDCL solver have a state $(\alpha, \gamma, \Delta, \mathcal{E})$ with current assignment $\alpha = \emptyset$, choice $\gamma = \emptyset$, learned clause store $\Delta = \{(a \lor b), (\neg c \lor d \lor e)\}$ and explanation function $\mathcal{E}$ the empty function. Let $\pi = (a\ c)(\neg a\ \neg c)(b\ d)(\neg b\ \neg d)$ be a syntactical symmetry of the input formula φ, and assume for the following exposition that no propagation happens from clauses in φ. As the current assignment is currently empty, the symmetrical learned clause store Θ is empty as well.

Suppose the CDCL algorithm chooses $\neg a$, so $\alpha = \gamma = \{\neg a\}$. During unit propagation, the CDCL algorithm propagates b, so $\alpha = \{\neg a, b\}$, $\gamma = \{\neg a\}$ and $\mathcal{E}(b) = a \vee b$. By Algorithm 1, SEL adds $\pi(\mathcal{E}(b)) = c \vee d$ to Θ, so $\Theta = \{c \vee d\}$. No further unit propagation is possible, and $c \vee d$ is not unit or conflicting, so the solver enters a new decision phase.

We let the solver choose $\neg d$, so $\alpha = \{\neg a, b, \neg d\}$, $\gamma = \{\neg a, \neg d\}$. Still, no unit propagation on clauses from φ or from the learned clause store Δ is possible. However, $c \vee d$ in Θ is unit, so SEL adds $c \vee d$ to Δ.

Now unit propagation is reinitiated, leading to the propagation of c with reason $\mathcal{E}(c) = c \vee d$ and e with reason $\mathcal{E}(e) = \neg c \vee d \vee e$, so $\alpha = \{\neg a, b, \neg d, c, e\}$. As both $\pi(\mathcal{E}(c)) = a \vee b$ and $\pi(\mathcal{E}(e)) = \neg a \vee d \vee e$ are satisfied by α, they are not added to Θ. No further propagation is possible, ending the propagation loop. ▲

Note that if a symmetry π is a syntactic symmetry of a formula φ, SEL will never learn a symmetrical clause $\pi(c)$ from a clause $c \in \varphi$, as $\pi(c) \in \varphi$ already, and has propagation priority on any other $\pi(c)$ constructed by SEL. Moreover, due to technical optimizations, $\pi(c)$ will not even be constructed by SEL, as it is satisfied due to unit propagation from φ's clauses. From another perspective, any clause learned by SEL is the symmetrical image of some previously learned clause.

Also note that SEL is able to learn symmetrical clauses of symmetry compositions $\pi' \circ \pi$, with π and π' two symmetries of the input formula. This happens when at a certain point, c is an explanation clause and $\pi(c)$ an unsatisfied unit clause, and at some later moment during search, $\pi(c)$ is an explanation clause and $\pi'(\pi(c))$ an unsatisfied unit clause.

3.1   Complexity of SEL

Assuming a two-watched literal implementation for checking the symmetrical clause store Θ on conflict or unit clauses, the computationally most intensive step for SEL is filling Θ with symmetrical explanation clauses during unit propagation. Worst case, for each propagated literal l, SEL constructs $\pi(\mathcal{E}(l))$ for each π in the set of input symmetries $\mathcal{P}$. Assuming k to be the size of the largest clause in φ or Δ, this incurs a polynomial $O(|\mathcal{P}|k)$ time overhead at each propagation. As for memory overhead, SEL must maintain a symmetrical clause store containing $O(|\mathcal{P}||\alpha|)$ clauses, with α the solver's current assignment.

Of course, as with any symmetrical learning approach, SEL might flood the learned clause store with many symmetrical clauses. In effect, as only symmetrical explanation clauses are added to the learned clause store if they propagate or are conflicting, an upper bound on the number of symmetrical clauses added is the number of propagations performed by the solver, which can be huge. Aggressive learned clause store cleaning strategies might be required to maintain efficiency.

4 Related Work

In this section, we describe the relation of SEL and symmetric learning to other SAT solving techniques from literature.

4.1 SEL and SLS

One proposed way to restrict the number of clauses generated by symmetrical learning is the *symmetrical learning scheme* (SLS) [6]. Given an input set of symmetries $\mathcal{P}$, SLS only learns $\pi(c)$ for each $\pi \in \mathcal{P}$, and for each clause c learned by resolution after a conflict. If c contains only one literal, the set of symmetrical learned clauses from c is extended to the orbit of c under the group generated by $\mathcal{P}$.

A disadvantage of SLS is that not all symmetrical learned clauses are guaranteed to contribute to the search by propagating a literal at least once. This might result in lots of useless clauses being learned, which do not actively avoid a symmetrical part of the search space. It also is possible that some clauses learned by this scheme already belong to the set of learned clauses, since most SAT solvers do not perform an expensive check for duplicate learned clauses.

In Sect. 5, we give experimental results with an implementation of SLS.

4.2 SEL and SP

Another way to restrict the number of learned symmetrical clauses is given by *symmetry propagation* (SP) [15]. SP also learns symmetrical clauses only when they are unit or conflicting, but it uses the notion of *weak activity* to derive which symmetrical clauses it will learn.

Definition 1. Let φ be a formula and $(\alpha, \gamma, \Delta, \mathcal{E})$ the state of a CDCL solver. A symmetry π of φ is *weakly active* for assignment α and choice literals γ if $\pi(\gamma) \subseteq \alpha$.

Weak activity is a is a refinement of *activity*; the latter is a technique used in dynamic symmetry handling approaches for constraint programming [18, 26].

Now, if a symmetry π of a formula φ is weakly active in the current solver state $(\alpha, \gamma, \Delta, \mathcal{E})$, then SP's implementation guarantees that for propagated literals $l \in \alpha \setminus \gamma$, $\pi(\mathcal{E}(l))$ is unit [15]. Then, SP adds any unsatisfied unit clauses $\pi(\mathcal{E}(l))$ to the learned clause store, and uses these to propagate $\pi(l)$.[2]

As SEL checks whether $\pi(\mathcal{E}(l))$ is unit for any input symmetry π, regardless of whether π is weakly active or not, SEL detects at least as many symmetrical clauses that are unit and unsatisfied as SP. Note that in Example 1, after making the choice $\neg a$, π is not weakly active, as $\neg a \in \alpha$ but $\pi(\neg a) = \neg c \notin \alpha$.

---

[2] SP focuses its presentation on propagating symmetrical literals $\pi(l)$ for weakly active symmetries, hence the name symmetry *propagation*. We present SP from a symmetrical learning point of view, using the fact that SP employs $\pi(\mathcal{E}(l))$ as a valid explanation clause for $\pi(l)$'s propagation.

Furthermore, after propagation of b and making the choice $\neg d$, SEL does learn the symmetrical explanation clause $\pi(\mathcal{E}(b)) = (c \vee d)$, propagating $\pi(a) = c$ in the process. This shows that SEL learns *strictly* more symmetrical clauses, performs more propagation than SP, and closes an increasing number of symmetrical search branches over time.

4.3   Compatibility of Symmetrical Learning and Preprocessing Techniques

As modern SAT solvers employ several preprocessing techniques [8] to transform an input formula φ to a smaller, hopefully easier, equisatisfiable formula φ', we should argue the soundness of SEL combined with those techniques. We do this by giving a sufficient condition of the preprocessed formula for which symmetrical learning remains a sound extension of a SAT solver's proof system.

Theorem 1. *Let φ and φ' be two formulas over vocabulary Σ, and let π be a symmetry of φ. Also, let φ' be*

1. *a logical consequence of φ and*
2. *equisatisfiable to φ.*

If clause c is a logical consequence of φ then $\varphi' \wedge \pi(c)$ is

1. *a logical consequence of φ and*
2. *equisatisfiable to φ.*

Proof. As c is a logical consequence of φ, $\pi(c)$ is as well, by Proposition 1. Hence, $\varphi' \wedge \pi(c)$ remains a logical consequence of φ, since both φ' and $\pi(c)$ hold in all models of φ, proving 1.

This also means that any satisfying assignment to φ is a satisfying assignment to $\varphi' \wedge \pi(c)$, so if φ is satisfiable, $\varphi' \wedge \pi(c)$ is satisfiable too. As the addition of an extra clause to a formula only reduces the number of satisfying assignments, if φ' is unsatisfiable, $\varphi' \wedge \pi(c)$ is unsatisfiable too. Since φ' is equisatisfiable with φ, $\varphi' \wedge \pi(c)$ is unsatisfiable if φ is unsatisfiable. This proves 2.

Corollary 1. *Let φ be a formula and π be a symmetry of φ. Symmetrical learning with symmetry π is sound for CDCL SAT solvers over a preprocessed formula φ' if φ' is a logical consequence of φ and if φ' is equisatisfiable to φ.*

Proof. Any clause c learned by resolution or symmetry application on clauses from φ' or logical consequences of φ is a logical consequence of φ. Hence, by Theorem 1, it is sound to learn the symmetrical clause $\pi(c)$ when solving for φ'.

In other words, if a preprocessing technique satisfies the conditions from Theorem 1, it is sound to symmetrically learn clauses in a CDCL SAT solver, as is done by the SEL algorithm.

This is not a trivial, but also not a strict requirement. For instance, common variable and clause elimination techniques based pioneered by SATELITE [16] and still employed by i.a. GLUCOSE satisfy this requirement. One exception, ironically, is static symmetry breaking, as the added symmetry breaking clauses are not logical consequences of the original formula. Also, a preprocessing technique that introduces new variables does not satisfy the above requirements, and risks to combine unsoundly with symmetrical learning.

4.4 Symmetrical Learning Does Not Break Symmetry

The earliest techniques to handle symmetry constructed formulas that removed symmetrical solutions from a problem specification, a process that *breaks* the symmetry in the original problem specification. Ever since, *handling* symmetry seems to have become eponymous with *breaking* it, even though a symmetrical learning based technique such as SEL only infers logical consequences of a formula, and hence does not a priori remove any solutions. In this paper, we tried to consistently use the term *symmetry handling* where appropriate.

An advantage of non-breaking symmetry handling approaches is that it remains possible for a solver to obtain any solution to the original formula. For instance, non-breaking symmetry handling approaches such as SEL can be used to generate solutions to a symmetric formula, which are evaluated under an asymmetric objective function [2]. Similarly, approaches such as SEL can be used in a #SAT solver [9].

5   Experiments

In this section, we present experiments gauging SEL's performance. We implemented SEL in the state-of-the-art SAT solver GLUCOSE 4.0 [4] and made our implementation available online [12]. Symmetry was detected by running BREAKID, which internally uses SAUCY as graph automorphism detector.

All experiments have a 5000 s time limit and a 16 GB memory limit. The hardware was an Intel Core i5-3570 CPU with 32 GiB of RAM and Ubuntu 14.04 Linux as operating system. Detailed results and benchmark instances have been made available online[3].

We first present a preliminary experiment on *row interchangeability* – a particular form of symmetry detected by BREAKID – in Subsect. 5.1, and give our main result in Subsect. 5.2.

5.1   Row Interchangeability

Row interchangeability is a particular form of symmetry where a subset of the literals are arranged as a matrix with k rows and m columns, and any permutation of the rows induces a symmetry [14]. This type of symmetry is common, occurring often when objects in some problem domain are interchangeable. For example, the interchangeability of pigeons in a pigeonhole problem, or the interchangeability of colors in a graph coloring problem lead to row interchangeability

---

[3] bitbucket.org/krr/sat_symmetry_experiments.

at the level of a propositional specification. This type of symmetry can be broken completely by static symmetry breaking formulas of polynomial size, resulting in exponential search space reduction in e.g. pigeonhole problem specifications [14].

While the dynamic symmetry breaking solver SYMCHAFF specializes in row interchangeability, it is unclear how SEL (or symmetrical learning in general) can efficiently handle this type of symmetry. As SEL only supports an input set of simple symmetries as defined in Sect. 2, we currently use a set of generators as a representation for a row interchangeability symmetry group. We investigate two possible representations for a row interchangeability group with k rows:

- a *linear* representation, containing $k - 1$ symmetries that swap consecutive rows, as well as the one symmetry swapping the first and last row.
- a *quadratic* representation, containing $k(k-1)/2$ symmetries that swap any two rows, including non-consecutive ones.

To experimentally verify the effectiveness of both approaches, we generated two benchmark sets. The first consists of unsatisfiable **pigeonhole** formulas, which assign k pigeons to $k-1$ holes. We only provided the row interchangeability symmetry group stemming from interchangeable pigeons to the symmetry handling routines. We shuffled the variable order of the formulas, to minimize lucky guesses by GLUCOSE's heuristic. The number of pigeons in the instances ranges between 10 and 100.

The second benchmark set consists of 110 both satisfiable and unsatisfiable graph **coloring** problems, where we try to color graphs with k or $k - 1$ colors, where k is the input graph's chromatic number[4]. We only provided the row interchangeability symmetry group stemming from interchangeable colors to the symmetry handling routines, ignoring any potential symmetry in the input graph. Input graphs are taken from Michael Trick's web page [32].

As a baseline, we use GLUCOSE 4.0 coupled with the static symmetry breaking preprocessor BREAKID, whose symmetry detection routine is disabled and only gets to break the row interchangeability groups mentioned previously. The results for **pigeonhole** are given in Table 1, and for **coloring** are given in Table 2. Solving time needed by BREAKID's symmetry detection is ignored, as this was always less than 2.1 s, and the same for any approach.

On **pigeonhole**, the quadratic approach outperforms the linear approach, solving instances with up to 30 pigeons opposed to only 16 pigeons. However, even the quadratic approach does not reach the speedup exhibited by BREAKID, easily handling instances with 100 pigeons. On **coloring**, the quadratic approach does manage to outperform both the linear approach and BREAKID.

Based on this experiment, we default SEL to use a quadratic amount of row-swapping symmetry generators to represent a row interchangeability symmetry group.

---

[4] For the few instances where the chromatic number was not known, we made an educated guess based on the graph's name.

Table 1. Solving time in seconds of BREAKID and SEL with different generator sets for **pigeonhole** interchangeability. A "-" means the time limit of 5000 s was reached.

# pigeons	BREAKID	SEL–quadratic	SEL–linear
10	0	0	0.04
11	0	0.06	0.3
12	0	0.15	0.08
13	0	0.93	0.81
14	0	0.01	4.97
15	0	0.03	791.26
16	0	0.04	4766.31
17	0	0.53	-
18	0	0.25	-
19	0	3.73	-
20	0	42.85	-
25	0.01	0.9	-
30	0.02	277.57	-
40	0.05	-	-
50	0.13	-	-
70	0.41	-	-
100	1.47	-	-

Table 2. Total number of graph coloring instances solved within 5000 s for BREAKID and SEL with different generator sets for graph **coloring** interchangeability.

# instances	BREAKID	SEL–quadratic	SEL–linear
110	70	**87**	74

5.2 Evaluation of SEL

We evaluate SEL by comparing the following five solver configurations:

- GLUCOSE: pure GLUCOSE 4.0, without any symmetry detection or handling routines.
- BREAKID: GLUCOSE 4.0 coupled with the BREAKID symmetry breaking preprocessor in its default settings.
- SEL: our implementation of SEL in GLUCOSE 4.0, taking as input the symmetries detected by BREAKID in its default settings. This includes any row interchangeability symmetry detected by BREAKID, which is interpreted as a quadratic set of row swapping symmetries.
- SP: the existing implementation of Symmetry propagation [13] in the classic SAT solver MINISAT [17], using its optimal configuration [15]. We slightly extended this implementation to take BREAKID's symmetry output as input,

and interpreted any row interchangeability symmetry as a quadratic set of symmetries.

– SLS: our implementation of SLS in GLUCOSE 4.0, taking as input the symmetries detected by BREAKID in its default settings. This includes any row interchangeability symmetry detected by BREAKID. Contrary to the other solvers, we interpret interchangeability here as the *linear* amount of generators. The reason for this is that we noticed in preliminary testing that SLS simply couldn't handle a quadratic number of generators: with this it almost always ran out of memory.

Our benchmark instances are partitioned in five benchmark sets:

– **app14**: the application benchmarks of the 2014 SAT competition. 300 instances; BREAKID detected some symmetry for 160 of these.
– **hard14**: the hard-combinatorial benchmarks of the 2014 SAT competition. 300 instances; BREAKID detected some symmetry for 107 of these.
– **app16**: the application benchmarks of the 2016 SAT competition. 299 instances; BREAKID detected some symmetry for 131 of these.
– **hard16**: the hard-combinatorial benchmarks of the 2016 SAT competition. 200 instances; BREAKID detected some symmetry for 131 of these.
– **highly**: an eclectic set of highly symmetric instances collected over the years. 204 instances; BREAKID detected some symmetry for 202 of these. Instance families include graph coloring, pigeonhole, Ramsey theorem, channel routing, planning, counting, logistics and Urquhart's problems.

We reiterate that detailed experimental results, benchmark instances and source code of the employed systems have been made available online [11–13].

Table 3 lists the number of successfully solved instances for each of the five solver configurations and each of the five benchmark sets. Except for GLUCOSE, BREAKID's symmetry detection and breaking time are accounted in the total solving time.

Table 3. Total number of successfully solved instances for each of the five solver configurations and each of the five benchmark sets.

Benchmark set	GLUCOSE	BREAKID	SEL	SP	SLS
app14 (300/160)	**222**	220	215	179	187
hard14 (300/107)	174	**189**	188	172	175
app16 (299/131)	150	**151**	**151**	127	141
hard16 (200/131)	52	**93**	80	48	70
highly (204/202)	106	151	**152**	151	113

First and foremost, BREAKID jumps out of the pack as the best all-around configuration: performing well on the strongly symmetric **highly** instances, as well as on the more challenging SAT competition instances, of which especially **app14** and **app16** feature very large instances.

Second, SEL seems quite competitive. In **hard14**, **app16** and **highly**, SEL and BREAKID trade blows. Only on **hard16**, SEL's performance seems significantly inferior to BREAKID.

Third, GLUCOSE performs badly on **highly**, which can be expected given the strong symmetry properties of those instances. **hard16**'s low success rate is due to 35 pigeonhole instances and 38 highly symmetric *tseitingrid* instances.

Fourth, SP performs very well on **highly**, but is dead last in all other benchmark sets. This might be due to its embedding in the older MINISAT, or due to the overhead of keeping track of weakly inactive symmetries.

Finally, SLS is not able to clinch the lead in any of the benchmark sets, and especially the bad results on **highly** are surprising for a symmetry exploiting technique. We conjecture that highly symmetric instances lead to an uncontrolled symmetrical clause generation, choking SLS's learned clause store.

For this, it is worth looking at the number of instances where a solver exceeded the 16 GB memory limit (a *memout*). These results are given in Table 4.

Table 4. Total number of instances that exceeded the 16 GB memory limit.

Benchmark set	GLUCOSE	BREAKID	SEL	SP	SLS
app14 (300/160)	0	0	18	23	41
hard14 (300/107)	0	0	1	1	13
app16 (299/131)	0	0	2	21	47
hard16 (200/131)	0	0	14	0	49
highly (204/202)	0	0	5	3	50

From Table 4, it is clear that all symmetrical learning approaches (SEL, SP, SLS) struggle with heavy memory consumption. SLS in particular is unable to solve many symmetrical instances due to memory constraints. SEL on the other hand, has relatively few memouts, concentrated mainly in the benchmark sets **app14** and **hard16** – the same as those were it had to give BREAKID the lead.

We conclude that SEL is a viable symmetrical learning, and by extension, dynamic symmetry handling approach. However, care must be taken that not too many symmetrical clauses are learned, filling up all available memory.

6 Conclusion

In this paper, we presented symmetric explanation learning (SEL), a form of symmetrical learning based on learning symmetric images of explanation clauses for unit propagations performed during search. A key idea is that these symmetric clauses are only learned when they would restrict the current search state, i.e., when they are unit or conflicting.

We related SEL to symmetry propagation (SP) and the symmetrical learning scheme (SLS), and gave a sufficient condition on when symmetrical learning can be combined with common SAT preprocessing techniques.

We further provided a working implementation of SEL and SLS embedded in GLUCOSE, and experimentally evaluated SEL, SLS, SP, GLUCOSE and the symmetry breaking preprocessor BREAKID on more than 1300 benchmark instances. Our conclusion is that SEL outperforms other symmetrical learning approaches, and functions as an effective general purpose dynamic symmetry handling technique, almost closing the gap with static symmetry breaking.

For future work, we expect that the efficiency of our implementation can still be improved. Specifically, investigating how to reduce SEL's memory overhead, perhaps by aggressive learned clause deletion techniques, has definite potential.

On the theoretical front of symmetrical learning, much work remains to be done on effectiveness guarantees similar to those provided by *complete* static symmetry breaking. Informally, a symmetry breaking formula ψ is complete for a given symmetry group if no two symmetric solutions satisfy ψ [33]. For instance, BREAKID guarantees that its symmetry breaking formulas are complete for row interchangeability symmetry, resulting in very fast pigeonhole solving times. Maybe a similar guarantee can be given for some form of symmetrical learning?

References

1. Aloul, F., Ramani, A., Markov, I., Sakallah, K.: Solving difficult SAT instances in the presence of symmetry. In: 39th Design Automation Conference, 2002 Proceedings, pp. 731–736 (2002)
2. Aloul, F.A., Ramani, A., Markov, I.L., Sakallah, K.A.: Dynamic symmetry-breaking for boolean satisfiability. Ann. Math. Artif. Intell. **57**(1), 59–73 (2009)
3. Aloul, F.A., Sakallah, K.A., Markov, I.L.: Efficient symmetry breaking for boolean satisfiability. IEEE Trans. Comput. **55**(5), 549–558 (2006)
4. Audemard, G., Simon, L.: Predicting learnt clauses quality in modern SAT solvers. In: Boutilier, C. (ed.) IJCAI, pp. 399–404 (2009)
5. Benhamou, B., Nabhani, T., Ostrowski, R., Saïdi, M.R.: Dynamic symmetry breaking in the satisfiability problem. In: Proceedings of the 16th International Conference on Logic for Programming, Artificial Intelligence, and Reasoning, LPAR-16, Dakar, Senegal, 25 April –1 May 2010
6. Benhamou, B., Nabhani, T., Ostrowski, R., Saïdi, M.R.: Enhancing clause learning by symmetry in SAT solvers. In: Proceedings of the 2010 22nd IEEE International Conference on Tools with Artificial Intelligence, ICTAI 2010, vol. 01, pp. 329–335 (2010). http://dx.doi.org/10.1109/ICTAI.2010.55
7. Benhamou, B., Saïs, L.: Tractability through symmetries in propositional calculus. J. Autom. Reason. **12**(1), 89–102 (1994). http://dx.doi.org/10.1007/BF00881844
8. Biere, A.: Preprocessing and inprocessing techniques in SAT. In: Eder, K., Lourenço, J., Shehory, O. (eds.) HVC 2011. LNCS, vol. 7261, p. 1. Springer, Heidelberg (2012). doi:10.1007/978-3-642-34188-5_1
9. Birnbaum, E., Lozinskii, E.L.: The good old Davis-Putnam procedure helps counting models. J. Artif. Intell. Res. **10**, 457–477 (1999)

10. Crawford, J.M., Ginsberg, M.L., Luks, E.M., Roy, A.: Symmetry-breaking predicates for search problems. In: Principles of Knowledge Representation and Reasoning, pp. 148–159. Morgan Kaufmann (1996)
11. Devriendt, J.: Binaries, experimental results and benchmark instances for "symmetric explanation learning: effective dynamic symmetry handling for SAT". bitbucket.org/krr/sat_symmetry_experiments
12. Devriendt, J.: An implementation of symmetric explanation learning in Glucose on Bitbucket. bitbucket.org/krr/glucose-sel
13. Devriendt, J.: An implementation of symmetry propagation in MiniSat on Github. github.com/JoD/minisat-SPFS
14. Devriendt, J., Bogaerts, B., Bruynooghe, M., Denecker, M.: Improved static symmetry breaking for SAT. In: Creignou, N., Le Berre, D. (eds.) SAT 2016. LNCS, vol. 9710, pp. 104–122. Springer, Cham (2016). doi:10.1007/978-3-319-40970-2_8
15. Devriendt, J., Bogaerts, B., De Cat, B., Denecker, M., Mears, C.: Symmetry propagation: improved dynamic symmetry breaking in SAT. In: IEEE 24th International Conference on Tools with Artificial Intelligence, ICTAI 2012, Athens, Greece, pp. 49–56. IEEE Computer Society, 7–9 November 2012. http://dx.doi.org/10.1109/ICTAI.2012.16
16. Eén, N., Biere, A.: Effective preprocessing in SAT through variable and clause elimination. In: Bacchus, F., Walsh, T. (eds.) SAT 2005. LNCS, vol. 3569, pp. 61–75. Springer, Heidelberg (2005). doi:10.1007/11499107_5
17. Eén, N., Sörensson, N.: An extensible SAT-solver. In: Giunchiglia, E., Tacchella, A. (eds.) SAT 2003. LNCS, vol. 2919, pp. 502–518. Springer, Heidelberg (2004). doi:10.1007/978-3-540-24605-3_37
18. Gent, I.P., Smith, B.M.: Symmetry breaking in constraint programming. In: Proceedings of ECAI-2000, pp. 599–603. IOS Press (2000)
19. Haken, A.: The intractability of resolution. Theor. Comput. Sci. **39**, 297–308 (1985). Third Conference on Foundations of Software Technology and Theoretical Computer Science. http://www.sciencedirect.com/science/article/pii/0304397585901446
20. Heule, M., Keur, A., Maaren, H.V., Stevens, C., Voortman, M.: CNF symmetry breaking options in conflict driven SAT solving (2005)
21. Junttila, T., Kaski, P.: Engineering an efficient canonical labeling tool for large and sparse graphs. In: Applegate, D., Brodal, G.S., Panario, D., Sedgewick, R. (eds.) Proceedings of the Ninth Workshop on Algorithm Engineering and Experiments and the Fourth Workshop on Analytic Algorithms and Combinatorics, pp. 135–149. SIAM (2007)
22. Katebi, H., Sakallah, K.A., Markov, I.L.: Symmetry and satisfiability: an update. In: Strichman, O., Szeider, S. (eds.) SAT 2010. LNCS, vol. 6175, pp. 113–127. Springer, Heidelberg (2010). doi:10.1007/978-3-642-14186-7_11
23. Krishnamurthy, B.: Short proofs for tricky formulas. Acta Inf. **22**(3), 253–275 (1985). http://dl.acm.org/citation.cfm?id=4336.4338
24. Marques-Silva, J.P., Sakallah, K.A.: GRASP: a search algorithm for propositional satisfiability. IEEE Trans. Comput. **48**(5), 506–521 (1999)
25. McKay, B.D., Piperno, A.: Practical graph isomorphism, II. J. Symbolic Comput. **60**, 94–112 (2014). http://www.sciencedirect.com/science/article/pii/S0747717113001193
26. Mears, C., García de la Banda, M., Demoen, B., Wallace, M.: Lightweight dynamic symmetry breaking. Constraints **19**(3), 195–242 (2014). http://dx.doi.org/10.1007/s10601-013-9154-2

27. Moskewicz, M., Madigan, C., Zhao, Y., Zhang, L., Malik, S.: Chaff: engineering an efficient SAT solver. In: DAC 2001, pp. 530–535. ACM (2001)
28. Pipatsrisawat, K., Darwiche, A.: On the power of clause-learning SAT solvers as resolution engines. Artif. Intell. **175**(2), 512–525 (2011). http://dx.doi.org/10.1016/j.artint.2010.10.002
29. Sabharwal, A.: SymChaff: exploiting symmetry in a structure-aware satisfiability solver. Constraints **14**(4), 478–505 (2009). http://dx.doi.org/10.1007/s10601-008-9060-1
30. Schaafsma, B., van Heule, M.J.H., Maaren, H.: Dynamic symmetry breaking by simulating zykov contraction. In: Kullmann, O. (ed.) SAT 2009. LNCS, vol. 5584, pp. 223–236. Springer, Heidelberg (2009). doi:10.1007/978-3-642-02777-2_22
31. Sörensson, N., Eén, N.: MiniSat v1.13 - a Sat solver with conflict-clause minimization. 2005. sat-2005 poster. Technical report (2005)
32. Trick, M.: Network resources for coloring a graph (1994). mat.gsia.cmu.edu/COLOR/color.html
33. Walsh, T.: Symmetry breaking constraints: Recent results. CoRR abs/1204.3348 (2012)

An Adaptive Prefix-Assignment Technique for Symmetry Reduction

Tommi Junttila, Matti Karppa$^{(\boxtimes)}$, Petteri Kaski, and Jukka Kohonen

Department of Computer Science, Aalto University, Espoo, Finland
{tommi.junttila,matti.karppa,petteri.kaski,jukka.kohonen}@aalto.fi

Abstract. This paper presents a technique for symmetry reduction that adaptively assigns a prefix of variables in a system of constraints so that the generated prefix-assignments are pairwise nonisomorphic under the action of the symmetry group of the system. The technique is based on McKay's canonical extension framework [J. Algorithms 26 (1998), no. 2, 306–324]. Among key features of the technique are (i) adaptability—the prefix sequence can be user-prescribed and truncated for compatibility with the group of symmetries; (ii) parallelisability—prefix-assignments can be processed in parallel independently of each other; (iii) versatility—the method is applicable whenever the group of symmetries can be concisely represented as the automorphism group of a vertex-colored graph; and (iv) implementability—the method can be implemented relying on a canonical labeling map for vertex-colored graphs as the only nontrivial subroutine. To demonstrate the tentative practical applicability of our technique we have prepared a preliminary implementation and report on a limited set of experiments that demonstrate ability to reduce symmetry on hard instances.

1 Introduction

Symmetry Reduction. Systems of constraints often have substantial symmetry. For example, consider the following system of Boolean clauses:

$$(x_1 \vee x_2) \wedge (x_1 \vee \bar{x}_3 \vee \bar{x}_5) \wedge (x_2 \vee \bar{x}_4 \vee \bar{x}_6). \tag{1}$$

The associative and commutative symmetries of disjunction and conjunction induce symmetries between the variables of (1), a fact that can be captured by stating that the group Γ generated by the two permutations $(x_1\ x_2)(x_3\ x_4)(x_5\ x_6)$ and $(x_4\ x_6)$ consists of all permutations of the variables that map (1) to itself. That is, Γ is the *automorphism group* of the system (1), cf. Sect. 2.

Known symmetry in a constraint system is a great asset from the perspective of solving the system, in particular since symmetry enables one to disregard partial solutions that are *isomorphic* to each other under the action of Γ on the space of partial solutions. Techniques for such *isomorph rejection*[1] [42] (alternatively, *symmetry reduction* or *symmetry breaking*) are essentially mandatory if

---

[1] A term introduced by J.D. Swift [42]; cf. Hall and Knuth [20] for a survey on early work on exhaustive computer search and combinatorial analysis.

© Springer International Publishing AG 2017
S. Gaspers and T. Walsh (Eds.): SAT 2017, LNCS 10491, pp. 101–118, 2017.
DOI: 10.1007/978-3-319-66263-3_7

one desires an exhaustive traversal of the (pairwise nonisomorphic) solutions of a highly symmetric system of constraints, or if the system is otherwise difficult to solve, for example, with many "dead-end" partial solutions compared with the actual number of solutions.

A prerequisite to symmetry reduction is that the symmetries are known. In many cases it is possible to automatically discover and compute these symmetries to enable practical and automatic symmetry reduction. In this context the dominant computational approach for combinatorial systems of constraints is to represent Γ via the automorphism group of a vertex-colored graph that captures the symmetries in the system. Carefully engineered tools for working with symmetries of vertex-colored graphs [15,27,35,37] and permutation group algorithms [10,40] then enable one to perform symmetry reduction. For example, for purposes of symmetry computations we may represent (1) as the following vertex-colored graph:

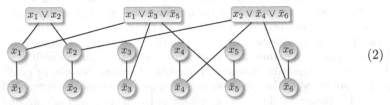

$$(2)$$

In particular, the graph representation (2) enables us to discover and reduce symmetry to avoid redundant work when solving the underlying system (1).

Our Contribution. The objective of this paper is to document a novel technique for symmetry reduction on systems of constraints. The technique is based on adaptively assigning values to a prefix of the variables so that the obtained prefix-assignments are pairwise nonisomorphic under the action of Γ. The technique can be seen as an instantiation of McKay's [36] influential canonical extension framework for isomorph-free exhaustive generation.

To give a brief outline of the technique, suppose we are working with a system of constraints over a finite set U of variables that take values in a finite set R. Suppose furthermore that $\Gamma \leq \mathrm{Sym}(U)$ is the automorphism group of the system. Select k distinct variables $u_1, u_2, \ldots, u_k$ in U. These k variables form the *prefix sequence* considered by the method. The technique works by assigning values in R to the variables of the prefix, in prefix-sequence order, with u_1 assigned first, then u_2, then u_3, and so forth, so that at each step the partial assignments so obtained are pairwise nonisomorphic under the action of Γ. For example, in (1) the partial assignments $x_1 \mapsto 0$, $x_2 \mapsto 1$ and $x_1 \mapsto 1$, $x_2 \mapsto 0$ are isomorphic since $(x_1\ x_2)(x_3\ x_4)(x_5\ x_6) \in \Gamma$ maps one assignment onto the other; in total there are three nonisomorphic assignments to the prefix x_1, x_2 in (1), namely (i) $x_1 \mapsto 0$, $x_2 \mapsto 0$, (ii) $x_1 \mapsto 0$, $x_2 \mapsto 1$, and (iii) $x_1 \mapsto 1$, $x_2 \mapsto 1$. Each partial assignment that represents an isomorphism class can then be used to reduce redundant work when solving the underlying system by standard techniques—in the nonincremental case, the system is augmented with a symmetry-breaking predicate requiring that one of the nonisomorphic partial assignments holds, while in the incremental setting [22, 43] the partial assignments can be solved independently or even in parallel.

Our contribution in this paper lies in how the isomorph rejection is implemented at the level of isomorphism classes of partial assignments by careful reduction to McKay's [36] isomorph-free exhaustive generation framework. The key technical contribution is that we observe how to generate the partial assignments in a normalized form that enables both *adaptability* (that is, the prefix $u_1, u_2, \ldots, u_k$ can be arbitrarily selected to match the structure of Γ) and precomputation of the extending variable-value orbits along a prefix.

Among further key features of the technique are:

1. *Implementability.* The technique can be implemented by relying on a canonical labeling map for vertex-colored graphs (cf. [27, 35, 37]) as the only nontrivial subroutine that is invoked once for each partial assignment considered.
2. *Versatility.* The method is applicable whenever the group of symmetries can be concisely represented as a vertex-colored graph; cf. (1) and (2). This is useful in particular when the underlying system has symmetries that are not easily discoverable from the final constraint encoding, for example, due to the fact that the constraints have been compiled or optimized[2] from a higher-level representation in a symmetry-obfuscating manner. A graphical representation can represent such symmetry directly and independently of the compiled/optimized form of the system.
3. *Parallelisability.* As a corollary of implementing McKay's [36] framework, the technique does not need to store representatives of isomorphism classes in memory to perform isomorph rejection, which enables easy parallelisation since the partial assignments can be processed independently of each other.

The main technical contribution of this paper is developed in Sect. 4 where we present the prefix-assignment technique. The required mathematical preliminaries on symmetry and McKay's framework are reviewed in Sects. 2 and 3, respectively. Our development in Sect. 4 relies on an abstract group Γ, with the understanding that a concrete implementation can be designed e.g. in terms of a vertex-colored graph representation, as will be explored in Sect. 5.

Preliminary Implementation and Experiments. To demonstrate the tentative practical applicability of our technique we have prepared a preliminary experimental implementation.[3] The implementation is structured as a preprocessor that works with an explicitly given graph representation and utilizes the *nauty* [35, 37] canonical labeling software for vertex-colored graphs as a subroutine to prepare an exhaustive collection of nonisomorphic prefix assignments relative to a user-supplied or heuristically selected prefix of variables. In Sect. 6 we

---

[2] For a beautiful illustration, we refer to Knuth's [31, Sect. 7.1.2, Fig. 10] example of optimum Boolean chains for 5-variable symmetric Boolean functions—from each optimum chain it is far from obvious that the chain represents a symmetric Boolean function. (See also Example 2.)

[3] This implementation can be found at https://github.com/pkaski/reduce/.

report on a limited set of experiments with solving systems of Boolean polynomial equations that demonstrate the ability to reduce symmetry on hard instances and give an initial favorable comparison with earlier techniques.

Earlier Work. A classical way to exploit symmetry in a system of constraints is to augment the system with so-called symmetry-breaking predicates (SBP) that eliminate either some or all symmetric solutions [5,14,19,39]. Such constraints are typically lexicographic leader (lex-leader) constraints that are derived from a generating set for the group of symmetries Γ. Among recent work in this area, Devriendt et al. [16] extend the approach by presenting a more compact way for expressing SBPs and a method for detecting "row interchangeabilities". Itzhakov and Codish [26] present a method for finding a set of symmetries whose corresponding lex-leader constraints are enough to completely break symmetries in search problems on small (10-vertex) graphs; this approach is extended by Codish et al. [12] by adding pruning predicates that simulate the first iterations of the equitable partition refinement algorithm of *nauty* [35,37]. Heule [23] shows that small complete symmetry-breaking predicates can be computed by considering arbitrary Boolean formulas instead of lex-leader formulas.

Our present technique can be seen as a method for producing symmetry-breaking predicates by augmenting the system of constraints with the disjunction of the nonisomorphic partial assignments. The main difference to the related work above is that our technique does not produce the symmetry-breaking predicate from a set of generators for Γ but rather the predicate is produced recursively, and with the possibility for parallelization, by classifying orbit representatives up to isomorphism using McKay's [36] framework. As such our technique breaks all symmetry with respect to the prescribed prefix, but comes at the cost of additional invocations of graph-automorphism and canonical-labeling tools. This overhead and increased symmetry reduction in particular means that our technique is best suited for constraint systems with hard combinatorial symmetry that is not easily capturable from a set of generators, such as symmetry in combinatorial classification problems [28]. In addition to McKay's [36] canonical extension framework, other standard frameworks for isomorph-free exhaustive generation in this context include *orderly algorithms* due to Faradžev [18] and Read [38], as well as the homomorphism principle for group actions due to Kerber and Laue [30].

It is also possible to break symmetry within a constraint solver during the search by dynamically adding constraints that rule out symmetric parts of the search space (cf. [11,19] and the references therein). If we use the nonisomorphic partial assignments produced by our technique as assumption sequences (cubes) in the incremental cube-and-conquer approach [22,43], our technique can be seen as a restricted way of breaking the symmetries in the beginning of the search, with the benefit—as with cube-and-conquer—that the portions of the search space induced by the partial assignments can be solved in parallel, either with complete independence or with appropriate sharing of information (such as conflict clauses) between the parallel nodes executing the search.

2 Preliminaries on Group Actions and Symmetry

This section reviews relevant mathematical preliminaries and notational conventions for groups, group actions, symmetry, and isomorphism for our subsequent development. (Cf. [10,17,25,28,29,40] for further reference.)

Groups and Group Actions. Let Γ be a finite group and let Ω be a finite set (the domain) on which Γ acts. For two groups Λ and Γ, let us write $\Lambda \leq \Gamma$ to indicate that Λ is a subgroup of Γ. We use exponential notation for group actions, and accordingly our groups act from the right. That is, for an object $X \in \Omega$ and $\gamma \in \Gamma$, let us write X^γ for the object in Ω obtained by acting on X with γ. Accordingly, we have $X^{(\beta\gamma)} = (X^\beta)^\gamma$ for all $\beta, \gamma \in \Gamma$ and $X \in \Omega$. For a finite set V, let us write $\mathrm{Sym}(V)$ for the group of all permutations of V with composition of mappings as the group operation.

Suppose that Γ acts on two sets, Ω and Σ. We extend the action to the Cartesian product $\Omega \times \Sigma$ elementwise by defining $(X, S)^\gamma = (X^\gamma, S^\gamma)$ for all $(X, S) \in \Omega \times \Sigma$ and $\gamma \in \Gamma$. Isomorphism extends accordingly; for example, we say that (X, S) and (Y, T) are isomorphic and write $(X, S) \cong (Y, T)$ if there exists a $\gamma \in \Gamma$ with $Y = X^\gamma$ and $T = S^\gamma$. Suppose that Γ acts on a set U. We extend the action of Γ on U to an elementwise action of Γ on subsets $W \subseteq U$ by setting $W^\gamma = \{w^\gamma : w \in W\}$ for all $\gamma \in \Gamma$ and $W \subseteq U$.

Orbit and Stabilizer, Automorphisms, Isomorphism. For an object $X \in \Omega$ let us write $X^\Gamma = \{X^\gamma : \gamma \in \Gamma\}$ for the *orbit* of X under the action of Γ and $\Gamma_X = \{\gamma \in \Gamma : X^\gamma = X\} \leq \Gamma$ for the *stabilizer* subgroup of X in Γ. Equivalently we say that Γ_X is the *automorphism group* of X and write $\mathrm{Aut}(X) = \Gamma_X$ whenever Γ is clear from the context; if we want to stress the acting group we write $\mathrm{Aut}_\Gamma(X)$.

We write $\Omega/\Gamma = \{X^\Gamma : X \in \Omega\}$ for the set of all orbits of Γ on Ω. For $\Lambda \leq \Gamma$ and $\gamma \in \Gamma$, let us write $\Lambda^\gamma = \gamma^{-1}\Lambda\gamma = \{\gamma^{-1}\lambda\gamma : \lambda \in \Lambda\} \leq \Gamma$ for the *γ-conjugate* of Λ. For all $X \in \Omega$ and $\gamma \in \Gamma$ we have $\mathrm{Aut}(X^\gamma) = \mathrm{Aut}(X)^\gamma$. That is, the automorphism groups of objects in an orbit are conjugates of each other.

We say that two objects are *isomorphic* if they are on the same orbit of Γ in Ω. In particular, $X, Y \in \Omega$ are isomorphic if and only if there exists an *isomorphism* $\gamma \in \Gamma$ from X to Y that satisfies $Y = X^\gamma$. An isomorphism from an object to itself is an *automorphism*. Let us write $\mathrm{Iso}(X, Y)$ for the set of all isomorphisms from X to Y. We have that $\mathrm{Iso}(X, Y) = \mathrm{Aut}(X)\gamma = \gamma\mathrm{Aut}(Y)$ where $\gamma \in \mathrm{Iso}(X, Y)$ is arbitrary. Let us write $X \cong Y$ to indicate that X and Y are isomorphic. If we want to stress the group Γ under whose action isomorphism holds, we write $X \cong_\Gamma Y$.

Canonical Labeling and Canonical Form. A function $\kappa : \Omega \to \Gamma$ is a *canonical labeling map* for the action of Γ on Ω if

(K)   for all $X, Y \in \Omega$ it holds that $X \cong Y$ implies $X^{\kappa(X)} = Y^{\kappa(Y)}$ (canonical labeling).

For $X \in \Omega$ we say that $X^{\kappa(X)}$ is the *canonical form* of X in Ω. From (K) it follows that isomorphic objects have identical canonical forms, and the canonical labeling map gives an isomorphism that takes an object to its canonical form.

We assume that the act of computing $\kappa(X)$ for a given X produces as a side-effect a set of generators for the automorphism group $\mathrm{Aut}(X)$.

3   McKay's Canonical Extension Method

This section reviews McKay's [36] canonical extension method for isomorph-free exhaustive generation. Mathematically it will be convenient to present the method so that the isomorphism classes are captured as orbits of a group action of a group Γ, and extension takes place in one step from "seeds" to "objects" being generated, with the understanding that the method can be applied inductively in multiple steps so that the "objects" of the current step become the "seeds" for the next step. We stress that all material in this section is well known (Cf. [28].).

Objects and Seeds. Let Ω be a finite set of *objects* and let Σ be a finite set of *seeds*. Let Γ be a finite group that acts on Ω and Σ. Let κ be a canonical labeling map for the action of Γ on Ω.

Extending Seeds to Objects. Let us connect the objects and the seeds by means of a relation $e \subseteq \Omega \times \Sigma$ that indicates which objects can be built from which seeds by extension. For $X \in \Omega$ and $S \in \Sigma$ we say that X *extends* S and write XeS if $(X, S) \in e$. We assume the relation e satisfies

(E1) e is a union of orbits of Γ, that is, $e^\Gamma = e$ (invariance), and

(E2)   for every object $X \in \Omega$ there exists a seed $S \in \Sigma$ such that XeS (completeness).

For a seed $S \in \Sigma$, let us write $e(S) = \{X \in \Omega : XeS\}$ for the set of all objects that extend S.

Canonical Extension. Next let us associate with each object a particular isomorphism-invariant extension by which we want to extend the object from a seed. A function $M : \Omega \to \Sigma$ is a *canonical extension map* if

(M1)   for all $X \in \Omega$ it holds that $(X, M(X)) \in e$ (extension), and

(M2)   for all $X, Y \in \Omega$ we have that $X \cong Y$ implies $(X, M(X)) \cong (Y, M(Y))$ (canonicity).

That is, (M1) requires that X is in fact an extension of $M(X)$ and (M2) requires that isomorphic objects have isomorphic canonical extensions. In particular, $X \mapsto (X, M(X))$ is a well-defined map from Ω/Γ to e/Γ.

Generating Objects from Seeds. Let us study the following procedure, which is invoked for exactly one representative $S \in \Sigma$ from each orbit in Σ/Γ:

(P)   Let $S \in \Sigma$ be given as input. Iterate over all $X \in e(S)$. Perform zero or more isomorph rejection tests on X and S. If the tests indicate we should accept X, visit X.

Let us first consider the case when there are no isomorph rejection tests. Here and in what follows we indicate with the "†"-symbol that a proof of a claim can be found in the full version of this conference abstract.[4]

Lemma 1 (†). *The procedure (P) visits every isomorphism class of objects in Ω at least once.*

Let us next modify procedure (P) so that any two visits to the same isomorphism class of objects originate from the same procedure invocation. Let $M : \Omega \to \Sigma$ be a canonical extension map. Whenever we construct X by extending S in procedure (P), let us visit X if and only if

(T1) $(X, S) \cong (X, M(X))$.

Lemma 2 (†). *The procedure (P) equipped with the test (T1) visits every isomorphism class of objects in Ω at least once. Furthermore, any two visits to the same isomorphism class must (i) originate by extension from the same procedure invocation on input S, and (ii) belong to the same $\mathrm{Aut}(S)$-orbit of this seed S.*

Let us next observe that the outcome of test (T1) is invariant on each $\mathrm{Aut}(S)$-orbit of extensions of S.

Lemma 3 (†). *For all $\alpha \in \mathrm{Aut}(S)$ we have that (T1) holds for (X, S) if and only if (T1) holds for (X^α, S).*

Lemma 3 in particular implies that we obtain complete isomorph rejection by combining the test (T1) with a further test that ensures complete isomorph rejection on $\mathrm{Aut}(S)$-orbits. Towards this end, let us associate an arbitrary order relation on every $\mathrm{Aut}(S)$-orbit on $e(S)$. Let us perform the following further test:

(T2) $X = \min X^{\mathrm{Aut}(S)}$.

The following lemma is immediate from Lemmas 2 and 3.

Lemma 4. *The procedure (P) equipped with the tests (T1) and (T2) visits every isomorphism class of objects in Ω exactly once.*

A Template for Canonical Extension Maps. We conclude this section by describing a template of how to use an arbitrary canonical labeling map $\kappa : \Omega \to \Gamma$ to construct a canonical extension map $M : \Omega \to \Sigma$.

---

[4] Available at https://arxiv.org/abs/1706.08325.

For $X \in \Omega$ construct the canonical form $Z = X^{\kappa(X)}$. Using the canonical form Z only, identify a seed T with ZeT. In particular, such a seed must exist by (E2). (Typically this identification can be carried out by studying Z and finding an appropriate substructure in Z that qualifies as T. For example, T may be the minimum seed in Σ that satisfies ZeT. Cf. Lemma 9.) Once T has been identified, set $M(X) = T^{\kappa(X)^{-1}}$.

Lemma 5 (†). *The map $X \mapsto M(X)$ above is a canonical extension map.*

4   Generation of Partial Assignments

This section describes an instantiation of McKay's method that generates partial assignments of values to a set of variables U one variable at a time following a *prefix sequence* at the level of isomorphism classes given by the action of a group Γ on U. Let R be a finite set where the variables in U take values.

Partial Assignments, Isomorphism, Restriction. For a subset $W \subseteq U$ of variables, let us say that a *partial assignment* of values to W is a mapping $X : W \to R$. Isomorphism for partial assignments is induced by the following group action.[5] Let $\gamma \in \Gamma$ act on $X : W \to R$ by setting $X^\gamma : W^\gamma \to R$ where X^γ is defined for all $u \in W^\gamma$ by

$$X^\gamma(u) = X(u^{\gamma^{-1}}). \tag{3}$$

Lemma 6 (†). *The action (3) is well-defined.*

For an assignment $X : W \to R$, let us write $\underline{X} = W$ for the underlying set of variables assigned by X. Observe that the underline map is a homomorphism of group actions in the sense that $\underline{X^\gamma} = \underline{X}^\gamma$ holds for all $\gamma \in \Gamma$ and $X : W \to R$. For $Q \subseteq \underline{X}$, let us write $X|_Q$ for the restriction of X to Q.

The Prefix Sequence and Generation of Normalized Assignments. We are now ready to describe the generation procedure. Let us begin by prescribing the prefix sequence. Let $u_1, u_2, \ldots, u_k$ be k distinct elements of U and let $U_j = \{u_1, u_2, \ldots, u_j\}$ for $j = 0, 1, \ldots, k$. In particular we observe that

$$U_0 \subseteq U_1 \subseteq \cdots \subseteq U_k$$

with $U_j \setminus U_{j-1} = \{u_j\}$ for all $j = 1, 2, \ldots, k$.

---

[5] For conciseness and accessibility, the present conference version develops the method only for variable symmetries, and accordingly the group action (3) acts only on the variables in U and not on the values in R. However, the method does extend to capture symmetries on both variables in U and values in R (essentially by consideration of the Cartesian product $U \times R$ in place of U), and such an extension will be developed in a full version of this paper.

For $j = 0, 1, \ldots, k$ let Ω_j consist of all partial assignments $X : W \to R$ with $W \cong U_j$. Or what is the same, using the underline notation, Ω_j consists of all partial assignments X with $\underline{X} \cong U_j$.

We rely on canonical extension to construct exactly one object from each orbit of Γ on Ω_j, using as seeds exactly one object from each orbit of Γ on Ω_{j-1}, for each $j = 1, 2, \ldots, k$. We assume the availability of canonical labeling maps $\kappa : \Omega_j \to \Gamma$ for each $j = 1, 2, \ldots, k$.

Our construction procedure will work with objects that are in a normal form to enable precomputation for efficient execution of the subsequent tests for isomorph rejection. Towards this end, let us say that $X \in \Omega_j$ is *normalized* if $\underline{X} = U_j$. It is immediate from our definition of Ω_j and (3) that each orbit in Ω_j/Γ contains at least one normalized object.

Let us begin with a high-level description of the construction procedure, to be followed by the details of the isomorph rejection tests and a proof of correctness. Fix $j = 1, 2, \ldots, k$ and study the following procedure, which we assume is invoked for exactly one normalized representative $S \in \Omega_{j-1}$ from each orbit in Ω_{j-1}/Γ:

(P') Let a normalized $S \in \Omega_{j-1}$ be given as input. For each $p \in u_j^{\mathrm{Aut}(U_{j-1})}$ and each $r \in R$, construct the assignment

$$X : U_{j-1} \cup \{p\} \to R$$

defined by $X(p) = r$ and $X(u) = S(u)$ for all $u \in U_{j-1}$. Perform the isomorph rejection tests (T1') and (T2') on X and S. If both tests accept, visit $X^{\nu(p)}$ where $\nu(p) \in \mathrm{Aut}(U_{j-1})$ normalizes X.

From an implementation perspective it is convenient to precompute the orbit $u_j^{\mathrm{Aut}(U_{j-1})}$ together with group elements $\nu(p) \in \mathrm{Aut}(U_{j-1})$ for each $p \in u_j^{\mathrm{Aut}(U_{j-1})}$ that satisfy $p^{\nu(p)} = u_j$. Indeed, a constructed X with $\underline{X} = U_{j-1} \cup \{p\}$ can now be normalized by acting with $\nu(p)$ on X to obtain a normalized $X^{\nu(p)}$ isomorphic to X.

The Isomorph Rejection Tests. Let us now complete the description of procedure (P') by describing the two isomorph rejection tests (T1') and (T2'). This paragraph only describes the tests with an implementation in mind, the correctness analysis is postponed to the following paragraph.

Let us assume that the elements of U have been arbitrarily ordered and that $\kappa : \Omega_j \to \Gamma$ is a canonical labeling map. Suppose that X has been constructed by extending a normalized S with $\underline{X} = \underline{S} \cup \{p\} = U_{j-1} \cup \{p\}$. The first test is:

(T1') Subject to the ordering of U, select the minimum $q \in U$ such that $q^{\kappa(X)^{-1}\nu(p)} \in u_j^{\mathrm{Aut}(U_j)}$. Accept if and only if $p \cong_{\mathrm{Aut}(X)} q^{\kappa(X)^{-1}}$.

From an implementation perspective we observe that we can precompute the orbit $u_j^{\mathrm{Aut}(U_j)}$. Furthermore, the only computationally nontrivial part of the test is the computation of $\kappa(X)$ since we assume that we obtain generators for

Aut(X) as a side-effect of this computation. Indeed, with generators for Aut(X) available, it is easy to compute the orbits $U/\text{Aut}(X)$ and hence to test whether $p \cong_{\text{Aut}(X)} q^{\kappa(X)^{-1}}$. Let us now describe the second test:

(T2')   Accept if and only if $p = \min p^{\text{Aut}(S)}$ subject to the ordering of U.

From an implementation perspective we observe that since S is normalized we have Aut$(S) \leq \text{Aut}(\underline{S}) = \text{Aut}(U_{j-1})$ and thus the orbit $u_j^{\text{Aut}(U_{j-1})}$ considered by procedure (P') partitions into one or more Aut(S)-orbits. Furthermore, generators for Aut(S) are readily available (due to S itself getting accepted in the test (T1') at an earlier level of recursion), and thus the orbits $u_j^{\text{Aut}(U_{j-1})}/\text{Aut}(S)$ and their minimum elements are cheap to compute. Thus, a fast implementation of procedure (P') will in most cases execute the test (T2') before the more expensive test (T1').

Correctness. This section establishes the correctness of procedure (P') together with the tests (T1') and (T2') by reduction to McKay's framework and Lemma 4. Fix $j = 1, 2, \ldots, k$. Let us start by defining the extension relation $e \subseteq \Omega_j \times \Omega_{j-1}$ for all $X \in \Omega_j$ and $S \in \Omega_{j-1}$ by setting XeS if and only if

there exists a $\gamma \in \Gamma$ such that $\underline{X}^\gamma = U_j, \underline{S}^\gamma = U_{j-1}$, and $X^\gamma|_{U_{j-1}} = S^\gamma$.   (4)

This relation is well-defined in the context of McKay's framework:

Lemma 7 (†). *The relation (4) satisfies (E1) and (E2).*

The following lemma establishes that the iteration in procedure (P') constructs exactly the objects $X \in e(S)$; cf. procedure (P).

Lemma 8 (†). *Let $S \in \Omega_{j-1}$ be normalized. For all $X \in \Omega_j$ we have XeS if and only if there exists a $p \in u_j^{\text{Aut}(U_{j-1})}$ with $\underline{X} = U_{j-1} \cup \{p\}$ and $X|_{U_{j-1}} = S$.*

Next we show the correctness of the test (T1') by establishing that it is equivalent with the test (T1) for a specific canonical extension function M. Towards this end, let us use the assumed canonical labeling map $\kappa : \Omega_j \rightarrow \Gamma$ to build a canonical extension function M using the template of Lemma 5. In particular, given an $X \in \Omega_j$ as input with $\underline{X} = U_{j-1} \cup \{p\}$, first construct the canonical form $Z = X^{\kappa(X)}$. In accordance with (T1'), select the minimum $q \in U$ such that $q^{\kappa(X)^{-1}}\nu(p) \in u_j^{\text{Aut}(U_j)}$. Now construct $M(X)$ from X by deleting the value of $q^{\kappa(X)^{-1}}$.

Lemma 9 (†). *The mapping $X \mapsto M(X)$ is well-defined and satisfies both (M1) and (M2).*

To complete the equivalence between (T1') and (T1), observe that since X and p determine S by $X|_{\underline{X}\setminus\{p\}} = S$, and similarly X and $q^{\kappa(X)^{-1}}$ determine $M(X)$ by $X|_{\underline{X}\setminus\{q^{\kappa(X)^{-1}}\}} = M(X)$, the test (T1) is equivalent to testing whether

$(X, p) \cong (X, q^{\kappa(X)^{-1}})$ holds, that is, whether $p \cong_{\mathrm{Aut}(X)} q^{\kappa(X)^{-1}}$ holds. Observe that this is exactly the test (T1').

It remains to establish the equivalence of (T2') and (T2). We start with a lemma that captures the $\mathrm{Aut}(S)$-orbits considered by (T2).

Lemma 10 (†). *For a normalized $S \in \Omega_{j-1}$ the orbits in $e(S)/\mathrm{Aut}(S)$ are in a one-to-one correspondence with the elements of $(u_j^{\mathrm{Aut}(U_{j-1})}/\mathrm{Aut}(S)) \times R$.*

Now order the elements $X \in e(S)$ based on the lexicographic ordering of the pairs $(p, X(p)) \in u_j^{\mathrm{Aut}(U_{j-1})} \times R$. Since the action (3) fixes the values in R elementwise, we have that (T2') holds if and only if (T2) holds for this ordering of $e(S)$. The correctness of procedure (P') equipped with the tests (T1') and (T2') now follows from Lemma 4.

Selecting a Prefix. This section gives a brief discussion on how to select the prefix. Let $U_k = \{u_1, u_2, \ldots, u_k\}$ be the set of variables in the prefix sequence. It is immediate that there exist $|R|^k$ distinct partial assignments from U_k to R. Let us write R^{U_k} for the set of these assignments. The group Γ now partitions R^{U_k} into orbits via the action (3), and it suffices to consider at most one representative from each orbit to obtain an exhaustive traversal of the search space, up to isomorphism. Our goal is thus to select the prefix U_k so that the setwise stabilizer Γ_{U_k} has comparatively few orbits on R^{U_k} compared with the total number of such assignments. In particular, the ratio of the number of orbits $|R^{U_k}/\Gamma_{U_k}|$ to the total number of mappings $|R|^k$ can be viewed as a proxy for the achieved symmetry reduction and as a rough[6] proxy for the speedup factor obtained compared with no symmetry reduction at all.

Subroutines. By our assumption, the canonical labeling map κ produces as a side-effect a set of generators for the automorphism group $\mathrm{Aut}(X)$ for a given input X. We also assume that generators for the groups $\mathrm{Aut}(U_j)$ for $j = 0, 1, \ldots, k$ can be precomputed by similar means. This makes the canonical labeling map essentially the only nontrivial subroutine needed to implement procedure (P'). Indeed, the orbit computations required by tests (T1') and (T2') are implementable by elementary permutation group algorithms [10, 40]. The next section describes how to implement κ by reduction to vertex-colored graphs.[7]

---

[6] Here it should be noted that executing the symmetry reduction carries in itself a nontrivial computational cost. That is, there is a tradeoff between the potential savings in solving the system gained by symmetry reduction versus the cost of performing symmetry reduction. For example, if the instance has no symmetry and Γ is a trivial group, then symmetry reduction merely makes it more costly to solve the system.

[7] Reduction to vertex-colored graphs is by no means the only possibility to obtain the canonical labeling map to enable (P'), (T1'), and (T2'). Another possibility would be to represent Γ directly as a permutation group and use dedicated permutation-group algorithms [33, 34]. Our present choice of vertex-colored graphs is motivated by easy availability of carefully engineered implementations for working with vertex-colored graphs.

5 Representation Using Vertex-Colored Graphs

This section describes one possible approach to represent the group of symmetries $\Gamma \leq \mathrm{Sym}(U)$ of a system of constraints over a finite set of variables U taking values in a finite set R. Our representation of choice will be vertex-colored graphs over a fixed finite set of vertices V. In particular, isomorphisms between such graphs are permutations $\gamma \in \mathrm{Sym}(V)$ that map edges onto edges and respect the colors of the vertices; that is, every vertex in V maps to a vertex of the same color under γ. It will be convenient to develop the relevant graph representations in steps, starting with the representation of the constraint system and then proceeding to the representation of setwise stabilizers and partial assignments. These representations are folklore (see e.g. [28]) and are presented here for completeness of exposition only.

Representing the Constraint System. To capture $\Gamma \cong \mathrm{Aut}(G)$ via a vertex-colored graph G with vertex set V, it is convenient to represent the variables U directly as a subset of vertices $U \subseteq V$ such that no vertex in $V \setminus U$ has a color that agrees with a color of a vertex in U. We then seek a graph G such that $\mathrm{Aut}(G) \leq \mathrm{Sym}(U) \times \mathrm{Sym}(V \setminus U)$ projected to U is exactly Γ. In most cases such a graph G is concisely obtainable by encoding the system of constraints with additional vertices and edges joined to the vertices representing the variables in U. We discuss two examples.

Example 1. Consider the system of clauses (1) and its graph representation (2). The latter can be obtained as follows. First, introduce a blue vertex for each of the six variables of (1). These blue vertices constitute the subset U. Then, to accommodate negative literals, introduce a red vertex joined by an edge to the corresponding blue vertex representing the positive literal. These edges between red and blue vertices ensure that positive and negative literals remain consistent under isomorphism. Finally, introduce a green vertex for each clause of (1) with edges joining the clause with each of its literals. It is immediate that we can reconstruct (1) from (2) up to labeling of the variables even after arbitrary color-preserving permutation of the vertices of (2). Thus, (2) represents the symmetries of (1).

Let us next discuss an example where it is convenient to represent the symmetry at the level of original constraints rather than at the level of clauses.

Example 2. Consider the following system of eight cubic equations over 24 variables taking values modulo 2:

$$x_{11}y_{11}z_{11} + x_{12}y_{12}z_{12} + x_{13}y_{13}z_{13} = 0 \qquad x_{21}y_{11}z_{11} + x_{22}y_{12}z_{12} + x_{23}y_{13}z_{13} = 0$$
$$x_{11}y_{11}z_{21} + x_{12}y_{12}z_{22} + x_{13}y_{13}z_{23} = 0 \qquad x_{21}y_{11}z_{21} + x_{22}y_{12}z_{22} + x_{23}y_{13}z_{23} = 1$$
$$x_{11}y_{21}z_{11} + x_{12}y_{22}z_{12} + x_{13}y_{23}z_{13} = 1 \qquad x_{21}y_{21}z_{11} + x_{22}y_{22}z_{12} + x_{23}y_{23}z_{13} = 1$$
$$x_{11}y_{21}z_{21} + x_{12}y_{22}z_{22} + x_{13}y_{23}z_{23} = 1 \qquad x_{21}y_{21}z_{21} + x_{22}y_{22}z_{22} + x_{23}y_{23}z_{23} = 1$$

This system seeks to decompose a $2 \times 2 \times 2$ tensor (whose elements appear on the right hand sides of the equations) into a sum of three rank-one tensors.

The symmetries of addition and multiplication modulo 2 imply that the symmetries of the system can be represented by the following vertex-colored graph:

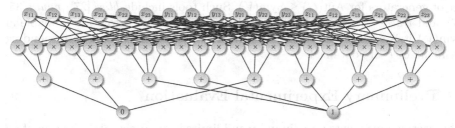

Indeed, we encode each monomial in the system with a product-vertex, and group these product-vertices together by adjacency to a sum-vertex to represent each equation, taking care to introduce two uniquely colored constant-vertices to represent the right-hand side of each equation.

The representation built directly from the system of polynomial equations in Example 2 concisely captures the symmetries in the system independently of the final encoding of the system (e.g. as CNF) for solving purposes. In particular, building the graph representation from such a final CNF encoding (cf. Example 1) results in a less compact graph representation and obfuscates the symmetries of the original system, implying less efficient symmetry reduction.

Representing the Values. In what follows it will be convenient to assume that the graph G contains a uniquely colored vertex for each value in R. (Cf. the graph in Example 2.) That is, we assume that $R \subseteq V \setminus U$ and that $\mathrm{Aut}(G)$ projected to R is the trivial group.

Representing Setwise Stabilizers in the Prefix Chain. To enable procedure (P') and the tests (T1') and (T2'), we require generators for $\mathrm{Aut}(U_j) \leq \Gamma$ for each $j = 0, 1, \ldots, k$. More generally, given a subset $W \subseteq U$, we seek to compute a set of generators for the setwise stabilizer $\mathrm{Aut}_\Gamma(W) = \Gamma_W = \{\gamma \in \Gamma : W^\gamma = W\}$, with $W^\gamma = \{w^\gamma : w \in W\}$. Assuming we have available a vertex-colored graph G that represents Γ by projection of $\mathrm{Aut}_{\mathrm{Sym}(V)}(G)$ to U, let us define the graph $G \uparrow W$ by selecting one vertex $r \in R$ and joining each vertex $w \in W$ with an edge to the vertex r. It is immediate that $\mathrm{Aut}_{\mathrm{Sym}(V)}(G \uparrow W)$ projected to U is precisely $\mathrm{Aut}_\Gamma(W)$.

Representing Partial Assignments. Let $X : W \to R$ be an assignment of values in R to variables in $W \subseteq U$. Again to enable procedure (P') together with the tests (T1') and (T2'), we require a canonical labeling $\kappa(X)$ and generators for the automorphism group $\mathrm{Aut}(X)$. Again assuming we have a vertex-colored graph G that represents Γ, let us define the graph $G \uparrow X$ by joining each vertex $w \in W$ with an edge to the vertex $X(w) \in R$. It is immediate that $\mathrm{Aut}_{\mathrm{Sym}(V)}(G \uparrow X)$ projected to U is precisely $\mathrm{Aut}_\Gamma(X)$. Furthermore, a canonical labeling $\kappa(X)$ can be recovered from $\kappa(G \uparrow X)$ and the canonical form $(G \uparrow X)^{\kappa(G \uparrow X)}$.

Using Tools for Vertex-Colored Graphs. Given a vertex-colored graph G as input, practical tools exist for computing a canonical labeling $\kappa(G) \in \text{Sym}(V)$ and a set of generators for $\text{Aut}(G) \leq \text{Sym}(V)$. Such tools include *bliss* [27], *nauty* [35, 37], and *traces* [37]. Once the canonical labeling and generators are available in $\text{Sym}(V)$ it is easy to map back to Γ by projection to U so that corresponding elements of Γ are obtained.

6 Preliminary Experimental Evaluation

This section documents a preliminary and limited experimental evaluation of an implementation of the adaptive prefix-assignment technique. The implementation is written in C and structured as a preprocessor that works with an explicitly given graph representation and utilizes the *nauty* [35, 37] canonical labeling software for vertex-colored graphs as a subroutine.

As instances we study systems of polynomial equations aimed at discovering the tensor rank of a small $m \times m \times m$ tensor $T = (t_{ijk})$ modulo 2, with $t_{ijk} \in \{0, 1\}$ and $i, j, k = 1, 2, \ldots m$. Computing the rank of a given tensor is NP-hard [21].[8] In precise terms, we seek to find the minimum r such that there exist three $m \times r$ matrices $A, B, C \in \{0, 1\}^{m \times r}$ such that for all $i, j, k = 1, 2, \ldots, m$ we have

$$\sum_{\ell=1}^{r} a_{i\ell} b_{j\ell} c_{k\ell} = t_{ijk} \pmod{2}. \tag{5}$$

Such instances are easily compilable into CNF with A, B, C constituting three matrices of Boolean variables so that the task becomes to find the minimum r such that the compiled CNF instance is satisfiable. Independently of the target tensor T such instances have a symmetry group of order at least $r!$ due to the fact that the columns of the matrices A, B, C can be arbitrarily permuted so that (5) maps to itself. In our experiments we select the entries of T uniformly at random so that the number of 1s in T is exactly n.

Hardware and Software Configuration. The experiments were performed on a cluster of Dell PowerEdge C4130 compute nodes, each equipped with two Intel Xeon E5-2680v3 CPUs and 128 GiB of main memory, running Linux version `3.10.0514.10.2.el7.x86_64`. All experiments were executed by allocating a single core on a single CPU of a compute node. Other unrelated compute load was in general present on the nodes during the experiments. A time-out of five hours of CPU time was applied.

Symmetry Reduction Tools and SAT Solvers. We report on three methods for symmetry reduction on aforementioned tensor-rank instances: (1) no reduction

---

[8] Yet considerable interest exists to determine tensor ranks of small tensors, in particular tensors that encode and enable fast matrix multiplication algorithms; cf. [1–4, 8, 9, 13, 24, 32, 41, 44].

("raw"), (2) breakid version 2.2 [16], (3) our technique ("reduce") with a user-selected prefix consisting of $2r$ variables that constitute the first two rows of the matrix A, and a graph representation of (5) as in Example 2. Three different SAT solvers were used in the experiments: lingeling and ilingeling version bbc-9230380 [7], and glucose version 4.1 [6]. We use the incremental solver ilingeling together with the incremental CNF output of reduce.

Results. Table 1 shows the results of a tensor rank computation modulo 2 for two random tensors T with $m = 5$, $n = 16$ and $m = 5$, $n = 18$. In both cases we observe that the rank is 9. The running times displayed in the table are in seconds, with "t/o" indicating a time-out. For both tensors we observe decreased running time due to symmetry reduction, with reduce performing better than breakid, but also taking more time to perform the preprocessing (indicated in columns labelled with "prep.") due to repeated calls to canonical labeling.

We would like to highlight that the comparison between reduce and breakid in particular illustrates the relevance of a graph representation of the symmetries in (5), which are not easily discoverable from the compiled CNF. In our experiments reduce receives as input the graph representation of the system (5) constructed as in Example 2, whereas breakid works on the compiled CNF input alone. This demonstrates the serendipity and versatility of using an auxiliary graph to represent the symmetries so that the symmetries are easily accessible. In fact, if we rely on the standard representation (2) built from the compiled CNF, the standard representation does *not* represent all the symmetries in (5).

Table 1. Computing tensor rank for two $5 \times 5 \times 5$ tensors

m	r	n	raw		prep. breakid	breakid		prep. reduce	reduce			Sat?
			glucose	lingeling		glucose	lingeling		glucose	lingeling	ilingeling	
5	6	16	60.02	114.14	0.06	36.01	81.86	0.29	3.70	9.67	7.28	UNSAT
5	7	16	t/o	t/o	0.19	t/o	t/o	1.07	372.43	614.04	249.10	UNSAT
5	8	16	t/o	t/o	0.12	t/o	t/o	2.56	t/o	t/o	15316.01	UNSAT
5	9	16	3.20	23.78	0.07	1.20	158.44	5.03	0.72	149.82	1.16	SAT
5	10	16	0.99	12.56	0.07	0.69	21.49	9.06	30.64	166.37	0.80	SAT
5	11	16	0.16	33.17	0.28	0.09	52.71	16.24	1.37	6.36	1.48	SAT
5	6	18	14.37	31.34	0.23	11.05	23.96	0.53	2.79	2.90	4.25	UNSAT
5	7	18	4606.30	t/o	0.18	3604.87	t/o	1.01	114.77	116.80	94.89	UNSAT
5	8	18	t/o	t/o	0.06	t/o	t/o	3.23	6901.79	t/o	3867.00	UNSAT
5	9	18	395.75	444.84	0.06	2.68	209.66	5.93	40.32	39.21	1.32	SAT
5	10	18	13.05	22.03	0.07	3.79	356.18	10.46	7.75	11.67	1.45	SAT
5	11	18	19.47	43.68	0.22	27.29	859.40	17.11	31.86	112.62	1.42	SAT

Acknowledgments. The research leading to these results has received funding from the European Research Council under the European Union's Seventh Framework Programme (FP/2007-2013)/ERC Grant Agreement 338077 "Theory and Practice of Advanced Search and Enumeration" (M.K., P.K., and J.K.). We gratefully acknowledge the use of computational resources provided by the Aalto Science-IT project at Aalto University. We thank Tomi Janhunen for useful discussions.

References

1. Alekseev, V.B.: On bilinear complexity of multiplication of 5 × 2 matrix by 2×2 matrix. Physics and mathematics, Uchenye Zapiski Kazanskogo Universiteta. Seriya Fiziko-Matematicheskie Nauki, vol. 156, pp. 19–29. Kazan University, Kazan (2014)
2. Alekseev, V.B.: On bilinear complexity of multiplication of $m \times 2$ and 2×2 matrices. Chebyshevskii Sb. **16**(4), 11–27 (2015)
3. Alekseev, V.B., Smirnov, A.V.: On the exact and approximate bilinear complexities of multiplication of 4 × 2 and 2 × 2 matrices. Proc. Steklov Inst. Math. **282**(1), 123–139 (2013)
4. Alekseyev, V.B.: On the complexity of some algorithms of matrix multiplication. J. Algorithms **6**(1), 71–85 (1985)
5. Aloul, F.A., Sakallah, K.A., Markov, I.L.: Efficient symmetry breaking for boolean satisfiability. In: Proceedings of the IJCAI 2003, pp. 271–276. Morgan Kaufmann (2003)
6. Audemard, G., Simon, L.: Extreme cases in SAT problems. In: Creignou, N., Le Berre, D. (eds.) SAT 2016. LNCS, vol. 9710, pp. 87–103. Springer, Cham (2016). doi:10.1007/978-3-319-40970-2_7
7. Biere, A.: Splatz, Lingeling, Plingeling, Treengeling, YalSAT entering the SAT Competition 2016. In: Proceedings of SAT Competition 2016 - Solver and Benchmark Descriptions. Department of Computer Science Series of Publications B, vol. B-2016-1, pp. 44–45. University of Helsinki (2016)
8. Bläser, M.: Lower bounds for the multiplicative complexity of matrix multiplication. Comput. Complex. **8**(3), 203–226 (1999)
9. Bläser, M.: On the complexity of the multiplication of matrices of small formats. J. Complex. **19**(1), 43–60 (2003)
10. Butler, G. (ed.): Fundamental Algorithms for Permutation Groups. LNCS, vol. 559. Springer, Heidelberg (1991)
11. Chu, G., de la Banda, M.G., Mears, C., Stuckey, P.J.: Symmetries, almost symmetries, and lazy clause generation. Constraints **19**(4), 434–462 (2014)
12. Codish, M., Gange, G., Itzhakov, A., Stuckey, P.J.: Breaking symmetries in graphs: the nauty way. In: Rueher, M. (ed.) CP 2016. LNCS, vol. 9892, pp. 157–172. Springer, Cham (2016). doi:10.1007/978-3-319-44953-1_11
13. Courtois, N.T., Hulme, D., Mourouzis, T.: Multiplicative complexity and solving generalized Brent equations with SAT solvers. In: Proceedings of the Third International Conference on Computational Logics, Algebras, Programming, Tools, and Benchmarking (COMPUTATION TOOLS 2012), pp. 22–27 (2012)
14. Crawford, J.M., Ginsberg, M.L., Luks, E.M., Roy, A.: Symmetry-breaking predicates for search problems. In: Proceedings of the KR 1996, pp. 148–159. Morgan Kaufmann (1996)
15. Darga, P.T., Liffiton, M.H., Sakallah, K.A., Markov, I.L.: Exploiting structure in symmetry detection for CNF. In: Proceedings of the DAC 2004, pp. 530–534. ACM (2004)
16. Devriendt, J., Bogaerts, B., Bruynooghe, M., Denecker, M.: Improved static symmetry breaking for SAT. In: Creignou, N., Le Berre, D. (eds.) SAT 2016. LNCS, vol. 9710, pp. 104–122. Springer, Cham (2016). doi:10.1007/978-3-319-40970-2_8
17. Dixon, J.D., Mortimer, B.: Permutation Groups. Graduate Texts in Mathematics, vol. 163. Springer, New York (1996)

18. Faradžev, I.A.: Constructive enumeration of combinatorial objects. In: Problèmes Combinatoires et Théorie des Graphes, pp. 131–135. No. 260 in Colloq. Internat. CNRS, CNRS, Paris (1978)
19. Gent, I.P., Petrie, K.E., Puget, J.: Symmetry in constraint programming. In: Handbook of Constraint Programming. Foundations of Artificial Intelligence, vol. 2, pp. 329–376. Elsevier (2006)
20. Hall Jr., M., Knuth, D.E.: Combinatorial analysis and computers. Amer. Math. Monthly **72**(2, Part 2), 21–28 (1965)
21. Håstad, J.: Tensor rank is NP-complete. J. Algorithms **11**(4), 644–654 (1990)
22. Heule, M.J.H., Kullmann, O., Wieringa, S., Biere, A.: Cube and conquer: guiding CDCL SAT solvers by lookaheads. In: Eder, K., Lourenço, J., Shehory, O. (eds.) HVC 2011. LNCS, vol. 7261, pp. 50–65. Springer, Heidelberg (2012). doi:10.1007/978-3-642-34188-5_8
23. Heule, M.J.H.: The quest for perfect and compact symmetry breaking for graph problems. In: Proceedings of the SYNASC 2016, pp. 149–156. IEEE Computer Society (2016)
24. Hopcroft, J.E., Kerr, L.R.: On minimizing the number of multiplications necessary for matrix multiplication. SIAM J. Appl. Math. **20**(1), 30–36 (1971)
25. Humphreys, J.F.: A Course in Group Theory. Oxford University Press, Oxford (1996)
26. Itzhakov, A., Codish, M.: Breaking symmetries in graph search with canonizing sets. Constraints **21**(3), 357–374 (2016)
27. Junttila, T., Kaski, P.: Engineering an efficient canonical labeling tool for large and sparse graphs. In: Proceedings of the ALENEX 2007. SIAM (2007)
28. Kaski, P., Östergård, P.R.J.: Classification Algorithms for Codes and Designs. Algorithms and Computation in Mathematics, vol. 15. Springer, Heidelberg (2006)
29. Kerber, A.: Applied Finite Group Actions. Algorithms and Combinatorics, vol. 19, 2nd edn. Springer, Heidelberg (1999)
30. Kerber, A., Laue, R.: Group actions, double cosets, and homomorphisms: unifying concepts for the constructive theory of discrete structures. Acta Appl. Math. **52**(1–3), 63–90 (1998)
31. Knuth, D.E.: The Art of Computer Programming. Combinatorial Algorithms, vol. 4A, Part 1. Addison-Wesley, Reading (2011)
32. Laderman, J.D.: A noncommutative algorithm for multiplying 3×3 matrices using 23 multiplications. Bull. Amer. Math. Soc. **82**, 126–128 (1976)
33. Leon, J.S.: Permutation group algorithms based on partitions, I: theory and algorithms. J. Symbol. Comput. **12**(4–5), 533–583 (1991)
34. Leon, J.S.: Partitions, refinements, and permutation group computation. In: Groups and Computation, II, pp. 123–158. No. 28 in DIMACS Series in Discrete Mathematics and Theoretical Computer Science, American Mathematical Society (1997)
35. McKay, B.D.: Practical graph isomorphism. Congressus Numerantium **30**, 45–87 (1981)
36. McKay, B.D.: Isomorph-free exhaustive generation. J. Algorithms **26**(2), 306–324 (1998)
37. McKay, B.D., Piperno, A.: Practical graph isomorphism. II. J. Symb. Comput. **60**, 94–112 (2014)
38. Read, R.C.: Every one a winner; or, how to avoid isomorphism search when cataloguing combinatorial configurations. Ann. Discrete Math. **2**, 107–120 (1978)

39. Sakallah, K.A.: Symmetry and satisfiability. In: Handbook of Satisfiability. Frontiers in Artificial Intelligence and Applications, vol. 185, pp. 289–338. IOS Press (2009)
40. Seress, Á.: Permutation Group Algorithms. Cambridge University Press, Cambridge (2003)
41. Strassen, V.: Gaussian elimination is not optimal. Numer. Math. **13**(4), 354–356 (1969)
42. Swift, J.D.: Isomorph rejection in exhaustive search techniques. In: Combinatorial Analysis, pp. 195–200. American Mathematical Society (1960)
43. Wieringa, S.: The icnf file format (2011). http://www.siert.nl/icnf/. Accessed April 2017
44. Winograd, S.: On multiplication of 2 × 2 matrices. Linear Algebra Appl. **4**(4), 381–388 (1971)

An Empirical Study of Branching Heuristics Through the Lens of Global Learning Rate

Jia Hui Liang[1(✉)], Hari Govind V.K.[2], Pascal Poupart[1], Krzysztof Czarnecki[1], and Vijay Ganesh[1]

[1] University of Waterloo, Waterloo, Canada
jliang@gsd.uwaterloo.ca
[2] College Of Engineering, Thiruvananthapuram, India

Abstract. In this paper, we analyze a suite of 7 well-known branching heuristics proposed by the SAT community and show that the better heuristics tend to generate more learnt clauses per decision, a metric we define as the global learning rate (GLR). Like our previous work on the LRB branching heuristic, we once again view these heuristics as techniques to solve the *learning rate* optimization problem. First, we show that there is a strong positive correlation between GLR and solver efficiency for a variety of branching heuristics. Second, we test our hypothesis further by developing a new branching heuristic that *maximizes GLR* greedily. We show empirically that this heuristic achieves very high GLR and interestingly very low literal block distance (LBD) over the learnt clauses. In our experiments this greedy branching heuristic enables the solver to solve instances faster than VSIDS, when the branching time is taken out of the equation. This experiment is a good proof of concept that a branching heuristic maximizing GLR will lead to good solver performance modulo the computational overhead. Third, we propose that machine learning algorithms are a good way to cheaply approximate the greedy GLR maximization heuristic as already witnessed by LRB. In addition, we design a new branching heuristic, called SGDB, that uses a stochastic gradient descent online learning method to dynamically order branching variables in order to maximize GLR. We show experimentally that SGDB performs on par with the VSIDS branching heuristic.

1 Introduction

Searching through a large, potentially exponential, search space is a reoccurring problem in many fields of computer science. Rather than reinventing the wheel and implementing complicated search algorithms from scratch, many researchers in fields as diverse as software engineering [7], hardware verification [9], and AI [16] have come to rely on SAT solvers as a general purpose tool to efficiently search through large spaces. By reducing the problem of interest down to a Boolean formula, engineers and scientists can leverage off-the-shelf SAT solvers to solve their problems without needing expertise in SAT or developing special-purpose algorithms. Modern conflict-driven clause-learning (CDCL) SAT solvers

© Springer International Publishing AG 2017
S. Gaspers and T. Walsh (Eds.): SAT 2017, LNCS 10491, pp. 119–135, 2017.
DOI: 10.1007/978-3-319-66263-3_8

can solve a wide-range of practical problems with surprising efficiency, thanks to decades of ongoing research by the SAT community. Two notable milestones that are key to the success of SAT solvers are the Variable State Independent Decaying Sum (VSIDS) branching heuristic (and its variants) [23] and conflict analysis techniques [22]. The VSIDS branching heuristic has been the dominant branching heuristic since 2001, evidenced by its presence in most competitive solvers such as Glucose [4], Lingeling [5], and CryptoMiniSat [26].

One of the challenges in designing branching heuristics is that it is not clear what constitutes a good decision variable. We proposed one solution to this issue in our LRB branching heuristic paper [19], which is to frame branching as an optimization problem. We defined a computable metric called *learning rate* and defined the objective as maximizing the learning rate. Good decision variables are ones with high learning rate. Since learning rate is expensive to compute a priori, we used a multi-armed bandit learning algorithm to estimate the learning rate on-the-fly as the basis for the LRB branching heuristic [19].

In this paper, we deepen our previous work and our starting point remains the same, namely, branching heuristics should be designed to solve the optimization problem of maximizing learning rate. In LRB, the learning rate metric is defined per variable. In this paper, we define a new metric, called the *global learning rate* (GLR) to measure the solver's overall propensity to generate conflicts, rather than the variable-specific metric we defined in the case of LRB. Our experiments demonstrate that GLR is an excellent objective to maximize.

1.1 Contributions

1. **A new objective for branching heuristic optimization:** In our previous work with LRB, we defined a metric that measures learning rate per variable. In this paper, we define a metric called the global learning rate (GLR), that measures the number of learnt clauses generated by the solver per decision, which intuitively is a better metric to optimize since it measures the solver as a whole. We show that the objective of maximizing GLR is consistent with our knowledge of existing branching heuristics, that is, the faster branching heuristics tend to achieve higher GLR. We perform extensive experiments over 7 well-known branching heuristics to establish the correlation between high GLR and better solver performance (Sect. 3).

2. **A new branching heuristic to greedily maximize GLR:** To further scientifically test the conjecture that GLR maximization is a good objective, we design a new branching heuristic that greedily maximizes GLR by always selecting decision variables that cause immediate conflicts. It is greedy in the sense that it optimizes for causing immediate conflicts, and it does not consider future conflicts as part of its scope. Although the computational overhead of this heuristic is very high, the variables it selects are "better" than VSIDS. More precisely, if we ignore the computation time to compute the branching variables, the greedy branching heuristic generally solves more instances faster than VSIDS. Another positive side-effect of the greedy branching heuristic is that relative to VSIDS, it has lower learnt clause literal

block distance (LBD) [3], a sign that it is learning higher quality clauses. The combination of learning faster (due to higher GLR) and learning better (due to lower LBD) clauses explains the power of the greedy branching heuristic. Globally optimizing the GLR considering all possible future scenarios a solver can take is simply too prohibitive. Hence, we limited our experiments to the greedy approach. Although this greedy branching heuristic takes too long to select variables in practice, it gives us a gold standard of what we should aim for. We try to approximate it as closely as possible in our third contribution (Sect. 4).

3. **A new machine learning branching heuristic to maximize GLR:** We design a second heuristic, called stochastic gradient descent branching (SGDB), using machine learning to approximate our gold standard, the greedy branching heuristic. SGDB trains an online logistic regression model by observing the conflict analysis procedure as the CDCL algorithm solves an instance. As conflicts are generated, SGDB will update the model to better fit its observations. Concurrently, SGDB also uses this model to rank variables based on their likelihood to generate conflicts if branched on. We show that in practice, SGDB is on par with the VSIDS branching heuristic over a large and diverse benchmark but still shy of LRB. However, more work is required to improve the learning in SGDB (Sect. 5).

2   Background

Clause Learning: Clause learning produces a new clause after each conflict to prevent the same or similar conflicts from reoccurring [22]. This requires maintaining an implication graph where the nodes are assigned literals and edges are implications forced by Boolean constraint propagation (BCP). When a clause is falsified, the CDCL solver invokes *conflict analysis* to produce a *learnt clause* from the conflict. It does so by *cutting* the implication graph, typically at the first-UIP [22], into the *reason side* and the *conflict side* with the condition that the decision variables appear on the reason side and the falsified clause appears on the conflict side. A new learnt clause is constructed by negating the reason side literals incident to the cut. Literal block distance (LBD) is a popular metric for measuring the "quality" of a learnt clause [3]. The lower the LBD the better.

Supervised Learning: Suppose there exists some function $f : Input \rightarrow Output$ that we do not have the code for. However, we do have *labeled* training data in the form of $\langle Input_i, f(Input_i) \rangle$ pairs. Given a large set of these labeled training data, also called a training set, there exists machine learning algorithms that can infer a new function $\tilde{f}$ that approximates f. These types of machine learning algorithms are called *supervised learning* algorithms. If everything goes well, $\tilde{f}$ will return the correct output with a high probability when given inputs from the training set, in which case we say $\tilde{f}$ fits the training set. Ideally, $\tilde{f}$ will also return the correct output for inputs that are not in the training set, in which case we say the function generalizes.

Most supervised learning algorithms require the input data to be represented as a vector of numbers. Feature extraction solves this issue by transforming each input data into a vector of real numbers, called a feature vector, that summarizes the input datum. During training, the feature vectors are used for training in place of the original input, hence learning the function $\tilde{f} : \mathbb{R}^n \to Output$ where $\mathbb{R}^n$ is the feature vector's type. Deciding which features to extract has a large impact on the learning algorithm's success.

In this paper, we only consider a special subclass of supervised learning called binary classification. In other words, the function we want to learn has the type $f : Input \to \{1, 0\}$, hence f maps every input to either the class 1 or the class 0.

We use logistic regression [10], a popular technique for binary classification, to learn a function $\tilde{f}$ that cheaply approximates f. The function learned by logistic regression has the type $\tilde{f} : \mathbb{R}^n \to [0, 1]$ where $\mathbb{R}^n$ is from the feature extraction and the output is a probability in $[0, 1]$ that the input is in class 1. Logistic regression defines the function $\tilde{f}$ as follows.

$$\tilde{f}([x_1, x_2, ..., x_n]) := \sigma(w_0 + w_1 x_1 + w_2 x_2 + ... + w_n x_n), \quad \sigma(z) := \frac{1}{1 + e^{-z}}$$

The weights $w_i \in \mathbb{R}$ measure the significance of each feature. The learning algorithm is responsible for finding values for these weights to make $\tilde{f}$ approximate f as closely as possible. The sigmoid function σ simply squeezes the linear function to be between 0 and 1. Hence $\tilde{f}$ outputs a real number between 0 and 1, which is expected since it is a probability.

The learning algorithm we use to set the weights is called stochastic gradient descent (SGD) [6], which is a popular algorithm for logistic regression. SGD minimizes the misclassification rate by taking a step in the opposite direction of the gradient with respect to each data point. The misclassification rate of a data point can be computed by the following error function:

$$Err(\mathbf{x}, y, \mathbf{W}) = y(1 - \tilde{f}(\mathbf{x}; \mathbf{W})) + (1 - y)(\tilde{f}(\mathbf{x}; \mathbf{W}))$$

where $\mathbf{x}$ is the input of a data point, y is the corresponding target class (0 or 1) for this data point and $\mathbf{W}$ is a vector weights. SGD takes a step in the opposite direction of the gradient as follows:

$$\mathbf{W}' \leftarrow \mathbf{W} - \alpha \frac{\partial Err(\mathbf{x}, y, \mathbf{W})}{\partial \mathbf{W}}$$

Here α is the step length (also known as the learning rate, not to be confused with the unrelated definition of learning rate in LRB). Under normal conditions, $\tilde{f}$ with the new weights $\mathbf{W}'$ will fit the training set better than with the old weights $\mathbf{W}$. If training time is not an issue, then SGD can be applied repeatedly until a fixed point is reached. The parameter $0 < \alpha < 1$ controls how aggressively the technique converges.

A common problem with machine learning in general is overfitting, where the trained function $\tilde{f}$ predicts correctly for the inputs it has seen in the training set, but works poorly for inputs it has not seen. We use a common technique called L2 regularization [24] to mitigate overfitting. L2 regularization introduces a new term in the error function that favors small weights

$$Err(\mathbf{x}, y, \mathbf{W}) = y(1 - \tilde{f}(\mathbf{x}; \mathbf{W})) + (1 - y)(\tilde{f}(\mathbf{x}; \mathbf{W})) + \lambda \|\mathbf{W}\|_2^2$$

Here λ is a parameter that determines the importance of the regularization penalty. How this prevents overfitting is beyond the scope of this paper.

SGD is also commonly used in an online fashion. Each time new data comes in, SGD is applied to this new data to update the weights, then the data is discarded. This has two advantages. Discarding the data keeps the memory usage low, especially useful when data is abundant. Additionally, the distribution in which the data is created can change over time. Online stochastic gradient does not assume the distribution is fixed and adjusts the weights accordingly after enough time. These two advantages are critical in our use of SGD.

3   GLR Maximization as a Branching Heuristic Objective

We framed the branching heuristic as an optimization problem in our earlier work [19], and we will continue to do so here. Formalizing the problem as an optimization problem opens up the problem to a wide range of existing optimization algorithms, and we exploited this very idea to develop the LRB [19] branching heuristic. The big difference between our previous papers and the current one is that the objective function for optimization in our previous work was learning per variable, whereas here we define it as the global learning rate (GLR) discussed below.

The first step to solving an optimization problem is to define the objective. Ideally the objective of the branching heuristic is to minimize the total running time. However, it is infeasible to calculate the running time a priori, which makes it unsuitable as an objective for branching. Instead, we target an easy to compute feature that correlates with solving time.

We define the *global learning rate* (GLR) of a solver as $GLR := \frac{\# \ of \ conflicts}{\# \ of \ decisions}$. Our goal is to construct a new branching heuristic to maximize the GLR. We assume that one clause is learnt per conflict. Learning multiple clauses per conflict has diminishing returns since they block the same conflict. But before we present our branching heuristic, let us justify why maximizing GLR is a reasonable objective for a branching heuristic. Past research concludes that clause learning is the most important feature for good performance in a CDCL solver [15], so perhaps it is not surprising that increasing the rate at which clauses are learnt is a reasonable objective. In our experiments, we assume the learning scheme is first-UIP since it is universally used by all modern CDCL solvers.

3.1   GLR vs Solving Time

We propose the following hypothesis: *for a given instance, the branching heuristic that achieves higher GLR tends to solve that instance faster than heuristics with lower GLR*. We provide empirical evidence in support of the hypothesis.

In the following experiment, we tested the above hypothesis on 7 branching heuristics: LRB [19], CHB [18], VSIDS (MiniSat [11] variation of VSIDS), CVSIDS (Chaff [23] variation of VSIDS), Berkmin [13], DLIS [21], and Jeroslow-Wang [14]. We created 7 versions of MapleSAT [1], one for each branching heuristic, keeping the code unrelated to the branching heuristic untouched. We ran all 7 branching heuristics on each application and hard combinatorial instance from every SAT Competition and SAT Race held between 2009 and 2016 with duplicate instances removed. At the end of each run, we recorded the solving time, GLR at termination, and the average LBD of clauses learnt. All experiments in this paper were conducted on StarExec [28], a platform purposefully designed for evaluating SAT solvers. For each instance, the solver was given 1800 seconds of CPU time and 8GB of RAM. The code for our experiments can be found on the MapleSAT website [2].

The results are presented in Table 1. Note that sorting by GLR in decreasing order, sorting by instances solved in decreasing order, sorting by LBD in increasing order, and sorting by average solving time in increasing order produces essentially the same ranking. This gives credence to our hypothesis that GLR correlates with branching heuristic effectiveness. Additionally, the experiment shows that high GLR correlates with low LBD.

Table 1. The GLR, number of instances solved, and average solving time for 7 different branching heuristics, sorted by the number of solved instances. Timed out runs have a solving time of 1800 s in the average solving time.

Heuristic	Avg LBD	Avg GLR	# Instances solved	Avg solving time (s)
LRB	10.797	0.533	1552	905.060
CHB	11.539	0.473	1499	924.065
VSIDS	17.163	0.484	1436	971.425
CVSIDS	19.709	0.406	1309	1043.971
BERKMIN	27.485	0.382	629	1446.337
DLIS	20.955	0.318	318	1631.236
JW	176.913	0.173	290	1623.226

To better understand the correlation between GLR and solving time, we ran a second experiment where for each instance, we computed the Spearman's rank correlation coefficient [27] (Spearman correlation for short) between the 7 branching heuristics' GLR and solving time. We then averaged all the instances' Spearman correlations by applying the Fisher transformation [12] to these correlations, then computing the mean, then applying the inverse Fisher transformation. This is a standard technique in statistics to average over correlations. This second experiment was performed on all the application and hard combinatorial instances from SAT Competition 2013 using the StarExec platform with a 5400 s timeout and 8 GB of RAM. For this benchmark, the average Spearman correlation is -0.3708, implying a negative correlation between GLR and solving time, or in other words, a high (resp. low) GLR tends to have low (resp. high) solving time as we hypothesized. Table 2 shows the results of the same correlation experiment with different solver configurations. The results show that the correlations remain moderately negative for all the configurations we tried.

Table 2. The Spearman correlation relating GLR to solving time between the 7 heuristics. The experiment is repeated with different solver configurations. MapleSAT is the default configuration which is essentially MiniSat [11] with phase saving [25], Luby restarts [20], and rapid clause deletion [3] based on LBD [3]. Clause activity based deletion is the scheme implemented in vanilla MiniSat.

Configuration	Spearman correlation
MapleSAT	-0.3708
No phase saving	-0.4492
No restarting	-0.3636
Clause deletion based on clause activity	-0.4235
Clause deletion based on LBD	-0.3958
Rapid clause deletion based on clause activity	-0.3881

Maximizing GLR also makes intuitive sense when viewing the CDCL solver as a proof system. Every conflict generates a new lemma in the proof. Every decision is like a new "case" in the proof. Intuitively, the solver wants to generate lemmas quickly using as few cases as possible, or in other words, maximize conflicts with as few decisions as possible. This is equivalent to maximizing GLR. Of course in practice, not all lemmas/learnt clauses are of equal quality, so the quality is also an important objective. We will comment more on this in later sections.

4   Greedy Maximization of GLR

Finding the globally optimal branching sequence that maximizes GLR is intractable in general. Hence we tackle a simpler problem to maximize GLR greedily instead. Although this is too computationally expensive to be effective in practice, it provides a proof of concept for GLR maximization and a gold standard for subsequent branching heuristics.

We define the function $c : PA \rightarrow \{1, 0\}$ that maps partial assignments to either class 1 or class 0. Class 1 is the "conflict class" which means that applying BCP to the input partial assignment with the current clause database would encounter a conflict once BCP hits a fixed-point. Otherwise the input partial assignment is given the class 0 for "non-conflict class". Note that c is a mathematical function with no side-effects, that is applying it does not alter the state of the solver. The function c is clearly decidable via one call to BCP, although it is quite costly when called too often.

Algorithm 1. Pseudocode for the *GGB* heuristic using the function c to greedily maximize GLR. Note that GGB is a meta-heuristic, it takes an existing branching heuristic (VSIDS in the following pseudocode) and makes it greedier by causing conflicts whenever possible. In general, VSIDS can be replaced with any other branching heuristic.

```
1: function PHASESAVING(Var)                          ▷ Return the variable plus a sign.
2:        return mkLit(Var, Var_savedPolarity)
3:
4: function VSIDS(Vars)        ▷ Return the variable with highest VSIDS activity plus a sign.
5:        return PhaseSaving(argmax_{v∈Vars} v_activity)
6:
7: function GGB
8:        CPA ← CurrentPartialAssignment
9:        V ← UnassignedVariables
10:       oneClass ← {v ∈ V  |  c(CPA ∪ {PhaseSaving(v)}) = 1}
11:       zeroClass ← V \ oneClass
12:       if oneClass ≠ ∅ then                      ▷ Next BCP will cause a conflict.
13:              return VSIDS(oneClass)
14:       else                                    ▷ Next BCP will not cause a conflict.
15:              return VSIDS(zeroClass)
```

The greedy GLR branching (GGB) heuristic is a branching heuristic that maximizes GLR greedily. When it comes time to branch, the branching heuristic is responsible for appending a decision variable (plus a sign) to the current partial assignment. GGB prioritizes decision variables where the new partial assignment falls in class 1 according to the function c. In other words, GGB branches on decision variables that cause a conflict during the subsequent call to BCP, if such variables exist. See Algorithm 1 for the implementation of GGB.

Unfortunately, GGB is very computationally expensive due to the numerous calls to the c function every time a new decision variable is needed. However, we show that GGB significantly increases the GLR relative to the base branching heuristic VSIDS. Additionally, we show that if the time to compute the decision variables was ignored, then GGB would be a more efficient heuristic than VSIDS. This suggests we need to cheaply approximate GGB to avoid the heavy computation. A cheap and accurate approximation of GGB would in theory be a better branching heuristic than VSIDS.

4.1   Experimental Results

In this section, we show that GGB accomplishes its goal of increasing the GLR and solving instances faster. Experiments were performed with MapleSAT using the StarExec platform with restarts and clause deletion turned off to minimize the effects of external heuristics. For each of the 300 instances in the SAT Competition 2016 application category, MapleSAT was ran twice, the first run configured with VSIDS and the second run configured with GGB. The run with VSIDS used a timeout of 5000 s. The run with GGB used a timeout of 24 h to account for the heavy computational overhead. We define *effective time* as the solving time minus the time spent by the branching heuristic selecting variables. Figure 1 shows the results of effective time between the two heuristics. Only *comparable* instances are plotted. An instance is comparable if either both heuristics

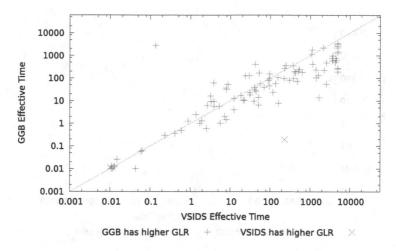

Fig. 1. GGB vs VSIDS. Each point in the plot is a comparable instance. Note that the axes are in log scale. GGB has a higher GLR for all but 2 instances. GGB has a mean GLR of 0.74 for this benchmark whereas VSIDS has a mean GLR of 0.59.

solved the instance or one heuristic solved the instance with an effective time of x seconds while the other heuristic timed out with an effective time greater than x seconds.

Of the comparable instances, GGB solved 69 instances with a lower effective time than VSIDS and 29 instances with a higher effective time. Hence if the branching was free, then GGB would solve instances faster than VSIDS 70% of the time. GGB achieves a higher GLR than VSIDS for all but 2 instances, hence it does a good job increasing GLR as expected. Figure 2 shows the same experiment except the points are colored by the average LBD of all clauses learnt from start until termination. GGB has a lower LBD than VSIDS for 72 of the 98 comparable instances. We believe this is because GGB by design causes conflicts earlier when the decision level is low, which keeps the LBD small since LBD cannot exceed the current decision level.

5   Stochastic Gradient Descent Branching Heuristic

GGB is too expensive in practice due to the computational cost of computing the c function. In this section, we describe a new branching heuristic called the stochastic gradient descent branching (SGDB) heuristic that works around this issue by cheaply approximating $c : PA \to \{1, 0\}$.

We use online stochastic gradient descent to learn the logistic regression function $\tilde{c} : \mathbb{R}^n \to [0, 1]$ where $\mathbb{R}^n$ is the partial assignment's feature vector and $[0, 1]$ is the probability the partial assignment is in class 1, the conflict class. Online training is a good fit since the function c we are approximating is non-stationary due to the clause database changing over time. For an instance with

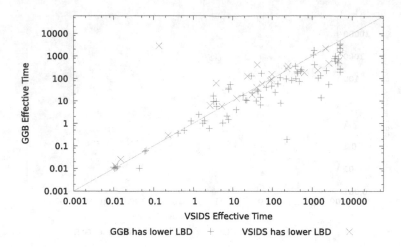

Fig. 2. GGB vs VSIDS. GGB has a lower average LBD for 72 of the 98 comparable instances. GGB has a mean average LBD of 37.2 for this benchmark whereas VSIDS has a mean average LBD of 61.1.

n Boolean variables and a partial assignment PA, we introduce the features $x_1, ..., x_n$ defined as follows: $x_i = 1$ if variable $i \in PA$, otherwise $x_i = 0$.

Recall that $\tilde{c} := \sigma(w_0 + w_1 x_1 + w_2 x_2 + ... + w_n x_n)$ is parameterized by the weights w_i, and the goal of SGDB is to find good weights dynamically as the solver roams through the search space. At the start of the search all weights are initialized to zero since we assume no prior knowledge.

To train these weights, SGDB needs to generate training data of the form $PA \times \{1, 0\}$ where 1 signifies the conflicting class, that is, applying BCP on PA with the current clause database causes a conflict. We leverage the existing conflict analysis procedure in the CDCL algorithm to create this data. Whenever the solver performs conflict analysis, SGDB creates a partial assignment PA_1 by concatenating the literals on the conflict side of conflict analysis with the negation[1] of the literals in the learnt clause and gives this partial assignment the label 1. Clearly applying BCP to PA_1 with the current clause database leads to a conflict, hence it is assigned to the conflict class. SGDB creates another partial assignment PA_0 by concatenating all the literals in the current partial assignment excluding the variables in the current decision level and excluding the variables in PA_1. Applying BCP to PA_0 does not lead to a conflict with the current clause database, because if it did, the conflict would have occurred at an earlier level. Hence PA_0 is given the label 0. In summary, SGDB creates two data points at every conflict, one for each class (the conflict class and the non-conflict class) guaranteeing a balance between the two classes.

---

[1] Recall that the learnt clause is created by negating some literals in the implication graph, this negation here is to un-negate them.

During conflict, two data points are created. SGDB then applies one step of stochastic gradient descent on these two data points to update the weights. Since we are training in an online fashion, the two data points are discarded after the weights are updated. To reduce the computation cost, regularization is performed lazily. Regularization, if done eagerly, updates the weights of every variable on every step of stochastic gradient descent. With lazy updates, only the weights of non-zero features are updated. As is typical with stochastic gradient descent, we gradually decrease the learning rate α over time until it reaches a fixed limit. This helps to rapidly adjust the weights at the start of the search.

When it comes time to pick a new decision variable, SGDB uses the $\tilde{c}$ function to predict the decision variable that maximizes the probability of creating a partial assignment in class 1, the conflict class. More precisely, it selects the following variable: $argmax_{v \in UnassignedVars}\tilde{c}(CPA \cup PhaseSaving(v))$ where CPA is the current partial assignment and $PhaseSaving(v)$ returns v plus the sign which the phase saving heuristic assigns to v if it were to be branched on. However, the complexity of the above computation is linear to the number of unassigned variables. Luckily this can be simplified by the following reasoning:

$$argmax_{v \in UnassignedVars}\tilde{c}(CPA \cup PhaseSaving(v))$$

$$= argmax_{v \in UnassignedVars}\sigma(w_0 + w_v + \sum_{l \in vars(CPA)} w_l)$$

Note that σ is a monotonically increasing function.

$$= argmax_{v \in UnassignedVars}(w_0 + w_v + \sum_{l \in vars(CPA)} w_l)$$

Remove the terms common to all the iterations of argmax.

$$= argmax_{v \in UnassignedVars}w_v$$

Hence it is equivalent to branching on the unassigned variable with the highest weight. By storing the weights in a max priority queue, the variable with the highest weight can be retrieved in time logarithmic to the number of unassigned variables, a big improvement over linear time. The complete algorithm is presented in Algorithm 2.

Differences with VSIDS: The SGDB branching heuristic presented thus far has many similarities with VSIDS. During each conflict, VSIDS increments the activities of the variables in PA_1 by 1 whereas SGDB increases the weights of the variables in PA_1 by a gradient. Additionally, the VSIDS decay multiplies every activity by a constant between 0 and 1, the L2 regularization in stochastic gradient descent also multiplies every weight by a constant between 0 and 1. SGDB decreases the weights of variables in PA_0 by a gradient, VSIDS does not have anything similar to this.

Sparse Non-conflict Extension: The AfterConflictAnalysis procedure in Algorithm 2 takes time proportional to $|PA_1|$ and $|PA_0|$. Unfortunately in practice, $|PA_0|$ is often quite large, about 75 times the size of $|PA_1|$ in our experiments. To shrink the size of PA_0, we introduce the sparse non-conflict extension.

Algorithm 2. Pseudocode for the $SGDB$ heuristic.

```
 1: function PHASESAVING(Var)                                      ▷ return the variable plus a sign
 2:     return mkLit(Var, Var_{SavedPolarity})
 3:
 4: procedure INITIALIZE
 5:     for all v ∈ Vars do
 6:         α ← 0.8, λ ← 0.1 × α, w_v ← 0
 7:         r_v ← 0                                      ▷ Stores the last time v was lazily regularized.
 8:     conflicts ← 0                                    ▷ The number of conflicts occurred so far.
 9:
10: function GETPA1(learntClause, conflictSide)
11:     return {¬l  |  l ∈ learntClause} ∪ conflictSide
12:
13: function GETPA0(PA_1)
14:     return {v ∈ AssignedVars  |  DecisionLevel(v) < currentDecisionLevel} \ PA_1
15:
16: procedure AFTERCONFLICTANALYSIS(learntClause, conflictSide)         ▷ Called after a learnt
        clause is generated from conflict analysis.
17:     if α > 0.12 then
18:         α ← α − 2 × 10^{-6}, λ ← 0.1 × α
19:     conflicts ← conflicts + 1
20:     PA_1 ← GetPA1(learntClause, conflictSide)
21:     PA_0 ← GetPA0(PA_1)
22:     for all v ∈ vars(PA_1 ∪ PA_0) do                              ▷ Lazy regularization.
23:         if conflicts − r_v > 1 then
24:             w_v ← w_v × (1 − \frac{αλ}{2})^{conflicts−r_v−1}
25:         r_v ← conflicts
26:     error_1 ← σ(w_0 + \sum_{i∈vars(PA_1)} w_i)                    ▷ Compute the gradients and descend.
27:     error_0 ← σ(w_0 + \sum_{i∈vars(PA_0)} w_i)
28:     w_0 ← w_0 × (1 − \frac{αλ}{2}) − \frac{α}{2}(error_1 + error_2)
29:     for all v ∈ vars(PA_1) do
30:         w_v ← w_v × (1 − \frac{αλ}{2}) − \frac{α}{2}(error_1)
31:     for all v ∈ vars(PA_0) do
32:         w_v ← w_v × (1 − \frac{αλ}{2}) − \frac{α}{2}(error_0)
33:
34: function SGDB
35:     d ← argmax_{v∈UnassignedVars} w_v
36:     while conflicts − r_d > 0 do                                  ▷ Lazy regularization.
37:         w_d ← w_d × (1 − \frac{αλ}{2})^{conflicts−r_d}
38:         r_d ← conflicts
```

With this extension PA_0 is constructed by randomly sampling one assigned literal for each decision level less than the current decision level. Then the literals in PA_1 are removed from PA_0 as usual. This construction bounds the size of PA_0 to be less than the number of decision levels. See Algorithm 3 for the pseudocode.

Reason-Side Extension: SGDB constructs the partial assignment PA_1 by concatenating the literals in the conflict side and the learnt clause. Although PA_1 is sufficient for causing the conflict, the literals on the reason side are the reason why PA_1 literals are set in the first place. Inspired by the LRB branching heuristic with a similar extension, the reason-side extension takes the literals on the reason side adjacent to the learnt clause in the implication graph and adds them to PA_1. This lets the learning algorithm associate these variables with the conflict class. See Algorithm 4 for the pseudocode.

Algorithm 3. Pseudocode for the sparse non-conflict extension. Only the `GetPA0` code is modified, the rest remains the same as SGDB.

```
1: function SAMPLE(level)
2:     C ← {v ∈ Vars  |  DecisionLevel(v) = level}
3:     return a variable sampled uniformly at random from C
4:
5: function GETPA0(PA₁)
6:     return (⋃_{i∈{1,2,...,currentDecisionLevel−1}} Sample(i)) \ PA₁
```

Algorithm 4. Pseudocode for the reason-side extension. Only the `GetPA1` code is modified, the rest remains the same as SGDB.

```
1: function GETPA1(learntClause, conflictSide)
2:     adjacent ← ⋃_{lit∈learntClause} Reason(¬lit)
3:     return {¬l  |  l ∈ learntClause} ∪ conflictSide ∪ adjacent
```

5.1   Experimental Results

We ran MapleSAT configured with 6 different branching heuristics (LRB, VSIDS, SGDB with four combinations of the two extensions) on all the application and hard combinatorial instances from SAT Competitions 2011, 2013, 2014, and 2016. At the end of each run, we recorded the elapsed time, the GLR at termination, and the average LBD of all clauses learnt from start to finish. Table 3 and Fig. 3 show the effectiveness of each branching heuristic in solving the instances in the benchmark. The reason-side extension (resp. sparse non-conflict extension) increases the number of solved instances by 97 (resp. 155). The two extensions together increase the number of solved instances by 219, and in total solve just 12 instances fewer than VSIDS. LRB solves 93 more instances than VSIDS. Table 4 shows the GLR and the average LBD achieved by the branching heuristics. Both extensions individually increased the GLR and decreased the LBD. The extensions combined increased the GLR and decreased the LBD even further. The best performing heuristic, LRB, achieves the highest GLR and lowest LBD in this experiment. It should not be surprising that LRB has high GLR, our goal when designing LRB was to generate lots of conflicts by branching on variables likely to cause conflicts. By design, LRB tries to achieve high GLR albeit indirectly by branching on variables with high learning rate.

6   Threats to Validity

1. **Did we overfit?** One threat is the possibility that the parameters are overtuned for the benchmarks and overfit them, and hence work poorly for untested benchmarks. To avoid overtuning parameters, we chose $\frac{\alpha}{2}$ in SGD to be the same as the step-size in LRB from our previous paper [19] and also chose $(1 - \frac{\alpha\lambda}{2})$ to be the same as the locality extension penalty factor in LRB from the same paper. We fixed these parameters from the start and never tuned them. Also, note that the training is online per instance.
2. **What about optimizing for quality of learnt clauses?** This remains a challenge. We did notice that when we maximize GLR we get a very nice

Table 3. # of solved instances by various configurations of SGD, VSIDS, and LRB.

Benchmark	Status	SGDB + No Ext	SGDB + Reason Ext	SGDB + Sparse Ext	SGDB + Both Ext	VSIDS	LRB
2011 Application	SAT	84	89	96	93	95	103
	UNSAT	87	87	96	94	99	98
	BOTH	171	176	192	187	194	201
2011 Hard Combinatorial	SAT	85	92	91	97	88	93
	UNSAT	36	50	43	51	48	64
	BOTH	121	142	134	148	136	157
2013 Application	SAT	91	92	108	112	127	132
	UNSAT	75	75	86	81	86	95
	BOTH	166	167	194	193	213	227
2013 Hard Combinatorial	SAT	107	109	118	118	115	116
	UNSAT	57	88	60	99	73	111
	BOTH	164	197	178	217	188	227
2014 Application	SAT	79	86	100	107	105	116
	UNSAT	65	62	79	73	94	76
	BOTH	144	148	179	180	199	192
2014 Hard Combinatorial	SAT	82	82	91	86	91	91
	UNSAT	41	61	56	73	59	89
	BOTH	123	143	147	159	150	180
2016 Application	SAT	52	55	62	62	60	61
	UNSAT	52	50	55	57	63	65
	BOTH	104	105	117	119	123	126
2016 Hard Combinatorial	SAT	5	7	6	7	3	6
	UNSAT	19	29	25	26	42	25
	BOTH	24	36	31	33	45	31
TOTAL (no duplicates)	SAT	585	612	672	682	684	718
	UNSAT	432	502	500	554	564	623
	BOTH	1017	1114	1172	1236	1248	1341

Table 4. GLR and average LBD of various configurations of SGD, VSIDS, and LRB on the entire benchmark with duplicate instances removed. LRB not solves the most instances and achieves the highest GLR and lowest average LBD in our experiments.

Metric	Status	SGDB + No Ext	SGDB + Reason Ext	SGDB + Sparse Ext	SGDB + Both Ext	VSIDS	LRB
Mean GLR	SAT	0.324501	0.333763	0.349940	0.357161	0.343401	0.375181
	UNSAT	0.515593	0.518362	0.542679	0.545567	0.527546	0.557765
	BOTH	0.403302	0.409887	0.429420	0.434854	0.419337	0.450473
Mean Avg LBD	SAT	22.553479	20.625091	19.470764	19.242937	28.833872	16.930723
	UNSAT	17.571518	16.896552	16.249930	15.832730	22.281780	13.574527
	BOTH	20.336537	18.965914	18.037512	17.725416	25.918232	15.437237

side-effect of low LBD. Having said that, in the future we plan to explore other notions of quality and integrate that into a multi-objective optimization problem view of branching heuristics.

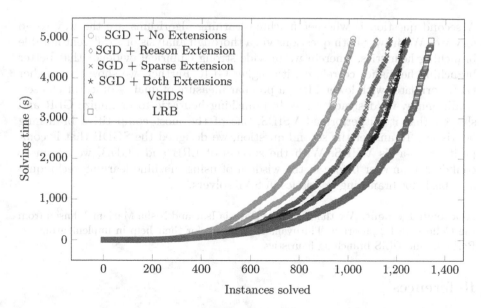

Fig. 3. A cactus plot of various configurations of SGD, VSIDS, and LRB on the entire benchmark with duplicate instances removed.

7   Related Work

The VSIDS branching heuristic, currently the most widely implemented branching heuristic in CDCL solvers, was introduced by the authors of the Chaff solver in 2001 [23] and later improved by the authors of the MiniSat solver in 2003 [11]. Carvalho and Marques-Silva introduced a variation of VSIDS in 2004 where the bump value is determined by the learnt clause length and backjump size [8] although their technique is not based on machine learning. Lagoudakis and Littman introduced a new branching heuristic in 2001 that dynamically switches between 7 different branching heuristics using reinforcement learning to guide the choice [17]. Liang et al. introduced two branching heuristics, CHB and LRB, in 2016 where a stateless reinforcement learning algorithm selects the branching variables themselves. CHB does not view branching as an optimization problem, whereas LRB, GGB, SGDB do. As stated earlier, LRB optimizes for learning rate, a metric defined with respect to variables. GGB and SGDB optimize for global learning rate, a metric defined with respect to the solver.

8   Conclusion and Future Work

Finding the optimal branching sequence is nigh impossible, but we show that using the simple framework of optimizing GLR has merit. The crux of the question since the success of our LRB heuristic is whether solving the learning rate optimization problem is indeed a good way of designing branching heuristics.

A second question is whether machine learning algorithms are the way to go forward. We answer both questions via a thorough analysis of 7 different notable branching heuristics, wherein we provide strong empirical evidence that better branching heuristics correlate with higher GLR. Further, we show that higher GLR correlates with lower LBD, a popular measure of quality of learnt clauses. Additionally, we designed a greedy branching heuristic to maximize GLR and showed that it outperformed VSIDS, one of the most competitive branching heuristics. To answer the second question, we designed the SGDB that is competitive vis-a-vis VSIDS. With the success of LRB and SGDB, we are more confident than ever before in the wisdom of using machine learning techniques as a basis for branching heuristics in SAT solvers.

Acknowledgement. We thank Sharon Devasia Isac and Nisha Mariam Johnson from the College Of Engineering, Thiruvananthapuram, for their help in implementing the Berkmin and DLIS branching heuristics.

References

1. https://sites.google.com/a/gsd.uwaterloo.ca/maplesat/
2. https://sites.google.com/a/gsd.uwaterloo.ca/maplesat/sgd
3. Audemard, G., Simon, L.: Predicting learnt clauses quality in modern SAT solvers. In: Proceedings of the 21st International Joint Conference on Artificial Intelligence, IJCAI 2009, pp. 399–404. Morgan Kaufmann Publishers Inc., San Francisco (2009)
4. Audemard, G., Simon, L.: Glucose 2.3 in the SAT 2013 Competition. In: Proceedings of SAT Competition 2013, pp. 42–43 (2013)
5. Biere, A.: Lingeling, Plingeling, PicoSAT and PrecoSAT at SAT Race 2010. FMV Report Series Technical Report 10(1) (2010)
6. Bottou, L.: On-line Learning in Neural Networks. On-line Learning and Stochastic Approximations, pp. 9–42. Cambridge University Press, New York (1998)
7. Cadar, C., Ganesh, V., Pawlowski, P.M., Dill, D.L., Engler, D.R.: EXE: automatically generating inputs of death. In: Proceedings of the 13th ACM Conference on Computer and Communications Security, CCS 2006, pp. 322–335. ACM, New York (2006)
8. Carvalho, E., Silva, J.P.M.: Using rewarding mechanisms for improving branching heuristics. In: Online Proceedings of The Seventh International Conference on Theory and Applications of Satisfiability Testing, SAT 2004, 10–13 May 2004, Vancouver, BC, Canada, (2004)
9. Clarke, E., Biere, A., Raimi, R., Zhu, Y.: Bounded model checking using satisfiability solving. Formal Meth. Syst. Des. **19**(1), 7–34 (2001)
10. Cox, D.R.: The regression analysis of binary sequences. J. Roy. Stat. Soc.: Ser. B (Methodol.) **20**(2), 215–242 (1958)
11. Eén, N., Sörensson, N.: An extensible SAT-solver. In: Giunchiglia, E., Tacchella, A. (eds.) SAT 2003. LNCS, vol. 2919, pp. 502–518. Springer, Heidelberg (2004). doi:10.1007/978-3-540-24605-3_37
12. Fisher, R.A.: Frequency distribution of the values of the correlation coefficient in samples from an indefinitely large population. Biometrika **10**(4), 507–521 (1915)
13. Goldberg, E., Novikov, Y.: BerkMin: a fast and robust SAT-solver. Discrete Appl. Math. **155**(12), 1549–1561 (2007)

14. Jeroslow, R.G., Wang, J.: Solving propositional satisfiability problems. Ann. Math. Artif. Intell. **1**(1–4), 167–187 (1990)
15. Katebi, H., Sakallah, K.A., Marques-Silva, J.P.: Empirical study of the anatomy of modern SAT solvers. In: Sakallah, K.A., Simon, L. (eds.) SAT 2011. LNCS, vol. 6695, pp. 343–356. Springer, Heidelberg (2011). doi:10.1007/978-3-642-21581-0_27
16. Kautz, H., Selman, B.: Planning as satisfiability. In: Proceedings of the 10th European Conference on Artificial Intelligence, ECAI 1992, pp. 359–363. Wiley Inc., New York (1992)
17. Lagoudakis, M.G., Littman, M.L.: Learning to select branching rules in the DPLL procedure for satisfiability. Electron. Notes Discrete Math. **9**, 344–359 (2001)
18. Liang, J.H., Ganesh, V., Poupart, P., Czarnecki, K.: Exponential recency weighted average branching heuristic for SAT solvers. In: Proceedings of the Thirtieth AAAI Conference on Artificial Intelligence, AAAI 2016, pp. 3434–3440. AAAI Press (2016)
19. Liang, J.H., Ganesh, V., Poupart, P., Czarnecki, K.: Learning rate based branching heuristic for SAT solvers. In: Proceedings of the 19th International Conference on Theory and Applications of Satisfiability Testing, SAT 2016, Bordeaux, France, 5–8 July 2016, pp. 123–140 (2016)
20. Luby, M., Sinclair, A., Zuckerman, D.: Optimal speedup of Las Vegas algorithms. Inf. Process. Lett. **47**(4), 173–180 (1993)
21. Marques-Silva, J.P.: The impact of branching heuristics in propositional satisfiability algorithms. In: Barahona, P., Alferes, J.J. (eds.) EPIA 1999. LNCS, vol. 1695, pp. 62–74. Springer, Heidelberg (1999). doi:10.1007/3-540-48159-1_5
22. Marques-Silva, J.P., Sakallah, K.A.: GRASP-a new search algorithm for satisfiability. In: Proceedings of the 1996 IEEE/ACM International Conference on Computer-aided Design, ICCAD 1996, pp. 220–227. IEEE Computer Society, Washington, D.C. (1996)
23. Moskewicz, M.W., Madigan, C.F., Zhao, Y., Zhang, L., Malik, S.: Chaff: engineering an efficient SAT solver. In: Proceedings of the 38th Annual Design Automation Conference, DAC 2001, pp. 530–535. ACM, New York (2001)
24. Murphy, K.P.: Machine Learning: A Probabilistic Perspective. The MIT Press, Cambridge (2012)
25. Pipatsrisawat, K., Darwiche, A.: A lightweight component caching scheme for satisfiability solvers. In: Marques-Silva, J., Sakallah, K.A. (eds.) SAT 2007. LNCS, vol. 4501, pp. 294–299. Springer, Heidelberg (2007). doi:10.1007/978-3-540-72788-0_28
26. Soos, M.: CryptoMiniSat v4. SAT Competition, p. 23 (2014)
27. Spearman, C.: The proof and measurement of association between two things. Am. J. Psychol. **15**(1), 72–101 (1904)
28. Stump, A., Sutcliffe, G., Tinelli, C.: StarExec: a cross-community infrastructure for logic solving. In: Demri, S., Kapur, D., Weidenbach, C. (eds.) IJCAR 2014. LNCS, vol. 8562, pp. 367–373. Springer, Cham (2014). doi:10.1007/978-3-319-08587-6_28

Coverage-Based Clause Reduction Heuristics for CDCL Solvers

Hidetomo Nabeshima[1(✉)] and Katsumi Inoue[2]

[1] University of Yamanashi, Kofu, Japan
nabesima@yamanashi.ac.jp
[2] National Institute of Informatics, Tokyo, Japan
inoue@nii.ac.jp

Abstract. Many heuristics, such as decision, restart, and clause reduction heuristics, are incorporated in CDCL solvers in order to improve performance. In this paper, we focus on learnt clause reduction heuristics, which are used to suppress memory consumption and sustain propagation speed. The reduction heuristics consist of evaluation criteria, for measuring the usefulness of learnt clauses, and a reduction strategy in order to select clauses to be removed based on the criteria. LBD (literals blocks distance) is used as the evaluation criteria in many solvers. For the reduction strategy, we propose a new concise schema based on the coverage ratio of used LBDs. The experimental results show that the proposed strategy can achieve higher coverage than the conventional strategy and improve the performance for both SAT and UNSAT instances.

1 Introduction

Many heuristics, such as decision, phase selection, restart, and clause reduction heuristics, are used in CDCL solvers in order to improve performance. For example, Katebi et al., show that decision and restart heuristics have resulted in significant performance improvement in their paper evaluating the components of CDCL solvers [5]. In this paper, we focus on clause reduction heuristics, which remove useless learnt clauses in order to suppress memory consumption and sustain propagation speed. Clause reduction is practically required since CDCL solvers learn a large number of clauses while solving. The reduction heuristics consist of evaluation criteria to measure the usefulness of learnt clauses and the reduction strategy for selecting clauses to be removed based on the criteria.

As the former evaluation criteria, LBD (literals blocks distance) [1] is implemented in many solvers. LBD is an excellent measure to identify learnt clauses that are likely to be used frequently. In this paper, we present experimental evidence of the identification power of LBD in a wide range of instances. Moreover, we show that an appropriate threshold for LBD, which are used to decide if clauses should be maintained or not, is determined depending on a given instance. However, a certain fixed threshold of LBD is often used in latter reduction strategies. In this paper, we propose a new reduction strategy based on the coverage of used LBDs, which dynamically computes an appropriate LBD

© Springer International Publishing AG 2017
S. Gaspers and T. Walsh (Eds.): SAT 2017, LNCS 10491, pp. 136–144, 2017.
DOI: 10.1007/978-3-319-66263-3_9

threshold in order to cover most propagations and conflicts. The experimental results show that our schema effectively maintains the used clauses and achieves performance improvement for both SAT and UNSAT instances.

The rest of this paper is organized as follows: Sect. 2 reviews clause reduction heuristics used in CDCL solvers. In Sects. 3 and 4, we provide experimental results, in order to clarify the property of LBD, and to point out some issues in the LBD-based reduction strategy, respectively. Our proposed reduction strategy is described in Sect. 5. Section 6 shows the experimental results and Sect. 7 concludes this paper.

2   Clause Reduction Heuristics

We briefly review the CDCL algorithm [3,6]. We assume that the reader is familiar with notions of propositional satisfiability (propositional variable, literal, clause, unit clause, unit propagation, and so on). The CDCL algorithm repeats the following two operations until a conflict occurs.

1. *Unit propagation*: the unassigned literal in each unit clause is assigned as true to satisfy the clause. This operation repeats until there is no unit clause.
2. *Decision*: when no unit clauses exist, an unassigned literal is selected and a truth value (true or false) is assigned to it.

For each assigned literal l, the *decision level of l* is defined as the number of decision literals on and before assigning l. By $dl(l)$, we denote the decision level of the literal l. When a conflict (falsified clause) occurs in the first step, the algorithm analyzes a cause of the conflict and *learns a clause* from the cause in order to prevent repeating the same conflict. The learnt clause is added to the clause database; then, the algorithm backjumps to the appropriate decision level computed from the clause.

CDCL solvers learn a large number of clauses during the search process of a given SAT instance. Hence, solvers should reduce the clause database periodically in order to suppress memory consumption and sustain propagation speed. In this section, we introduce reduction heuristics based on LBD, which was firstly introduced in Glucose solver. First, we present the evaluation criteria LBD in order to sort learnt clauses according to usefulness.

Definition 1 (Literals Blocks Distances (LBD) [1]). *The LBD of a clause C is defined as $|\{dl(l) \mid l \in C\} \cap \mathbb{N}|$, where $\mathbb{N}$ is the set of all natural numbers including 0; that is, the number of kinds of decision levels of literals in C.*

By $lbd(C)$, we denote the LBD of clause C. When a learnt clause is generated, the LBD of the clause is computed based on the current assignment. Additionally, the LBD is updated when the clause is used in unit propagations and the new LBD is smaller than the old one[1]. Literals with the same decision level have

---

[1] In Glucose 3.0 or later, the LBD update is executed only for clauses used in unit propagations on and after the first UIP in conflict analysis.

a possibility to be assigned at the same time in the future. Hence, small LBD clauses have a high possibility, which will be used in unit propagations and conflicts. We show the experimental evidence in the next section.

Next, we review the clause reduction strategy used in Glucose. The clause database reduction is executed every $l_{first} + l_{inc} \times x$ conflicts[2], where x is the number of reduction calls (initially $x = 0$). On each reduction, clauses are reduced according to the following policy that contains two exceptional conditions:

- Half of learnt clauses are removed in descending order of LBD except the following clauses.
  - keeps clauses whose LBDs ≤ 2 (these are called *glue clauses*).
  - keeps clauses used after the last reduction and updated LBD is less than a certain threshold.

In the following, we refer the above base policy and two exceptional conditions as **BP**, **E1** and **E2**, respectively. This Glucose reduction strategy and its derivatives are used in many solvers. For example, Lingeling dynamically selects either Glucose-based or classical activity-based strategies [4]. If the standard deviation of the LBDs of learnt clauses is too small or too large, then, the activity-based strategy is selected. MapleCOMSPS uses the reduction strategy combining both. This keeps clauses whose LBDs ≤ 6, while others are managed by the activity-based strategy. In addition, clauses with LBDs of 4 to 6, which have not been used for a while, are managed by the activity-based strategy [8]. In Sect. 4, we show some issues in the Glucose reduction strategy.

3   Experimental Evaluation of LBD

LBD is a well-known criterion for identifying learnt clauses that are likely to be used frequently. In this section, we present the experimental evidence of LBD usefulness in a wide variety of instances. From the results, we design our reduction strategy based on LBD. Throughout the paper, we use 755 instances, excluding duplicates, from the application instances used in competitions over the last 3 years,[3] as benchmark instances. All experiments were conducted on a Core i5 (1.4 GHz) with 4 GB memory. We set the timeout for solvers to 5,000 CPU seconds. We used our SAT solver GlueMiniSat 2.2.10. The main difference with Glucose is that GlueMiniSat uses the lightweight in-processing techniques [7]. Detailed results and the source code of GlueMiniSat can be found at http:// www.kki.yamanashi.ac.jp/~nabesima/sat2017/.

Figure 1 shows the distributions of LBDs of learnt clauses (left) and used LBDs (right) for each instance. In this experiment, learnt clause reduction was disabled; that is, the solver held every learnt clause. The numbers in the legend represent LBDs. The red line in the left graph will be explained in Sect. 4. Each

---

[2] In Glucose 3.0 or later, l_{first} and l_{inc} are 2000 and 300 respectively [2].

[3] SAT 2014 competition, SAT-Race 2015 and SAT 2016 competition.

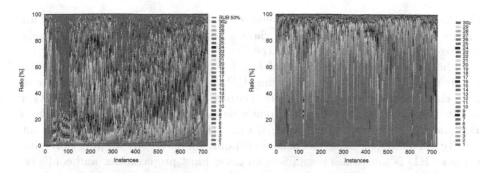

Fig. 1. Distribution of LBDs of learnt clauses (left) and used LBDs (right). Instances are sorted by ascending order of CPU time. (Color figure online)

stacked bar in the left graph represents the ratio distribution of LBDs, for all learnt clauses, after the solver stopped (e.g., if there are 10% learnt clauses whose LBDs are 2, the height of LBD 2 bar is 10%). In the right graph, each bar represents the ratio distribution of clause LBDs, which caused propagations or conflicts (e.g., if 50% of propagations or conflicts are caused by LBD 2 clauses, then the height of the LBD 2 bar is 50%).

The left graph shows that learnt clauses have various LBDs. In easy instances at the left end of the graph, small LBDs are somewhat dominant; however, the other instances have many different LBDs. On the other hand, from the right graph, it is clear that most propagations or conflicts are caused by small LBD clauses. This strongly supports the identification power of the LBD criterion.

4   Issues of Glucose Reduction Strategy

The Glucose reduction schema consists of **BP**, **E1** and **E2**, described in Sect. 2. In this section, we discuss the issues of the schema. We consider the base policy **BP**. Suppose that L_k is the number of learnt clauses after k-th reductions ($k \geq 1$) and r is the residual ratio ($0 \leq r < 1$, 0.5 in Glucose). L_k is defined as follows:

$$L_k = \begin{cases} rl_{\text{first}} & (k = 1) \\ r(L_{k-1} + l_{\text{first}} + (k-1)l_{\text{inc}}) & (k > 1) \end{cases} \tag{1}$$

We consider the difference $d_k = L_k - L_{k-1}$, which can be represented as $d_k = rd_{k-1} + rl_{\text{inc}}$. For this equation, when we add $-\frac{r}{1-r}l_{\text{inc}}$ to both sides, it represents a geometric progression with initial value $rl_{\text{first}} - \frac{r}{1-r}l_{\text{inc}}$ and common ratio r. Hence, we can get the following relationship:

$$L_k - L_{k-1} = (rl_{\text{first}} - \frac{r}{1-r}l_{\text{inc}}))r^{k-1} + \frac{r}{1-r}l_{\text{inc}} \tag{2}$$

The difference between L_k and L_{k-1} represents the number of clauses that can be newly held. The limit of $L_k - L_{k-1}$ as k approaches ∞ is a constant:

$$\lim_{k \to \infty}(L_k - L_{k-1}) = \frac{r}{1-r}l_{\text{inc}} \tag{3}$$

On the other hand, the interval between reductions increases exponentially. This means that the ratio of number of clauses that can be newly held, gradually approaches 0 as k increases. This is the first issue regarding **BP**.

The red line in the left graph of Fig. 1 represents the upper bound of the number of clauses held when following **BP**. In most instances, the number of glue clauses exceeds the bound. Glue clauses are never removed by **E1**. As a result, the solver can not hold new non-glue clauses at all, as long as it follows **BP** and **E1**. Even with a high residual ratio, the increment becomes a constant by (3); therefore, the issue essentially remains. Moreover, by keeping only glue clauses (**E1**) is sometimes insufficient to cover most propagations and conflicts. In the right graph of Fig. 1, the purple and green bars at the bottom represent the ratio of LBD 1 and 2. This indicates that the appropriate upper bound of LBD depends on a given instance. These are the second issue regarding **E1**.

In Glucose, clauses used after the last reduction and with LBD less than, or equal to 30, are not removed (**E2**). The right graph in Fig. 1 shows that this threshold can cover most propagations and conflicts; however, it may be overly retentive. This is the third issue related to **E2**.

In the next section, we propose a new concise reduction strategy to address the above mentioned concerns.

5   Coverage-Based Reduction Strategy

Most propagations and conflicts are caused by small LBD clauses. We propose a reduction strategy to dynamically compute the upper bound of LBD in order to cover most propagations and conflicts.

Let c be the specified coverage ($0 \leq c \leq 1$) and f_k be the number of times that LBD k clauses are used, where we call that a clause is *used* when it causes a unit propagation or a conflict, that is, when the clause becomes a unit or a falsified clause in a unit propagation process. We define the cumulative frequency up to k as $f_k^{\mathrm{cum}} = \sum_{i=1}^{k} f_i$ and the total frequency as $f^{\mathrm{tot}} = f_{|V|}^{\mathrm{cum}}$, where V is the set of variables at a given instance. The LBD threshold $lbd - thld(c)$ is defined as the minimum LBD l such that f_l^{cum} achieves the cover rate c of f^{tot} uses, that is,

$$lbd - thld(c) = l \quad \text{s.t.} \quad (f_{l-1}^{\mathrm{cum}} < cf^{\mathrm{tot}}) \wedge (cf^{\mathrm{tot}} \leq f_l^{\mathrm{cum}}). \tag{4}$$

This does not guarantee that the rate c of used clauses in the future will be covered by holding clauses with LBD $\leq lbd - thld(c)$, because a discarded clause may be required in order to propagate a clause with LBD $\leq lbd - thld(c)$. Nevertheless, we will present experimental results that our approach can achieve high coverage rate.

Next, we consider the trade-off between coverage and number of kept clauses. A high coverage requires the retention of a large number of clauses. Figure 2 exhibits the holding ratio, which is the number of maintained clauses at the termination of solver divided by the total number of learnt clauses, when we

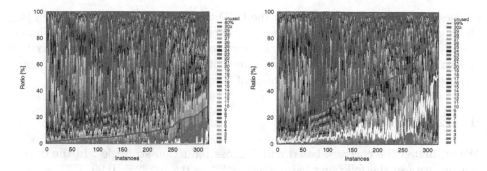

Fig. 2. Holding ratio of coverage of 80% (left) and 99% (right) for unsolved instances. Instances are sorted by ascending order of holding ratio. (Color figure online)

specify the coverage as 80% (left) and 99% (right). The red line represents the holding ratio and the white line denotes the ratio of unused and held clauses[4]. At the right end of the left graph, the red line is 30% and the white line is 10%. This means that a maximum of 30% of learnt clauses are required to cover at least 80% of uses of learnt clauses, and that a maximum of 10% of clauses are unused. The right graph at Fig. 2 shows that it is necessary to keep the majority of clauses in order to cover almost all uses, in the worst case, and that approximately half of them are not used.

In order to suppress the number of held clauses while covering propagations and conflicts as much as possible, we classify learnt clauses into three types: *core*, *support*, and *other* clauses. We make core clauses cover most uses (e.g. 80%), and support clauses cover the remains (e.g. 99%). We give support clauses a short lifetime since their number can be enormous; we give core clauses a longer lifetime, where the *lifetime* n of clause C means that C will be removed when it is unused while n reductions occur. We provide the formal definition of core, support, and other clauses. Let C be a clause, and c^{core} and c^{supp} be the specified coverage of core and support clauses ($c^{\mathrm{core}} \leq c^{\mathrm{supp}}$), respectively.

– C is a *core clause* if $lbd(C) \leq lbd - thld(c^{\mathrm{core}})$.
– C is a *support clause* if $lbd - thld(c^{\mathrm{core}}) < lbd(C) \leq lbd - thld(c^{\mathrm{supp}})$.
– Otherwise, C is an *other clause*.

The coverage-based clause reduction is executed every $l_{\mathrm{first}} + l_{\mathrm{inc}} \times x$ conflicts, same as in the Glucose schema, where x is the number of reduction calls (initially $x = 0$). For each reduction, we compute the core LBD threshold $lbd - thld(c^{\mathrm{core}})$ and the support LBD threshold $lbd - thld(c^{\mathrm{supp}})$ based on the frequency distribution of used LBDs. The lifetime of a core and support clause is the specified value l^{core} and l^{supp} ($l^{\mathrm{core}} \geq l^{\mathrm{supp}}$), respectively. Other clauses are removed at the next reduction (that is, the lifetime is 0). The computational cost of the coverage-based reduction strategy is $O(n)$, where n is the number of learnt

---

[4] A clause is unused if it does not produce any propagation or conflict, except for the UIP propagation immediately after being learn it.

clauses. Because the computation of the LBD threshold (4) is $O(m)$, where m is the maximum LBD, the removal of clauses exceeding the threshold is $O(n)$, and usually $m \ll n$. The Glucose reduction strategy requires $O(n \log n)$ since it needs to sort learnt clauses by their LBDs. Note that our reduction strategy does not impose the upper bound to the number of clauses (**BP**).

6   Experimental Results

We evaluated the coverage-based and Glucose reduction strategies. In the evaluation, we use the following parameters: $c^{\text{core}} = 0.8$, $c^{\text{supp}} = 0.99$, $l^{\text{core}} = 10$, $l^{\text{supp}} = 1$, $l_{\text{first}} = 2000$ and $l_{\text{inc}} = 300$. The first 4 parameters were determined by preliminary experiments. l_{first} and l_{inc} are the same as in Glucose. We also compared our approach with Glucose 4.0, MapleCOMSPS, and Lingeling. The latter two solvers use the Glucose-style schema as part of the reduction strategy, as described in Sect. 2. These solvers took the first and third place in the main track of the SAT 2016 competition, respectively[5].

Table 1. Solved instances, where "X (Y + Z)" denotes the number of solved instances (X), solved satisfiable instances (Y) and solved unsatisfiable instances (Z), respectively.

Solver	Solved instances
GlueMiniSat 2.2.10 (Glucose schema)	510 (255 + 255)
GlueMiniSat 2.2.10 (Coverage schema)	**524** (259 + 265)
Glucose 4.0	484 (244 + 240)
MapleCOMSPS	519 (**276** + 243)
Lingeling bbc	522 (249 + **273**)
Virtual best solver	597 (305 + 292)

Table 1 shows the number of instances solved by each solver and Fig. 3 is the cactus plot of these results. GlueMiniSat, with the Glucose schema, has better performance than Glucose. The coverage schema can further improve performance for both SAT and UNSAT instances. MapleCOMSPS and Lingeling show the superior results for SAT and UNSAT instances, respectively. GlueMiniSat, with the coverage schema, shows that the well-balanced result and total number of solved instances are comparable with these state of the art solvers.

Table 2 is the comparison of statistics between Glucose and coverage schema. Each value in the first 5 lines denotes the average for commonly solved 494 instances of both strategies. The first two lines in the table show that the Glucose schema reduces more clauses than the coverage schema; hence, the Glucose schema shows higher propagation speed. On the other hand, the coverage schema

---

[5] We exclude Riss 6, which ranked 2nd in the competition. Because it uses Linux-specific APIs, we could not compile it in our computing environment (Mac OS X).

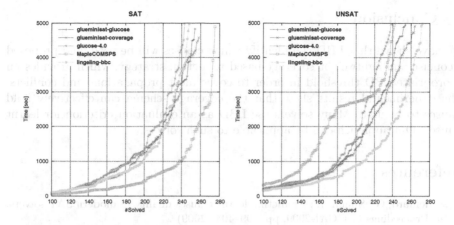

Fig. 3. Time to solve instances

Table 2. Comparison of two reduction strategies in GlueMiniSat.

Solver	Glucose schema	Coverage schema
Removed learnt [%]	76.4	72.9
Propagation speed [literals/sec]	1749748	1716494
CPU time [sec]	812.0	757.5
Conflicts	2519806	2325005
Reduction time [sec]	6.9	2.5
Coverage [%]	73.3	85.0
Precision [%]	67.0	74.1
Recall [%]	33.9	44.1

requires shorter CPU time and less conflicts in order to solve instances. It shows that the coverage schema can hold more useful clauses than the Glucose schema. In the coverage schema, the reduction of learnt clauses is slightly faster since the computational cost is $O(n)$ while the Glucose schema is $O(n \log n)$.

The last three lines in Table 2 are the results of different experiments, in which each solver does not actually remove learnt clauses to calculate coverage, precision, and recall. Each value indicates the average for commonly solved 406 instances. Coverage in Table 2 is the ratio of the number of used clauses that are caused only by maintained clauses to the total number of used clauses that are caused by all clauses. Precision is the ratio of used and held clauses to held clauses; recall is the ratio of used and held clauses to used clauses. In the coverage schema, these values have improved. This indicates that the coverage schema can better identify which clauses will be used.

7   Conclusion

We have shown that LBD can identify which clauses will be used, and proposed a concise and lightweight coverage-based reduction strategy, which provides an appropriate LBD threshold in order to cover most propagations and conflicts. The experimental results show that the coverage schema can effectively hold clauses to be used. Many solvers use LBD as an evaluation criterion for learnt clauses. Our approach can be applicable to such solvers.

References

1. Audemard, G., Simon, L.: Predicting learnt clauses quality in modern SAT solvers. In: Proceedings of IJCAI-2009, pp. 399–404 (2009)
2. Audemard, G., Simon, L.: Glucose 3.1 in the SAT 2014 competition (2014). http:// satcompetition.org/edacc/sc14/solver-description-download/118. SAT Competition 2014 Solver Description
3. Bayardo Jr., R.J., Schrag, R.: Using CSP look-back techniques to solve real-world SAT instances. In: Proceedings of the 14th National Conference on Artificial Intelligence (AAAI 1997), pp. 203–208 (1997)
4. Biere, A.: Lingeling and Friends at the SAT Competition 2011 (2011). http://fmv. jku.at/papers/biere-fmv-tr-11-1.pdf. SAT Competition 2011 Solver Description
5. Katebi, H., Sakallah, K.A., Marques-Silva, J.P.: Empirical study of the anatomy of modern sat solvers. In: Sakallah, K.A., Simon, L. (eds.) SAT 2011. LNCS, vol. 6695, pp. 343–356. Springer, Heidelberg (2011). doi:10.1007/978-3-642-21581-0_27
6. Marques-Silva, J.P., Sakallah, K.A.: GRASP: a search algorithm for propositional satisfiability. IEEE Trans. Comput. **48**(5), 506–521 (1999)
7. Nabeshima, H., Iwanuma, K., Inoue, K.: On-the-fly lazy clause simplification based on binary resolvents. In: ICTAI, pp. 987–995. IEEE (2013)
8. Oh, C.: Between SAT and UNSAT: the fundamental difference in CDCL SAT. In: Heule, M., Weaver, S. (eds.) SAT 2015. LNCS, vol. 9340, pp. 307–323. Springer, Cham (2015). doi:10.1007/978-3-319-24318-4_23

Maximum Satisfiability and Minimal Correction Sets

(I Can Get) Satisfaction: Preference-Based Scheduling for Concert-Goers at Multi-venue Music Festivals

Eldan Cohen[✉], Guoyu Huang, and J. Christopher Beck

Department of Mechanical and Industrial Engineering,
University of Toronto, Toronto, Canada
{ecohen,jcb}@mie.utoronto.ca, guoyu.huang@mail.utoronto.ca

Abstract. With more than 30 million attendees each year in the U.S. alone, music festivals are a fast-growing source of entertainment, visited by both fans and industry professionals. Popular music festivals are large-scale events, often spread across multiple venues and lasting several days. The largest festivals exceed 600 shows per day across dozens of venues. With many artists performing at overlapping times in distant locations, preparing a personal schedule for a festival-goer is a challenging task. In this work, we present an automated system for building a personal schedule that maximizes the utility of the shows attended based on the user preferences, while taking into account travel times and required breaks. Our system leverages data mining and machine learning techniques together with combinatorial optimization to provide optimal personal schedules in real time, over a web interface. We evaluate MaxSAT and Constraint Programming formulations on a large set of real festival timetables, demonstrating that MaxSAT can provide optimal solutions in about 10 s on average, making it a suitable technology for such an online application.

1 Introduction

In recent years, music festival have been growing in popularity, generating significant revenue [14–16]. In the U.S. alone, over 30 million people attend music festivals each year, with more than 10 million attending more than one festival each year [17].

Modern music festivals are large-scale events consisting of a set of musical shows, scheduled over the course of a few days at several different venues. Preparing a personal schedule for a music festival is a challenging task due to the existence of time conflicts between shows and travel times between venues. Festival-goers often spend a significant amount of time deciding which shows to attend, while trying to account for their musical preferences, travel times, and breaks for eating and resting. This problem is often discussed in the entertainment media:

© Springer International Publishing AG 2017
S. Gaspers and T. Walsh (Eds.): SAT 2017, LNCS 10491, pp. 147–163, 2017.
DOI: 10.1007/978-3-319-66263-3_10

"The majority of the major conflicts come late in each day—will you dance to HAIM or Flume on Sunday? Will you opt for the upbeat melodies of St. Lucia or Grimes on Saturday?" [12].

"Just when Coachella is upon us and you couldn't be more excited, a cloud enters – the set times are out, and there are heartbreaking conflicts. Difficult decisions must be made. Do you pass over an artist you love because an artist you love even more is playing all the way across the fest?" [13].

In this work, we address the problem of building an optimal schedule based on user preferences. We present a system that uses combinatorial optimization and machine learning techniques to learn the user musical preferences and generate a schedule that maximizes the user utility, while taking into account travel times and required breaks. Our system is implemented over a web interface and is able to generate optimal schedules in less than 10 s.

2   Problem Definition

The problem we address consists of two subproblems that need to be solved sequentially. The preference learning subproblem consists of predicting the user's musical preferences based on a small sample of preferences provided by the user. The scheduling subproblem then consists of finding an optimal personal schedule based on the user's preferences.

We first define the problem parameters and then the two subproblems.

2.1   Parameters

Shows. We consider a set of n festival shows $S := \{s_1, s_2, ..., s_n\}$, each associated with one of the performing artists (or bands) in the festival and taking place in one of the festival venues $V := \{v_1, v_2, ..., v_{|V|}\}$. Each show $s_i \in S$ has a fixed start time, t_i^s, and a fixed end time, t_i^e, such that the show length is $t_i^l = t_i^e - t_i^s$.

Travel Times. We consider an $n \times n$ travel time matrix TT, such that TT_{ij} is the travel time between the venue of show s_i and the venue of show s_j. We do not restrict TT to be symmetric, however, we assume it satisfies the triangle inequality.

Show Preferences. To represent the user's musical preferences, we consider the tuple $\langle f_p, M, N \rangle$. $f_p : S \to \mathbb{Z}^+ \cup \{\perp\}$ is a mapping from a show to either an integer score or the special value $\perp$ indicating that the user did not provide a score for the show. $M := \{m_1, m_2, ..., m_{|M|}\}$ is a set of show groups, $m_i \subseteq S$, such that the user must attend at least one show in each group. These groups can be used to model a simple list of shows the user has to attend (i.e., if each group is a singleton), as well as more sophisticated musical preferences such as seeking a diversity of musical styles by grouping shows based on style. Finally, $N \subseteq S$ is a set of shows the user is not interested in attending.

Break Preferences. We consider a set of l required breaks $B := \{b_1, b_2, ..., b_l\}$, such that for each $b_k \in B$, w_k^s and w_k^e represent the start and end of a time window in which the break should be scheduled and w_k^t represents the required break length. We assume that breaks are ordered temporally by their index, the time windows are non-overlapping, and at most one break can be scheduled between each pair of consecutive scheduled shows. The breaks are not allocated to a specific venue: the user can choose to enjoy a break at any location (e.g., at one of the venues or on the way to the next venue). The purpose of scheduling the breaks is to guarantee that sufficient free time is reserved for each b_k during the requested time window.

2.2 Preference Learning Subproblem

Given a function $f_p : S \to \mathbb{Z}^+ \cup \{\bot\}$ that maps shows to scores, our preference learning problem consists of replacing each $\bot$ value by an integer value to produce a full mapping f_p^* that is consistent with the user's preferences. Formally, our problem consists of finding a function $g : S \to \mathbb{Z}^+$ that minimizes the mean squared error [10], a common measure of fit, over the set of shows for which a score was provided $Q = \{s_i \mid f_p(s_i) \neq \bot\}$:

$$min \; \frac{1}{|Q|} \sum_{s_i \in Q} (g(s_i) - f_p(s_i))^2$$

The function g will then be used to predict the missing scores:

$$f_p^*(s) = \begin{cases} f_p(s), & \text{if } f_p(s) \neq \bot \\ g(s), & \text{if } f_p(s) = \bot \end{cases}$$

2.3 Scheduling Subproblem

Our scheduling subproblem consists of finding an assignment of values for a set of boolean variables $\{x_i \mid i \in [1..n]\}$, representing whether or not the festival-goer attends show s_i, and a set of integer values $\{y_j \mid j \in [1..l]\}$, specifying the start time of break b_j. The assignment has to satisfy the user preference w.r.t. M, N, and B (i.e., groups, shows not attended, and break time-windows). Our objective is to maximize the sum of the user-specified scores for the attended shows:

$$max \sum_{s_i \in S} x_i \cdot f_p^*(s_i)$$

3 System Architecture

The proposed system architecture is illustrated in Fig. 1. Our system implements a web interface, accessible using any web-enabled device.

Given an input of user preferences (f_p, M, N, B), provided over a web interface, we start by populating the missing scores in f_p using our preference learning

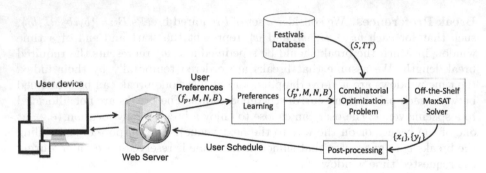

Fig. 1. The system architecture

algorithm. Then, we formulate the scheduling problem as a MaxSAT problem and solve it using an off-the-shelf MaxSAT solver. The details of the shows, S, and the travel time matrix, TT, for all festivals are stored in a database on the server. The results are processed and a schedule is produced and displayed over the web interface.

In Sect. 4 we present our MaxSAT model for the scheduling problem, followed in Sect. 5 with an alternative constraint programming (CP) model for the same problem. In Sect. 6 we describe our preference learning algorithm. Section 7 describes an empirical evaluation of the system, in which we compare the MaxSAT and the CP model, and compare our preference learning method to a simple baseline.

4   MaxSAT Model

In this section, we present our MaxSAT formulation of the problem. We first describe a boolean formulation and then provide a *weighted partial MaxSAT* encoding of the problem.

4.1   A Boolean Formulation

Consider the scheduling subproblem defined in Sect. 2.3. The variables x_i that represent show attendance are boolean while the variables y_k that represent the start time of each break are integer.

A time-indexed formulation of the problem would consist of replacing each y_k variable with a set of $y_{k,p}$ boolean variables such that $y_{k,p}=1 \iff y_k=p$. Assuming a time horizon of H time units, we need $H \times |B|$ variables, to represent the breaks. This size may not be unreasonable but here we develop an equivalent model that does not scale with the horizon length.

For any feasible solution for our problem, we can show that there exists a feasible solution with the same objective value, in which if there exists a break b_k between the shows s_i and s_j, it is scheduled in one of the following positions:

1. Immediately after s_i.
2. At the beginning of b_k's time window.

We prove this observation starting with the introduction of required notation and definitions.

We start with z, an assignment of values to the x_i and y_k variables of our problem, and $b_k \in B$, an arbitrary break scheduled at start time y_k. We use $\texttt{before}_z(k)$ to denote the end time of the latest scheduled show before break b_k and $\texttt{after}_z(k)$ to denote the start time of the earliest scheduled show after break b_k, according to the assignment z. We also use $\texttt{tt}_z(k)$ to denote the travel time between the location of the user at time $\texttt{before}_z(k)$ and the location the user needs to be at time $\texttt{after}_z(k)$.

Definition 1 (Feasible Assignment). *Let z be an assignment of values to variables x_i and y_k. We consider z to be a feasible solution if*

1. *All attended shows are non-overlapping.*
2. *The scheduled breaks and the attended shows do not overlap.*
3. *All scheduled breaks are within their time windows, i.e., $y_k \geq w_k^s$ and $y_k + w_k^t \leq w_k^e$.*
4. *For every break b_k that is scheduled between the attended shows s_i and s_j, there is enough time to travel between the show venues and have the break: $\texttt{after}_z(k) - \texttt{before}_z(k) \geq \texttt{tt}_z(k) + w_k^t$.*

Definition 2 (Earliest-break Assignment). *Let z be a feasible assignment of values to variables x_i and y_k. We consider z^*, to be the earliest-break assignment of z if:*

$$x_i^* = x_i \ \forall s_i \in S$$

$$y_k^* = \begin{cases} \texttt{before}_z(k), & \text{if } \texttt{before}_z(k) \geq w_k^s \\ w_k^s, & \text{if } \texttt{before}_z(k) < w_k^s \end{cases}$$

Figure 2 demonstrates the two possible locations of *earliest-break* assignments. Note that by definition of z^*, $y_k^* \leq y_k$ (otherwise y_k is not feasible).

Lemma 1. *There exists a feasible assignment z if and only if there exists a feasible earliest-break assignment z^*.*

Proof. **Direction** $\Longrightarrow$: We show that if z is feasible, we can construct a z^* that satisfies all the requirements of a feasible solution.

1. Since z is feasible, the attended shows $\{s_i | x_i = true\}$ are not overlapping (Definition 1). Since z^* is an earliest-break assignment of z, $x_i^* = x_i \ \forall s_i \in S$ (Definition 2). Therefore, the attended shows $\{s_i | x_i^* = true\}$ do not overlap (z^* satisfies requirement 1).
2. Since z^* is an earliest-break assignment of z, $\texttt{before}_z(k) \leq y_k^* \leq y_k$ (Definition 2). Consequently, $y_k^* + w_k^t \leq \texttt{after}_z(k)$. Since the shows and breaks do not overlap in z, they do not overlap in z^* (z^* satisfies requirement 2).

3. Since z^* is an earliest-break assignment of z, $w_k^s \le y_k^* \le y_k$ (Definition 2). Consequently, $y_k^* + w_k^t \le y_k + w_k^t$. Therefore, all breaks in z^* are within their time windows (z^* satisfies requirement 3).
4. $\texttt{before}_z(k)$, $\texttt{after}_z(k)$, $tt_z(k)$, and w_k^t remain in z^* as they were in z. Therefore, z^* satisfies requirement 4.

Direction $\Longleftarrow$: If exists an earliest-break assignment z^* that is feasible, then $z = z^*$ is a feasible assignment.   □

Note that because the cost function depends only on the x_i variables and $x_i^* = x_i$ for all i, z^* has the same objective value.

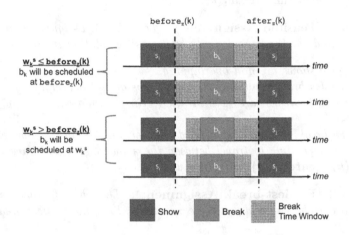

Fig. 2. The two possible locations of *earliest-break* assignments.

We can, therefore, translate the integer start time of a break to a set of boolean variables representing the two possible locations for each break: $q_{k,i}$ that represents whether break b_k is scheduled immediately after show s_i and r_k that represents whether break b_k is scheduled at the beginning of its time window.

4.2   Weighted Partial MaxSAT Encoding

We present a weighted partial MaxSAT model in Fig. 3, using *soft clauses* to model the objective function by attaching a weight for each show that corresponds to the user preferences. In our model, (c, w) denotes a clause c with a weight w, while (c, ∞) denotes that c is a *hard-clause*.

We use $R(k)$ to denote the set of all pairs of shows $s_i, s_j \in S$, such that the break b_k cannot be scheduled between the s_i and s_j, due to the necessary transition time. Formally, $R(k) = \{(i, j) \mid t_i^s \le t_j^s \wedge t_j^s - t_i^e \le TT_{ij} + w_k^t\}$.

We use the following boolean decision variables in our MaxSAT formulation:

$x_i :=$ show $s_i \in S$ is attended.

$q_{k,i} :=$ break b_k is scheduled immediately after show s_i.

$r_k :=$ break b_k is scheduled at the beginning of its time window.

$$(\neg x_i \vee \neg x_j, \infty) \qquad \forall s_i, s_j \in S : t_i^s \leq t_j^s \wedge t_i^e + TT_{ij} \geq t_j^s \quad (1)$$

$$\left(\bigvee_{x_i \in m_i} x_i, \infty \right) \qquad \forall m_i \in M \quad (2)$$

$$(\neg x_i, \infty) \qquad \forall s_i \in N \quad (3)$$

$$(\neg q_{k,i}, \infty) \qquad \forall s_i \in S, b_k \in B : t_i^e + w_k^t \geq w_k^e \quad (4)$$

$$(\neg q_{k,i}, \infty) \qquad \forall s_i \in S, k \in B : t_i^e \leq w_k^s \quad (5)$$

$$(\neg x_i \vee \neg r_k, \infty) \qquad \forall s_i \in S, b_k \in B : t_i^s \leq w_k^s \leq t_i^e \quad (6)$$

$$(\neg x_i \vee \neg r_k, \infty) \qquad \forall s_i \in S, b_k \in B : t_i^s \leq w_k^s + w_k^t \leq t_i^e \quad (7)$$

$$(\neg x_i \vee \neg r_k, \infty) \qquad \forall s_i \in S, b_k \in B : w_k^s \leq t_i^s \wedge w_k^s + w_k^t \geq t_i^e \quad (8)$$

$$(x_i \vee \neg q_{k,i}, \infty) \qquad \forall i \in S, b_k \in B \quad (9)$$

$$(\neg x_i \vee \neg x_j \vee \neg q_{k,i}, \infty) \qquad \forall b_k \in B, (i,j) \in R(k) \quad (10)$$

$$(\neg x_i \vee \neg x_j \vee \neg r_k, \infty) \qquad \forall b_k \in B, (i,j) \in R(k) : t_i^e \leq w_k^s \leq t_j^s \quad (11)$$

$$(r_k \vee q_{k,1} \vee q_{k,2} \vee ... \vee q_{k,n}, \infty) \qquad \forall b_k \in B \quad (12)$$

$$(x_i, f_p(s_i)) \qquad \forall s_i \in S \quad (13)$$

Fig. 3. MaxSAT model

Constraint (1) makes sure that for every pair of shows that cannot both be attended (either because they are overlapping or because of insufficient travel time), at most one will be attended. Constraint (2) ensures the shows that the user must attend are part of the schedule and constraint (3) ensures the shows the user is not interested in attending are not part of the schedule.

Constraints (4) and (5) ensure that a break is not scheduled *immediately after* a show, if it means that the break start time or end time is not inside the break's time window.

Constraints (6), (7) and (8) ensure that a break is not scheduled at the beginning of its time window if it overlaps an attended show. Constraints (9) ensures that a break is not scheduled *immediately after* a show that is not attended.

Constraint (10) and (11) ensure that a break is not scheduled in any of the two possible locations between two attended shows, if there is not sufficient time.

Constraint (12) ensures that all breaks are scheduled at least once (if a break can be scheduled more than once while maintaining optimality we arbitrarily select one). Constraint (13) is the only soft constraint and is used for optimization. Each clause corresponds to a show with weight equal to the show's score.

5   Constraint Programming Model

We present a CP model in Fig. 4. For this model, we utilize optional interval variables, which are decision variables whose possible values are a convex interval: $\{\perp\} \cup \{[s, e) \mid s, e \in \mathbb{Z}, s \leq e\}$. s and e are the start and end times of the interval and $\perp$ is a special value indicating the interval is not present in the solution [9]. The variable $\mathtt{pres}(var)$ is 1 if interval variable var is present in the solution, and 0 otherwise. Model constraints are only enforced on interval variables that are present in the solution. $\mathtt{start}(var)$ and $\mathtt{end}(var)$ return the integer start time and end time of the interval variable var, respectively.

We use the following decision variables in our CP formulation:

$x_i := (interval)$ present if the user attends show s_i and absent otherwise,
$y_k := (interval)$ always present interval representing b_k such that $\mathtt{start}(y_k)$ and $\mathtt{end}(y_k)$ represent the start and end time of break b_k, respectively.

$$\max \quad \sum_{s_i \in S} f_p(s_i) \cdot \mathtt{pres}(x_i) \tag{1}$$

$$\text{s.t.} \quad \mathtt{start}(x_i) = t_i^s \qquad\qquad\qquad \forall s_i \in S \tag{2}$$

$$\mathtt{end}(x_i) = t_i^e \qquad\qquad\qquad\quad \forall s_i \in S \tag{3}$$

$$\sum_{s_i \in m_j} \mathtt{pres}(x_i) \geq 1 \qquad\qquad \forall m_j \in M \tag{4}$$

$$\mathtt{pres}(x_i) = 0 \qquad\qquad\qquad\quad\; \forall s_i \in N \tag{5}$$

$$\mathtt{start}(y_k) \geq w_k^s \qquad\qquad\qquad \forall b_k \in B \tag{6}$$

$$\mathtt{end}(y_k) \leq w_k^e \qquad\qquad\qquad\; \forall b_k \in B \tag{7}$$

$$\mathtt{end}(y_k) - \mathtt{start}(y_k) = w_k^t \qquad\quad \forall b_k \in B \tag{8}$$

$$\mathtt{noOverlap}([x_1, x_2, ..., x_n], TT) \tag{9}$$

$$\mathtt{pres}(x_i) \wedge \mathtt{pres}(x_j) \Rightarrow \qquad\qquad \forall b_k \in B; (i,j) \in R(k) \tag{10}$$
$$\quad (\mathtt{end}(y_k) \leq \mathtt{start}(x_i)) \vee (\mathtt{start}(y_k) \geq \mathtt{end}(x_j))$$

$$\mathtt{start}(y_k) = w_k^s \vee \mathtt{start}(y_k) \in \{t_i^s \mid s_i \in S, w_k^s \leq t_i^s \leq w_k^e\} \quad \forall b_k \in B \tag{11}$$

Fig. 4. Constraint programming model

Objective (1) maximizes the user's utility, by summing the utility values of the attended shows. Constraints (2) and (3) set the shows' interval variables, of fixed duration and at fixed times, based on the festival schedule. Constraint (4) ensures that for every group in M, at least one show is attended and constraint (5) ensures the shows the user is not interested in attending are not attended.

Constraints (6) and (7) define the time window for each break based on the user's request by setting a lower bound on the start time and an upper bound on the end time of each break. Constraint (8) defines the length of the break based on the requested break length.

Constraint (9) is responsible for enforcing the feasibility of attending the chosen shows. We use the `noOverlap` global constraint which performs efficient domain filtering on the interval variables with consideration for transition times [9]. Here, the `noOverlap` global constraint ensures that the attended shows do not overlap in time, with consideration for the transition time between the different venues. For example, two consecutive shows can be attended if they are in the same venue. However, if the transition time between the venues is larger than the time difference between the end time of the first show and the start time of the following show, it is impossible to attend both shows.

Constraint (10) ensures the consistency of the break schedule with the attended shows. It verifies that a break is not scheduled between a pair of attended shows that does not have sufficient time difference to allow traveling and taking the break. For every triplet of two attended shows and a break such that the break cannot be scheduled between the two shows due to time (i.e., the time difference between the two shows is smaller than the sum of the transition time and the break time), we make sure the break is either scheduled before the earlier show or after the later show. Note that simple `noOverlap` constraints are not sufficient as we need to ensure the break will be scheduled either before the earlier show or after the later one and there is no required transition time between the end (resp., start) of a show and the start (resp. end) of a break.

Constraint (11) reduces the domains of each break variable, based on Lemma 1. While it does not enforce an *earliest-break assignment* (not required), it does reduce the domains of the break interval variables by allowing each break to be scheduled only at the beginning of the break's time window, or immediately after shows that end in the break's time window.

6   Learning User Preferences

The user cannot be expected to know all the performing artists in a festival or to spend time assigning a score to every artist. Therefore, we employ a learning-based approach to populate the missing scores based on the user's assignment of scores to a subset of the shows. We leverage the availability of large on-line music datasets to mine the features to predict the user's score.

As every show is associated with a performing artist, we assume the score assigned to each show reflects the user preference for the musical style of the performing artist. Therefore, our approach is to use the tags assigned to each artist on Last.fm,[1] a popular music website, as a feature set for a regression model that predicts the user score. The tags typically describe the artist musical style and origin (e.g., *pop, indie rock, punk, australian, spanish*, etc.). Tag-based features have been shown to be successful in recommending music [6,11]. In this work, we use linear regression model, due to its simplicity and its success in tag-based recommendation systems [22].

Given a training set of K inputs of $(\overrightarrow{o_i}, p_i)$ for $1 \leq i \leq K$ where $\overrightarrow{o_i}$ if a vector of F features and $p_i \in \mathbb{R}$ is a real value, the regression problem finds a set of

---

[1] http://www.last.fm.

weights, β_i, that expresses the value as a linear combination of the features and an error ε_i: $p_i = \beta_0 + \beta^1 o_i^1 + \beta^2 o_i^2 + ... + \beta^K o_i^K + \varepsilon_i$, such that the mean squared error over the training set is minimized [21]:

$$min \ \frac{1}{K} \sum_{i=1}^{K} \varepsilon_i^2$$

We collect the tags associated with all performing artists in the festival using Last.fm API.[2] For each artist, we construct a binary feature vector describing whether a tag applies to an artist. We train our model based on the subset of artists for which a score has been provided by the user. We then predict scores for the rest of performing artists, based on their associated tags. The predicted scores are rounded to integers as per the problem definition.

Due to the properties of this problem, notably a large number of features compared to a small training set, we choose to use Elastic Nets [26]. Elastic Nets employ a convex combination of L_1 and L_2 regularization using a parameter α, such that $\alpha = 1$ corresponds to ridge regression [8] and $\alpha = 0$ corresponds to lasso regression [24]. We use 5-fold cross-validation on the training set to choose the α value from a set of 10 values in $[0, 1]$. To reduce training time, we start by performing a univariate feature selection based on the F-test for regression [4], i.e., for each feature we calculate the F-test value for the hypothesis that the regression coefficient is zero, and select the top 75 features.

For comparison, we also consider Support Vector Regression (SVR). In our empirical evaluation we compare the linear regression based on Elastic Nets, to a linear SVR model. Linear SVR are much faster to train and to generate predictions than nonlinear SVR, and often give competitive results [7], making them an interesting candidate for our application. The regularization parameter and the loss function are chosen using a 5-fold cross validation on the training set.

7   Empirical Evaluation

In this section we present an empirical evaluation of our system. As our work consists of two parts – preference learning and scheduling, we evaluate each part separately.

A thorough evaluation of our application would require performing an experiment with real users to evaluate their satisfaction of the system. Unfortunately, such experiment is outside the scope of this research at this time. Instead, we leverage the existence of on-line datasets to empirically evaluate our system using real data.

First, we perform an experiment that evaluates the success of our preference learning method in predicting the musical taste of real users. Then, we perform an empirical evaluation of the scheduling system based on the timetables of real music festivals.

---

[2] http://www.last.fm/api.

7.1 Preference Learning Evaluation

In this section we present an experiment designed to evaluate the success of our preference learning method in predicting the musical taste of a user. The preference learning method we develop in Sect. 6 takes in a training set of tuples $(user, artist, score)$ and predicts the score for $(user, artist)$ tuples using tags from Last.fm. To evaluate this method without real users, we need a way to generate potential user scores on a subset of artists as an input to our learning algorithm and then a way to find a set of "ground truth" scores for the rest of the artists to compare against our predicted scores.

We could not find a dataset of this form that we can directly apply our method on. Instead, we take an existing dataset describing the listening habits of users and manipulate it into the required form of $(user, artist, score)$. We then split it to a training set used to train our learning algorithm and a test set on which we can compare our predicted score.

Data Preparation. The chosen dataset r is a collection of tuples $(user, artist, \#plays)$ describing the listening habits of 360,000 users on Last.fm, collected by Celma [3]. Each tuple describes the number of times each user listened to songs of each artist. For this experiment, we sample a set of 1000 users, U, each with at least 50 tuples. On average, each user in our sample has tuples for 57.85 artists.

In order to transform this dataset to the required form, we need to substitute the $\#plays$ column with a score column. To do so, we sort each user's records based on $\#plays$, and provide a score between 1 to 8, based on the corresponding quantile of $\#plays$. The artists with the lowest $\#plays$ will have a score of 1, and the artists with the highest $\#plays$ will have a score of 8. For convenience, we refer to the transformed dataset as a set of 1000 *user-specific datasets*. Given a user $u \in U$, we use r_u to denote the dataset that consists of $(artist, score)$ tuples that correspond to the user's tuples in r, $r_i = \Pi_{\langle artist,score \rangle}(\sigma_{user=u}(r)) \ \ \forall u \in U$.

Experiment Setup. To evaluate the preference learning method, we split each user's transformed dataset into a training set (60%) and a test set (40%). Given approximately 58 tuples per user, we get an average training set of approximately 35 records. We consider this to be a reasonable number of input scores to expect from a music festival attendee.

For each user, we train the model described in Sect. 6 based on the training set and then measure the *Minimum Squared Error* (MSE) and *Minimum Absolute Error* (MAE) in predicting the score of the records in the test set. We then calculate the median value of MSE and MAE across all users.

We compare our preference learning method, based on Elastic Net, with a linear SVR model and a baseline that consists of substituting all the missing scores with the mean score.

Table 1. Median MSE, median MAE and average time across a dataset of 1000 users

Algorithm	Median MSE	Median MAE	Mean runtime (sec)
Elastic Net	3.45	1.56	0.94
Linear SVR	3.79	1.64	0.35
Mean score baseline	6.24	2.25	0.24

Experiment Results. Table 1 shows the median MSE, the median MAE, and the mean runtime (train + test) across all r_u, for the Elastic Net, the linear SVR, and the mean baseline. It is clear that the learning based methods significantly outperform the baseline. The Elastic Net method yields more accurate results, however it requires longer runtime. Note that we expect the runtime in our system to be longer due to a larger set of shows for which we must predict scores (all shows in the festival).

7.2 Schedule Evaluation

In this section we present an empirical evaluation of the scheduling system. All experiments were run on a dual-core i5 (2.7 GHz) machine with 16 GB of RAM running Mac OS X Sierra. For our MaxSAT model, we used MaxHS v2.9, that employs a hybrid SAT and MIP approach [5]. For our CP model, we used CP Optimizer from the IBM ILOG CPLEX Optimization Studio version 12.6.3, single-threaded with default search and inference settings. We use a 10-minute run-time limit for each experiment.

Problem Instances. We consider 34 instances based on the real timetables of seven popular music festivals in the recent years as shown in Table 2. The instances have a large range of sizes defined by the number of shows, $|S|$, the number of venues, $|V|$, the number of must-attend groups, $|M|$, the number of shows the user is not interested in attending, $|N|$, and the number of required breaks, $|B|$. The parameters S and V are based on the real festival timetable. The travel time matrix TT is randomly generated in a range based on the estimated travel time between the real festival venues and it satisfies the triangle inequality. The breaks in B are generated in two configurations: either 2 breaks of 30 min or one break of 60 min. The shows in M and N were arbitrarily chosen, discarding infeasible instances. Each M has a mix of singletons and groups with multiple items. Random scores between 1–10 were assigned to a subset of the artists that ranged between 50% for the smallest festival to 15% for the largest one. The missing scores are predicted using our Elastic Net model. The preference learning runtime for each instance ranges between 2 to 6 s per user.

Table 2. Description of the problem instances: number of shows $|S|$, number of venues $|V|$, number of must-attend groups $|M|$, number of unattended shows $|N|$, and the time it takes to train the learning model and predict the missing scores.

Timetable source	$|S|$	$|V|$	$|M|$	$|N|$	$|B|$	Learning time (sec)
Pitchfork'16 Saturday	17	3	1	1	2	5.72
Pitchfork'16 Sunday	16	3	2	2	1	3.27
Pitchfork'17 Saturday	14	3	1	1	2	2.64
Pitchfork'17 Sunday	14	3	2	2	1	3.38
Lollapalooza Chile'17 Saturday	34	6	1	1	2	2.92
Lollapalooza Chile'17 Sunday	34	6	3	3	1	2.97
Primavera'16 Thursday	35	6	1	1	1	2.65
Primavera'16 Friday	35	6	3	3	2	2.52
Primavera'17 Thursday	35	6	1	1	2	2.42
Primavera'17 Friday	35	6	3	3	1	2.68
Osheaga'15 Friday	40	6	1	1	2	2.73
Osheaga'15 Saturday	38	6	3	3	1	2.29
Osheaga'15 Sunday	38	6	5	5	2	2.63
Osheaga'16 Friday	38	6	1	1	1	3.25
Osheaga'16 Saturday	38	6	3	3	2	2.36
Osheaga'16 Sunday	37	6	5	5	1	2.54
Glastonbury'15 Friday	90	10	1	1	2	3.21
Glastonbury'15 Saturday	92	10	3	3	1	2.91
Glastonbury'15 Sunday	94	10	5	5	2	2.84
Glastonbury'16 Friday	95	10	1	1	1	2.40
Glastonbury'16 Saturday	90	10	3	3	2	2.65
Glastonbury'16 Sunday	90	10	5	5	1	2.60
Tomorrowland'14 #1 Friday	149	15	1	1	2	2.76
Tomorrowland'14 #1 Saturday	139	15	3	3	1	2.51
Tomorrowland'14 #1 Sunday	118	14	5	5	2	2.73
Tomorrowland'14 #2 Friday	149	15	1	1	1	2.58
Tomorrowland'14 #2 Saturday	139	15	3	3	2	2.96
Tomorrowland'14 #2 Sunday	129	15	5	5	1	2.56
SXSW'15 Thursday	685	100	1	1	1	4.21
SXSW'15 Friday	706	102	3	3	2	4.04
SXSW'15 Saturday	715	99	5	5	1	4.19
SXSW'17 Thursday	644	96	1	1	2	3.60
SXSW'17 Friday	564	94	3	3	1	3.62
SXSW'17 Saturday	566	77	5	5	2	3.66

Mean learning time: 3.03

Table 3. Time to find and prove optimal solution and time to find optimal solution for MaxSAT and CP (results that are at least 10 times better are in bold).

Instance	Objective	Find+Prove Opt. (sec)		Find Opt. (sec)	
		MaxSAT	CP	MaxSAT	CP
Pitchfork'16 Saturday	40	0.02	0.02	0.00	0.00
Pitchfork'16 Sunday	30	0.02	0.02	0.00	0.00
Pitchfork'17 Saturday	43	0.02	0.01	0.00	0.00
Pitchfork'17 Sunday	39	0.02	0.02	0.00	0.00
Lollapalooza Chile'17 Saturday	34	0.04	0.39	0.00	0.01
Lollapalooza Chile'17 Sunday	33	0.03	0.04	0.00	0.00
Primavera'16 Thursday	50	0.03	0.06	0.00	0.01
Primavera'16 Friday	60	**0.04**	0.48	**0.01**	0.19
Primavera'17 Thursday	36	0.03	0.05	0.00	0.00
Primavera'17 Friday	38	**0.03**	0.39	0.00	0.06
Osheaga'15 Friday	52	0.12	0.91	0.00	0.01
Osheaga'15 Saturday	38	0.02	0.05	0.00	0.01
Osheaga'15 Sunday	33	0.02	0.04	0.00	0.00
Osheaga'16 Friday	51	0.05	0.39	0.00	0.00
Osheaga'16 Saturday	26	0.04	0.09	0.01	0.02
Osheaga'16 Sunday	38	0.03	0.04	0.00	0.00
Glastonbury'15 Friday	64	**0.59**	91.27	**0.38**	15.28
Glastonbury'15 Saturday	67	**0.13**	3.13	0.03	0.20
Glastonbury'15 Sunday	53	0.06	0.20	0.00	0.07
Glastonbury'16 Friday	85	**0.06**	110.78	**0.01**	1.33
Glastonbury'16 Saturday	70	**0.06**	1.90	**0.01**	0.12
Glastonbury'16 Sunday	61	0.18	1.07	0.01	0.05
Tomorrowland'14 #1 Friday	67	**1.81**	532.46	**1.71**	211.99
Tomorrowland'14 #1 Saturday	49	**0.39**	4.08	0.02	0.12
Tomorrowland'14 #1 Sunday	45	0.06	0.34	**0.00**	0.13
Tomorrowland'14 #2 Friday	58	**11.13**	408.23	**0.02**	0.42
Tomorrowland'14 #2 Saturday	48	**0.09**	2.68	**0.01**	0.99
Tomorrowland'14 #2 Sunday	42	0.37	1.15	0.30	0.08
SXSW'15 Thursday	133	**8.00**	T/O	**4.67**	61.10
SXSW'15 Friday	119	**3.00**	T/O	**1.11**	16.41
SXSW'15 Saturday	116	**9.88**	218.35	6.59	8.72
SXSW'17 Thursday	98	**6.82**	T/O	2.73	25.18
SXSW'17 Friday	101	**2.76**	80.04	**0.19**	5.93
SXSW'17 Saturday	125	**4.90**	T/O	**0.57**	96.04
Mean run-time		**1.50**	113.49	**0.54**	13.07

Numerical Results. Table 3 shows the time it takes to find and prove an optimal solution. The table is ordered in increasing size of the festivals. For the smaller instances, both MaxSAT and CP find and prove an optimal solution in a short time (usually less than 1 s). For the medium to large instances (starting from Glastonbury), CP struggles to find and prove optimal solutions, especially for the less constrained instances (i.e., smaller $|M|$ and $|N|$). For the largest music festival (SXSW), CP times-out on some of the instances (denoted "T/O"), failing to prove optimality (although it does *find* an optimal solution in all cases). MaxSAT, however, seems to be able to scale well, with the hardest instance taking approximately 12 s to find an optimal solution and prove its optimality.

Table 3 also shows the time it took to find the optimal solution without proving its optimality. MaxSAT still demonstrates better results in most cases, however in almost all cases, CP manages to find the optimal solution in less than one minute.

8 Related Work

Personal-level scheduling has received little attention in recent optimization literature. Refanidis and Yorke-Smith [20] presented a CP model for the problem of automating the management of an individual's time, noting the problem's difficulty due to the variety of tasks, constraints, utilities, and preference types involved. Alexiadis and Refanidis [1,2] presented a post-optimization approach, in which an existing personal schedule is optimized using local search. They developed a bundle of transformation methods to explore the neighborhood of a solution using either hill climbing or simulated annealing and achieved more than 6% improvement on average.

Closely-related problems, such as conference scheduling, have only been addressed from the perspective of building the *event schedule* with the objective of either meeting the presenters' or attendees' preferences [23]. An example for a presenter-based perspective approach can be found in Potthoff and Brams's integer programming formulation for conference scheduling w.r.t. presenters' availability [18]. Examples for attendee-based perspective approaches can be found in Quesnelle and Steffy's work on minimizing attendee conflicts [19], using an integer programming model, and Vangerven et al.'s work on maximizing attendance using a hierarchical optimization approach [25]. We are not aware of any work that directly addresses music festival scheduling nor of any work which takes the event schedule as input and optimizes for the individual attendee.

9 Conclusions and Future Work

We present a preference-based scheduling system for concert-goers at multi-venue music festivals. We utilize data mining and machine learning techniques to learn the user preferences and reduce the required user input. We use MaxSAT to efficiently find and prove an optimal schedule that maximizes the user utility, while taking into consideration the travel times between venues and the user's break preferences. Our system implements a web interface in which the user provides the required inputs and accesses the resulting schedule.

Our empirical evaluation shows that the use of preference learning allows us to provide more accurate results and the use of a MaxSAT model allows us to provide an efficient online service, with most instances taking less than 5 s and the hardest instances reaching 15 s for learning and optimization.

We believe this system can easily be adapted to other kinds of multi-venue events, such as conferences and large sporting events. For example, in the context of a conference, the preference learning can rely on the keywords of each talk and generate a preference-based personal schedule of talks to attend.

Another potential extension of this work is to explore ways to provide the users with alternative schedules. In this work the preference learning method is aimed at finding a schedule that is consistent with the user preferences. However, as some visitors often use the festival to expand their musical horizons, investigating ways to generate schedules that introduce the users to music they are not familiar with is an interesting direction of research.

References

1. Alexiadis, A., Refanidis, I.: Optimizing individual activity personal plans through local search. AI Commun. **29**(1), 185–203 (2015)
2. Alexiadis, A., Refanidis, J.: Post-optimizing individual activity plans through local search. In: Proceedings of the 8th Workshop on Constraint Satisfaction Techniques for Planning and Scheduling Problems (COPLAS 2013), pp. 7–15 (2013)
3. Celma, Ò.: Music Recommendation and Discovery: The Long Tail, Long Fail, and Long Play in the Digital Music Space. Springer, Heidelberg (2010)
4. Chatterjee, S., Hadi, A.S.: Regression Analysis by Example. Wiley, Hoboken (2015)
5. Davies, J., Bacchus, F.: Exploiting the power of MIP solvers in MaxSAT. In: Järvisalo, M., Van Gelder, A. (eds.) SAT 2013. LNCS, vol. 7962, pp. 166–181. Springer, Heidelberg (2013). doi:10.1007/978-3-642-39071-5_13
6. Firan, C.S., Nejdl, W., Paiu, R.: The benefit of using tag-based profiles. In: Web Conference 2007. LA-WEB 2007. Latin American, pp. 32–41. IEEE (2007)
7. Ho, C.H., Lin, C.J.: Large-scale linear support vector regression. J. Mach. Learn. Res. **13**, 3323–3348 (2012)
8. Hoerl, A.E., Kennard, R.W.: Ridge regression: biased estimation for nonorthogonal problems. Technometrics **42**(1), 80–86 (2000)
9. Laborie, P.: IBM ILOG CP optimizer for detailed scheduling illustrated on three problems. In: Hoeve, W.-J., Hooker, J.N. (eds.) CPAIOR 2009. LNCS, vol. 5547, pp. 148–162. Springer, Heidelberg (2009). doi:10.1007/978-3-642-01929-6_12
10. Lehmann, E.L., Casella, G.: Theory of Point Estimation. Springer Science & Business Media, Heidelberg (2006)
11. Levy, M., Sandler, M.: Music information retrieval using social tags and audio. IEEE Trans. Multimedia **11**(3), 383–395 (2009)
12. Long, Z.: Start planning your weekend with the lollapalooza 2016 schedule, May 2016. https://www.timeout.com/chicago/blog/start-planning-your-weekend-with-the-lollapalooza-2016-schedule-050916. Accessed 9 May 2016

13. Lynch, J.: Coachella 2016: 10 heartbreaking set time conflicts (and how to handle them), May 2016. http://www.billboard.com/articles/columns/music-festivals/7333891/coachella-2016-set-time-schedule-conflicts. Accessed 14 April 2016
14. McIntyre, H.: America's top five music festivals sold $183 million in tickets in 2014, March 2015. http://www.forbes.com/sites/hughmcintyre/2015/03/21/americas-top-five-music-festivals-sold-183-million-in-tickets-in-2014. Accessed 21 Mar 2015
15. McIntyre, H.: New York City's music festival market is becoming increasingly crowded, June 2016. http://www.forbes.com/sites/hughmcintyre/2016/06/21/new-york-citys-music-festival-market-is-becoming-increasingly-crowded. 21 Jun 2016
16. Mintel: Music concerts and festivals market is star performer in the UK leisure industry as sales grow by 45% in 5 years, December 2015. http://www.mintel.com/press-centre/leisure/music-concerts-and-festivals-market-is-star-performer-in-the-uk-leisure-industry-as-sales-grow-by-45-in-5-years. Accessed 9 Dec 2015
17. Nielsen: For music fans, the summer is all a stage, April 2015. http://www.nielsen.com/us/en/insights/news/2015/for-music-fans-the-summer-is-all-a-stage.html. Accessed 14 Apr 2015
18. Potthoff, R.F., Brams, S.J.: Scheduling of panels by integer programming: results for the 2005 and 2006 New Orleans meetings. Public Choice **131**(3–4), 465–468 (2007)
19. Quesnelle, J., Steffy, D.: Scheduling a conference to minimize attendee preference conflicts. In: Proceedings of the 7th Multidisciplinary International Conference on Scheduling: Theory and Applications (MISTA), pp. 379–392 (2015)
20. Refanidis, I., Yorke-Smith, N.: A constraint-based approach to scheduling an individual's activities. ACM Trans. Intell. Syst. Technol. (TIST) **1**(2), 12 (2010)
21. Sen, A., Srivastava, M.: Regression Analysis: Theory, Methods, and Applications. Springer Texts in Statistics. Springer, New York (1990). http://cds.cern.ch/record/1611847
22. Sen, S., Vig, J., Riedl, J.: Tagommenders: connecting users to items through tags. In: Proceedings of the 18th International Conference on World Wide Web, pp. 671–680. ACM (2009)
23. Thompson, G.M.: Improving conferences through session scheduling. Cornell Hotel Restaur. Adm. Q. **43**(3), 71–76 (2002)
24. Tibshirani, R.: Regression shrinkage and selection via the lasso. J. R. Stat. Soc. Ser. B (Methodol.) 267–288 (1996)
25. Vangerven, B., Ficker, A., Goossens, D., Passchyn, W., Spieksma, F., Woeginger, G.: Conference scheduling: a personalized approach. In: Burke, E., Di Gaspero, L., Özcan, E., McCollum, B., Schaerf, A. (eds.) Proceedings of the 11th International Conference on the Practice and Theory of Automated Timetabling, PATAT, pp. 385–401 (2016)
26. Zou, H., Hastie, T.: Regularization and variable selection via the elastic net. J. Roy. Stat. Soc. B Ser. B **67**, 301–320 (2005)

On Tackling the Limits of Resolution
in SAT Solving

Alexey Ignatiev[1,2]($\boxtimes$), Antonio Morgado[1], and Joao Marques-Silva[1]

[1] LASIGE, Faculty of Science, University of Lisbon, Lisbon, Portugal
{aignatiev,ajmorgado,jpms}@ciencias.ulisboa.pt
[2] ISDCT SB RAS, Irkutsk, Russia

Abstract. The practical success of Boolean Satisfiability (SAT) solvers stems from the CDCL (Conflict-Driven Clause Learning) approach to SAT solving. However, from a propositional proof complexity perspective, CDCL is no more powerful than the resolution proof system, for which many hard examples exist. This paper proposes a new problem transformation, which enables reducing the decision problem for formulas in conjunctive normal form (CNF) to the problem of solving maximum satisfiability over Horn formulas. Given the new transformation, the paper proves a polynomial bound on the number of MaxSAT resolution steps for pigeonhole formulas. This result is in clear contrast with earlier results on the length of proofs of MaxSAT resolution for pigeonhole formulas. The paper also establishes the same polynomial bound in the case of modern core-guided MaxSAT solvers. Experimental results, obtained on CNF formulas known to be hard for CDCL SAT solvers, show that these can be efficiently solved with modern MaxSAT solvers.

1 Introduction

Boolean Satisfiability (SAT) solvers have made remarkable progress over the last two decades. Unable to solve formulas with more than a few hundred variables in the early 90s, SAT solvers are now capable of routinely solving formulas with a few million variables [13,17]. The success of SAT solvers is supported by the CDCL (Conflict-Driven Clause Learning) [17, Chap. 04] paradigm, and the ability of SAT solvers to learn clauses from induced conflicts [17]. Nevertheless, being no more powerful than the general resolution proof system [60], CDCL SAT solvers are also known not to scale for specific formulas, which are hard for resolution [23,39,69]. Recent work has considered different forms of extending CDCL with techniques adapted from more powerful proof systems as well as others [7,12,16,22,40,68], with success in some settings. Nevertheless, for pigeonhole formulas [27], and with the exception of the lingeling SAT solver [14] on specific encodings, modern CDCL SAT solvers are unable to prove unsatisfiability even for a fairly small numbers of pigeons.

This work was supported by FCT funding of post-doctoral grants SFRH/BPD/103609/2014, SFRH/BPD/120315/2016, and LASIGE Research Unit, ref. UID/CEC/00408/2013.

© Springer International Publishing AG 2017
S. Gaspers and T. Walsh (Eds.): SAT 2017, LNCS 10491, pp. 164–183, 2017.
DOI: 10.1007/978-3-319-66263-3_11

This paper proposes an alternative path to tackle the difficulties of the resolution proof system, by developing an approach that aims to complement existing SAT solvers, and which also builds upon efficient CDCL SAT solvers. The motivation is to transform the original problem, from one clausal form to another, the latter encoding a restricted maximum satisfiability problem, but in such a way that CDCL SAT solvers can still be exploited. Given any CNF formula $\mathcal{F}$, the paper shows how to encode the problem as Horn Maximum Satisfiability (HornMaxSAT), more concretely by requiring a given cost on the HornMaxSAT formulation. This enables solving the modified problem with either a MaxSAT solver or with a dedicated HornMaxSAT solver. The paper then shows that for propositional encodings of the pigeonhole principle [27], transformed to Horn-MaxSAT, there exists a polynomially time bounded sequence of MaxSAT resolution steps which enables deriving a number of falsified clauses that suffices for proving unsatisfiable the original PHP formula. Similarly, the paper also proves that for modern core-guided MaxSAT solvers there exist sequences of unsatisfiable cores that enable reaching the same conclusion in polynomial time. This in turn suggests that MaxSAT algorithms [55] can be effective in practice when applied to such instances.

Experimental results, obtained on different encodings of the pigeonhole principle, but also on other instances that are well-known to be hard for resolution [69], confirm the theoretical result. Furthermore, a recently-proposed family of MaxSAT solvers [29,64], based on iterative computation of minimum hitting sets, is also shown to be effective in practice and on a wider range of classes of instances.

The paper is organized as follows[1]. Section 2 introduces the definitions and notation used throughout the paper. Section 3 develops a simple encoding from SAT into HornMaxSAT. Section 4 derives a polynomial bound on the number and size of MaxSAT-resolution steps to establish the unsatisfiability of propositional formulas encoding the pigeonhole principle transformed into HornMaxSAT. The section also shows that there are executions of core-guided MaxSAT solvers that take polynomial time to establish a lower bound of the cost of the MaxSAT solution which establishes the unsatisfiability of the original CNF formula. Experimental results on formulas encoding the pigeonhole principle, but also on other formulas known to be hard for CDCL SAT solvers [69] are analyzed in Sect. 5.2. The paper concludes in Sect. 6.

2   Preliminaries

The paper assumes definitions and notation standard in propositional satisfiability (SAT) and maximum satisfiability (MaxSAT) [17]. Propositional variables are taken from a set $X = \{x_1, x_2, \ldots\}$. A Conjunctive Normal Form (CNF) formula is defined as a conjunction of disjunctions of literals, where a literal is a variable or its complement. CNF formulas can also be viewed as sets of

---

[1] An extended version of the paper containing additional detail and proofs can be found in [42].

sets of literals, and are represented with calligraphic letters, $\mathcal{A}$, $\mathcal{F}$, $\mathcal{H}$, etc. A truth assignment is a map from variables to $\{0, 1\}$. Given a truth assignment, a clause is satisfied if at least one of its literals is assigned value 1; otherwise it is falsified. A formula is satisfied if all of its clauses are satisfied; otherwise it is falsified. If there exists no assignment that satisfies a CNF formula $\mathcal{F}$, then $\mathcal{F}$ is referred to as *unsatisfiable*. (Boolean) Satisfiability (SAT) is the decision problem for propositional formulas, i.e. to decide whether a given propositional formula is satisfiable. Since the paper only considers propositional formulas in CNF, throughout the paper SAT refers to the decision problem for propositional formulas in CNF.

To simplify modeling with propositional logic, one often represents more expressive constraints. Concrete examples are cardinality constraints and pseudo-Boolean constraints [17]. A cardinality constraint of the form $\sum x_i \leq k$ is referred to as an AtMost$k$ constraint, whereas a cardinality constraint of the form $\sum x_i \geq k$ is referred to as an AtLeast$k$ constraint. The study of propositional encodings of cardinality and pseudo-Boolean constraints is an area of active research [1,4,5,9,10,17,25,34,59,66,70].

A clause is Horn if it contains at most one positive literal. A Horn clause is a *goal* clause if it has no positive literals; otherwise it is a *definite* clause. The decision problem for Horn formulas is well-known to be in P, with linear time algorithms since the 80s [32,53]. A number of function problems defined on Horn formulas can be solved in polynomial time [49]. These include computing the lean kernel, finding a minimal unsatisfiable subformula and finding a maximal satisfiable subformula.

2.1   Propositional Encodings of the Pigeonhole Principle

The propositional encoding of the pigeonhole hole principle is well-known [27].

Definition 1 (Pigeonhole Principle, PHP [27]). *The pigeonhole principle states that if $m + 1$ pigeons are distributed by m holes, then at least one hole contains more than one pigeon. A more formal formulation is that there exists no injective function mapping from $\{1, 2, ..., m + 1\}$ to $\{1, 2, ..., m\}$, for $m \geq 1$.*

Propositional formulations of PHP encode the negation of the principle, and ask for an assignment such that the $m + 1$ pigeons are placed into m holes. The propositional encoding of the PHP_m^{m+1} problem can be derived as follows. Let the variables be x_{ij}, with $1 \leq i \leq m + 1, 1 \leq j \leq m$, with $x_{ij} = 1$ iff the ith pigeon is placed in the jth hole. The constraints are that each pigeon must be placed in at least one hole, and each hole must not have more than one pigeon.

$$\bigwedge_{i=1}^{m+1} \text{AtLeast1}(x_{i1}, \dots, x_{im}) \wedge \bigwedge_{j=1}^{m} \text{AtMost1}(x_{1j}, \dots, x_{m+1,j}) \tag{1}$$

An AtLeast1 constraint can be encoded with a single clause. For the AtMost1 constraint there are different encodings, including [17,34,66]. For example, the pairwise encoding [17] of $\mathsf{AtMost1}(x_{1j}, \ldots, x_{m+1,j})$ uses no auxiliary variables and the clauses $\wedge_{r=2}^{m+1} \wedge_{s=1}^{r-1} (\neg x_{rj} \vee \neg x_{sj})$. It is well-known that resolution has an exponential lower bound for PHP [11, 39, 62].

2.2   MaxSAT, MaxSAT Resolution and MaxSAT Algorithms

MaxSAT. For unsatisfiable formulas, the maximum satisfiability (MaxSAT) problem is to find an assignment that maximizes the number of satisfied clauses (given that not all clauses can be satisfied). There are different variants of the MaxSAT problem [17, Chap. 19]. Partial MaxSAT allows for *hard* clauses (which must be satisfied) and *soft* clauses (which represent a preference to satisfy those clauses). There are also weighted variants, in which soft clauses are given a weight, and for which hard clauses (if any) have a weight of $\top$ (meaning clauses that must be satisfied). The notation (c, w) will be used to represent a clause c with w denoting the cost of falsifying c. The paper considers partial MaxSAT instances, with hard clauses, for which $w = \top$, and soft clauses, for which $w = 1$. $\langle \mathcal{H}, \mathcal{S} \rangle$ is used to denote partial MaxSAT problems with sets of hard ($\mathcal{H}$) and soft ($\mathcal{S}$) clauses. In the paper, a MaxSAT *solution* represents either a maximum cardinality set of satisfied soft clauses or an assignment that satisfies all hard clauses and also maximizes (minimizes) the number of satisfied (falsified, resp.) soft clauses.

MaxSAT Resolution [18, 47]. In contrast with SAT, the MaxSAT resolution operation requires the introduction of additional clauses other than the resolvent, and resolved clauses cannot be resolved again. Let $(x \vee A, u)$ and $(\neg x \vee B, w)$ be two clauses, and let $m \triangleq \min(u, w)$, $u \ominus w \triangleq (u == \top) ? \top : u - w$, with $u \geq w$. The (non-clausal) MaxSAT resolution step [47] is shown in Table 1. (We could have used the clausal formulation [18], but it is more verbose and unnecessary for the purposes of the paper. It suffices to mention that clausal MaxSAT resolution adds at most $2n$ clauses at each resolution step, where the number of variables is n and the number of literals in each clause does not exceed n.) It is well-known that MaxSAT-resolution is unlikely to improve propositional resolution [18]. For the original PHP_m^{m+1} formulas, there are known exponential lower bounds on the size of deriving one empty clause by MaxSAT-resolution (given that the remaining clauses are satisfiable) [18, Corollary 18].

Table 1. Example MaxSAT-resolution steps

Clause 1	Clause 2	Derived Clauses
$(x \vee A, u)$	$(\neg x \vee B, w)$	$(A \vee B, m)$, $(x \vee A, u \ominus m)$, $(\neg x \vee B, w \ominus m)$, $(x \vee A \vee \neg B, m)$, $(\neg x \vee \neg A \vee B, m)$
$(x \vee A, 1)$	$(\neg x, \top)$	$(A, 1)$, $(\neg x, \top)$, $(\neg x \vee \neg A, 1)$

MaxSAT Algorithms. Many algorithms for MaxSAT have been proposed over the years [17, Chap. 19]. The most widely investigated can be broadly organized into branch and bound [17, Chap. 19], iterative-search [12,36,46], core-guided [2,36,50,51,54,55,57], and minimum hitting sets [29,64]. In most proposed algorithms, core-guided and minimum hitting sets MaxSAT algorithms iteratively détermine formulas to be unsatisfiable, until satisfiability is reached for a formula that relaxes clauses of minimum cost. This paper analyzes the operation of core-guided MaxSAT algorithms, concretely the MSU3 algorithm [50][2]. Moreover, and to our best knowledge, the relationship between core-guided MaxSAT algorithms and MaxSAT resolution was first investigated in [57].

2.3   Related Work

The complexity of resolution on pigeonhole formulas has been studied by different authors, e.g. see [11,27,39,58,62] and references therein, among others. It is well-known that for other proof systems, including cutting planes and extended resolution, PHP has polynomial proofs [6,20,21,26,28,65]. Different authors have looked into extending CDCL (and so resolution) with the goal of solving formulas for which resolution has known exponential lower bounds [7,12,13,16,37,38,40,44,45,68]. Some SAT solvers apply pattern matching techniques [14], but these are only effective for specific propositional encodings. Furthermore, there has been limited success in applying cutting planes and extended resolution in practical SAT solvers.

3   Reducing SAT to HornMaxSAT

The propositional satisfiability problem for CNF formulas can be reduced to HornMaxSAT, more concretely to the problem of deciding whether for some target Horn formula there exists an assignment that satisfies a given number of soft clauses.

Let $\mathcal{F}$ be a CNF formula, with N variables $\{x_1 \ldots, x_N\}$ and M clauses $\{c_1, \ldots, c_M\}$. Given $\mathcal{F}$, the reduction creates a Horn MaxSAT problem with hard clauses $\mathcal{H}$ and soft clauses $\mathcal{S}$, $\langle \mathcal{H}, \mathcal{S} \rangle = \mathsf{HEnc}(\mathcal{F})$. For each variable $x_i \in X$, create new variables p_i and n_i, where $p_i = 1$ iff $x_i = 1$, and $n_i = 1$ iff $x_i = 0$. Thus, we need a hard clause $(\neg p_i \lor \neg n_i)$, to ensure that we do not simultaneously assign $x_i = 1$ and $x_i = 0$. (Observe that the added clause is Horn.) This set of hard Horn clauses is referred to as $\mathcal{P}$. For each clause c_j, we require c_j to be satisfied, by requiring that one of its literals *not* to be falsified. For each literal x_i use $\neg n_i$, and for each literal $\neg x_i$ use $\neg p_i$. Thus, c_j is encoded with a new (hard) clause c'_j with the same number of literals as c_j, but with only negative literals on the p_i and n_i variables, and so the resulting clause is also Horn. The set of soft clauses $\mathcal{S}$ is given by (p_i) and (n_i) for each of the original variables x_i. If the resulting Horn formula has a HornMaxSAT solution with at least N variables assigned

---

[2] Different implementations of the MSU3 have been proposed over the years [2,50,51, 55], which often integrate different improvements. A well-known implementation of MSU3 is OpenWBO [51], one of the best MaxSAT solvers in the MaxSAT Evaluations since 2014.

value 1, then the original formula is satisfiable; otherwise the original formula is unsatisfiable. (Observe that, by construction, the HornMaxSAT solution cannot assign value 1 to more than N variables. Thus, unsatisfiability implies being unable to satisfy more than $N - 1$ soft clauses.) Clearly, the encoding outlined in this section can be the subject of different improvements.

The transformation proposed above can be related with the well-known dual-rail encoding, used in different settings [19,43,48,61,63]. To our best knowledge, the use of this encoding for deriving a pure Horn formula has not been proposed in earlier work.

Lemma 1. *Given $\langle \mathcal{H}, \mathcal{S} \rangle = \mathsf{HEnc}(\mathcal{F})$, there can be no more than N satisfied soft clauses.*

Lemma 2. *Let $\mathcal{F}$ have a satisfying assignment ν. Then, there exists an assignment that satisfies $\mathcal{H}$ and N soft clauses in $\langle \mathcal{H}, \mathcal{S} \rangle = \mathsf{HEnc}(\mathcal{F})$.*

Lemma 3. *Let ν' be an assignment that satisfies the clauses in $\mathcal{H}$ and N clauses in $\mathcal{S}$. Then there exists an assignment ν that satisfies $\mathcal{F}$.*

Theorem 1. *$\mathcal{F}$ is satisfiable if and only if there exists an assignment that satisfies $\mathcal{H}$ and N clauses in $\mathcal{S}$.*

Example 1 (Pigeonhole Principle). The reduction of SAT into HornMaxSAT can also be applied to the PHP_m^{m+1} problem. With each variable x_{ij}, $1 \leq i \leq m + 1, 1 \leq j \leq m$, we associate two new variables: n_{ij} and p_{ij}. The set of clauses $\mathcal{P}$ prevents a variable x_i from being assigned value 0 and 1 simultaneously: $\mathcal{P} = \{(\neg n_{ij} \vee \neg p_{ij}) \mid 1 \leq i \leq m + 1, 1 \leq j \leq m\}$. $\mathcal{L}_i$ represents the encoding of each AtLeast1 constraint, concretely $\mathcal{L}_i = (\neg n_{i1} \vee \ldots \vee \neg n_{im})$. $\mathcal{M}_j$ represents the encoding of each AtMost1 constraint, which will depend on the encoding used. The soft clauses $\mathcal{S}$ are given by,

$$\{(n_{11}), \ldots, (n_{1m}), \ldots, (n_{m+1\,1}), \ldots, (n_{m+1\,m}),$$
$$(p_{11}), \ldots, (p_{1m}), \ldots, (p_{m+1\,1}), \ldots, (p_{m+1\,m})\}$$

with $|\mathcal{S}| = 2m(m + 1)$. Thus, the complete reduction of PHP into MaxSAT becomes:

$$\mathsf{HEnc}\left(\mathrm{PHP}_m^{m+1}\right) \triangleq \langle \mathcal{H}, \mathcal{S} \rangle = \langle \wedge_{i=1}^{m+1} \mathcal{L}_i \wedge \wedge_{j=1}^{m} \mathcal{M}_j \wedge \mathcal{P}, \mathcal{S} \rangle \qquad (2)$$

Clearly, given $\mathcal{P}$, one cannot satisfy more the $m(m + 1)$ soft clauses. By Theorem 1, PHP_m^{m+1} is satisfiable if and only if there exists an assignment that satisfies the hard clauses $\mathcal{H}$ and $m(m + 1)$ soft clauses from $\mathcal{S}$.

4   Short MaxSAT Proofs for PHP

This section shows that the reduction of PHP_m^{m+1} to HornMaxSAT based on a dual-rail encoding enables both existing core-guided MaxSAT algorithms and also MaxSAT resolution, to prove in polynomial time that the original problem formulation[3] is unsatisfiable. Recall from Theorem 1, that PHP_m^{m+1} is satisfiable if and only if, given (2), there exists an assignment that satisfies $\mathcal{H}$ and $m(m+1)$ soft clauses in $\mathcal{S}$. This section shows that for both core-guided algorithms and for MaxSAT resolution, we can conclude in polynomial time that satisfying $\mathcal{H}$ requires falsifying at least $m(m+1)+1$ soft clauses, thus proving PHP_m^{m+1} to be unsatisfiable.

The results in this section should be contrasted with earlier work [18], which proves that MaxSAT resolution requires an exponentially large proof to produce an empty clause, this assuming the *original* propositional encoding for PHP_m^{m+1}.

4.1   A Polynomial Bound on Core-Guided MaxSAT Algorithms

This section shows that a core-guided MaxSAT algorithm will conclude in polynomial time that more than $m(m+1)$ clauses must be falsified, when the hard clauses are satisfied, thus proving the original PHP_m^{m+1} to be unsatisfiable. The analysis assumes the operation of basic core-guided algorithm, MSU3 [50], but similar analyses could be carried out for other families of core-guided algorithms[4].

The following observations about (2) are essential to prove the bound on the run time. First, the clauses in the $\mathcal{L}_i$ constraints do not share variables in common with the clauses in the $\mathcal{M}_j$ constraints. Second, each constraint $\mathcal{L}_i$ is of the form $(\neg n_{i1} \vee \ldots \vee \neg n_{im})$ and so its variables are disjoint from any other $\mathcal{L}_k$, $k \neq i$. Third, assuming a pairwise encoding, each constraint $\mathcal{M}_j$ is of the form $\wedge_{r=2}^{m+1} \wedge_{s=1}^{r-1} (\neg p_{rj} \vee \neg p_{sj})$, and so its variables are disjoint from any other $\mathcal{M}_l$, $l \neq j$. Since the sets of variables for each constraint are disjoint from the other sets of variables, we can exploit this partition of the clauses, and run a MaxSAT solver *separately* on each one. (Alternatively, we could assume the MSU3 MaxSAT algorithm to work with disjoint unsatisfiable cores.)

Table 2 summarizes the sequence of unit propagation steps that yields a lower bound on the number of falsified clauses larger than $m(m+1)$.

For each $\mathcal{L}_i$, the operation is summarized in the second row of Table 2. Unit propagation yields a conflict between m soft clauses and the corresponding hard clause. This means that at least one of these soft clauses must be falsified. Since there are $m+1$ constraints $\mathcal{L}_i$, defined on disjoint sets of variables, then each will contribute at least one falsified soft clause, which puts the lower bound on the number of falsified clauses at $m+1$.

For each $\mathcal{M}_j$ the operation is summarized in rows 3 to last of Table 2. Each row indicates a sequence of unit propagation steps that produces a conflict, each on a distinct set of soft clauses. Observe that each soft clause (p_{kj}), $k \geq 2$, induces

---

[3] This section studies the original *pairwise* encoding of PHP_m^{m+1}. However, a similar argument can be applied to PHP_m^{m+1} provided any encoding of AtMost1 constraints $\mathcal{M}_j$, as confirmed by the experimental results in Sect. 5.2.

[4] Basic knowledge of core-guided MaxSAT algorithms is assumed. The reader is referred to recent surveys for more information [2,55].

Table 2. Partitioned core-guided unit propagation steps

Con-straint	Hard clause(s)	Soft clause(s)	Relaxed clauses	Updated AtMost$k$ Constraints	LB increase
$\mathcal{L}_i$	$(\neg n_{i1} \vee \ldots \vee \neg n_{im})$	$(n_{i1}), \ldots, (n_{im})$	$(r_{il} \vee n_{i1}),$ $1 \leq l \leq m$	$\sum_{l=1}^{m} r_{il} \leq 1$	1
$\mathcal{M}_j$	$(\neg p_{1j} \vee \neg p_{2j})$	$(p_{1j}), (p_{2j})$	$(r_{1j} \vee p_{1j}),$ $(r_{2j} \vee p_{2j})$	$\sum_{l=1}^{2} r_{lj} \leq 1$	1
$\mathcal{M}_j$	$(\neg p_{1j} \vee \neg p_{3j}),$ $(\neg p_{2j} \vee \neg p_{3j}),$ $(r_{1j} \vee p_{1j}),$ $(r_{2j} \vee p_{2j}),$ $\sum_{l=1}^{2} r_{lj} \leq 1$	(p_{3j})	$(r_{3j} \vee p_{3j})$	$\sum_{l=1}^{3} r_{lj} \leq 2$	1
		$\ldots$			
$\mathcal{M}_j$	$(\neg p_{1j} \vee \neg p_{m+1j}), \ldots,$ $(\neg p_{mj} \vee \neg p_{m+1j}),$ $(r_{1j} \vee p_{1j}), \ldots,$ $(r_{mj} \vee p_{mj}),$ $\sum_{l=1}^{m} r_{lj} \leq m-1$	(p_{m+1j})	$(r_{m+1j} \vee p_{m+1j})$	$\sum_{l=1}^{m+1} r_{lj} \leq m$	1

a sequence of unit propagation steps, that causes the AtMost$\{k-1\}$ constraint to become inconsistent. Concretely, for iteration k (where row 3 corresponds to iteration 1), the sequence of unit propagation steps is summarized in Table 3[5].

Table 3. Analysis of $\mathcal{M}_j$, iteration k

Clauses	Unit Propagation
(p_{k+1j})	$p_{k+1j} = 1$
$(\neg p_{1j} \vee \neg p_{k+1j}), \ldots, (\neg p_{kj} \vee \neg p_{k+1j})$	$p_{1j} = \ldots = p_{kj} = 0$
$(r_{1j} \vee p_{1j}), \ldots, (r_{kj} \vee p_{kj})$	$r_{1j} = \ldots = r_{kj} = 1$
$\sum_{l=1}^{k} r_{lj} \leq k-1$	$\left(\sum_{l=1}^{k} r_{lj} \leq k-1\right) \vdash_1 \bot$

Since there are m such rows, then each $\mathcal{M}_j$ contributes at least m falsified soft clauses. Moreover, the number of $\mathcal{M}_j$ constraints is m, and so the $\mathcal{M}_j$ constraints increase the bound by $m \cdot m$.

Given the above, in total we are guaranteed to falsify at least $m+1+m \cdot m = m(m+1)+1$ clauses, thus proving that one cannot satisfy $m(m+1)$ soft clauses if the hard clauses are satisfied. In turn, this proves that the PHP$_m^{m+1}$ problem is unsatisfiable.

We can also measure the run time of the sequence of unit propagation steps. For each $\mathcal{L}_i$, the run time is $\mathcal{O}(m)$, and there will be m such unit propagation

---

[5] The notation $\Phi \vdash_1 \bot$ indicates that inconsistency (i.e. a falsified clause) is derived by unit propagation on the propositional encoding of Φ. This is the case with existing encodings of AtMost$k$ constraints.

steps, for a total $\mathcal{O}(m^2)$. For each $\mathcal{M}_j$ there will be m unit propagation steps, with run time between $\mathcal{O}(1)$ and $\mathcal{O}(m)$. Thus, the run time of the sequence of unit propagation steps for each $\mathcal{M}_j$ is $\mathcal{O}(m^2)$. Since there are m constraints $\mathcal{M}_j$, then the total run time is $\mathcal{O}(m^3)$.

Proposition 1. *Given* (2), *and for a core-guided MSU3-like MaxSAT solver, there is a sequence of unit propagation steps such that a lower bound of $m(m+1) + 1$ is computed in $\mathcal{O}(m^3)$ time.*

Moreover, it is important to observe that the unit propagation steps considered in the analysis above avoid the clauses in $\mathcal{P}$, i.e. only the clauses in $\mathcal{L}_i$, $\mathcal{M}_j$, $\mathcal{S}$, and relaxed clauses, are used for deriving the lower bound of $m(m+1) + 1$ on the minimum number of falsified soft clauses. As shown in Sect. 5.2, and for the concrete case of PHP, the clauses in $\mathcal{P}$ are unnecessary and actually impact negatively the performance of core-guided MaxSAT solvers. Finally, and although the proof above assumes an MSU3-like core-guided algorithm, similar ideas could be considered in the case of other variants of core-guided MaxSAT algorithms [2,36,55,57].

4.2 A Polynomial Bound on MaxSAT Resolution

We can now exploit the intuition from the previous section to identify the sequence of MaxSAT resolution steps that enable deriving $m(m+1) + 1$ empty clauses, thereby proving that *any* assignment that satisfies the hard clauses must falsify at least $m(m+1) + 1$ soft clauses, and therefore proving that the propositional encoding of PHP is unsatisfiable. As before, we assume that the pairwise encoding is used to encode each constraint $\mathcal{M}_j$. As indicated earlier in Sect. 2.2, we consider a simplified version of MaxSAT resolution [47], which is non-clausal. As explained below, this is not problematic, as just a few clauses are of interest. For the clausal version of MaxSAT resolution, the other clauses, which our analysis ignores, are guaranteed to be linear in the number of variables at each step, and will *not* be considered again.

Table 4 summarizes the essential aspects of the MaxSAT resolution steps used to derive $m(m+1) + 1$ empty clauses. (Also, Sect. 4.1 clarifies that the formula can be partitioned if $\mathcal{P}$ is ignored.) Similarly to the previous section, the $\mathcal{L}_i$ constraints serve to derive $m+1$ empty clauses, whereas each $\mathcal{M}_j$ constraint serves to derive m empty clauses. In total, we derive $m(m+1)+1$ empty clauses, getting the intended result.

As shown in Table 4, for each constraint $\mathcal{L}_i$, start by applying MaxSAT resolution between the hard clause $\mathcal{L}_i \triangleq (\neg n_{i1} \vee \ldots \vee \neg n_{im})$ and soft clause (n_{i1}) to get soft clause $(\neg n_{i2} \vee \ldots \vee \neg n_{im})$, and a few other clauses (which are irrelevant for our purposes). Next, apply $m - 1$ additional MaxSAT resolution steps, resolving soft clause $(\neg n_{ik} \vee \ldots \vee \neg n_{im})$ with soft clause (n_{ik}) to get soft clause $(\neg n_{ik+1} \vee \ldots \vee \neg n_{im})$. Clearly, the final MaxSAT resolution step will yield an empty clause. Therefore, over all $m + 1$ $\mathcal{L}_i$ constraints, we derive $m + 1$ empty clauses. Table 4 also illustrates the application of the MaxSAT resolution steps to the pairwise encoding of $\mathcal{M}_j$. At iteration i, with $2 \leq i \leq m + 1$, we apply

Table 4. Simplified MaxSAT resolution steps

Constraint	Clauses	Resulting clause(s)
$\mathcal{L}_i$	$(\neg n_{i1} \vee \ldots \vee \neg n_{im}, \top),$ $(n_{i1}, 1)$	$(\neg n_{i2} \vee \ldots \vee \neg n_{im}, 1),\ldots$
$\mathcal{L}_i$	$(\neg n_{i2} \vee \ldots \vee \neg n_{im}, 1),$ $(n_{i2}, 1)$	$(\neg n_{i3} \vee \ldots \vee \neg n_{im}, 1),\ldots$
	$\ldots$	
$\mathcal{L}_i$	$(\neg n_{im}, 1),$ $(n_{im}, 1)$	$(\bot, 1),\ldots$
$\mathcal{M}_j$	$(\neg p_{1j} \vee \neg p_{2j}, \top),$ $(p_{1j}, 1)$	$(\neg p_{2j}, 1), (\neg p_{1j} \vee \neg p_{2j}, \top), (p_{1j} \vee p_{2j}, 1)$
$\mathcal{M}_j$	$(\neg p_{2j}, 1),$ $(p_{2j}, 1)$	$(\bot, 1)$
$\mathcal{M}_j$	$(\neg p_{1j} \vee \neg p_{3j}, \top),$ $(p_{1j} \vee p_{2j}, 1)$	$(p_{2j} \vee \neg p_{3j}, 1), (\neg p_{1j} \vee \neg p_{3j}, \top),$ $(\neg p_{1j} \vee \neg p_{3j} \vee \neg p_{2j}, 1), (p_{1j} \vee p_{2j} \vee p_{3j}, 1)$
$\mathcal{M}_j$	$(\neg p_{2j} \vee \neg p_{3j}, \top),$ $(p_{2j} \vee \neg p_{3j}, 1)$	$(\neg p_{3j}, 1), (\neg p_{2j} \vee \neg p_{3j}, \top)$
$\mathcal{M}_j$	$(\neg p_{3j}, 1),$ $(p_{3j}, 1)$	$(\bot, 1)$
	$\ldots$	
$\mathcal{M}_j$	$(\neg p_{1j} \vee \neg p_{m+1j}, \top),$ $(p_{1j} \vee \ldots \vee p_{mj}, 1)$	$(p_{2j} \cdots p_{mj} \vee \neg p_{m+1j}, 1),\ldots$
$\mathcal{M}_j$	$(\neg p_{2j} \vee \neg p_{m+1j}, \top),$ $(p_{2j} \vee \ldots \vee p_{mj} \vee$ $\neg p_{m+1j}, 1)$	$(p_{3j} \cdots p_{mj} \vee \neg p_{m+1j}, 1),\ldots$
	$\ldots$	
$\mathcal{M}_j$	$(\neg p_{mj} \vee \neg p_{m+1j}, \top),$ $(p_{mj} \vee \neg p_{m+1j}, 1)$	$(\neg p_{m+1j}, 1),\ldots$
$\mathcal{M}_j$	$(p_{m+1j}, 1),$ $(\neg p_{m+1j}, 1)$	$(\bot, 1)$

i MaxSAT resolution steps to derive another empty clause. In total, we derive m empty clauses for each $\mathcal{M}_j$. An essential aspect is selecting the initial clause from which each sequence of MaxSAT resolution steps is executed. These reused clauses are highlighted in Table 4, and are crucial for getting the right sequence of MaxSAT resolution steps. For each $\mathcal{M}_j$, the MaxSAT resolution steps can be organized in m phases, each yielding an empty soft clause. For phase l, the previous phase $l - 1$ produces the clause $(p_{1j} \vee p_{2j} \vee \ldots \vee p_{lj}, 1)$, which is then iteratively simplified, using unit soft clauses, until the empty soft clause for phase l is derived. It should be noted that the first phase uses two unit soft clauses to produce $(p_{1j} \vee p_{2j}, 1)$, which is used in the second phase. As in Sect. 4.1, is immediate that each soft clause is *never* reused.

Regarding the run time complexity, observe that each MaxSAT resolution step runs in time linear on the number of literals in the clauses. The clauses in the problem formulation have no more than $\mathcal{O}(m)$ literals. This also holds true as MaxSAT resolution steps are applied. By analogy with the analysis of the core-guided algorithm, a total of $\mathcal{O}(m^2)$ empty soft clauses will be derived. From the analysis above, summarized in Table 4, deriving the $\mathcal{O}(m^2)$ empty clauses requires a total of $\mathcal{O}(m^3)$ MaxSAT resolution steps. For non-clausal MaxSAT resolution, since the number of generated (non-clausal) terms is constant for each MaxSAT resolution step, then the run time is $\mathcal{O}(m^3)$. In contrast, for clausal MaxSAT resolution [18, Definition 1], since the number of literals for each resolution step is $\mathcal{O}(m^2)$, then the run time becomes $\mathcal{O}(m^5)$.

Proposition 2. *For the HornMaxSAT encoding of* PHP_m^{m+1}, *there exists a polynomial sequence of MaxSAT resolution steps, each producing a number of constraints polynomial in the size of the problem formulation, that produces* $m(m+1)+1$ *soft empty clauses.*

4.3   Integration in SAT Solvers

This section shows that off-the-shelf MaxSAT solvers, which build on CDCL SAT solvers, can solve PHP_m^{m+1} in polynomial time, provided the right order of conflicts is chosen. This motivates integrating core-guided MaxSAT reasoning into SAT solvers. Similarly, one could consider integrating MaxSAT resolution (or a mix of both [57]) but, like resolution, MaxSAT resolution is harder to implement in practice. The proposed transformation can be applied on demand, and the operation of CDCL can be modified to integrate some form of core-guided reasoning. In contrast to other attempts at extending CDCL, the use of MaxSAT reasoning, will build also on CDCL itself.

MaxHS-like Horn MaxSAT. The reduction to Horn MaxSAT also motivates the development of dedicated MaxSAT solvers. One approach is to build upon MaxHS-solvers [29, 64], since in this case the SAT checks can be made to run in linear time, e.g. using LTUR [53]. Similar technique can possibly be integrated into SAT solvers.

Handling $\mathcal{P}$ clauses. The $\mathcal{P}$ clauses prevent assigning a variable simultaneously value 0 and value 1. As the analysis for the PHP instances suggests, and the experimental results confirm, these clauses can be responsible for non-polynomial run times. One can envision attempting to solve problems without considering $\mathcal{P}$ clauses, and adding these clauses on demand, as deemed necessary to block non-solutions. The operation is similar to the counterexample-guided abstraction refinement paradigm (CEGAR) [24].

5 Experimental Evaluation

5.1 Experimental Setup

To illustrate the main points of the paper, a number of solvers were tested. However and in order to save space, the results are detailed below only for some of the tested competitors.[6] The families of the evaluated solvers as well as the chosen representatives for the families are listed in Table 5. The family of CDCL SAT solvers comprises MiniSat 2.2 (*minisat*) and Glucose 3 (*glucose*) while the family of SAT solvers strengthened with the use of other powerful techniques (e.g. Gaussian elimination, GA and/or cardinality-based reasoning, CBR) includes lingeling (*lgl*) and CryptoMiniSat (*crypto*). The MaxSAT solvers include the known tools based on implicit minimum-size hitting set enumeration, i.e. MaxHS (*maxhs*) and LMHS (*lmhs*), and also a number of core-guided solvers shown to be best for industrial instances in a series of recent MaxSAT Evaluations[7], e.g. MSCG (*mscg*), OpenWBO16 (*wbo*) and WPM3 (*wpm3*), as well as the recent MaxSAT solver Eva500a (*eva*) based on MaxSAT resolution. Other competitors considered include CPLEX (*lp*), OPB solvers cdcl-cuttingplanes (*cc*) and Sat4j (*sat4j*) as well as a solver based on ZBDDs called ZRes (*zres*).

Table 5. Families of solvers considered in the evaluation (their best performing representatives are written in *italics*). *SAT+* stands for SAT strengthened with other techniques, *IHS MaxSAT* is for implicit hitting set based MaxSAT, *CG MaxSAT* is for core-guided MaxSAT, *MRes* is for MaxSAT resolution, *MIP* is for mixed integer programming, *OPB* is for pseudo-Boolean optimization, *BDD* is for binary decision diagrams.

SAT		SAT+		IHS MaxSAT		CG MaxSAT			MRes	MIP	OPB		BDD
minisat	*glucose*	*lgl*	crypto	*maxhs*	*lmhs*	*mscg*	*wbo*	wpm3	*eva*	*lp*	*cc*	sat4j	*zres*
[33]	[8]	[14,15]	[67,68]	[29–31]	[64]	[56]	[52]	[3]	[57]	[41]	[35]	[12]	[22]

Note that three configurations of CPLEX were tested: (1) the default configuration and the configurations used in (2) MaxHS and (3) LMHS. Given the overall performance, we decided to present the results for one best performing configuration, which turned out to be the default one. Also, the performance of CPLEX was measured for the following two types of LP instances: (1) the instances encoded to LP directly from the original CNF formulas (see *lp-cnf*) and (2) the instances obtained from the HornMaxSAT formulas (*lp-wcnf*). A similar remark can be made with regard to the cc solver: it can deal with the original CNF formulas as well as their OPB encodings (the corresponding configurations of the solver are *cc-cnf* and *cc-opb*[8], respectively).

---

[6] The discussion focuses on the results of the best performing *representatives*. Solvers that are missing in the discussion are meant to be "dominated" by their representatives.

[7] http://www.maxsat.udl.cat.

[8] The two tested versions of *cc-opb* behave almost identically with a minor advantage of linear search. As a result, *cc-opb* stands for the linear search version of the solver.

Regarding the IHS-based MaxSAT solvers, both MaxHS and LMHS implement the *Eq-Seeding* constraints [30]. Given that all soft clauses constructed by the proposed transformation are *unit* and that the set of all variables of Horn-MaxSAT formulas is *covered* by the soft clauses, these eq-seeding constraints replicate the complete MaxSAT formula on the MIP side. As a result, after all disjoint unsatisfiable cores are enumerated by MaxHS or LMHS, only one call to an MIP solver is needed to compute the optimum solution. In order to show the performance of an IHS-based MaxSAT solver with this feature disabled, we considered another configuration of LMHS (*lmhs-nes*).[9]

All the conducted experiments were performed in Ubuntu Linux on an Intel Xeon E5-2630 2.60 GHz processor with 64 GByte of memory. The time limit was set to 1800 s and the memory limit to 10 GByte for each individual process to run.

5.2   Experimental Results

The efficiency of the selected competitors was assessed on the benchmark suite consisting of 3 sets: (1) pigeonhole formulas (PHP) [27], (2) Urquhart formulas (*URQ*) [69], and (3) their combinations (*COMB*).

Pigeonhole principle benchmarks. The set of PHP formulas contains 2 families of benchmarks differing in the way AtMost1 constraints are encoded: (1) standard pairwise-encoded (*PHP-pw*) and (2) encoded with sequential counters [66] (*PHP-sc*). Each of the families contains 46 CNF formulas encoding the pigeonhole principle for 5 to 100 pigeons. Figure 1[10] shows the performance of the solver on sets PHP-pw and PHP-sc. As can be seen, the MaxSAT solvers (except *eva* and *wbo*) and also *lp-** are able to solve all instances. As expected, CDCL SAT solvers perform poorly for PHP with the exception of lingeling, which in some cases detects cardinality constraints in PHP-pw. However, disabling cardinality constraints reasoning or considering the PHP-sc benchmarks impairs its performance tremendously. Also note that we were unable to reproduce the performance of *zres* applied to PHP reported in [22].

On discarding $\mathcal{P}$ clauses. To confirm the conjecture that the $\mathcal{P}$ clauses can hamper a MaxSAT solver's ability to get *good* unsatisfiable cores, we also considered both PHP-pw and PHP-sc instances *without* the $\mathcal{P}$ clauses. Figure 2 compares the performance of the solvers working on PHP formulas w/ and w/o the $\mathcal{P}$ clauses. The lines with *(no P)* denote solvers working on the formulas w/o $\mathcal{P}$ (except *maxhs* and *lmhs* whose performance is not affected by removal of $\mathcal{P}$). As in Fig. 2b, the efficiency of *wbo* is improved by a few orders of magnitude if $\mathcal{P}$ clauses are discarded. Also, *mscg* gets about an order of magnitude performance improvement outperforming other solvers.

---

[9] We chose LMHS (not MaxHS) because it has a command-line option to disable eq-seeding.

[10] Note that all the shown cactus plots below scale the Y axis logarithmically.

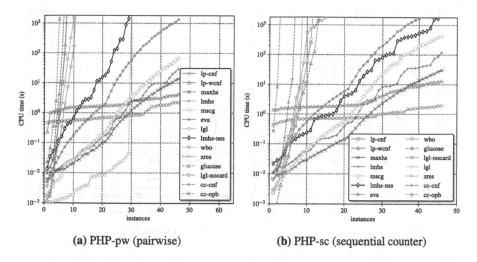

(a) PHP-pw (pairwise)   **(b)** PHP-sc (sequential counter)

Fig. 1. Performance of the considered solvers on pigeonhole formulas.

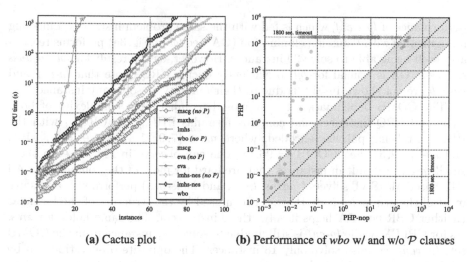

(a) Cactus plot   **(b)** Performance of *wbo* w/ and w/o $\mathcal{P}$ clauses

Fig. 2. Performance of MaxSAT solvers on PHP-pw $\cup$ PHP-sc w/ and w/o $\mathcal{P}$ clauses.

Urquhart benchmarks and combined instances. The URQ instances are known to be hard for resolution [69], but not for BDD-based reasoning [22]. Here, we follow the encoding of [22] to obtain the formulas of varying size given the parameter n of the encoder. In the experiments, we generated 3 CNF formulas for each n from 3 to 30 (i.e. URQ$_{n,i}$ for $n \in \{3, \ldots, 30\}$ and $i \in \{1, 2, 3\}$), which resulted in 84 instances. As expected, the best performance on the URQ instances is demonstrated by zres. Both *maxhs* and *lmhs* are not far behind. Note that both *maxhs* and *lmhs* do exactly 1 call to CPLEX (due to eq-seeding) after enumerating disjoint unsatisfiable cores. This contrasts sharply with the poor

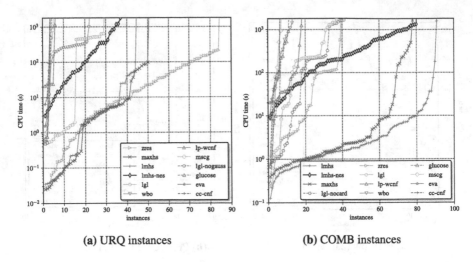

(a) URQ instances   (b) COMB instances

Fig. 3. Performance of the considered solvers on URQ and combined formulas.

performance of *lp-wcnf*, which is fed with the same problem instances. Lingeling if augmented with Gaussian elimination (GA, see *lgl* in Fig. 3a) performs reasonably well being able to solve 29 instances. However, as the result for *lgl-nogauss* suggests, GA is crucial for *lgl* to efficiently decide URQ. Note that *lp-cnf* and *cc-opb* are not shown in Fig. 3a due to their inability to solve any instance.

The COMB benchmark set is supposed to inherit the complexity of both PHP and URQ instances and contains formulas $PHP_m^{m+1} \vee URQ_{n,i}$ with the PHP part being pairwise-encoded, where $m \in \{7, 9, 11, 13\}$, $n \in \{3, \ldots, 10\}$, and $i \in \{1, 2, 3\}$, i.e. $|COMB| = 96$. As one can observe in Fig. 3b, even these small m and n result in instances that are hard for most of the competitors. All IHS-based MaxSAT solvers (*maxhs*, *lmhs*, and *lmhs-nes*) perform well and solve most of the instances. Note that *lgl* is confused by the structure of the formulas (neither CBR nor GA helps it solve these instances). The same holds for *zres*. As for CPLEX, while *lp-cnf* is still unable to solve any instance from the COMB set, *lp-wcnf* can also solve only 18 instances. The opposite observation can be made for *cc-cnf* and *cc-opb*.

Summary. As shown in Table 6, given all the considered benchmarks sets, the proposed problem transformation and the follow-up IHS-based MaxSAT solving can cope with by far the largest number of instances overall (see the data for *maxhs*, *lmhs*, and *lmhs-nes*). The core-guided and also resolution based MaxSAT solvers generally perform well on the pigeonhole formulas (except *wbo*, and this has to be investigated further), which supports the theoretical claims of papers. However, using them does not help solving the URQ and also COMB benchmarks. Also, as shown in Fig. 2, the $\mathcal{P}$ clauses can be harmful for MaxSAT solvers. As expected, SAT solvers cannot deal with most of the considered formulas as long as they do not utilize more powerful reasoning (e.g. GA or CBR). However, and as the COMB instances demonstrate, it is easy to construct instances that

Table 6. Number of solved instances per solver.

	glucose	lgl	lgl-no[a]	maxhs	lmhs	lmhs-nes	mscg	wbo	eva	lp-cnf	lp-wcnf	cc-cnf	cc-opb	zres
PHP-pw (46)	7	29	7	**46**	**46**	29	**46**	10	**46**	**46**	**46**	6	5	10
PHP-sc (46)	13	11	11	**46**	**46**	45	**46**	15	40	**46**	**46**	6	2	8
URQ (84)	3	29	4	50	44	37	5	22	3	0	6	3	0	**84**
COMB (96)	11	37	41	78	**91**	80	7	13	6	0	18	6	0	39
Total (272)	34	106	63	220	**227**	191	104	60	95	92	116	21	7	141

[a]This represents *lgl-nogauss* for URQ and *lgl-nocard* for PHP-pw, PHP-sc, and COMB.

are hard for the state-of-the-art SAT solvers strengthened with GA and CBR. Finally, one should note the performance gap between *maxhs* (also *lmhs*) and *lp-wcnf* given that they solve the same instances by one call to the same MIP solver with the only difference being the disjoint cores precomputed by *maxhs* and *lmhs*.

6   Conclusions and Research Directions

Resolution is at the core of CDCL SAT solving, but it also represents its Achilles' heel. Many crafted formulas are known to be hard for resolution, with pigeonhole formulas representing a well-known example [27]. More importantly, some of these examples can occur naturally in some practical settings. In the context of MaxSAT, researchers have proposed a dedicated form of resolution, i.e. MaxSAT resolution [18,47], which was also shown not to be more powerful than propositional resolution [18] for the concrete case of pigeonhole formulas [27].

This paper proposes a general transformation for CNF formulas, by encoding the SAT decision problem as a MaxSAT problem over Horn formulas. The transformation is based on the well-known dual-rail encoding, but it is modified such that all clauses are Horn. More importantly, the paper shows that, on this modified formula, MaxSAT resolution can identify in polynomial time a large enough number of empty soft clauses such that this number implies the unsatisfiability of the original pigeonhole formula. Furthermore, the paper shows that the same argument can be used to prove a polynomial run time for the well-known class of core-guided MaxSAT solvers [55]. Experimental results, obtained on formulas known to be hard for SAT solvers, show that different families of MaxSAT solvers perform far better than the best performing SAT solvers, and also ILP solvers, on these instances. Future work will investigate effective mechanisms for integrating Horn MaxSAT problem transformation and MaxSAT reasoning techniques into SAT solvers. In contrast to cutting planes or extended resolution, MaxSAT algorithms already build on CDCL SAT solvers; this is expected to facilitate integration. Another research direction is to investigate similar transformations for the many other examples for which resolution has exponential lower bounds, but also when to opt to apply such transformations.

References

1. Abío, I., Nieuwenhuis, R., Oliveras, A., Rodríguez-Carbonell, E.: BDDs for pseudo-boolean constraints – revisited. In: Sakallah, K.A., Simon, L. (eds.) SAT 2011. LNCS, vol. 6695, pp. 61–75. Springer, Heidelberg (2011). doi:10.1007/978-3-642-21581-0_7

2. Ansótegui, C., Bonet, M.L., Levy, J.: SAT-based MaxSAT algorithms. Artif. Intell. **196**, 77–105 (2013)

3. Ansótegui, C., Didier, F., Gabàs, J.: Exploiting the structure of unsatisfiable cores in MaxSAT. In: IJCAI, pp. 283–289 (2015)

4. Asín, R., Nieuwenhuis, R., Oliveras, A., Rodríguez-Carbonell, E.: Cardinality networks and their applications. In: Kullmann, O. (ed.) SAT 2009. LNCS, vol. 5584, pp. 167–180. Springer, Heidelberg (2009). doi:10.1007/978-3-642-02777-2_18

5. Asín, R., Nieuwenhuis, R., Oliveras, A., Rodríguez-Carbonell, E.: Cardinality networks: a theoretical and empirical study. Constraints **16**(2), 195–221 (2011)

6. Atserias, A., Kolaitis, P.G., Vardi, M.Y.: Constraint propagation as a proof system. In: Wallace, M. (ed.) CP 2004. LNCS, vol. 3258, pp. 77–91. Springer, Heidelberg (2004). doi:10.1007/978-3-540-30201-8_9

7. Audemard, G., Katsirelos, G., Simon, L.: A restriction of extended resolution for clause learning SAT solvers. In: AAAI (2010)

8. Audemard, G., Lagniez, J.-M., Simon, L.: Improving glucose for incremental SAT solving with assumptions: application to MUS extraction. In: Järvisalo, M., Van Gelder, A. (eds.) SAT 2013. LNCS, vol. 7962, pp. 309–317. Springer, Heidelberg (2013). doi:10.1007/978-3-642-39071-5_23

9. Bailleux, O., Boufkhad, Y.: Efficient CNF encoding of boolean cardinality constraints. In: Rossi, F. (ed.) CP 2003. LNCS, vol. 2833, pp. 108–122. Springer, Heidelberg (2003). doi:10.1007/978-3-540-45193-8_8

10. Bailleux, O., Boufkhad, Y., Roussel, O.: New encodings of pseudo-boolean constraints into CNF. In: Kullmann, O. (ed.) SAT 2009. LNCS, vol. 5584, pp. 181–194. Springer, Heidelberg (2009). doi:10.1007/978-3-642-02777-2_19

11. Beame, P., Pitassi, T.: Simplified and improved resolution lower bounds. In: FOCS, pp. 274–282 (1996)

12. Berre, D.L., Parrain, A.: The Sat4j library, release 2.2. JSAT **7**(2–3), 59–64 (2010)

13. Biere, A.: Picosat essentials. JSAT **4**(2–4), 75–97 (2008)

14. Biere, A.: Lingeling, plingeling and treengeling entering the SAT competition 2013. In: Balint, A., Belov, A., Heule, M., Järvisalo, M. (eds.) Proceedings of SAT Competition 2013, vol. B-2013-1, pp. 51–52. Department of Computer Science Series of Publications B, University of Helsinki (2013)

15. Biere, A.: Lingeling essentials, a tutorial on design and implementation aspects of the SAT solver lingeling. In: Pragmatics of SAT Workshop, p. 88 (2014)

16. Biere, A., Berre, D., Lonca, E., Manthey, N.: Detecting cardinality constraints in CNF. In: Sinz, C., Egly, U. (eds.) SAT 2014. LNCS, vol. 8561, pp. 285–301. Springer, Cham (2014). doi:10.1007/978-3-319-09284-3_22

17. Biere, A., Heule, M., van Maaren, H., Walsh, T. (eds.): Handbook of Satisfiability. Frontiers in Artificial Intelligence and Applications, vol. 185. IOS Press, Amsterdam (2009)

18. Bonet, M.L., Levy, J., Manyà, F.: Resolution for Max-SAT. Artif. Intell. **171**(8–9), 606–618 (2007)

19. Bryant, R.E., Beatty, D., Brace, K., Cho, K., Sheffler, T.: COSMOS: a compiled simulator for MOS circuits. In: DAC, pp. 9–16 (1987)

20. Buss, S.R.: Polynomial size proofs of the propositional pigeonhole principle. J. Symb. Log. **52**(4), 916–927 (1987)
21. Buss, S.R., Turán, G.: Resolution proofs of generalized pigeonhole principles. Theor. Comput. Sci. **62**(3), 311–317 (1988)
22. Chatalic, P., Simon, L.: Multiresolution for SAT checking. Int. J. Artif. Intell. Tools **10**(4), 451–481 (2001)
23. Chvátal, V., Szemerédi, E.: Many hard examples for resolution. J. ACM **35**(4), 759–768 (1988)
24. Clarke, E.M., Grumberg, O., Jha, S., Lu, Y., Veith, H.: Counterexample-guided abstraction refinement for symbolic model checking. J. ACM **50**(5), 752–794 (2003)
25. Codish, M., Zazon-Ivry, M.: Pairwise cardinality networks. In: Clarke, E.M., Voronkov, A. (eds.) LPAR 2010. LNCS (LNAI), vol. 6355, pp. 154–172. Springer, Heidelberg (2010). doi:10.1007/978-3-642-17511-4_10
26. Cook, S.A.: A short proof of the pigeon hole principle using extended resolution. ACM SIGACT News **8**(4), 28–32 (1976)
27. Cook, S.A., Reckhow, R.A.: The relative efficiency of propositional proof systems. J. Symb. Log. **44**(1), 36–50 (1979)
28. Cook, W.J., Coullard, C.R., Turán, G.: On the complexity of cutting-plane proofs. Discrete Appl. Math. **18**(1), 25–38 (1987)
29. Davies, J., Bacchus, F.: Solving MAXSAT by solving a sequence of simpler SAT instances. In: Lee, J. (ed.) CP 2011. LNCS, vol. 6876, pp. 225–239. Springer, Heidelberg (2011). doi:10.1007/978-3-642-23786-7_19
30. Davies, J., Bacchus, F.: Exploiting the power of MIP solvers in MAXSAT. In: Järvisalo, M., Van Gelder, A. (eds.) SAT 2013. LNCS, vol. 7962, pp. 166–181. Springer, Heidelberg (2013). doi:10.1007/978-3-642-39071-5_13
31. Davies, J., Bacchus, F.: Postponing optimization to speed up MAXSAT solving. In: Schulte, C. (ed.) CP 2013. LNCS, vol. 8124, pp. 247–262. Springer, Heidelberg (2013). doi:10.1007/978-3-642-40627-0_21
32. Dowling, W.F., Gallier, J.H.: Linear-time algorithms for testing the satisfiability of propositional Horn formulae. J. Log. Program. **1**(3), 267–284 (1984)
33. Eén, N., Sörensson, N.: An extensible SAT-solver. In: Giunchiglia, E., Tacchella, A. (eds.) SAT 2003. LNCS, vol. 2919, pp. 502–518. Springer, Heidelberg (2004). doi:10.1007/978-3-540-24605-3_37
34. Eén, N., Sörensson, N.: Translating pseudo-Boolean constraints into SAT. JSAT **2**(1–4), 1–26 (2006)
35. Jan Elffers' personal webpage. http://www.csc.kth.se/~elffers
36. Fu, Z., Malik, S.: On solving the partial MAX-SAT problem. In: Biere, A., Gomes, C.P. (eds.) SAT 2006. LNCS, vol. 4121, pp. 252–265. Springer, Heidelberg (2006). doi:10.1007/11814948_25
37. Goldberg, E.: Testing satisfiability of CNF formulas by computing a stable set of points. In: Voronkov, A. (ed.) CADE 2002. LNCS (LNAI), vol. 2392, pp. 161–180. Springer, Heidelberg (2002). doi:10.1007/3-540-45620-1_15
38. Goldberg, E.: Testing satisfiability of CNF formulas by computing a stable set of points. Ann. Math. Artif. Intell. **43**(1), 65–89 (2005)
39. Haken, A.: The intractability of resolution. Theor. Comput. Sci. **39**, 297–308 (1985)
40. Huang, J.: Extended clause learning. Artif. Intell. **174**(15), 1277–1284 (2010)
41. IBM ILOG: CPLEX optimizer 12.7.0 (2016). http://www-01.ibm.com/software/commerce/optimization/cplex-optimizer
42. Ignatiev, A. Morgado, A., Marques-Silva, J.: On tackling the limits of resolution in SAT solving. CoRR, abs/1705.01477 (2017). https://arxiv.org/abs/1705.01477

43. Jabbour, S., Marques-Silva, J., Sais, L., Salhi, Y.: Enumerating prime implicants of propositional formulae in conjunctive normal form. In: Fermé, E., Leite, J. (eds.) JELIA 2014. LNCS (LNAI), vol. 8761, pp. 152–165. Springer, Cham (2014). doi:10.1007/978-3-319-11558-0_11

44. Jovanović, D., Moura, L.: Cutting to the chase solving linear integer arithmetic. In: Bjørner, N., Sofronie-Stokkermans, V. (eds.) CADE 2011. LNCS (LNAI), vol. 6803, pp. 338–353. Springer, Heidelberg (2011). doi:10.1007/978-3-642-22438-6_26

45. Jovanovic, D., de Moura, L.M.: Cutting to the chase - solving linear integer arithmetic. J. Autom. Reason. 51(1), 79–108 (2013)

46. Koshimura, M., Zhang, T., Fujita, H., Hasegawa, R.: QMaxSAT: a partial MaxSAT solver. JSAT 8(1/2), 95–100 (2012)

47. Larrosa, J., Heras, F., de Givry, S.: A logical approach to efficient Max-SAT solving. Artif. Intell. 172(2–3), 204–233 (2008)

48. Manquinho, V.M., Flores, P.F., Silva, J.P.M., Oliveira, A.L.: Prime implicant computation using satisfiability algorithms. In: ICTAI, pp. 232–239 (1997)

49. Marques-Silva, J., Ignatiev, A., Mencía, C., Peñaloza, R.: Efficient reasoning for inconsistent horn formulae. In: Michael, L., Kakas, A. (eds.) JELIA 2016. LNCS (LNAI), vol. 10021, pp. 336–352. Springer, Cham (2016). doi:10.1007/978-3-319-48758-8_22

50. Marques-Silva, J., Planes, J.: On using unsatisfiability for solving maximum satisfiability. CoRR, abs/0712.1097 (2007). https://arxiv.org/abs/0712.1097

51. Martins, R., Joshi, S., Manquinho, V., Lynce, I.: Incremental cardinality constraints for MaxSAT. In: O'Sullivan, B. (ed.) CP 2014. LNCS, vol. 8656, pp. 531–548. Springer, Cham (2014). doi:10.1007/978-3-319-10428-7_39

52. Martins, R., Manquinho, V., Lynce, I.: Open-WBO: a modular MaxSAT solver. In: Sinz, C., Egly, U. (eds.) SAT 2014. LNCS, vol. 8561, pp. 438–445. Springer, Cham (2014). doi:10.1007/978-3-319-09284-3_33

53. Minoux, M.: LTUR: a simplified linear-time unit resolution algorithm for Horn formulae and computer implementation. Inf. Process. Lett. 29(1), 1–12 (1988)

54. Morgado, A., Dodaro, C., Marques-Silva, J.: Core-guided MaxSAT with soft cardinality constraints. In: O'Sullivan, B. (ed.) CP 2014. LNCS, vol. 8656, pp. 564–573. Springer, Cham (2014). doi:10.1007/978-3-319-10428-7_41

55. Morgado, A., Heras, F., Liffiton, M.H., Planes, J., Marques-Silva, J.: Iterative and core-guided MaxSAT solving: a survey and assessment. Constraints 18(4), 478–534 (2013)

56. Morgado, A., Ignatiev, A., Marques-Silva, J.: MSCG: robust core-guided MaxSAT solving. JSAT 9, 129–134 (2015)

57. Narodytska, N., Bacchus, F.: Maximum satisfiability using core-guided MaxSAT resolution. In: AAAI, pp. 2717–2723 (2014)

58. Nordström, J.: On the interplay between proof complexity and SAT solving. SIGLOG News 2(3), 19–44 (2015)

59. Ogawa, T., Liu, Y., Hasegawa, R., Koshimura, M., Fujita, H.: Modulo based CNF encoding of cardinality constraints and its application to MaxSAT solvers. In: ICTAI, pp. 9–17 (2013)

60. Pipatsrisawat, K., Darwiche, A.: On the power of clause-learning SAT solvers as resolution engines. Artif. Intell. 175(2), 512–525 (2011)

61. Previti, A. Ignatiev, A. Morgado, A., Marques-Silva, J.: Prime compilation of nonclausal formulae. In: IJCAI, pp. 1980–1988 (2015)

62. Razborov, A.A.: Proof complexity of pigeonhole principles. In: Kuich, W., Rozenberg, G., Salomaa, A. (eds.) DLT 2001. LNCS, vol. 2295, pp. 100–116. Springer, Heidelberg (2002). doi:10.1007/3-540-46011-X_8

63. Roorda, J.-W., Claessen, K.: A new SAT-based algorithm for symbolic trajectory evaluation. In: Borrione, D., Paul, W. (eds.) CHARME 2005. LNCS, vol. 3725, pp. 238–253. Springer, Heidelberg (2005). doi:10.1007/11560548_19
64. Saikko, P., Berg, J., Järvisalo, M.: LMHS: a SAT-IP hybrid MaxSAT solver. In: Creignou, N., Le Berre, D. (eds.) SAT 2016. LNCS, vol. 9710, pp. 539–546. Springer, Cham (2016). doi:10.1007/978-3-319-40970-2_34
65. Segerlind, N.: The complexity of propositional proofs. Bul. Symb. Logic **13**(4), 417–481 (2007)
66. Sinz, C.: Towards an optimal CNF encoding of boolean cardinality constraints. In: Beek, P. (ed.) CP 2005. LNCS, vol. 3709, pp. 827–831. Springer, Heidelberg (2005). doi:10.1007/11564751_73
67. Soos, M.: Enhanced Gaussian elimination in DPLL-based SAT solvers. In: POS@SAT, pp. 2–14 (2010)
68. Soos, M., Nohl, K., Castelluccia, C.: Extending SAT solvers to cryptographic problems. In: Kullmann, O. (ed.) SAT 2009. LNCS, vol. 5584, pp. 244–257. Springer, Heidelberg (2009). doi:10.1007/978-3-642-02777-2_24
69. Urquhart, A.: Hard examples for resolution. J. ACM **34**(1), 209–219 (1987)
70. Warners, J.P.: A linear-time transformation of linear inequalities into conjunctive normal form. Inf. Process. Lett. **68**(2), 63–69 (1998)

Improving MCS Enumeration via Caching

Alessandro Previti[1(✉)], Carlos Mencía[2], Matti Järvisalo[1],
and Joao Marques-Silva[3]

[1] HIIT, Department of Computer Science, University of Helsinki, Helsinki, Finland
alessandro.previti@helsinki.fi
[2] Department of Computer Science, University of Oviedo, Gijón, Spain
[3] LASIGE, Faculty of Science, University of Lisbon, Lisbon, Portugal

Abstract. Enumeration of minimal correction sets (MCSes) of conjunctive normal form formulas is a central and highly intractable problem in infeasibility analysis of constraint systems. Often complete enumeration of MCSes is impossible due to both high computational cost and worst-case exponential number of MCSes. In such cases partial enumeration is sought for, finding applications in various domains, including axiom pinpointing in description logics among others. In this work we propose caching as a means of further improving the practical efficiency of current MCS enumeration approaches, and show the potential of caching via an empirical evaluation.

1 Introduction

Minimal correction sets (MCSes) of an over-constrained system are subset-minimal sets of constraints whose removal restores the consistency of the system [6]. In terms of unsatisfiable conjunctive normal form (CNF) propositional formulas, the focus of this work, MCSes are hence minimal sets of clauses such that, once removed, the rest of the formula is satisfiable. Due to the generality of the notion, MCSes find applications in various domains where understanding infeasibility is a central problem, ranging from minimal model computation and model-based diagnosis to interactive constraint satisfaction and configuration [17], as well as ontology debugging and axiom pinpointing in description logics [1].

On a fundamental level, MCSes are closely related to other fundamental notions in infeasibility analysis. These include maximal satisfiable subsets (MSSes), which represent the complement notion of MCSes (sometimes referred to as co-MSSes [11]), and minimally unsatisfiable subsets (MUSes), with the well-known minimal hitting set duality providing a tight connection between MCSes and MUSes [4,6,26]. Furthermore, MCSes are strongly related to maximum satisfiability (MaxSAT), the clauses satisfied in an optimal MaxSAT solution being

A. Previti and M. Järvisalo were supported by Academy of Finland (grants 251170 COIN, 276412, and 284591) and the Research Funds of the University of Helsinki. C. Mencía was supported by grant TIN2016-79190-R. J. Marques-Silva was supported by FCT funding of LASIGE Research Unit, ref. UID/CEC/00408/2013.

© Springer International Publishing AG 2017
S. Gaspers and T. Walsh (Eds.): SAT 2017, LNCS 10491, pp. 184–194, 2017.
DOI: 10.1007/978-3-319-66263-3_12

the residual formula after removing a smallest (minimum-weight) MCS over the soft clauses. Not surprisingly, MCS extraction surpasses in terms of computational complexity the task of satisfiability checking, deciding whether a given subset of clauses of a CNF formula is an MCS being DP-complete [7]. Despite this, and on the other hand motivated by the various applications and fundamental connections, several algorithms for extracting an MCS of a given CNF formula have been recently proposed [3,11,17–20,22,24], iteratively using Boolean satisfiability (SAT) solvers as the natural choice for the underlying practical NP oracle.

In this work we focus on the computationally more challenging task of MCS enumeration. Complete enumeration of MCSes is often impossible due to both high computational cost and the worst-case exponential number of MCSes. In such cases partial enumeration is sought for, which finds many application domains, including axiom pinpointing in description logics [1] among others.

Instead of proposing a new algorithm for MCS enumeration, we propose the use of *caching* as a means of further improving the scalability of current state-of-the-art MCS enumeration algorithms. Caching (or memoization) is of course a well-known general concept, and has been successfully applied in speeding up procedures for other central problems related to satisfiability. A prime example is the use of subformula caching in the context of the #P-complete model counting problem [2,5,12,13,27,30]. Similarly, clause learning in CDCL SAT solvers [23, 28] can be viewed as a caching mechanism where learned clauses summarize and prevent previously identified conflicts.

In more detail, we propose caching unsatisfiable cores met during search within SAT-based MCS enumeration algorithms. Putting this idea into practice, we show that core caching has clear potential in scaling up MCS enumeration, especially for those instances whose extraction of a single MCS is not trivial. In terms of related work, to the best of our knowledge the use of caching to scale up MCS enumeration has not been previously proposed or studied. Partial MUS enumerators (e.g. [14,31]) store MUSes and MCSes in order to exploit hitting set duality and enumerate both. In contrast, we use caching to avoid potentially hard calls to a SAT solver.

The rest of this paper is organized as follows. In Sect. 2 we overview necessary preliminaries and notation used throughout, and in Sect. 3 provide an overview of MCS extraction and enumeration algorithms. We propose caching as a means of improving MCS enumeration in Sect. 4, and, before conclusions, present empirical results on the effects of using this idea in practice in Sect. 5.

2   Preliminaries

We consider propositional formulas in conjunctive normal form (CNF). A CNF formula $\mathcal{F}$ over a set of Boolean variables $X = \{x_1, ..., x_n\}$ is a conjunction of clauses $(c_1 \wedge ... \wedge c_m)$. A clause c_i is a disjunction of literals $(l_{i,1} \vee ... \vee l_{i,k_i})$ and a literal l is either a variable x or its negation $\neg x$. We refer to the set of literals appearing in $\mathcal{F}$ as $L(\mathcal{F})$. CNF formulas can be alternatively represented as sets of clauses, and clauses as sets of literals. Unless explicitly specified, formulas and clauses are assumed to be represented as sets.

A truth assignment, or interpretation, is a mapping $\mu : X \rightarrow \{0,1\}$. If each of the variables in X is assigned a truth value, μ is a *complete* assignment. Interpretations can be also seen as conjunctions or sets of literals. Truth valuations are lifted to clauses and formulas as follows: μ satisfies a clause c if it contains at least one of its literals, whereas μ falsifies c if it contains the complements of all its literals. Given a formula $\mathcal{F}$, μ satisfies $\mathcal{F}$ (written $\mu \models \mathcal{F}$) if it satisfies all its clauses, in which case μ is a *model* of $\mathcal{F}$.

Given two formulas $\mathcal{F}$ and $\mathcal{G}$, $\mathcal{F}$ entails $\mathcal{G}$ (written $\mathcal{F} \models \mathcal{G}$) if and only if each model of $\mathcal{F}$ is also a model of $\mathcal{G}$. A formula $\mathcal{F}$ is satisfiable ($\mathcal{F} \not\models \bot$) if it has a model, and otherwise unsatisfiable ($\mathcal{F} \models \bot$). SAT is the NP-*complete* [8] decision problem of determining the satisfiability of a given propositional formula.

The following definitions give central notions of subsets of an unsatisfiable formula $\mathcal{F}$ in terms of (set-wise) minimal unsatisfiability and maximal satisfiability [15,17].

Definition 1. $\mathcal{M} \subseteq \mathcal{F}$ *is a* minimally unsatisfiable subset *(MUS) of $\mathcal{F}$ if and only if $\mathcal{M}$ is unsatisfiable and $\forall c \in \mathcal{M}, \mathcal{M} \setminus \{c\}$ is satisfiable.*

Definition 2. $\mathcal{C} \subseteq \mathcal{F}$ *is a* minimal correction subset *(MCS) if and only if $\mathcal{F} \setminus \mathcal{C}$ is satisfiable and $\forall c \in \mathcal{C}, \mathcal{F} \setminus (\mathcal{C} \setminus \{c\})$ is unsatisfiable.*

Definition 3. $\mathcal{S} \subseteq \mathcal{F}$ *is a* maximal satisfiable subset *(MSS) if and only if $\mathcal{S}$ is satisfiable and $\forall c \in \mathcal{F} \setminus \mathcal{S}, \mathcal{S} \cup \{c\}$ is unsatisfiable.*

Note that an MSS is the set-complement of an MCS. MUSes and MCSes are closely related by the well-known hitting set duality [4,6,26,29]: Every MCS (MUS) is an irreducible hitting set of all MUSes (MCSes) of the formula. In the worst case, there can be an exponential number of MUSes and MCSes [15,25]. Besides, MCSes are related to the maximum satisfiability (MaxSAT) problem, which consists in finding an assignment satisfying as many clauses as possible; a smallest MCS (i.e., largest MSS) is the set of clauses left unsatisfied by some optimal MaxSAT solution.

Given the practical significance of handling soft constraints [21], we consider that formulas may be partitioned into sets of hard and soft clauses, i.e., $\mathcal{F} = \mathcal{F}_H \cup \mathcal{F}_S$. Hard clauses must be satisfied, while soft clauses can be relaxed if necessary. Thus, an MCS will be a subset of $\mathcal{F}_S$.

The following simple proposition will be useful in the remainder of this paper.

Proposition 1. *Let $\mathcal{M}$ be an unsatisfiable formula. Then, for any $\mathcal{M}' \supseteq \mathcal{M}$ we have that also $\mathcal{M}'$ is unsatisfiable.*

3   MCS Extraction and Enumeration

In this section we overview the state-of-the-art MCS enumeration algorithms. These algorithms work on a formula $\mathcal{F} = \mathcal{F}_H \cup \mathcal{F}_S$ partitioned into hard and soft clauses, $\mathcal{F}_H$ and $\mathcal{F}_S$, respectively. The hard clauses are added as such to a SAT solver. Each soft clause c_i is extended, or reified, with a fresh selector (or assumption) variable s_i, i.e., soft clause c_i results in the clause $(\neg s_i \vee c_i)$, before adding them to the SAT solver. The use of selector variables is a standard technique used to add and remove clauses, enabling incremental SAT solving. Selector variables are set as assumptions at the beginning of each call to the SAT solver in order to activate (add) or deactivate (remove) a clause. In particular, if s_i is set to 1, then the associated clause is activated. If s_i is set to 0, then c_i is deactivated. The addition of the selector variables makes $\mathcal{F}$ satisfiable (provided that $\mathcal{F}_H$ is satisfiable). When all the selector variables s_i are set to 1, the result is the original formula $\mathcal{F}$. MCS algorithms use the selector variables as assumptions for selecting subsets of $\mathcal{F}_S$ over which to check satisfiability together with the hard clauses. We will refer to the subset of soft clauses of $\mathcal{F}_S$ identified by a set of selector variables together with the hard clauses as the *induced formula*. In presenting the algorithms, we will avoid referring explicitly to selector variables; rather, we identify a formula $\mathcal{F}$ with all selector variables of the soft clauses in $\mathcal{F}_S$ being set to 1.

State-of-the-art MCS extraction algorithms rely on making a sequence of calls to a SAT solver that is used as a witness NP oracle. The solver is queried a number of times on subformulas of the unsatisfiable input formula $\mathcal{F}$. A SAT solver call is represented on line 5 by $\langle st, \mu, C \rangle \leftarrow \text{SAT}(\mathcal{F})$, where st is a Boolean value indicating whether the formula is satisfiable or not. If the formula is satisfiable, the SAT solver returns a model μ. Otherwise it returns an unsatisfiable core C over the soft clauses.

We overview in more detail a simple example of such algorithms: the basic linear search (BLS) approach, depicted in Algorithm 1. This algorithm maintains a partition of $\mathcal{F}$ in two disjoint sets during the computation of an MCS. The set S represents a satisfiable subformula of $\mathcal{F}$, i.e., the MSS under construction. The set $\mathcal{U}$ is formed by the clauses that still need to be checked. The initial assignment used to split $\mathcal{F}$ is a model μ of $\mathcal{F}_H$. All the clauses in $\mathcal{F}$ satisfied by μ are put in S, while the falsified clauses become part of $\mathcal{U}$. These operations are enclosed inside the function $\text{InitialAssignment}(\mathcal{F})$ on line 3. Then, iteratively until all the clauses in $\mathcal{U}$ have been checked, the algorithm picks a clause $c \in \mathcal{U}$ and checks the satisfiability of $S \cup \{c\}$. If it is satisfiable, c is added to S. Otherwise, c is known to belong to the MCS under construction and is added to $\mathcal{M}$. Upon termination, S represents an MSS and $\mathcal{M} = \mathcal{F} \setminus S$ represents an MCS of $\mathcal{F}$.

In linear search, the number of SAT solver calls necessary is linear in terms of the number of soft clauses in the input formula. Different alternatives and optimization techniques have been proposed in recent years, leading to substantial improvements, including FastDiag [10], dichotomic search [25], clause D (CLD) [17], relaxation search [3], and the CMP algorithm [11]. In addition, algorithms such as the literal-based extractor (LBX) [20] represent the current state-of-the-art for extracting a single MCS. Recently, algorithms such as LOPZ, UCD and UBS, which also target the extraction of a single MCS, have been proposed [18], requiring a sublinear number of SAT solver calls on the number

Algorithm 1. Basic linear search

```
1  Function BLS(F)
2      M ← ∅
3      (S, U) ← InitialAssignment(F)
4      foreach c ∈ U do
5          ⟨st, μ, C⟩ ← SAT(S ∪ {c})
6          if st then S ← S ∪ {c}
7          else M ← M ∪ {c}
8      return M                                    // MCS of F
```

of clauses. Optimization techniques include exploiting satisfying assignments, backbone literals, and disjoint unsatisfiable cores [17], among others, and are integrated into MCS extraction algorithms for improving efficiency, giving rise to, e.g., enhanced linear search (ELS) [17].

MCS enumeration relies on iteratively extracting an MCS $C \in F$ and blocking it by adding the hard clause $\bigvee_{l \in L(C)} l$ to F. This way, no superset of C will subsequently be considered during the enumeration. The process continues until F_H becomes unsatisfiable, at which time all MCSes have been enumerated.

To the best of our knowledge, the current state-of-the-art in MCS enumeration is represented by the algorithms implemented in the tool mcsls [17], specifically, ELS and CLD. These algorithms have been shown to be complementary to each other [16].

4   Caching for MCS Enumeration

We will now introduce caching as a way to improve MCS enumeration. For some intuition, when a formula has a large number of MCSes, many of the MCSes tend to share many clauses. This suggests that similar satisfiability problems are solved in the computation of several MCSes. Our proposal aims at making use of this observation by storing information that could lead to avoiding potentially time-consuming calls to the SAT solver on $S \cup \{c\}$.

The idea is to keep a global database which is updated and queried by the MCS extraction algorithm during the enumeration process. The only requirements for realizing the database are the two operations *store(C)* and *hasSubset(K)*, where C is an unsatisfiable core of F and $K \subseteq F$. The intent of the function *hasSubset(K)* is to check for a given subset K of F whether an unsatisfiable core C of F with $C \subseteq K$ has already been extracted. If this is the case, we know by Proposition 1 that K is unsatisfiable. Naturally, as the cache database queries should avoid the cost of calling a SAT solver on the actual instance, the functions *store(C)* and *hasSubset(K)* need to be fast to compute.

Considering these requirements, as well as ease of implementing the cache and queries to the cache, in this work we implement the database by means of a SAT solver, storing a formula referred to as D formula in Algorithm 2. Variables of this formula are the selector variables of the original formula F, while clauses represent unsatisfiable cores of $\mathcal{F}$. For an unsatisfiable core C of $\mathcal{F}$ the corresponding clause is given by $(\bigvee_{c_i \in C} \neg\sigma(c_i))$, where $\sigma(c_i)$ is a function which for a given clause c_i returns the associated selector variable s_i. As an example, suppose that $C = \{c_1, c_2, c_3\}$ is an unsatisfiable core of $\mathcal{F}$. The corresponding clause added to $\mathcal{D}$ is $(\neg s_1 \vee \neg s_2 \vee \neg s_3)$. Notice that in $\mathcal{D}$ all literals are pure, since no positive literal is part of any clause. The $\mathcal{D}$ formula is in fact *monotone*. Consequently, checking the satisfiability of the $\mathcal{D}$ formula under any assumptions can be done in polynomial time. From a theoretical point of view, this clearly shows an advantage compared to calling a SAT solver on the formula $\mathcal{F}$.

Algorithm 2 shows the BLS algorithm extended with caching. The algorithm is here presented for simplicity in terms of extracting a single (next) MCS. To avoid repetition, we assume that the formula $\mathcal{F}_H$ contains blocking clauses of all the previously computed MCSes. As a consequence, any initial assignment computed at line 3 is guaranteed to split the formula in two parts $\mathcal{S}$ and $\mathcal{U}$ such that for any MCS $\mathcal{M} \subseteq \mathcal{U}$, $\mathcal{M}$ is not an already computed MCS.

Proposition 2. *Let $\mathcal{G}$ be a formula and $\mathcal{A} = \{s_i | c_i \in \mathcal{G}\}$. If $\mathcal{D} \cup \mathcal{A} \models \bot$, then $\mathcal{G} \models \bot$.*

Proof. Recall that each clause in $\mathcal{D}$ represents an unsatisfiable core. For $\mathcal{D} \cup \mathcal{A} \models \bot$ to hold, there has to be a clause $c \in \mathcal{D}$ whose literals are all falsified. This can happen if and only if we have $c \subseteq \mathcal{A}'$, where $\mathcal{A}' = \{\neg s_i | s_i \in \mathcal{A}\}$. Since the formula induced by c is unsatisfiable and $c \subseteq \mathcal{A}'$, by Proposition 1 it follows that $\mathcal{G}$ is unsatisfiable. □

Proposition 2 is applied on line 6 of Algorithm 2. In case $\mathcal{D} \cup \mathcal{A}$ is unsatisfiable, the call to the SAT solver on line 10 becomes unnecessary and we can immediately

---

Algorithm 2. Basic linear search with caching

```
1  Function BLS-CACHING(F)
       Global: D
2      M ← ∅                                    // MCS under construction
3      (S,U) ← InitialAssignment(F)
4      foreach c ∈ U do
5          A ← {s_i | c_i ∈ S ∪ {c}}
6          ⟨st, μ, C⟩ ← SAT(D ∪ A)
7          if not st then
8              M ← M ∪ {c}
9              continue
10         ⟨st, μ, C⟩ ← SAT(S ∪ {c})
11         if not st then
12             D ← D ∪ {(⋁_{c_i ∈ C} ¬σ(c_i))}
13             M ← M ∪ {c}
14         else  S ← S ∪ {c}
15     return M                                  // MCS of F
```

add c to $\mathcal{M}$ and proceed with testing the next clause. Otherwise, we are forced to test the clause c on the original formula (line 10). If $\mathcal{S} \cup \{c\}$ is unsatisfiable, we add the unsatisfiable core C to the formula $\mathcal{D}$ and c to the MCS under construction $\mathcal{M}$. If instead the outcome returned by the call is satisfiable, we add the clause to $\mathcal{S}$, the MSS under construction. When all clauses have been tested, the MCS $\mathcal{M} = \mathcal{F} \setminus \mathcal{S}$ is returned.

Example 1. Assume that $\mathcal{F} = \{c_1, c_2, c_3, c_4, c_5\}$ is an unsatisfiable formula with $M_1 = \{c_1, c_2, c_3\}$ and $M_2 = \{c_1, c_4, c_5\}$ the only MUSes of $\mathcal{F}$. An example run of Algorithm 2 is shown in Table 1. First $\mathcal{D}$ is empty and a SAT call on the original formula is made to identify c_3 as part of the MCS under construction. The query returns UNSAT, c_3 is added to the MCS under construction, and the unsatisfiable core $\{c_1, c_2, c_3\}$ is added to $\mathcal{D}$. For testing $\mathcal{S} \cup \{c_5\} = \{c_1, c_4, c_2, c_5\}$, $\mathcal{D} \cup \mathcal{A}$ (represented in CNF as $(\neg s_1 \vee \neg s_2 \vee \neg s_3) \wedge s_1 \wedge s_4 \wedge s_2 \wedge s_5$) is satisfiable, so another SAT call on $\mathcal{F}$ is required. This adds the additional unsatisfiable core $\{c_1, c_4, c_5\}$ to $\mathcal{D}$ and c_5 to $\mathcal{M}_1$, which is now a complete MCS. Finally, when testing clauses c_3 and c_4 (for the next MCS), we have that in both cases $\mathcal{D} \cup \mathcal{A}$ is unsatisfiable and the two clauses are added to $\mathcal{M}_2$. This example shows that while two SAT solver calls are needed for determining the first MCS, for the second one it suffices to query the core database. ∎

Table 1. Example execution of Algorithm 2

$\mathcal{S} \cup \{c\}$	$\mathcal{D}$	Query	$\mathcal{M}_1$
$\{c_1, c_4, c_2\} \cup \{c_3\}$	$\emptyset$	$\mathcal{F}$: UNSAT	$\{c_3\}$
$\{c_1, c_4, c_2\} \cup \{c_5\}$	$\{c_1, c_2, c_3\}$	$\mathcal{F}$: UNSAT	$\{c_3, c_5\}$
$\mathcal{S} \cup \{c\}$	$\mathcal{D}$	Query	$\mathcal{M}_2$
$\{c_1, c_2, c_5\} \cup \{c_3\}$	$\{c_1, c_2, c_3\}, \{c_1, c_4, c_5\}$	$\mathcal{D} \cup \mathcal{A}$: UNSAT	$\{c_3\}$
$\{c_1, c_2, c_5\} \cup \{c_4\}$	$\{c_1, c_2, c_3\}, \{c_1, c_4, c_5\}$	$\mathcal{D} \cup \mathcal{A}$: UNSAT	$\{c_3, c_4\}$

5   Experimental Results

We implemented the proposed approach (Algorithm 2), mcscache-els, on top of the state-of-the-art MCS enumeration tool mcsls [17] in C++, extending the ELS algorithm to use a core database for caching, and using Minisat 2.2.0 [9] as a backend solver. We implemented two optimizations: we use (i) satisfying assignments obtained from satisfiable SAT solver calls to extend the set $\mathcal{S}$ with all clauses in $\mathcal{U}$ satisfied by the assignments, and (ii) disjoint unsatisfiable cores by computing a set of disjoint cores at the beginning of search, which can lead to avoiding some calls to the SAT solver during the computation of MCSes. The current implementation does not use the so-called backbone literals optimization [17] with the intuition that this would make the core database non-monotone and thereby queries to the cache potentially more time-consuming.

We compare mcscache-els with two state-of-the-art approaches: ELS, which is the basis of mcscache-els, and CLD [17]. Both ELS (mcsls-els) and CLD (mcsls-cld) are implemented in the tool mcsls [17]. All algorithms accept formulas with soft and hard clauses. As benchmarks, we used the 811 instances from [17] for which an MCS could be extracted. These instances were originally used for benchmarking algorithms for extracting only a single MCS. Note that, in terms of MCS enumeration, these are therefore hard instances, and we expect caching to be most beneficial on such hard instances. The experiments were run on a computing cluster running 64-bit Linux on 2-GHz processors using a per-instance memory limit of 8 GB and a time limit of 1800 s.

We compare the performance of the algorithms in terms of the number of MCSes enumerated within the per-instance time limit. A comparison of mcscache-els and mcsls-els is shown in Fig. 1. Using caching clearly and consistently improves performance: with only few exceptions, caching enables enumerating higher numbers of MCSes. Note that the only difference between mcscache-els and mcsls-els is that the first uses the caching approach proposed in this work. We also compare mcscache-els to mcsls-cld; in this comparison the base algorithms are different. As can be observed from Fig. 2, mcsls-cld exhibits better performance for instances on which a lower number of MCSes are enumerated. However, as the number of MCSes enumerated increases, the performance of mcsls-cld noticeably degrades compared to mcscache-els and mcscache-els starts to clearly dominate.

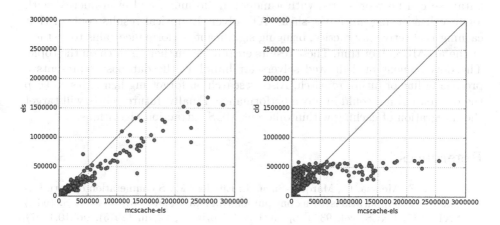

Fig. 1. mcscache-els vs mcsls-els   **Fig. 2.** mcscache-els vs mcsls-cld

Finally, we consider more statistics on the effects of caching. First, we observed that on a significant number of the instances the number of cache misses (cache queries which do not find the core queried for in the database) was very low, i.e., the success rate in querying the cache was high, as >90% of MCS clauses were often detected from the cache, without direct access to the original formula. On the other hand, querying the core database as currently implemented can still take a substantial amount of time on some instances. The query time seems to correlate with the average size of cores in the cache. In particular, on some instances cores can be very large (up to 200,000 clauses), which made the databases queries for the SAT solver time-consuming. In the present implementation, this seems to introduce an unnecessary overhead. This observation, together with the empirical results, motivates studying alternative ways of querying the database by taking into account the very simplistic form of the database, in order to mitigate the observed negative effects. Alternatively, heuristics aiming at removing unused or too large cores could also be a viable and practical solution.

6   Conclusions

Analysis of over-constrained sets of constraints finds a widening range of practical applications. A central task in this context is the enumeration of minimal correction sets of constraints, namely, MCSes. Best-performing algorithms for the highly intractable task of MCS enumeration make high numbers of increasingly hard SAT solver calls as the number of MCSes increases. Motivated by this, we developed caching mechanisms to speed-up MCS enumeration. By keeping a global database in which unsatisfiable cores found during the computation of MCSes are stored, future calls to the SAT solver in the computation of new MCSes can be substituted by a polynomial-time check. In particular, the global database can be represented with a monotone formula, and even queried with low overhead using an off-the-shelf SAT solver. Empirical results confirm that caching is effective in practice, bringing significant performance gains to a state-of-the-art MCS algorithm. These results encourage further research on the topic. The development of dedicated solvers for handling the database represents a promising line of future research. Also, research on forgetting heuristics to keep the database small could improve performance. Finally, future efforts will target the integration of caching within different MCS extraction algorithms.

References

1. Arif, M.F., Mencía, C., Marques-Silva, J.: Efficient MUS enumeration of horn formulae with applications to axiom pinpointing. In: Heule, M., Weaver, S. (eds.) SAT 2015. LNCS, vol. 9340, pp. 324–342. Springer, Cham (2015). doi:10.1007/978-3-319-24318-4_24
2. Bacchus, F., Dalmao, S., Pitassi, T.: Solving #SAT and Bayesian inference with backtracking search. J. Artif. Intell. Res. 34, 391–442 (2009)
3. Bacchus, F., Davies, J., Tsimpoukelli, M., Katsirelos, G.: Relaxation search: a simple way of managing optional clauses. In: Proceedings of AAAI, pp. 835–841. AAAI Press (2014)
4. Bailey, J., Stuckey, P.J.: Discovery of minimal unsatisfiable subsets of constraints using hitting set dualization. In: Hermenegildo, M.V., Cabeza, D. (eds.) PADL 2005. LNCS, vol. 3350, pp. 174–186. Springer, Heidelberg (2005). doi:10.1007/978-3-540-30557-6_14

5. Beame, P., Impagliazzo, R., Pitassi, T., Segerlind, N.: Formula caching in DPLL. Trans. Comput. Theor. **1**(3), 9:1–9:33 (2010)
6. Birnbaum, E., Lozinskii, E.L.: Consistent subsets of inconsistent systems: structure and behaviour. J. Exp. Theor. Artif. Intell. **15**(1), 25–46 (2003)
7. Chen, Z., Toda, S.: The complexity of selecting maximal solutions. Inf. Comput. **119**(2), 231–239 (1995)
8. Cook, S.A.: The complexity of theorem-proving procedures. In: Proceedings of STOC, pp. 151–158. ACM (1971)
9. Eén, N., Sörensson, N.: An extensible SAT-solver. In: Giunchiglia, E., Tacchella, A. (eds.) SAT 2003. LNCS, vol. 2919, pp. 502–518. Springer, Heidelberg (2004). doi:10.1007/978-3-540-24605-3_37
10. Felfernig, A., Schubert, M., Zehentner, C.: An efficient diagnosis algorithm for inconsistent constraint sets. Artif. Intell. Eng. Des. Anal. Manuf. **26**(1), 53–62 (2012)
11. Grégoire, É., Lagniez, J., Mazure, B.: An experimentally efficient method for (MSS, coMSS) partitioning. In: Proceedings of AAAI, pp. 2666–2673. AAAI Press (2014)
12. Kitching, M., Bacchus, F.: Symmetric component caching. In: Proceedings of IJCAI, pp. 118–124 (2007)
13. Kopp, T., Singla, P., Kautz, H.A.: Toward caching symmetrical subtheories for weighted model counting. In: Proceedings of AAAI Beyond NP Workshop. AAAI Workshops, vol. WS-16-05. AAAI Press (2016)
14. Liffiton, M.H., Previti, A., Malik, A., Marques-Silva, J.: Fast, flexible MUS enumeration. Constraints **21**(2), 223–250 (2016)
15. Liffiton, M.H., Sakallah, K.A.: Algorithms for computing minimal unsatisfiable subsets of constraints. J. Autom. Reasoning **40**(1), 1–33 (2008)
16. Malitsky, Y., O'Sullivan, B., Previti, A., Marques-Silva, J.: Timeout-sensitive portfolio approach to enumerating minimal correction subsets for satisfiability problems. In: Proceedings of ECAI. Frontiers in Artificial Intelligence and Applications, vol. 263, pp. 1065–1066. IOS Press (2014)
17. Marques-Silva, J., Heras, F., Janota, M., Previti, A., Belov, A.: On computing minimal correction subsets. In: Proceedings of IJCAI, pp. 615–622. AAAI Press (2013)
18. Mencía, C., Ignatiev, A., Previti, A., Marques-Silva, J.: MCS extraction with sublinear oracle queries. In: Creignou, N., Le Berre, D. (eds.) SAT 2016. LNCS, vol. 9710, pp. 342–360. Springer, Cham (2016). doi:10.1007/978-3-319-40970-2_21
19. Mencía, C., Marques-Silva, J.: Efficient relaxations of over-constrained CSPs. In: Proceedings of ICTAI, pp. 725–732. IEEE Computer Society (2014)
20. Mencía, C., Previti, A., Marques-Silva, J.: Literal-based MCS extraction. In: Proceedings of IJCAI, pp. 1973–1979. AAAI Press (2015)
21. Meseguer, P., Bouhmala, N., Bouzoubaa, T., Irgens, M., Sánchez, M.: Current approaches for solving over-constrained problems. Constraints **8**(1), 9–39 (2003)
22. Morgado, A., Liffiton, M.H., Marques-Silva, J.: MaxSAT-based MCS enumeration. In: Biere, A., Nahir, A., Vos, T. (eds.) HVC 2012. LNCS, vol. 7857, pp. 86–101. Springer, Heidelberg (2013). doi:10.1007/978-3-642-39611-3_13
23. Moskewicz, M.W., Madigan, C.F., Zhao, Y., Zhang, L., Malik, S.: Chaff: engineering an efficient SAT solver. In: Proceedings of DAC, pp. 530–535. ACM (2001)
24. Nöhrer, A., Biere, A., Egyed, A.: Managing SAT inconsistencies with HUMUS. In: Proceedings of VaMoS, pp. 83–91. ACM (2012)
25. O'Sullivan, B., Papadopoulos, A., Faltings, B., Pu, P.: Representative explanations for over-constrained problems. In: Proceedings of AAAI, pp. 323–328. AAAI Press (2007)

26. Reiter, R.: A theory of diagnosis from first principles. Artif. Intell. **32**(1), 57–95 (1987)
27. Sang, T., Bacchus, F., Beame, P., Kautz, H.A., Pitassi, T.: Combining component caching and clause learning for effective model counting. In: SAT Online Proceedings (2004)
28. Silva, J.P.M., Sakallah, K.A.: GRASP: a search algorithm for propositional satisfiability. IEEE Trans. Comput. **48**(5), 506–521 (1999)
29. Slaney, J.: Set-theoretic duality: a fundamental feature of combinatorial optimisation. In: Proceedings of ECAI. Frontiers in Artificial Intelligence and Applications, vol. 263, pp. 843–848. IOS Press (2014)
30. Thurley, M.: sharpSAT – counting models with advanced component caching and implicit BCP. In: Biere, A., Gomes, C.P. (eds.) SAT 2006. LNCS, vol. 4121, pp. 424–429. Springer, Heidelberg (2006). doi:10.1007/11814948_38
31. Zielke, C., Kaufmann, M.: A new approach to partial MUS enumeration. In: Heule, M., Weaver, S. (eds.) SAT 2015. LNCS, vol. 9340, pp. 387–404. Springer, Cham (2015). doi:10.1007/978-3-319-24318-4_28

Introducing Pareto Minimal Correction Subsets

Miguel Terra-Neves[(✉)], Inês Lynce, and Vasco Manquinho

INESC-ID/Instituto Superior Técnico, Universidade de Lisboa, Lisbon, Portugal
{neves,ines,vmm}@sat.inesc-id.pt

Abstract. A Minimal Correction Subset (MCS) of an unsatisfiable constraint set is a minimal subset of constraints that, if removed, makes the constraint set satisfiable. MCSs enjoy a wide range of applications, one of them being approximate solutions to constrained optimization problems. However, existing work on applying MCS enumeration to optimization problems focuses on the single-objective case.

In this work, a first definition of Pareto Minimal Correction Subsets (Pareto-MCSs) is proposed with the goal of approximating the Pareto-optimal solution set of multi-objective constrained optimization problems. We formalize and prove an equivalence relationship between Pareto-optimal solutions and Pareto-MCSs. Moreover, Pareto-MCSs and MCSs can be connected in such a way that existing state-of-the-art MCS enumeration algorithms can be used to enumerate Pareto-MCSs.

An experimental evaluation considers the multi-objective virtual machine consolidation problem. Results show that the proposed Pareto-MCS approach outperforms the state-of-the-art approaches.

1 Introduction

Given an unsatisfiable set of constraints F, a Minimal Correction Subset (MCS) is a minimal subset of constraints C such that, if all constraints in C are removed from F, then F becomes satisfiable. MCSs enjoy a wide range of applications, such as analysis of over-constrained systems [8,11], minimal model computation [3], interactive constraint satisfaction [17] and approximation of constrained combinatorial optimization problems [15]. Many efficient MCS enumeration algorithms have been proposed in the recent years [1,8,9,14,16], and MCS-based approximation algorithms are able to compute good quality approximations of optimal solutions efficiently [15]. However, the usage of MCSs has focused only on approximating single-objective problems.

In many scenarios, a decision maker may need to optimize multiple conflicting objectives [21]. In this case, multiple optimal solutions may exist, referred to as Pareto-optimal solutions [18], each of them favoring certain objectives at the expense of others. One such example is Virtual Machine Consolidation (VMC),

The authors would like to thank Salvador Abreu and Pedro Salgueiro from Universidade de Évora for granting the authors permission to use their cluster. This work was supported by national funds through Fundação para a Ciência e a Tecnologia (FCT) with references UID/CEC/50021/2013 and SFRH/BD/111471/2015.

S. Gaspers and T. Walsh (Eds.): SAT 2017, LNCS 10491, pp. 195–211, 2017.
DOI: 10.1007/978-3-319-66263-3_13

where a cloud provider is interested in finding a virtual machine placement over a set of servers that minimizes energy consumption while also minimizing the performance penalty incurred by the new placement. Current state-of-the-art solutions for multi-objective VMC are mostly based on meta-heuristics, such as genetic algorithms [22] and biogeography-based optimization [23], as is the case for many other Multi-Objective Combinatorial Optimization (MOCO) problems. Such approaches are non-deterministic and are known to be parameter sensitive, resulting in each different problem requiring a distinct configuration of the algorithm to achieve a competitive performance. Moreover, meta-heuristics are known to struggle as instances become more tightly constrained. On the other hand, it is widely accepted that constraint-based methods, compared to other approaches, usually thrive in tightly constrained problems, and some exact algorithms exist for computing the set of Pareto-optimal solutions of a MOCO instance [19]. However, such solutions are impractical for large scale problems.

The main contributions of this paper are as follows: (1) a first definition of Multi-MCSs and Pareto-MCSs, an extension of MCSs to constrained multi-objective optimization problems; (2) a proof of an equivalence relationship between Pareto-MCSs and Pareto-optimal solutions; (3) a proof of a relationship between Multi-MCSs and MCSs that allows one to use state-of-the-art MCS enumerators right off-the-shelf as Multi-MCS enumerators; (4) an extensive experimental evaluation on VMC instances from the Google Custer Data project which clearly shows the suitability of the Pareto-MCS approach for finding good quality approximations of the solution sets of MOCO instances.

The paper is organized as follows. In Sect. 2, the basic definitions are introduced along with the notation used in the remainder of the paper. In Sect. 3, a definition of Multi-MCSs and Pareto-MCSs is proposed and some of their properties are described and proven. Section 4 introduces the VMC problem and the corresponding instances are evaluated in Sect. 5 using the Pareto-MCS approach. Section 6 concludes this paper and suggests future work directions.

2   Preliminaries

In this section, we introduce the necessary definitions and notations that will be used throughout the rest of the paper. We start by describing Weighted Boolean Optimization (WBO) in Sect. 2.1. Next, Minimal Correction Subsets (MCSs) are defined and the ClauseD (CLD) algorithm for MCS enumeration is presented. Finally, Multi-Objective Combinatorial Optimization (MOCO) and the Guided Improvement Algorithm (GIA) for solving MOCO problems are described.

2.1   Weighted Boolean Optimization

Let $X = \{x_1, x_2, \ldots, x_n\}$ be a set of n Boolean variables. A literal is either a variable x_i or its complement $\neg x_i$. Given a set of m literals $l_1, l_2, \ldots, l_m$ and their respective coefficients $\omega_1, \omega_2, \ldots, \omega_m \in \mathbb{Z}$, a Pseudo-Boolean (PB) expression has the following form:

$$\sum \omega_i \cdot l_i. \tag{1}$$

Given an integer $k \in \mathbb{Z}$, a PB constraint is a linear inequality with the form:

$$\sum \omega_i \cdot l_i \bowtie k, \quad \bowtie \in \{\leq, \geq, =\}. \tag{2}$$

Given a set $F = \{c_1, c_2, \ldots, c_k\}$ of k PB constraints defined over a set of X Boolean variables, the Pseudo-Boolean Satisfiability (PBS) problem consists of deciding if there exists a complete assignment $\alpha : X \rightarrow \{0,1\}$, such that all PB constraints in F are satisfied. If that is the case, we say that F is satisfiable and α satisfies F, denoted $\alpha(F) = 1$. Otherwise, we say that F is unsatisfiable and $\alpha(F) = 0$ for any assignment α. Analogously, given a PB constraint c, $\alpha(c) = 1$ ($\alpha(c) = 0$) denotes that α satisfies (does not satisfy) c. A weighted PB constraint is a pair (c, ω), where c is a PB constraint and $\omega \in \mathbb{N}^+$ is the cost of not satisfying c. In WBO [13], given a formula $\mathbb{F} = (F_H, F_S)$, where F_H and F_S denote sets of hard and soft weighted PB constraints respectively, the goal is to compute a complete assignment that satisfies all of the constraints in F_H and minimizes the sum of the weights of the constraints in F_S that are not satisfied.

Example 1. Let $F_H = \{(x_1 + x_2 + x_3 \leq 2)\}$ be the set of hard PB constraints and $F_S = \{(x_1 \geq 1, 4), (x_2 \geq 1, 2), (x_3 \geq 1, 3)\}$ the set of weighted soft PB constraints of a WBO instance. $\alpha_1 = \{(x_1, 1), (x_2, 0), (x_3, 1)\}$ is an optimal assignment that does not satisfy only the second soft constraint ($x_2 \geq 1$), having a cost of 2. $\alpha_2 = \{(x_1, 0), (x_2, 1), (x_3, 1)\}$ is not an optimal assignment because it does not satisfy ($x_1 \geq 1$), which has a weight of 4. $\alpha_3 = \{(x_1, 1), (x_2, 1), (x_3, 1)\}$ is an invalid assignment because it does not satisfy F_H.

For simplicity reasons, given p PB constraints $c_1, c_2, \ldots, c_p$, the disjunction operator $\vee$ is used to represent the constraint that at least one of the PB constraints must be satisfied (e.g. $c_1 \vee c_2 \vee \cdots \vee c_p$). Note that such disjunctions can be easily converted to sets of PB constraints using auxiliary variables.

2.2 Minimal Correction Subsets

Given an unsatisfiable set of PB constraints F, a minimal correction subset (MCS) is a minimal subset $C \subseteq F$ such that $F \backslash C$ is satisfiable.

Definition 1. *Let F be an unsatisfiable set of PB constraints. A subset $C \subseteq F$ is an MCS of F if, and only if, $F \backslash C$ is satisfiable and $(F \backslash C) \cup \{c\}$ is unsatisfiable for all $c \in C$.*

Example 2. Consider the unsatisfiable set of PB constraints $F = \{(x_1 + x_2 = 1), (x_1 \geq 1), (x_2 \geq 1)\}$. F has three MCSs $C_1 = \{(x_1 \geq 1)\}$, $C_2 = \{(x_2 \geq 1)\}$ and $C_3 = \{(x_1 + x_2 = 1)\}$.

Several algorithms exist for finding MCSs [2,8,14–16]. For the purpose of this work, the state-of-the-art CLD algorithm was used [14]. CLD's pseudo-code is presented in Algorithm 1. It starts by initializing the sets S and C of satisfied and not satisfied PB constraints respectively (lines 1 and 2). Initially, all constraints

---

Algorithm 1. CLD algorithm for computing an MCS of a PB formula [14]

Input: F

1 $S \leftarrow \emptyset$
2 $C \leftarrow F$
3 $status \leftarrow SAT$
4 **while** $status = SAT$ **do**
5 | $\quad D \leftarrow (\bigvee_{c \in C} c)$
6 | $\quad (status, \alpha) \leftarrow \mathsf{PBS}(S \cup \{D\})$
7 | $\quad$ **if** $status = SAT$ **then**
8 | $\quad\quad$ | $S \leftarrow S \cup \bigcup_{c \in C, \alpha(c)=1} \{c\}$
9 | $\quad\quad$ | $C \leftarrow F \setminus S$

10 **return** C

---

are not satisfied. Then, the CLD algorithm repeatedly checks if it is possible to satisfy at least one of the constraints in C, while satisfying all constraints in S (lines 5 and 6). If so, then sets S and C are updated accordingly (lines 8 and 9). If not, then C is an MCS and is returned by the algorithm (line 10).

Algorithm 1 computes a single MCS C, but it can be used to find another MCS by incorporating the constraint $(\bigvee_{c \in C} c)$ in the initialization process of S in line 1. Such a constraint "blocks" MCS C from being identified again by the algorithm. Hence, the CLD algorithm can be used to enumerate all MCSs of F by blocking previous MCSs in subsequent invocations of the algorithm.

The following definition extends the notion of MCS to WBO instances. For simplicity, we assume that the set of hard PB constraints F_H of a WBO instance is always satisfiable. (This can be checked using a single call to a PBS solver.)

Definition 2. *Let* $\mathbb{F} = (F_H, F_S)$ *be a WBO instance, where* F_H *and* F_S *denote the hard and soft PB constraint sets respectively. A subset* $C \subseteq F_S$ *is an MCS of* $\mathbb{F}$ *if, and only if,* $F_H \cup (F_S \setminus C)$ *is satisfiable and* $F_H \cup (F_S \setminus C) \cup \{c\}$ *is unsatisfiable for all* $c \in C$.

Observe that Algorithm 1 can be used to enumerate MCSs of $\mathbb{F}$ by initializing S as F_H and C as F_S in lines 1 and 2. An MCS provides an approximation to a WBO optimal solution, and in some problems is faster to compute [10]. Actually, the WBO problem can be reduced to finding the MCS $C \subseteq F_S$ that minimizes the sum of the weights of the soft constraints in C [4].

Example 3. Consider the WBO instance given by $F_H = \{(x_1 + x_2 = 1)\}$ and $F_S = \{(x_1 \geq 1, 1), (x_2 \geq 1, 2), (\neg x_1 \geq 1, 4), (\neg x_2 \geq 1, 6)\}$. This instance has two MCSs $C_1 = \{(x_1 \geq 1, 1), (\neg x_2 \geq 1, 6)\}$ and $C_2 = \{(x_2 \geq 1, 2), (\neg x_1 \geq 1, 4)\}$. The sum of the weights of the constraints in C_1 and C_2 is 7 and 6 respectively. C_2 is the MCS that minimizes the sum of the weights of its constraints. Therefore, any assignment that satisfies $\{(x_1 + x_2 = 1), (x_1 \geq 1), (\neg x_2 \geq 1)\}$ is an optimal solution of the WBO instance.

2.3   Multi-Objective Combinatorial Optimization

A Multi-Objective Combinatorial Optimization (MOCO) [21] instance is composed of two sets: a set $F = \{c_1, c_2, \ldots, c_k\}$ of constraints that must be satisfied
and a set $O = \{f_1, f_2, \ldots, f_l\}$ of objective functions to minimize. In this work, we
focus on the special case where $c_1, c_2, \ldots, c_k$ are PB constraints and $f_1, f_2, \ldots, f_l$
are PB expressions over a set X of Boolean variables. Given an objective function $f \in O$ and a complete assignment $\alpha : X \to \{0, 1\}$, we denote as $f(\alpha)$ the
objective value of α for f.

Definition 3. *Let* $\mathbb{M} = (F, O)$ *be a MOCO instance, where F and O are the
constraint and objective function sets respectively. Let $\alpha, \alpha' : X \to \{0, 1\}$ be
two complete assignments such that $\alpha \neq \alpha'$ and $\alpha(F) = \alpha'(F) = 1$. We say
that α dominates α', written $\alpha \prec \alpha'$, if, and only if, $\forall_{f \in O} f(\alpha) \leq f(\alpha')$ and
$\exists_{f' \in O} f'(\alpha) < f'(\alpha')$.*

Definition 4. *Let* $\mathbb{M} = (F, O)$ *be a MOCO instance and $\alpha : X \to \{0, 1\}$ a
complete assignment. α is said to be Pareto-optimal if, and only if, $\alpha(F) = 1$
and no other complete assignment α' exists such that $\alpha'(F) = 1$ and $\alpha' \prec \alpha$.*

In MOCO, the goal is to find the set of Pareto-optimal [18] solutions, also
referred to as solution set.

Example 4. Let $F = \{(x_1 + x_2 \leq 1)\}$ be the set of PB constraints and $O = \{(x_1 +
2 \cdot \neg x_2), (\neg x_1)\}$ the set of objective functions of a MOCO instance. Table 1 shows
the objective values for each possible assignment. The lines that correspond
to Pareto-optimal solutions are highlighted in bold. Note that $\{(x_1, 1), (x_2, 1)\}$
violates the constraint in F. Hence, it is not a valid assignment. $\{(x_1, 0), (x_2, 0)\}$
is not Pareto-optimal because it is dominated by $\{(x_1, 0), (x_2, 1)\}$. However,
$\{(x_1, 0), (x_2, 1)\}$ and $\{(x_1, 1), (x_2, 0)\}$ are Pareto-optimal solutions because they
are not dominated by any other assignment that satisfies F.

Table 1. Possible assignments and objective values for the instance in Example 4.

x_1	x_2	$x_1 + 2 \cdot \neg x_2$	$\neg x_1$
0	0	2	1
0	**1**	**0**	**1**
1	**0**	**3**	**0**
1	1	-	-

Next, we review the Guided Improvement Algorithm (GIA) [19] for finding
the solution set of a MOCO instance. The GIA algorithm is implemented in the
optimization engine of Microsoft's SMT solver Z3 [5] for finding Pareto-optimal
solutions of SMT instances with multiple objective functions.

Algorithm 2. Guided Improvement Algorithm for MOCO problems [19]

Input: (F, O)

1 $F_W \leftarrow F$

2 $\alpha_{sol} \leftarrow \emptyset$

3 $(status, \alpha) \leftarrow \text{PBS}(F_W)$

4 **while** $status = SAT$ **do**

5    $F'_W \leftarrow F_W$

6   **while** $status = SAT$ **do**

7      $\alpha_{sol} \leftarrow \alpha$

8      $(v_1, v_2, \ldots, v_l) \leftarrow (f_1(\alpha), f_2(\alpha), \ldots, f_l(\alpha))$

9      $F'_W \leftarrow F'_W \cup \bigcup_{f_i \in O}\{(f_i \le v_i)\} \cup \{(\bigvee_{f_i \in O} f_i \le v_i - 1)\}$

10      $(status, \alpha) \leftarrow \text{PBS}(F'_W)$

11   **yield**(α_{sol})

12    $F_W \leftarrow F_W \cup \{(\bigvee_{f_i \in O} f_i \le v_i - 1)\}$

13    $(status, \alpha) \leftarrow \text{PBS}(F_W)$

The pseudo-code for the GIA algorithm is presented in Algorithm 2. It starts by building a working formula F_W, that is initialized to F (line 1), and checking if F_W is satisfiable (line 3). If so, then the algorithm enters a loop that enumerates the Pareto-optimal assignments (lines 4 to 13). At each iteration, the algorithm first builds a second working formula F'_W (line 5). Next, it searches for a single Pareto-optimal solution by repeatedly adding constraints to F'_W that force future assignments to dominate the assignment α obtained in the last PBS call (lines 8 and 9) and checking if F'_W is still satisfiable (line 10). The algorithm guarantees that it found a Pareto-optimal solution when F'_W becomes unsatisfiable.

After finding a Pareto-optimal assignment α, the algorithm adds new constraints to F_W that force at least one of the objective values of future assignments to be better than α's (line 12). Then, it checks if F_W remains satisfiable (line 13), in which case, more Pareto-optimal solutions exists. The loop in lines 4 to 13 is repeated until all Pareto-optimal assignments have been found.

3 Pareto Minimal Correction Subsets

This section introduces the novel concept of Multi-MCSs and explains how they can be used to approximate the solution set of a MOCO instance. First, the multi-objective version of WBO is introduced. Next, Multi-MCSs and Pareto-MCSs are defined. Finally, we describe some properties of Multi-MCSs and Pareto-MCSs. In particular, we prove a relationship between MCSs and Multi-MCSs that allows MCS enumerators to be used off-the-shelf as Multi-MCS enumerators.

3.1 Multi-Objective Weighted Boolean Optimization

A Multi-Objective Weighted Boolean Optimization (MOWBO) instance is composed of a set $F_H = \{c_1, c_2, \ldots, c_k\}$ of hard PB constraints and a set $O_S = \{F_{S1}, F_{S2}, \ldots, F_{Sl}\}$ of soft weighted PB constraint sets. Given an assignment α and a set $F_{Si} \in O_S$, let $w(F_{Si}, \alpha)$ denote the sum of the weights of the PB constraints in F_{Si} unsatisfied by α, i.e.,

$$w(F_{Si}, \alpha) = \sum_{(c,w)\in F_{Si},\, \alpha(c)=0} w. \tag{3}$$

Like in MOCO, the goal is to find the set of Pareto-optimal solutions. Dominance in MOWBO is defined as follows.

Definition 5. *Let* $\mathbb{W} = (F_H, O_S)$ *be a MOWBO instance, with* $O_S = \{F_{S1}, F_{S2}, \ldots, F_{Sl}\}$. *Let* $\alpha, \alpha' : X \to \{0, 1\}$ *be two complete assignments such that* $\alpha \neq \alpha'$ *and* $\alpha(F_H) = \alpha'(F_H) = 1$. *We say that* α *dominates* α' $(\alpha \prec \alpha')$ *if, and only if,* $\forall_{F_S \in O_S} w(F_S, \alpha) \leq w(F_S, \alpha')$ *and* $\exists_{F'_S \in O_S} w(F'_S, \alpha) < w(F'_S, \alpha')$.

Similarly to the reduction from the single-objective case to WBO [13], a MOCO instance $\mathbb{M} = (F, O)$ can be reduced to a MOWBO instance $\mathbb{W} = (F_H, O_S)$ as follows: (1) we set $F_H = F$; (2) for each $f \in O$, with $f = \omega_1 \cdot l_1 + \omega_2 \cdot l_2 + \cdots + \omega_m \cdot l_m$, we add a soft constraint set $F_S = \{(\neg l_1, \omega_1), (\neg l_2, \omega_2), \ldots, (\neg l_m, \omega_m)\}$ to O_S.

Example 5. Recall the MOCO instance from Example 4, with $F = \{(x_1 + x_2 \leq 1)\}$ and $O = \{(x_1 + 2 \cdot \neg x_2), (\neg x_1)\}$. An equivalent MOWBO instance has $F_H = \{(x_1 + x_2 \leq 1)\}$ and $O_S = \{F_{S1}, F_{S2}\}$, where $F_{S1} = \{(\neg x_1, 1), (x_2, 2)\}$ and $F_{S2} = \{(x_1, 1)\}$.

3.2 Multi and Pareto Minimal Correction Subsets

The definition of Multi-MCSs, an extension of MCSs to MOWBO formulas, builds upon the concept of MCS for PBS and WBO formulas (see Sect. 2.2).

Definition 6. *Let* $\mathbb{W} = (F_H, O_S)$ *be a MOWBO instance, with* $O_S = \{F_{S1}, F_{S2}, \ldots, F_{Sl}\}$. *Let* $\mathbb{C} = (C_1, C_2, \ldots, C_l)$ *be a tuple of sets such that* $C_i \subseteq F_{Si}$, $1 \leq i \leq l$. $\mathbb{C}$ *is a Multi-MCS of* $\mathbb{W}$ *if, and only if,* $F_H \cup \bigcup_{i=1}^{l}(F_{Si} \backslash C_i)$ *is satisfiable and* $F_H \cup \bigcup_{i=1}^{l}(F_{Si} \backslash C_i) \cup \{c\}$ *is unsatisfiable for all* $c \in \bigcup_{i=1}^{l} C_i$.

The dominance relation between two MOWBO solutions in Definition 5 can also be extended to pairs of Multi-MCSs as follows.

Definition 7. *Let* $\mathbb{W} = (F_H, O_S)$ *be a MOWBO instance. Let* $\mathbb{C} = (C_1, C_2, \ldots, C_l)$ *and* $\mathbb{C}' = (C'_1, C'_2, \ldots, C'_l)$ *be two Multi-MCSs of* $\mathbb{W}$. *We say that* $\mathbb{C}$ *dominates* $\mathbb{C}'$ $(\mathbb{C} \prec \mathbb{C}')$ *if, and only if,* $\forall_{1 \leq i \leq l} \sum_{(c,w)\in C_i} w \leq \sum_{(c',w')\in C'_i} w'$ *and* $\exists_{1 \leq j \leq l} \sum_{(c,w)\in C_j} w < \sum_{(c',w')\in C'_j} w'$.

Considering the definition of dominance for Multi-MCSs, the concept of Pareto-MCS can be formalized.

Definition 8. *Let* $W = (F_H, O_S)$ *be a MOWBO instance and* $\mathbb{C}$ *a Multi-MCS of* W. $\mathbb{C}$ *is a Pareto-MCS if, and only if, no other Multi-MCS* $\mathbb{C}'$ *exists such that* $\mathbb{C}' \prec \mathbb{C}$.

Example 6. Recall the MOWBO instance $W = (F_H, O_S)$ from Example 5, with $F_H = \{(x_1 + x_2 \leq 1)\}$ and $O_S = \{F_{S1}, F_{S2}\}$, where $F_{S1} = \{(\neg x_1, 1), (x_2, 2)\}$ and $F_{S2} = \{(x_1, 1)\}$. W has two Multi-MCSs $\mathbb{C}_1 = (\{\}, \{(x_1, 1)\})$ and $\mathbb{C}_2 = (\{(\neg x_1, 1), (x_2, 2)\}, \{\})$, which are also Pareto-MCSs of W.

3.3   Properties of Multi and Pareto Minimal Correction Subsets

Recall from Sect. 2.2 that an MCS provides an approximation to a WBO optimal solution. Moreover, the WBO problem can be reduced to finding an MCS that minimizes the sum of the weights of its soft constraints. Therefore, one could expect that a MOWBO instance can be reduced to finding the set of its Pareto-MCSs. The following results reveal an equivalence relationship between Pareto-MCSs and Pareto-optimal solutions.

Proposition 1. *Let* $W = (F_H, O_S)$ *be a MOWBO instance, with* $O_S = \{F_{S1}, F_{S2}, \ldots, F_{Sl}\}$, *and* α *a Pareto-optimal solution of* W. *Let* $\mathbb{C} = (C_1, C_2, \ldots, C_l)$, *where* $C_i = \{(c, \omega) : (c, \omega) \in F_{Si} \wedge \alpha(c) = 0\}$ *for all* $1 \leq i \leq l$. *Then,* $\mathbb{C}$ *is a Pareto-MCS of* W.

Proof. We prove this by contradiction. Suppose that $\mathbb{C}$ is not a Pareto-MCS. Then, there exists some Multi-MCS $\mathbb{C}' = (C'_1, C'_2, ..., C'_l)$ such that $\mathbb{C}' \prec \mathbb{C}$. Therefore, by Definition 7, the following holds:

$$\left(\forall_{1 \leq i \leq l} \sum_{(c', \omega') \in C'_i} \omega' \leq \sum_{(c, \omega) \in C_i} \omega \right) \wedge \left(\exists_{1 \leq j \leq l} \sum_{(c', \omega') \in C'_j} \omega' < \sum_{(c, \omega) \in C_j} \omega \right). \quad (4)$$

If $\mathbb{C}'$ is a Multi-MCS, then by Definition 6 we have that $F_H \cup \bigcup_{i=1}^{l}(F_{Si} \setminus C'_i)$ is satisfiable, and thus some complete assignment α' exists such that α' satisfies $F_H \cup \bigcup_{i=1}^{l}(F_{Si} \setminus C'_i)$. For all $1 \leq i \leq l$, α' satisfies all of the constraints in $F_{Si} \setminus C'_i$. Then, $w(F_{Si}, \alpha') \leq \sum_{(c', \omega') \in C'_i} \omega'$. Moreover, note that, by definition of $\mathbb{C}$, we have $w(F_{Si}, \alpha) = \sum_{(c, \omega) \in C_i} \omega$. Replacing in Eq. (4), we get

$$(\forall_{1 \leq i \leq l} \, w(F_{Si}, \alpha') \leq w(F_{Si}, \alpha)) \wedge (\exists_{1 \leq j \leq l} \, w(F_{Sj}, \alpha') < w(F_{Sj}, \alpha)). \quad (5)$$

By Definition 5, we have that $\alpha' \prec \alpha$, but α is a Pareto-optimal solution, thus we have a contradiction. □

Proposition 2. *Let* $W = (F_H, O_S)$ *be a MOWBO instance, with* $O_S = \{F_{S1}, F_{S2}, \ldots, F_{Sl}\}$, *and* $\mathbb{C} = (C_1, C_2, \ldots, C_l)$ *a Pareto-MCS of* W. *Then, any complete assignment* α *that satisfies* $F_H \cup \bigcup_{i=1}^{l}(F_{Si} \setminus C_i)$ *is a Pareto-optimal solution.*

Proof. Let α be a complete assignment that satisfies $F_H \cup \bigcup_{i=1}^{l}(F_{Si}\backslash C_i)$. Suppose that α is not Pareto-optimal. Then, there exists a complete assignment α' such that α' is Pareto-optimal and $\alpha' \prec \alpha$. Let $\mathbb{C}' = \{C_1', C_2', ..., C_l'\}$ such that $C_i' = \{(c,\omega) : (c,\omega) \in F_{Si} \wedge \alpha'(c) = 0\}$ for all $1 \leq i \leq l$. By Proposition 1, we have that $\mathbb{C}'$ is a Pareto-MCS. Since $\alpha' \prec \alpha$, the following holds:

$$(\forall_{1\leq i \leq l}\, w(F_{Si}, \alpha') \leq w(F_{Si}, \alpha)) \wedge (\exists_{1\leq j \leq l}\, w(F_{Sj}, \alpha') < w(F_{Sj}, \alpha)). \quad (6)$$

For all $1 \leq i \leq l$, since α satisfies $F_H \cup \bigcup_{i=1}^{l}(F_{Si}\backslash C_i)$, we also have that $w(F_{Si}, \alpha) \leq \sum_{(c,\omega)\in C_i}\omega$ and, by definition of $\mathbb{C}'$, $w(F_{Si}, \alpha') = \sum_{(c',\omega')\in C_i'}\omega'$. Replacing in Eq. (6), we have:

$$\left(\forall_{1\leq i \leq l}\sum_{(c',\omega')\in C_i'}\omega' \leq \sum_{(c,\omega)\in C_i}\omega\right) \wedge \left(\exists_{1\leq j\leq l}\sum_{(c',\omega')\in C_j'}\omega' < \sum_{(c,\omega)\in C_j}\omega\right). \quad (7)$$

By Definition 7, we have that $\mathbb{C}' \prec \mathbb{C}$, but $\mathbb{C}$ is a Pareto-MCS, thus we have a contradiction. $\qquad\square$

Propositions 1 and 2 show that for each Pareto-MCS there is at least one Pareto-optimal solution and that each Pareto-optimal solution has an associated Pareto-MCS. Therefore, MOWBO can be reduced to Pareto-MCS enumeration much in the same way that WBO can be reduced to MCS enumeration.

Proposition 3. *Let $\mathbb{W} = (F_H, O_S)$ be a MOWBO instance, with $O_S = \{F_{S1}, F_{S2}, \ldots, F_{Sl}\}$, and $\mathbb{C} = (C_1, C_2, \ldots, C_l)$ a Multi-MCS of $\mathbb{W}$. Then, $C = \bigcup_{i=1}^{l} C_i$ is an MCS of the WBO instance $\mathbb{F} = (F_H, \bigcup_{i=1}^{l} F_{Si})$.*

Proof. By Definition 6, we have that $F_H \cup \bigcup_{i=1}^{l}(F_{Si}\backslash C_i)$ is satisfiable and $F_H \cup \bigcup_{i=1}^{l}(F_{Si}\backslash C_i) \cup \{c\}$ is unsatisfiable for all $c \in C$. First, we show that the following equation holds:

$$F_H \cup \bigcup_{i=1}^{l}(F_{Si}\backslash C_i) = F_H \cup \left(\bigcup_{i=1}^{l}F_{Si}\backslash\bigcup_{i=1}^{l}C_i\right). \quad (8)$$

That is the case if at least one of the following is true: (1) $F_{S1}, F_{S2}, \ldots, F_{Sl}$ are disjoint; or (2) for all $1 \leq j, k \leq l$ such that $j \neq k$ and all $c \in F_{Sj} \cap F_{Sk}$, if $c \in C_j$ then $c \in C_k$. By definition, we have that, if $c \in C_j$ for some $1 \leq j \leq l$, then $F_H \cup \bigcup_{i=1}^{l}(F_{Si}\backslash C_i)$ is satisfiable and $F_H \cup \bigcup_{i=1}^{l}(F_{Si}\backslash C_i) \cup \{c\}$ is unsatisfiable. Therefore, if $c \in F_{Sk}$ for some $1 \leq k \leq l$ such that $k \neq j$, then $c \in C_k$, otherwise we would have $c \in F_{Sk}\backslash C_k$ and, consequently, $F_H \cup \bigcup_{i=1}^{l}(F_{Si}\backslash C_i) = F_H \cup \bigcup_{i=1}^{l}(F_{Si}\backslash C_i) \cup \{c\}$, which is a contradiction. Therefore, Eq. (8) holds. Due to (8), $F_H \cup \left(\bigcup_{i=1}^{l}F_{Si}\backslash\bigcup_{i=1}^{l}C_i\right)$ is satisfiable and $F_H \cup \left(\bigcup_{i=1}^{l}F_{Si}\backslash\bigcup_{i=1}^{l}C_i\right) \cup \{c\}$ is unsatisfiable for all $c \in C$. By Definition 2, then C is an MCS of $\mathbb{F}$. $\qquad\square$

Proposition 3 implies that Pareto-MCS enumeration of a MOWBO instance $W = (F_H, O_S)$ can be reduced to enumerating MCSs of the WBO instance $\mathbb{F} = (F_H, \bigcup_{F_S \in O_S} F_S)$ as follows: (1) build $\mathbb{F}$ from W; (2) enumerate MCSs C of $\mathbb{F}$ using an off-the-shelf MCS enumerator and convert them to Multi-MCSs $\mathbb{C} = (C_1, C_2, \ldots, C_l)$ of W, where $C_i = \{(c, \omega) : (c, \omega) \in C \cap F_{Si}\}$ for all $1 \leq i \leq l$; (3) filter out Multi-MCSs dominated by any other Multi-MCS using nondominated sorting [6]. Therefore, efficient state-of-the-art algorithms for MCS enumeration can be used to enumerate Pareto-MCSs right off-the-shelf.

4   Virtual Machine Consolidation

This section introduces a multi-objective formulation for the Virtual Machine Consolidation (VMC) problem. The VMC problem instances are later considered as a test case for the techniques proposed in the paper. The VMC problem occurs in the context of data centers management where the goal is to place a set of Virtual Machines (VMs) in a set of servers such that the quality of service contracted with the data center clients is achieved. Moreover, the minimization of multiple objectives are to be considered, in particular the minimization of energy usage, resource wastage and migration of VMs.

Consider a set $J = \{j_1, j_2, \ldots, j_m\}$ of m jobs in a data center, where each job is composed of several VMs. For each job $j \in J$, let $V_j = \{\nu_1, \nu_2, \ldots, \nu_{k_j}\}$ denote the set of k_j VMs in job j. Furthermore, let $V = \bigcup_{j \in J} V_j$ denote the set of all VMs and for each VM $\nu \in V$, $req_{CPU}(\nu)$ and $req_{RAM}(\nu)$ are the CPU and memory requirements of VM ν, respectively.

Let $S = \{s_1, s_2, \ldots, s_n\}$ denote the set of n servers in the data center. For each server $s \in S$, we denote as $cap_{CPU}(s)$ and $cap_{RAM}(s)$ the CPU and memory capacity of server s, respectively. A server s is said to be active if at least one VM is placed in s. Otherwise, it is inactive.

We denote as $ld_r(s)$ the total requirements of a resource $r \in \{CPU, RAM\}$ of the VMs placed in s. $nrm\_ld_r(s)$ denotes the normalized load of s for resource r, and is computed as follows:

$$nrm\_ld_r(s) = \frac{ld_r(s)}{cap_r(s)}. \tag{9}$$

The residual capacity of s for resource r, denoted $rsd\_cap_r(s)$, is given by:

$$rsd\_cap_r(s) = 1 - nrm\_ld_r(s). \tag{10}$$

The energy consumption of a server considers that energy consumption can be accurately described as a linear function of the server's CPU load [7,23]. Let $enr_{idle}(s)$ and $enr_{full}(s)$ denote, respectively, the energy consumed by s when there are no VMs placed in s and when s is at its full capacity. If $ld_{CPU}(s) = 0$, then we assume that s is inactive and consumes no energy. Otherwise, the energy consumption of s is given by the following equation:

$$energy(s) = (enr_{full}(s) - enr_{idle}(s)) \cdot nrm\_ld_{CPU}(s) + enr_{idle}(s). \tag{11}$$

Different VM placements may utilize the resources of a given server differently. If the remaining capacities of a server's resources are not well balanced, it may prevent future VMs from being placed in that server. For example, if a server s has very little memory left, but its CPU is far from being fully utilized, it is very likely that s will not be able to host any CPU intensive VMs anyway due to lack of memory. The resource wastage of s is quantified with the equation:

$$wastage(s) = |rsd\_cap_{CPU}(s) - rsd\_cap_{RAM}(s)|. \tag{12}$$

In a realistic scenario, one must consider that some VMs can already be placed in some server. If necessary, these already placed VMs can be migrated to a different server, but this process has an associated cost to the data center provider. Therefore, the VM set can be seen as partitioned into two sets, a set of placed VMs that can be migrated and a set of fresh VMs that have not yet been placed. We denote as M the existing placement of VMs in servers, i.e.

$$M = \{(\nu, s) : \nu \in V \wedge s \in S \wedge \nu \text{ is placed in } s\}. \tag{13}$$

The associated cost of each VM migration is proportional to the amount of resources the VM is using. Consider the migration of a VM ν to some other server. Based on the observation that the cost of a migration depends on its memory size and memory access patterns [20], we assume that the cost of migrating ν is equal to its memory requirement.

In the VMC problem the goal is to determine a placement of all VMs of V in the servers of S that simultaneously minimizes total energy consumption, resource wastage and migration costs. The VM placement is subject to the following constraints: (1) each VM $\nu \in V$ must be placed in exactly one server and (2) for each server $s \in S$, the sum of the CPU (memory) requirements of the VMs placed in s cannot exceed its CPU (memory) capacity. In order to achieve higher fault tolerance levels, some applications require that all its VMs have to be assigned to different servers. Hence, the VMC problem also considers anti-collocation constraints among VMs in the same job. Given a job $j \in J$ and two VMs $\nu, \nu' \in V_j$, then ν and ν' must be placed in different servers.

We also consider that the provider may enforce a migration budget constraint in order to prevent algorithms from producing placements that exceed a given memory budget b. The purpose of this constraint is to discard placements that require too many migrations, thus deteriorating the performance of the data center. The budget b is specified by the provider and depends on the size of the data center and the nature of its workload. For example, the provider may specify a percentile bp of the data center's total memory capacity. In this case, the budget is defined as $b = bp \cdot \sum_{s \in S} cap_{RAM}(s)$. We refer to bp as the migration budget percentile.

For each server $s_k \in S$, a Boolean variable y_k is introduced that indicates if s_k is active or not. For each VM-server pair $\nu_i \in V$ and $s_k \in S$, we introduce a Boolean variable $x_{i,k}$ that indicates whether ν_i is placed in server s_k. Given a resource r and a server s_k, $ld_r(s_k)$ denotes the load of server s_k for resource r

and is given by:

$$ld_r(s_k) = \sum_{\nu_i \in V} req_r(\nu_i) \cdot x_{i,k}. \tag{14}$$

$$\text{min:} \sum_{s_k \in S} [(enr_{full}(s_k) - enr_{idle}(s_k)) \cdot nrm\_ld_{CPU}(s_k) + enr_{idle}(s_k) \cdot y_k] \tag{15}$$

$$\text{min:} \sum_{s_k \in S} wastage(s_k) \tag{16}$$

$$\text{min:} \sum_{(\nu_i,s_k) \in M} req_{RAM}(\nu_i) \cdot \neg x_{i,k} \tag{17}$$

subject to:

$$ld_r(s_k) \leq y_k \times cap_r(s_k) \qquad k \in \{1,\ldots,n\}, r \in \{CPU, RAM\} \tag{18}$$

$$\sum_{\nu_i \in V_j} x_{i,k} \leq 1 \qquad j \in J, k \in \{1,\ldots,n\} \tag{19}$$

$$\sum_{s_k \in S} x_{i,k} = 1 \qquad i \in \{1,\ldots,m\} \tag{20}$$

$$\sum_{(\nu_i,s_k) \in M} req_{RAM}(\nu_i) \cdot \neg x_{i,k} \leq b \tag{21}$$

$$x_{i,k}, y_k \in \{0,1\} \qquad i \in \{1,\ldots,m\}, k \in \{1,\ldots,n\} \tag{22}$$

Fig. 1. Multi-objective VMC problem formulation.

Figure 1 presents the multi-objective VMC problem formulation, where n denotes the number of available servers in the data center and m the number of VMs to be placed. Objective functions (15), (16) and (17) represent, respectively, the energy consumption, resource wastage and migration costs to be minimized. The constraints in (18) ensure that the resource capacities of each server are not exceeded, while constraints in (19) correspond to the anti-collocation constraints. Constraints in (20) guarantee that each VM is placed on exactly one server. Finaly, the constraint in (21) corresponds to the migration budget constraint. Note that $nrm\_ld_{CPU}(s_k)$, $wastage(s_k)$ and $ld_r(s_k)$ correspond to the values obtained from the expressions introduced in (9), (12) and (14), respectively.

Observe that the VMC formulation is not a MOCO formulation as defined in Sect. 2.3 because the resource wastage objective function is not a PB expression. In order to apply the GIA and CLD algorithms, the modulus in Eq. (12) must be removed through the use of auxiliary variables and constraints. Moreover, the coefficients in Eqs. (15) and (16) might not be integers. In this case, a scaling factor is applied so that all coefficients are integer and the aforementioned algorithms can be applied.

5   Experimental Results

In this section, the performance of the Pareto-MCS based approach is evaluated on the multi-objective VMC use case described in Sect. 4. We refer to our approach for enumerating Pareto-MCSs as PCLD, since the CLD algorithm (Sect. 2.2) was used. First, we compare the performance of PCLD with GIA. Next, the performance of PCLD is compared with that of the state-of-the-art evolutionary algorithms for VMC. All algorithms were implemented in Java. Sat4j-PB [12] (version 2.3.4) was used as the PBS solver, and the evolutionary algorithms were implemented on top of the MOEA Framework[1] (version 2.9.1).

The VMC benchmarks used in this evaluation are based on subsets of workload traces randomly selected from the Google Cluster Data project[2]. The benchmark set includes instances with 32, 64 and 128 servers. For each instance, the sum of VM resource requirements is approximately 25%, 50%, 75% and 90% of the total capacity of the servers. The existing placements (set M) were generated by placing a subset of the VMs. Placements with a sum of requirements of the VMs comprising approximately 0% (no placements), 25%, 50%, 75% and 100% (all VMs placed) of the total VM resource requirements were used. Five different instances were generated for each number of servers, total VM resource requirement and mapping requirement percentile combination, amounting to a total of 300 benchmarks[3]. For each server, the energy consumption parameters enr_{idle} and enr_{full} were chosen from the ranges $[110, 300]$ and $[300, 840]$, respectively, depending on their resource capacities. These ranges are based on the energy consumption values of the Amazon EC2 dataset used in previous works [23].

In the VMC problem instances, it is impractical to find the full set of Pareto-optimal solutions within a reasonable amount of time. Therefore, the evaluation process considers approximations of the Pareto-optimal solution set that each algorithm is able to produce within the time limit of 1800 s. The hypervolume (HV) [24] provides a combined measure of convergence and diversity of a given approximation, thus being a quality indicator commonly used to compare the performance of multi-objective optimization algorithms. HV measures the volume of the objective space between the set of non-dominated solutions and a given reference point. The reference point depends on the benchmark, and is set to the worst possible objective values. Hence, larger values of HV mean that the solution set is composed of solutions of better quality and/or diversity. The evaluation was conducted on an AMD Opteron 6376 (2.3 GHz) running Debian jessie and each algorithm was executed with a memory limit of 4 GB. Evolutionary algorithms were executed with 10 different seeds for each instance, and the analysis is performed using the median values over all executions.

---

[1] http://moeaframework.org/.

[2] http://code.google.com/p/googleclusterdata/.

[3] http://sat.inesc-id.pt/dome.

5.1   Pareto-MCSs vs GIA

Table 2, compares the performance of PCLD with that of GIA, with migration budget percentiles of 100% (no migration restriction), 5%, 1% and 0.5%. Observe that migration budgets are not relevant for the 60 instances where $M = \emptyset$ (no VMs already placed in servers). Hence, these are not considered in the 5%, 1% and 0.5% rows. The 'Solved' column presents the number of instances for which at least one solution was found. The 'HV' columns show the number of instances for which the algorithm had the best hypervolume (HV) among both algorithms ('wins' column) and the average of the differences between the HV obtained by the algorithm and the best HV ('avg' column). For example, assuming that the best HV obtained for an instance is 0.9, and the algorithm obtained an HV of 0.8, then the difference for that instance is 0.1.

Table 2. Number of VMC instances solved and overall comparison of GIA and PCLD.

Algorithm	Budget percentile	Solved	HV	
			Wins	Avg.
GIA	100%	215	49	0.01212
PCLD		**217**	**178**	**0.00094**
GIA	5%	186	51	0.00909
PCLD		**187**	**152**	**0.00101**
GIA	1%	198	59	0.00704
PCLD		**199**	**157**	**0.00124**
GIA	0.5%	**199**	72	0.00658
PCLD		**199**	**147**	**0.00163**

Results in Table 2 clearly show that PCLD is able to find much better quality approximations of the solution set than those found by GIA. However, the quality difference becomes smaller as the budget percentile decreases. This happens because migration budget constraints reduce the search space of the problem, possibly resulting in instances that are easier to solve using constraint based methods. This claim is backed by the results in Table 2, where one can observe that both algorithms are able to find solutions for more instances when using smaller migration budgets.

5.2   Pareto-MCSs vs Evolutionary Algorithms

Table 3, shows the results PCLD and evolutionary algorithms VMPMBBO [23], MGGA [22] and NSGAII [6]. MGGA was adapted to consider migration costs instead of thermal dissipation and configured to use a population size of 12, and crossover rate and mutation rate as suggested by Xu and Fortes [22].

Table 3. Comparison of PCLD, VMPMBBO, MGGA and NSGAII on VMC instances.

Algorithm	Budget percentile	Solved		HV	
		Always	Median	Wins	Avg.
PCLD	100%	217	217	32	0.17980
VMPMBBO		**226**	**227**	**212**	**0.01777**
MGGA		220	225	0	0.17508
NSGAII		224	231	17	0.16518
PCLD	5%	**187**	**187**	**108**	0.05170
VMPMBBO		103	111	85	**0.00606**
MGGA		12	13	0	0.30225
NSGAII		13	14	0	0.28676
PCLD	1%	**199**	**199**	**182**	**0.00158**
VMPMBBO		76	83	21	0.05155
MGGA		0	0	-	-
NSGAII		0	0	-	-
PCLD	0.5%	**199**	**199**	**187**	**0.00060**
VMPMBBO		67	76	15	0.06063
MGGA		0	0	-	-
NSGAII		0	0	-	-

We note that VMPMBBO was originally designed for optimizing energy consumption and resource wastage. When $M = \emptyset$, no migrations occur and VMPMBBO was run with the configuration suggested by Zheng *et al.* [23]. However, when $M \neq \emptyset$, we have to consider migration costs. VMPMBBO's population is divided into subsystems and each subsystem optimizes a single objective function. The suggested configuration uses 4 subsystems, 2 per objective. When $M \neq \emptyset$, we use 6 subsystems instead to account for migration costs. NSGAII [6] is a general purpose genetic algorithm. It was configured with a population size, crossover rate and mutation rate of 100, 0.8 and 0.05 respectively, as suggested by our fine tuning experiments.

When compared with the remaining algorithms, VMPMBBO with a budget percentile of 100% has the best performance, finding solution sets for more instances and of better quality. However, considering large migration costs is not realistic for an active data center where live migrations may result in a large performance deterioration of running applications. Therefore, considering a limited budget for migrations of VMs is more realistic. In these cases, PCLD is, by far, the best performing algorithm when smaller budgets are used. MGGA's and NSGAII's performance deteriorates the most as the budget decreases. Both algorithms are barely able to find even a single solution for budgets of 5% or less. With a budget of 5%, VMPMBBO already solves less 75 instances than PCLD, despite being competitive in some instances. However, with budgets of 1% and

0.5%, VMPMBBO is completely outclassed by PCLD. Not only is PCLD able to find solutions for many more instances (116 and 122 more with budgets of 1% and 0.5% respectively), the corresponding solution sets are of far better quality than VMPMBBO's.

6 Conclusion and Future Work

This paper introduces the Pareto Minimal Correction Subset (Pareto-MCS) of a multi-objective constraint optimization problem. An equivalence relationship between Pareto-MCSs and Pareto-optimal solutions is proved, showing that Pareto-optimal solution enumeration can be reduced to Pareto-MCS enumeration. Additionally, we show that Pareto-MCS enumeration can be reduced to MCS enumeration, allowing for state-of-the-art efficient MCS enumeration algorithms to be used right off-the-shelf instead of developing entirely new algorithms for Pareto-MCS enumeration.

An experimental evaluation on instances of the Virtual Machine Consolidation (VMC) problem shows that Pareto-MCS enumeration clearly outperforms the state-of-the-art on a large set of problem instances, while remaining competitive on all instances. Not only is this new approach able to find solutions for more instances, but it is also able to find solution sets of far higher quality.

Finally, we note that there is still room for improvement in the Pareto-MCS enumeration procedure. Our approach does not yet account for weights of soft constraints in the search process. We plan to develop heuristics to help guiding the MCS enumeration process towards Pareto-optimal solutions more effectively.

References

1. Bacchus, F., Davies, J., Tsimpoukelli, M., Katsirelos, G.: Relaxation search: a simple way of managing optional clauses. In: 28th Conference on Artificial Intelligence, pp. 835–841. AAAI (2014)
2. Bailey, J., Stuckey, P.J.: Discovery of minimal unsatisfiable subsets of constraints using hitting set dualization. In: Hermenegildo, M.V., Cabeza, D. (eds.) PADL 2005. LNCS, vol. 3350, pp. 174–186. Springer, Heidelberg (2005). doi:10.1007/978-3-540-30557-6_14
3. Ben-Eliyahu, R., Dechter, R.: On computing minimal models. Ann. Math. Artif. Intell. **18**(1), 3–27 (1996)
4. Birnbaum, E., Lozinskii, E.L.: Consistent subsets of inconsistent systems: structure and behaviour. J. Exp. Theoret. Artif. Intell. **15**(1), 25–46 (2003)
5. Bjørner, N., Phan, A.-D., Fleckenstein, L.: νZ - an optimizing SMT solver. In: Baier, C., Tinelli, C. (eds.) TACAS 2015. LNCS, vol. 9035, pp. 194–199. Springer, Heidelberg (2015). doi:10.1007/978-3-662-46681-0_14
6. Deb, K., Agrawal, S., Pratap, A., Meyarivan, T.: A fast elitist non-dominated sorting genetic algorithm for multi-objective optimization: NSGA-II. In: Schoenauer, M., Deb, K., Rudolph, G., Yao, X., Lutton, E., Merelo, J.J., Schwefel, H.-P. (eds.) PPSN 2000. LNCS, vol. 1917, pp. 849–858. Springer, Heidelberg (2000). doi:10.1007/3-540-45356-3_83

7. Fan, X., Weber, W., Barroso, L.A.: Power provisioning for a warehouse-sized computer. In: 34th International Symposium on Computer Architecture, pp. 13–23 (2007)
8. Felfernig, A., Schubert, M., Zehentner, C.: An efficient diagnosis algorithm for inconsistent constraint sets. Artif. Intell. Eng. Des. Anal. Manuf. **26**(1), 53–62 (2012)
9. Grégoire, É., Lagniez, J., Mazure, B.: An experimentally efficient method for (mss, comss) partitioning. In: 28th Conference on Artificial Intelligence, pp. 2666–2673. AAAI (2014)
10. Ignatiev, A., Janota, M., Marques-Silva, J.: Towards efficient optimization in package management systems. In: 36th International Conference on Software Engineering, pp. 745–755 (2014)
11. Junker, U.: QUICKXPLAIN: preferred explanations and relaxations for over-constrained problems. In: 19th National Conference on Artificial Intelligence, 16th Conference on Innovative Applications of Artificial Intelligence, pp. 167–172 (2004)
12. Le Berre, D., Parrain, A.: The sat4j library, release 2.2. J. Satisfiab. Bool. Model. Comput. **7**(2–3), 59–64 (2010)
13. Manquinho, V., Marques-Silva, J.P., Planes, J.: Algorithms for weighted boolean optimization. In: Kullmann, O. (ed.) SAT 2009. LNCS, vol. 5584, pp. 495–508. Springer, Heidelberg (2009). doi:10.1007/978-3-642-02777-2_45
14. Marques-Silva, J., Heras, F., Janota, M., Previti, A., Belov, A.: On computing minimal correction subsets. In: 23rd International Joint Conference on Artificial Intelligence, IJCAI, pp. 615–622 (2013)
15. Mencía, C., Previti, A., Marques-Silva, J.: Literal-based MCS extraction. In: 24th International Joint Conference on Artificial Intelligence, IJCAI, pp. 1973–1979 (2015)
16. Nöhrer, A., Biere, A., Egyed, A.: Managing SAT inconsistencies with HUMUS. In: 6th International Workshop on Variability Modelling of Software-Intensive Systems, pp. 83–91 (2012)
17. O'Callaghan, B., O'Sullivan, B., Freuder, E.C.: Generating corrective explanations for interactive constraint satisfaction. In: Beek, P. (ed.) CP 2005. LNCS, vol. 3709, pp. 445–459. Springer, Heidelberg (2005). doi:10.1007/11564751_34
18. Pareto, V.: Manuale di Economia Politica, vol. 13. Societa Editrice, Milano (1906)
19. Rayside, D., Estler, H.C., Jackson, D.: The guided improvement algorithm for exact, general-purpose, many-objective combinatorial optimization. Technical report. MIT-CSAIL-TR-2009-033, MIT Massachusetts Institute of Technology (2009)
20. Salfner, F., Tröger, P., Polze, A.: Downtime analysis of virtual machine live migration. In: 4th International Conference on Dependability, IARIA, pp. 100–105 (2011)
21. Ulungu, E.L., Teghem, J.: Multi-objective combinatorial optimization problems: a survey. J. Multi-Crit. Dec. Anal. **3**(2), 83–104 (1994)
22. Xu, J., Fortes, J.A.B.: Multi-objective virtual machine placement in virtualized data center environments. In: IEEE/ACM International Conference on Green Computing and Communications, GreenCom, & International Conference on Cyber, Physical and Social Computing, CPSCom, pp. 179–188 (2010)
23. Zheng, Q., Li, R., Li, X., Shah, N., Zhang, J., Tian, F., Chao, K., Li, J.: Virtual machine consolidated placement based on multi-objective biogeography-based optimization. Future Gener. Comput. Syst. **54**, 95–122 (2016)
24. Zitzler, E., Thiele, L.: Multiobjective evolutionary algorithms: a comparative case study and the strength pareto approach. IEEE Trans. Evol. Comput. **3**(4), 257–271 (1999)

Parallel SAT Solving

A Distributed Version of Syrup

Gilles Audemard$^{(\boxtimes)}$, Jean-Marie Lagniez, Nicolas Szczepanski,
and Sébastien Tabary

Univ. Lille-Nord de France, CRIL/CNRS UMR 8188, Lens, France
{audemard,lagniez,szczepanski,tabary}@cril.fr

Abstract. A portfolio SAT solver has to share clauses in order to be efficient. In a distributed environment, such sharing implies additional problems: more information has to be exchanged and communications among solvers can be time consuming. In this paper, we propose a new version of the state-of-the-art SAT solver Syrup that is now able to run on distributed architectures. We analyze and compare different programming models of communication. We show that, using a dedicated approach, it is possible to share many clauses without penalizing the solvers. Experiments conducted on SAT 2016 benchmarks with up to 256 cores show that our solver is very effective and outperforms other approaches. This opens a broad range of possibilities to boost parallel solvers needing to share many data.

1 Introduction

Using SAT technology to solve problems coming from different areas is increasingly becoming a standard practice. As recent examples, one can cite the use of SAT solvers to solve mathematical conjectures [19,20]. These new application domains produce harder and harder SAT formulas. At the same time, it becomes difficult to improve SAT solvers. It is noticeable even by reading the contents of the last SAT conferences. Then, a way to solve such hard SAT formulas is to use parallel SAT solvers [19].

During the last decade, several parallel SAT solvers have been developed [5,7,9,13,18,32]. Even if it exists some distributed parallel SAT solvers, that is, SAT solvers working on different computers [5,9,13], most of them, are multi-core ones. In that case, solvers use multiple cores of only one computer [7,18,32]. One of the reasons is that parallel SAT solvers exchange data between sequential solvers, and communications are easier with a multi-core architecture rather than a distributed one. Nevertheless, computers with more than 32 cores are very expensive and remain limited to a given number of cores. A way to overcome this problem and thus, acquire many more cores, consists in considering several computers. This case is easy to obtain by using the tremendous number of computers available in grid or cloud computing.

Authors were partially supported by the SATAS ANR-15-CE40-0017.

S. Gaspers and T. Walsh (Eds.): SAT 2017, LNCS 10491, pp. 215–232, 2017.
DOI: 10.1007/978-3-319-66263-3_14

In distributed environment, one has to deal with two issues: the increase of search units and the exchange of learned clauses through the network. The former implies that systematically more clauses will be produced and the latter that more messages need to be exchanged between the units and thus, causing overheads. To confront the first problem, already present in multi-core architectures, solvers carefully select the clauses to exchange or manage each shared clause in a different way [3,7]. Concerning the second problem, to the best of our knowledge, the best approach consists in using topologies of message exchange as proposed in TOPOSAT [13].

The aim of this paper is to extend the parallel SAT solver SYRUP [7] to distributed environments. This solver is a portfolio based approach with clause sharing, initially designed for multi-core architectures. The key point of its efficiency comes from the way that clauses are carefully selected before being exchanged and a dedicated data structure is used to manage them. Both are necessary to avoid overloading search units.

Although SYRUP is scalable up to 32 cores (see Sect. 3), it was not initially designed to run over this number of cores. During the last SAT competition SYRUP used 24 cores out of the 48 available cores. In particular, the quantity of clauses exchanged between the search units, and thus the number of clauses a solver has to deal with, increases with the number of cores. To reduce the impact of exchanging this potentially huge quantity of information, the number of clauses that can be stored in the buffer, that is placed in the shared memory, was limited in the initial version of SYRUP. This limit depends on the number of available cores and was set empirically. Consequently, it is not obvious that parameters selected to tune SYRUP on less than 32 cores are always the best when the number of cores increases. Therefore, in Sect. 4, the impact of this parameter on each solver (taken individually) is empirically evaluated. The results show that sharing too many clauses dramatically slows down the solvers. Consequently, a new limit for the buffer size is proposed.

Another important point concerns the way the information is communicated between solvers. When running SYRUP on multi-core architecture information is exchanged by using the shared memory almost for free. Nevertheless, when considering distributed architectures, communication can be costly and not so trivial to implement. Indeed, as empirically demonstrated in Sect. 4, it exists several ways to communicate information between solvers and the used schema has a direct impact on the effectiveness of the solver. It is shown that if solvers do fewer communication cycles then they can share more clauses without affecting performances (namely the number of unit propagation per second does not dramatically decrease). To avoid these cycles, a new fully hybrid distributed implementation of SYRUP is proposed. This schema allows to exchange clauses faster and, as it is empirically shown in the end of the Sect. 4, sharing clauses, as soon as possible, among search units provides better results.

The rest of the paper is organized as follows. In the next section, basic notions about SAT as well as sequential and parallel CDCL SAT solvers are provided. In Sect. 3, we briefly describe the global architecture of the parallel SAT solver

SYRUP. In Sect. 4, we present and evaluate two existing distributed programming models which are already used for SAT. Section 5 introduces our fully hybrid version of SYRUP. Before concluding and providing original perspectives (Sect. 7), in Sect. 6 we compare the different versions of SYRUP itself and then, the fully hybrid version of SYRUP against two state-of-the-art distributed solvers.

2   Preliminaries

A *CNF formula* is a conjunction of *clauses* built on a finite set of Boolean variables where a clause is a disjunction of *literals*. A literal is either a Boolean variable (x) or the negation of a Boolean variable ($\neg x$). A *unit* (resp. *binary*) clause is a clause with only one literal (resp. two literals), called *unit literal* (resp. *binary clause*). An interpretation assigns a value from $\{0,1\}$ to every Boolean variable, and, following usual compositional rules, naturally extended to a CNF formula. A formula ϕ is consistent or satisfiable when it exists at least one interpretation that satisfies it. This interpretation is then called a model of ϕ and is represented by the set of literals that it satisfies.

SAT is the NP-complete problem that consists in checking whether or not a CNF is satisfiable, i.e. whether or not the CNF has a model. Several techniques have been proposed to tackle this problem in practice (see [14] for more details). In this paper, we focus on CDCL SAT solvers exploiting the so-called Conflict Driven Clause Learning features (see e.g. [12,24,25,34]) which are currently the most efficient complete SAT solvers. Let us recall briefly the global architecture of a CDCL SAT solver. CDCL solving is a branching search process, where at each step a literal is selected for branching. Usually, the variable is picked w.r.t. the VSIDS heuristic [25] and its value is taken from a vector, called polarity vector, in which the previous value assigned to the variable is stored [28]. Afterwards, Boolean constraint propagation is performed. When a literal and its opposite are propagated, a conflict is reached, a clause is learnt from this conflict [24] and a backjump is executed. These operations are repeated until a solution is found (satisfiable) or the empty clause is derived (unsatisfiable). CDCL SAT solvers can be enhanced by considering restart strategies [16] and deletion policies for learnt clauses [4,15]. Among the measures used to identify relevant clauses, the Literal Blocked Distance measure (LBD in short) proposed in GLUCOSE is one of the most efficient. Then, as experimentally shown by the authors of [6], clauses with smaller LBD should be considered more relevant.

Many attempts have been made recently to build SAT solvers on top of multi-core architectures. These solvers can partition the search space and so, work on subformula [22,32] or they can work on the entire formula using a portfolio approach [3,10,18]. Clause sharing plays a key role in such solvers. Each thread selects some of the learnt clauses and shares them using shared memory. Since sharing too many clauses can have a negative impact on the effectiveness of solvers (for instance by slowing down the Boolean constraint propagation), several strategies have been proposed in order to select which clauses have to be communicated [3,7,18]. In addition to this problem, communications between solvers can be prohibitive. Usually, one has to deal with critical section, shared memory and cache misses, involving an important bottleneck in communications (see [1] for a thorough experimental evaluation of parallel solvers on multi-core architectures).

Some SAT solvers are also implemented on distributed architectures [5,9,13]. When considering such distributed architectures using message passing such as MPI (Message Passing Interface), one has to take the cost brought from the data link (CPU time, memory and network speed) into account. Typically, there are two ways to tackle this problem. The first one consists in reducing the number of messages exchanged by sharing clauses carefully between solvers using graph topologies (one-to-one, ring, ...) [13]. Another way to reduce the communication cost, is by realizing a hybridization between Multi-Threading and Distributed architectures. When considering pure message passing architectures, such as those used in [5,13], messages between programs are exchanged as if processes are running on independent computers. Conversely, in hybrid models, processes running on the same computer communicate via shared memory and messages are used only to communicate between computers. As demonstrated in [9] for SAT and more generally in [2,30], using such hybridization drastically reduces the network congestion. The authors of [11] present an extended taxonomy of different distributed programming models. In Sect. 4, we will study in details the communication in distributed architectures.

3   Syrup: A Portfolio Approach

Syrup is a portfolio-based parallel SAT solver based on Glucose [7]. In Syrup, clauses are exchanged using the shared memory. Figure 1 gives an overview of how two (Glucose) solvers, $S1$ and $S2$, working on two different threads communicate. Let us present how clauses are imported and exported. When a solver decides to share a clause (represented by black dotted lines), then this clause is added into a buffer located in the shared memory. Similarly to its predecessors [3,10,18], Syrup tries to reduce the number of clauses shared between cores by only sending learnt clauses that are expected to be very useful. That is, unit clauses, binary and glue (LBD = 2) ones. The other clauses are only exported if they are seen at least twice during the conflict analysis. It is important to note that clauses are exported using a buffer of limited size (by default the size is set to 0.4 MB × the number of threads), and that a clause is removed from the buffer when all threads have imported it. Consequently, if the buffer is full, clauses that should have been exchanged are discarded and therefore not exported. This buffer prevents an overload of the shared memory.

In Syrup, imported clauses (represented by black plain lines) are considered by a solver only right after a restart occurs. In addition, imported clauses are not directly added into the solver. As a special feature, Syrup allows their importation in a lazy manner. Imported clauses are initially watched by only one literal which is sufficient to ensure that any conflicting clause will be detected during unit propagation. If an imported clause is conflicting, then it is treated as

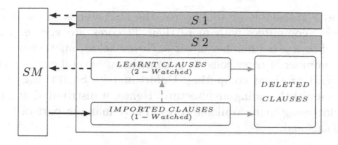

Fig. 1. Schematic overview of SYRUP. A black dotted (resp. plain) line means that a clause is exported to (resp. imported from) the shared memory (SM). A gray dotted line represents a promoted clause due to a conflict. The gray plain lines represent deleted clauses during the databases reduction.

a local learnt clause and becomes 2-watched (represented by gray dotted lines). This allows to dynamically select useful clauses to be imported, namely, the ones that are relevant to the current solver search space. Furthermore, using the 1-literal watched scheme limits the overhead of imported clauses. Since the number of imported clauses can become huge, SYRUP uses a dedicated strategy to periodically remove clauses that are supposed to be useless for the search (represented by gray plain lines).

In order to evaluate the scalability of SYRUP, we ran it on all hundred benchmarks of the SAT Race 2015 (parallel track) using 1 (which is the version 4.1 of GLUCOSE), 2, 4, 8, 16 and 32 cores. The wall clock time limit was set to 1,800 s. For these experiments a computer of 32 cores with 256 GB of RAM has been used (quad-processor Intel XEON X7550). As shown in Fig. 2, SYRUP is clearly scalable. The sequential GLUCOSE solves 26 benchmarks (15 SAT, 11 UNSAT), SYRUP with 8 cores solves 56 (31 SAT, 25 UNSAT) and with 32 cores it solves 70 benchmarks (42 SAT, 28 UNSAT).

#cores	SAT	UNS	Total
1	15	11	26
2	25	15	40
4	29	23	52
8	31	25	56
16	36	27	63
32	42	28	70

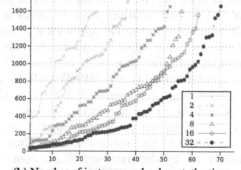

(a) Overview table   (b) Number of instances solved w.r.t. the time

Fig. 2. Scalability of SYRUP

Since SYRUP is scalable up to 32 cores, the next step is to use it with more cores. However, computers with more than 32 cores are very expensive and remain limited to a given number of cores. Nevertheless, another way to obtain an unlimited number of cores consists in considering several computers as those available in grid or cloud computing. Unfortunately, SYRUP was designed to only work on multi-threading architecture. Hence, a distributed architecture is appropriate for using more available cores. The remaining part of the paper is dedicated to this end.

4   Parallel SAT in Distributed Architecture

As aforementioned, parallel portfolio solvers exchange a tremendous quantity of information. When considering multi-core architectures this information can be efficiently collected by the different solvers using the shared memory. Since all processors share a single view of the data, the communication between the solvers can be as fast as memory accesses to the same location, which explains the need of such a centralized scheme. However, when considering a distributed architecture, using a centralized scheme can be prohibitive when the number of processes that need to communicate is huge. Indeed, as experimentally demonstrated in [5], the process in charge of collecting information and sharing them between the solvers becomes quickly congested. Consequently, using a centralized architecture to communicate information may not be the best solution when the aim is to use distributed architecture with a considerable number of computers.

To overcome the bottleneck induced by centralized architecture, each process will directly communicate with all others. However, communications have to be realized carefully to avoid deadlocks occurring when a process requests a resource that is already occupied by others. For instance, let us consider the case where each process is implemented so that it first realizes a blocking **send** and then it realizes a blocking **receive**. In such a situation, if two processes send a message at once, since each process has to wait that the message is received by the other to go on, then they are both blocked. More specifically, a blocking send is not completed until the corresponding receive is not executed. This problem is well known and several solutions exist to avoid and detect such deadlocks [8,21]. In SAT behavior, most of the approaches proposed in the literature [5,9,13] use non-blocking or/and collective communications to avoid this problem. Generally, the retained solution consists in waiting that all sender buffers can be safely accessed and modified (using MPI_Waitall) in order to finalize a communication cycle (and then avoid deadlock). We follow this approach for the distributed versions of SYRUP presented in this section.

Another important point concerns the way communications are performed between sequential SAT solvers. Two solutions are generally considered: (i) only one thread serially alternates search and communication steps (it is the solution retained in AMPHAROS [5]) or (ii) two distinct threads that separately realize these two tasks (it is the solution retained by the authors of HORDESAT [9] and TOPOSAT [13]). In the following, we decide to combine communicator threads devoted to exchange information (clauses and solution notification) and search threads dedicated to solve the problem. In such a solution, each communicator thread can be attached to one or several solver threads running on the same machine. Therefore, a communicator thread alternates between sharing clauses phase and sleeping phase (this thread is in standby during a given time). A communication cycle consists in one sharing clauses phase followed by one sleeping phase. During a communication cycle, and in order to avoid deadlocks, communicator threads use critical sections (mutex) and have to wait for sender buffers to be available twice (via MPI_Waitall): a first time to check if a solution has been found and a second time to exchange clauses.

In the rest of this section, we empirically evaluate a distributed version of SYRUP using communicator threads with communication cycles as aforementioned. In this evaluation we make the distinction between a pure message passing programming model, where couples (solver, communicator) threads are considered as computers (allows to only use a MPI library), and a partially hybrid programming model, where communication between solvers on the same machine are realized using shared memory (using both MPI and shared memory libraries).

4.1   Pure Message-Passing Programming Model

This programming model respects the communication protocol presented in introduction of this section. Precisely:

- it uses several processes per computer;
- each sequential SAT solver forms a process: it is able to perform search as well as communication by managing one or two threads (solver and communicator);
- it uses the message passing (*i.e.,* MPI library or equivalent) to exchange data between processes, even between processes running on the same computer;
- data are duplicated for each solver on both network and shared memory.

Figure 3 illustrates this programming model with two threads per solver process. The solver process ($PROCESS_1$), represented by a dashed square, embeds a search thread S_1 and its own communicator thread C_1. Data (*i.e.,* clauses) are shared between these two threads using the shared memory via a reception buffer called $DATA$. On this picture, we focus on data sent by C_1 to all other solvers. In fact, in pure message passing model, each solver behaves as an independent computer. The main drawback of this architecture is the data replication which decreases the communication speed. On the same computer, it requires the copy of the data between solvers without sharing a global address space (black arrows). And, when data have to be sent to solvers running on another computer, it requires sending one message for each sequential solver (gray arrows) rather than one per computer. As a result, the same data is duplicated as many times as the number of solvers and thus, data exchange may cause network congestion. However, the advantage of this programming scheme is the possibility to use only a MPI library or equivalent, thereby facilitating application development and their debugging. This means that, using this programming

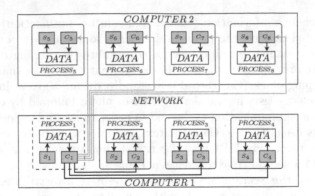

Fig. 3. Pure message passing programming model: a focus is performed on data (clauses) sent by the solver S_1 to other solvers with two computers.

model, a distributed solver can directly be created from a sequential SAT solver rather than a multi-core one.

The pure message-passing programming model is the solution retained by the authors of TOPOSAT to realize their communications. They have chosen 5 s for their communication cycles. A key point of TOPOSAT architecture is the use of non-blocking and point to point communication that allow building different topologies of message exchange (2-dimensional grid, medium-coupled, ...). These topologies reduce the network congestions caused by this model. However, since our aim is not to compare topologies but low level communication models, we only implement the topology where each process communicates with all others.

In order to compare the different programming models, we have developed a pure message-passing approach based on the sequential SAT solver GLUCOSE using communicator threads as presented in introduction of this section. This implementation directly provides a distributed version of SYRUP. In a first experiment, we have studied the impact of the reception buffer size (*DATA*) and the frequency of exchange on the global performance of each sequential solver (roughly, the number of unit propagation per second). To this end, 256 processes have been run on 32 computers of 8 cores during 100 s (wall clock time). We have selected 20 varied instances coming from the SAT Race 2015 (parallel track) and not solved by SYRUP within 300 s (results obtained in the Sect. 3 with 32 cores). In order to evaluate the impact of the reception buffer sizes and the communication cycles times on the solver effectiveness, three reception buffer sizes (20 MB, 100 MB, and 200 MB) and three different communication cycles times (every 0.5, 5, and 10 s) have been considered.

Table 1 reports for each combination the average number of received clauses noted "Received" in the table, the average number of received clauses really retained by solvers noted "Retained" (imported clauses through the network that effectively become 1-watched, see Sect. 3), the average number of retained clauses that lead to a conflict noted "Retained (conflicts)", the average number

Table 1. Pure message passing model. Impact of the reception buffer size (*DATA*) and the frequency of exchange on the global performances of solvers. All results correspond to the obtained average of all processes per second.

	20 MB			100 MB			200 MB		
Com. cycle	**0.5 s**	**5 s**	**10 s**	**0.5 s**	**5 s**	**10 s**	**0.5 s**	**5 s**	**10 s**
Received	8,270	30,989	38,820	9,920	28,369	39,890	11,007	28,296	34,346
Retained	82%	25%	0.8%	93%	51%	34%	93%	71%	62%
Ret. (conflicts)	43	55	26	54	51	16	72	33	12
Ret. (useful)	494	618	306	553	575	173	783	452	192
Propagation	100,379	489,289	525,738	91,103	479,145	540,286	69,912	469,021	443,696
Conflicts	216	823	1,101	254	734	1,069	275	721	891

of retained clauses used during the unit propagation process noted "Retained (useful)", the average number of unit propagation per second noted "Propagation" and the average number of conflicts per second noted "Conflicts". For all these measures the average is taken on all benchmarks.

Firstly, we can observe that whatever the *DATA* buffer size is, the worst results are obtained for a cycle of 0.5 s. Indeed the number of propagation and conflicts per second is drastically affected: 5 times less than when using cycles of 5 or 10 s. This is due to blocking functions (MPI_WaitAll and Mutex operations) used during the communication cycles. Indeed, the shorter the cycle is, the more these functions are called. When we increase the communication cycle to 5 s, the number of these blocking functions is reduced and consequently the global performance of the solver in term of propagations and conflicts is improved. However, we can observe that the number of retained clauses decreases when the time of the communication cycle increases. Indeed, more clauses have to be exchanged, consequently the shared memory buffer has more chance to become full. This is a potential drawback, because some of these discarded clauses can be useful ones.

We focus now on a cycle of 5 s. We observe that the percentage of retained clauses increases as the buffer size grows. The same amount of data is sent but fewer clauses are deleted. The number of retained clauses for useful and conflicted as well as the number of propagation and conflicts slightly decreases when the size of the buffer increases. This is explained by an overload of the solvers due to the management of the clauses database.

4.2   Partially Hybrid Programming Model

We propose a partially hybrid model as an alternative to overcome the drawbacks of the pure message passing scheme. This programming scheme is as follows:

- it embeds only one process per computer;
- several sequential SAT solvers form a process: it is a clustering of search threads associated with only one communicator thread;

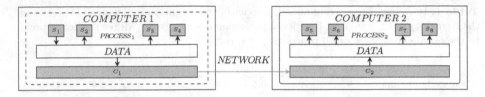

Fig. 4. Partially hybrid programming model: a focus is performed on data (clauses) sent by the solver S_1 to other solvers with two computers.

- two independent libraries are used to communicate: one to deal with shared memory and another one to deal with network communications;
- it pools data that have to be sent over the network.

The communicator thread is used as an interface to share data between solvers on distinct computers whereas the shared memory allows to exchange data between solvers on the same computer. Figure 4 depicts this schema. Each computer uses a single process that embeds all sequential solvers. In this figure, solver S_1 sends clauses to S_2, S_3, S_4 and its communicator C_1 using the buffer located in the shared memory (*DATA*). Communicator C_1 sends clauses through the network to the communicator C_2 which distributes them to S_5, S_6, S_7, S_8. Obviously, the main advantage of this approach is the mutualization of data sent over the network.

Table 2. Partially hybrid programming model. Impact of the buffer size (*DATA*) and the frequency of exchange on the global performances of solvers. All results correspond to the obtained average of all processes per second.

	20 MB			100 MB			200 MB		
Com. cycle	0.5 s	5 s	10 s	0.5 s	5 s	10 s	0.5 s	5 s	10 s
Received	28,202	38,448	41,826	26,980	36,505	36,370	26,836	34,643	35,427
Retained	84%	68%	43%	86%	85%	81%	86%	86%	86%
Ret. (conflicts)	342	318	209	325	280	190	326	258	176
Ret. (useful)	3,841	3,256	2,578	3,429	2,688	2,008	3,647	2,603	1,853
Propagation	504,540	636,850	658,177	528,529	577,653	603,685	511,709	559,477	595,084
Conflicts	976	1,364	1,283	928	1,253	1,248	929	1,184	1,203

The solver HORDESAT is based on this programming model. Its key feature is the management of sent messages through the communicator thread. It involves a collective communication of fixed buffer size of clauses (every 5 s). If the amount of data that has to be sent is not sufficient, some padding is added to obtain the desired size. Reversely, when the size limit is reached, clauses are simply skipped. To avoid this phenomena, the authors propose to adapt (for the next sending) the amount of clauses provided by the search threads to the demand.

We have developed our own partially hybrid solver, using communicator threads as presented in the beginning of this section, and based on the parallel SAT solver SYRUP (version 4.1). Note that in this programming model, one has to deal with two different buffers: the buffer located in the shared memory and used by SYRUP, and the buffer used to send clauses over the network. Nevertheless, we think that the collective communication of fixed size used by HORDESAT to send clauses on the network is an obstacle for the efficiency of distributed SAT solvers. Therefore, we propose to perform a collective communication that exchanges data of different sizes. This is allowed by MPI libraries. It discards sending useless integers and prevent discarding some clauses.

As in Sect. 4.1, we also investigate the impact of the buffer size *DATA* located in the shared memory and the frequency of exchange on the global performances of solvers for the partially hybrid programming model. Results are presented in Table 2. They show the same trend as for the pure model (Table 1). Indeed, we can also observe that short communication cycles penalize the performances of the sequential solvers and a large size of buffer reduces the number of useful clauses retained. An important point to note is that the partially hybrid model provides higher propagations and conflicts rates, showing that this approach seems more beneficial.

5 Fully Hybrid Programming Model

As suggested by the results reported in Table 2, it seems beneficial to share clauses between sequential solvers over the network as soon as possible. Indeed, we observed that the amount of useful retained clauses, whatever the buffer size (see lines Retained (conflicts) and Retained (useful)) is more important when the communication cycles frequency increases. However, when considering short cycle times, we pointed out that the partially hybrid model suffers from critical section problems that drastically reduce the effectiveness of solvers. In this section, we tackle this problem by introducing an algorithm that has fewer critical sections and sends clauses as soon as possible without communication cycle.

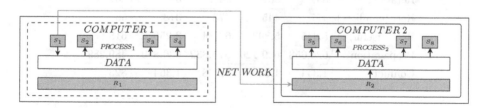

Fig. 5. Fully hybrid message passing programming model: a focus is performed on data sent by the solver S_1 to other solvers between two computers.

In order to communicate clauses directly, search threads bypass the communicator and send clauses themselves on the network. Consequently, the communicator thread of each computer becomes a receiver thread. Contrary to the partially hybrid model, this programming model allows several threads to communicate on the network at the same time. We call such a programming model fully hybrid.

Since there is no more communication cycle, one has to avoid deadlock in a different way. Thread safety has to be established to ensure that sender and listener threads never deadlock for any ordering of thread execution [8]. This is guaranteed due to the distributed mutual exclusion algorithms [23,27,29,33] which state that only one process is allowed to execute the critical section at any given time. The overhead for imposing thread safety for MPI have been largely studied [31], and most implementations do not have sophisticated locking mechanisms algorithms. However, the MPICH implementation [26] of MPI works fine with the fully hybrid programming model. In [17], the authors detail a new conceptual concurrency support mechanism for MPI to deal with this problem. To the best of our knowledge, no distributed SAT solver is based on this programming model.

Figure 5 depicts the fully hybrid programming model. Each computer uses a single process that embeds all solver threads and one receiver thread. In this picture, the solver S_1 sends data with S_2, S_3 and S_4 via the shared memory ($DATA$) and to the 2nd computer via the network. The receiver thread R_2 receives data (here a clause and not a pool of clauses) and sends it to S_5, S_6, S_7 and S_8 via the shared memory buffer.

Table 3. Fully hybrid model. Impact of the buffer size ($DATA$) on the global performances of solvers. All results correspond to the obtained average of all processes on 20 instances per second. Last two columns recall best compromise for pure and partially hybrid models.

Model	Fully			Partially	Pure
	20 MB	100 MB	200 MB	20 MB	20 MB
Received	35,776	32,840	33,343	38,448	30,989
Retained	83%	86%	86%	68%	25%
Ret. (conflicts)	477	445	473	318	55
Ret. (useful)	5,845	5,899	6,184	3,256	618
Propagation	626,658	629,126	614,832	636,850	489,289
Conflicts	1,344	1,246	1,268	1,364	823

Theoretically, the search thread is blocked until a sent clause is received by all other computers. However, in practice a buffer is used to allow the solver to send clauses without waiting the end of the communication. Concerning the receiver thread of each computer, it receives clauses one after another using blocking operations. Therefore, this thread performs a busy-waiting until the message is received. However, since many data are sent by all search threads, this thread always receives clauses and does not wait much. Furthermore, busy-waiting allows to receive clauses faster than non-blocking communication.

Using the same protocol as used in Sects. 4.1 and 4.2, we evaluate the impact of this programming model on the sequential solvers effectiveness. Table 3 displays the results. Since clauses are sent as soon as they are selected, there is no communication cycle. So, we just show the impact of the different sizes of the buffer (*DATA*). Here again, the best compromise seems to be a buffer of size 20 MB. Interestingly, we have a propagation and conflict rate comparable to the partially hybrid one. But, the number of useful retained clauses is more important. It seems that our goal is achieved: maintaining a good propagation rate while sharing many useful clauses. In the next section, we will show that this programming model effectively outperforms the two others.

6   Experiments

In the first part of this section we compare the different programming models on the benchmarks of the SAT competition 2015 in order to tune our solver. Then, we compare the best combination we found against the distributed SAT solvers TOPOSAT and HORDESAT on the benchmarks coming from the SAT competition 2016.

6.1   Comparison of the Different Programming Models

Initially, we compare the three programming models: pure, partially and fully hybrid on the 100 instances of SAT-Race 2015 of the parallel track. As in Sect. 4, we use 32 computers containing 8 cores each (32 bi-processors Intel XEON X7550 at 2 GHz with a gigabyte Ethernet controller), resulting in a solver with 256 cores. Again the wall time is set to 1,800 s. We use the best configuration for each model, namely the shared memory buffer size is set to 20 MB for all models, and communication cycles are done every 5 s for pure MPI and partially hybrid model. Results are reported in the cactus plot Fig. 6a. It is easy to see that the pure message programming model is the worst one. It is only able to solve 61 instances whereas the partially (resp. fully) hybrid model solves 84 (resp. 87) problems. The main reason is that the pure message passing model replicates too many data. Note that on the same hardware, SYRUP solves 57 instances with only 8 cores. Therefore, a distributed approach can become inefficient due to its communication cost! This explains why TOPOSAT (based on this programming model) tries to reduce the number of shared clauses in order to remain efficient. Now, if we perform a pairwise comparison between the partially and the fully hybrid models (see Fig. 6b), we observe that the fully hybrid model is significantly better. More precisely, we show that, except for some satisfiable instances, most of the instances are solved faster with the fully hybrid programming model.

All these results validate the experimental study held previously in Sects. 4 and 5. They also validate that the more retained clauses are present, the better the solver is. It is noticeable that in order to increase the number of useful shared clauses, a good practice is to share them as soon as possible.

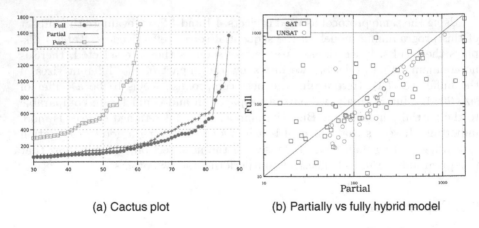

(a) Cactus plot   (b) Partially vs fully hybrid model

Fig. 6. Comparison of pure, partially and fully hybrid models on SAT'15 instances (parallel track). (a) shows the classical cactus plot and (b) compares the partially to the fully hybrid models. In the latter, each dot corresponds to an instance and dots below the diagonal are instances solved faster by the fully hybrid model.

6.2 Evaluation on SAT'16 Benchmarks

In this last section, we compare our fully hybrid version of SYRUP, called D-SYRUP, against the partially hybrid HORDESAT which embeds the solver PLIN-GELING and the pure TOPOSAT which embeds the solver GLUCOSE 3.0. All solvers use 256 threads (*i.e.*, 32 computers with 8 cores each). In order to be fair, we use new benchmarks coming from SAT'16 competition, application track that includes 300 instances. Due to limited resources, the wall time is now set to 900 s. To give a wider picture, we also add the sequential solver GLUCOSE, the parallel solver SYRUP with 8 cores and a version of D-SYRUP with 128 cores (16 computers with 8 cores) to see if, on this benchmarks set, D-SYRUP is also scalable.

Results are reported in Fig. 7. Clearly, on this test set, D-SYRUP outperforms HORDESAT and TOPOSAT. HORDESAT seems penalized first because its communication cycles are too short (1 s) thus leading to more access to critical sections and second because it uses a collective communication buffer of fixed size (see Sect. 4.2). Concerning TOPOSAT, it is clearly inefficient because it uses the pure message passing programming model. For a more detailed view, Fig. 8a displays a pairwise comparison between D-SYRUP and HORDESAT with 256 cores. It confirms that our solver outperforms HORDESAT for almost all instances.

When comparing D-SYRUP on both 128 and 256 cores, in Figs. 7b and 8b, we see that D-SYRUP scales well. Additionally, when comparing SYRUP on 8 cores to D-SYRUP, we observe a big improvement in terms of solved benchmarks. Note that the multi-core version of SYRUP on a 32 cores computer solves 154 benchmarks (62 SAT and 82 UNSAT). These results show that the SYRUP with its lazy strategy maintains its efficiency independently of the number of cores

Solver	#cores	SAT	UNS	Total
D-SYRUP	256	75	109	184
HORDESAT	256	70	84	154
TOPOSAT	256	69	58	127
D-SYRUP	128	73	104	177
SYRUP	8	56	75	131
GLUCOSE	1	49	57	106

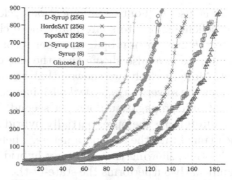

(a) Overview table (b) Number of instances solved w.r.t. the time

Fig. 7. Comparing D-SYRUP versus the portfolio distributed solvers and scalability evaluation of D-SYRUP with respect to 1 core (GLUCOSE), 8 cores (SYRUP) and 128 cores (D-SYRUP).

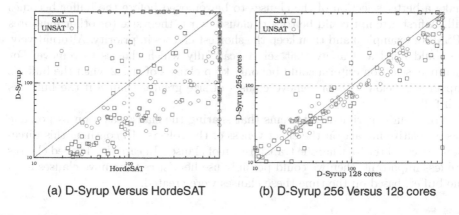

(a) D-Syrup Versus HordeSAT (b) D-Syrup 256 Versus 128 cores

Fig. 8. Scatter plots. For each plot, each dot represents an instance, dots below the diagonal correspond to instances solved faster by D-SYRUP (256 cores).

used. Indeed, recall that all clauses shared by SYRUP are also shared to all computers with D-SYRUP.

7 Conclusion and Future Work

In this paper we designed and evaluated a distributed version of the multi-core parallel portfolio SAT solver SYRUP. For this purpose, we developed several programming models to manage communications. In particular, we implemented the programming models that are currently the state-of-the-art distributed SAT solvers and used communication cycles. We have shown that the higher the frequency of the communication cycles is, the fewer unit propagations are done per second. Due to this observation, we proposed a fully hybrid model that is able to share clauses without communication cycles. We empirically demonstrated that, this scheme allows to share many more clauses without penalizing the solvers, since the number of propagation per second is less affected. In these first experiments, we have also shown that the size of the buffer used in SYRUP is important. More precisely, we showed that beyond a certain threshold, the number of clauses the solvers have to deal with increases that much that decreases the number of unit propagation done per second. Therefore, we designed three versions of SYRUP using the pure, partially hybrid and fully hybrid programming model, each tuned with the best parameters. Then, we compared these versions between each other, showing that the fully hybrid model is the best choice. This model significantly outperforms the state-of-the-art distributed SAT solvers TOPOSAT and HORDESAT on a wide set of benchmarks.

Our direct plan for the near future is to provide docker containers to simplify the deployment of the proposed distributed version of SYRUP in a distributed architecture and thus, facilitate its use. Another research direction, is to explore more advanced concepts that are related to clause managements, in order to make a better selection of the clauses to be removed before the buffer becomes full. A first attempt could be to sort clauses w.r.t. their size (or other measures (PSM for example)) and then keep the shortest clauses in priority. Alternatively, we could build a strategy that selects carefully which clauses to remove. For instance, selection criteria could be: only keep clauses of size 2 when the buffer's capacity is below 10% or remove clauses of size greater than 8 if the buffer is half full.

From our experiments, it seems that sharing the clauses as soon as possible has a positive impact on the effectiveness of the solvers. To continue this direction, we could try to delay the transmission of clauses. In case the delayed clauses are less important, then we could possibly use heuristics to remove clauses from the buffer, based on the time these clauses were created.

References

1. Aigner, M., Biere, A., Kirsch, C.M., Niemetz, A., Preiner, M.: Analysis of portfolio-style parallel SAT solving on current multi-core architectures. In: Fourth Pragmatics of SAT Workshop, a Workshop of the SAT 2013 Conference, POS 2013, pp. 28–40 (2013)
2. Amer, A., Lu, H., Balaji, P., Matsuoka, S.: Characterizing MPI and hybrid MPI+threads applications at scale: case study with BFS. In: 15th IEEE/ACM International Symposium on Cluster, Cloud and Grid Computing, CCGrid 2015, pp. 1075–1083 (2015)
3. Audemard, G., Hoessen, B., Jabbour, S., Lagniez, J.-M., Piette, C.: Revisiting clause exchange in parallel SAT solving. In: Cimatti, A., Sebastiani, R. (eds.) SAT 2012. LNCS, vol. 7317, pp. 200–213. Springer, Heidelberg (2012). doi:10.1007/978-3-642-31612-8_16
4. Audemard, G., Lagniez, J.-M., Mazure, B., Saïs, L.: On freezing and reactivating learnt clauses. In: Sakallah, K.A., Simon, L. (eds.) SAT 2011. LNCS, vol. 6695, pp. 188–200. Springer, Heidelberg (2011). doi:10.1007/978-3-642-21581-0_16

5. Audemard, G., Lagniez, J.-M., Szczepanski, N., Tabary, S.: An adaptive parallel SAT solver. In: Rueher, M. (ed.) CP 2016. LNCS, vol. 9892, pp. 30–48. Springer, Cham (2016). doi:10.1007/978-3-319-44953-1_3

6. Audemard, G., Simon, L.: Predicting learnt clauses quality in modern SAT solvers. In: Proceedings of the 21st International Joint Conference on Artificial Intelligence, IJCAI 2009, pp. 399–404 (2009)

7. Audemard, G., Simon, L.: Lazy clause exchange policy for parallel SAT solvers. In: Sinz, C., Egly, U. (eds.) SAT 2014. LNCS, vol. 8561, pp. 197–205. Springer, Cham (2014). doi:10.1007/978-3-319-09284-3_15

8. Balaji, P., Buntinas, D., Goodell, D., Gropp, W., Thakur, R.: Fine-grained multithreading support for hybrid threaded MPI programming. Int. J. High Perform. Comput. Appl. IJHPCA 24(1), 49–57 (2010)

9. Balyo, T., Sanders, P., Sinz, C.: HordeSat: a massively parallel portfolio SAT solver. In: Heule, M., Weaver, S. (eds.) SAT 2015. LNCS, vol. 9340, pp. 156–172. Springer, Cham (2015). doi:10.1007/978-3-319-24318-4_12

10. Biere, A.: (P)lingeling. http://fmv.jku.at/lingeling

11. Carribault, P., Pérache, M., Jourdren, H.: Enabling low-overhead hybrid MPI/OpenMP parallelism with MPC. In: Sato, M., Hanawa, T., Müller, M.S., Chapman, B.M., Supinski, B.R. (eds.) IWOMP 2010. LNCS, vol. 6132, pp. 1–14. Springer, Heidelberg (2010). doi:10.1007/978-3-642-13217-9_1

12. Eén, N., Sörensson, N.: An extensible SAT-solver. In: Giunchiglia, E., Tacchella, A. (eds.) SAT 2003. LNCS, vol. 2919, pp. 502–518. Springer, Heidelberg (2004). doi:10.1007/978-3-540-24605-3_37

13. Ehlers, T., Nowotka, D., Sieweck, P.: Communication in massively-parallel SAT solving. In: Proceedings of the 26th IEEE International Conference on Tools with Artificial Intelligence, ICTAI 2014, pp. 709–716 (2014)

14. Franco, J., Martin, J.: A history of satisfiability, Chap.1. In: Biere, A., Heule, M., van Maaren, H., Walsh, T. (eds.) Handbook of Satisfiability. Frontiers in Artificial Intelligence and Applications, vol. 185, pp. 3–74. IOS Press, Amsterdam (2009)

15. Goldberg, E., Novikov, Y.: BerkMin: a fast and robust sat-solver. Discrete Appl. Math. 155(12), 1549–1561 (2007)

16. Gomes, C.P., Selman, B., Kautz, H.A.: Boosting combinatorial search through randomization. In: Proceedings of the Fifteenth National Conference on Artificial Intelligence and Tenth Innovative Applications of Artificial Intelligence Conference, AAAI 1998, IAAI 1998, pp. 431–437 (1998)

17. Grant, R.E., Skjellum, A., Bangalore, P.V.: Lightweight threading with MPI using persistent communications semantics. National Nuclear Security Administration, USA (2015)

18. Hamadi, Y., Jabbour, S., Sais, L.: ManySat: a parallel SAT solver. J. Satisf. Boolean Model. Comput. JSAT 6(4), 245–262 (2009)

19. Heule, M.J.H., Kullmann, O., Marek, V.W.: Solving and verifying the boolean pythagorean triples problem via cube-and-conquer. In: Creignou, N., Le Berre, D. (eds.) SAT 2016. LNCS, vol. 9710, pp. 228–245. Springer, Cham (2016). doi:10.1007/978-3-319-40970-2_15

20. Konev, B., Lisitsa, A.: A SAT attack on the Erdős discrepancy conjecture. In: Sinz, C., Egly, U. (eds.) SAT 2014. LNCS, vol. 8561, pp. 219–226. Springer, Cham (2014). doi:10.1007/978-3-319-09284-3_17

21. Kshemkalyani, A.D., Singhal, M.: Efficient detection and resolution of generalized distributed deadlocks. IEEE Trans. Softw. Eng. TSE 20(1), 43–54 (1994)

22. Lanti, D., Manthey, N.: Sharing information in parallel search with search space partitioning. In: Nicosia, G., Pardalos, P. (eds.) LION 2013. LNCS, vol. 7997, pp. 52–58. Springer, Heidelberg (2013). doi:10.1007/978-3-642-44973-4_6

23. Lodha, S., Kshemkalyani, A.D.: A fair distributed mutual exclusion algorithm. IEEE Tran. Parallel Distrib. Syst. (TPDS) 11(6), 537–549 (2000)

24. Silva, J.P.M., Sakallah, K.A.: GRASP - a new search algorithm for satisfiability. In: Proceedings of the 1996 IEEE/ACM International Conference on Computer-aided Design, ICCAD 1996, pp. 220–227 (1996)

25. Moskewicz, M.W., Madigan, C.F., Zhao, Y., Zhang, L., Malik, S.: Chaff: engineering an efficient sat solver. In: Proceedings of the 38th Design Automation Conference, DAC 2001, pp. 530–535 (2001)

26. MPICH2. http://www.mcs.anl.gov/mpi/mpich2

27. Nishio, S., Li, K.F., Manning, E.G.: A resilient mutual exclusion algorithm for computer networks. IEEE Trans. Parallel Distrib. Syst. (TPDS) 1(3), 344–356 (1990)

28. Pipatsrisawat, K., Darwiche, A.: A lightweight component caching scheme for satisfiability solvers. In: Marques-Silva, J., Sakallah, K.A. (eds.) SAT 2007. LNCS, vol. 4501, pp. 294–299. Springer, Heidelberg (2007). doi:10.1007/978-3-540-72788-0_28

29. Singhal, M.: A taxonomy of distributed mutual exclusion. J. Parallel Distrib. Comput. 18(1), 94–101 (1993)

30. Smith, L., Bull, M.: Development of mixed mode MPI/OpenMP applications. Sci. Program. 9(2–3), 83–98 (2001)

31. Thakur, R., Gropp, W.: Test suite for evaluating performance of MPI implementations that support **MPI_THREAD_MULTIPLE**. In: Cappello, F., Herault, T., Dongarra, J. (eds.) EuroPVM/MPI 2007. LNCS, vol. 4757, pp. 46–55. Springer, Heidelberg (2007). doi:10.1007/978-3-540-75416-9_13

32. Tak, P., Heule, M.J.H., Biere, A.: Concurrent cube-and-conquer. In: Cimatti, A., Sebastiani, R. (eds.) SAT 2012. LNCS, vol. 7317, pp. 475–476. Springer, Heidelberg (2012). doi:10.1007/978-3-642-31612-8_42

33. Weigang, W., Zhang, J., Luo, A., Cao, J.: Distributed mutual exclusion algorithms for intersection traffic control. IEEE Trans. Parallel Distrib. Syst. (TPDS) 26(1), 65–74 (2015)

34. Zhang, L., Madigan, C., Moskewicz, M., Malik, S.: Efficient conflict driven learning in boolean satisfiability solver. In: Proceedings of the International Conference on Computer Aided Design (ICCAD), pp. 279–285 (2001)

PaInleSS: A Framework for Parallel SAT Solving

Ludovic Le Frioux[1,2,3(✉)], Souheib Baarir[1,2,3,4], Julien Sopena[2,3],
and Fabrice Kordon[2,3]

[1] LRDE, EPITA, Kremlin-Bicêtre, France
{ludovic.le.frioux,souheib.baarir}@lrde.epita.fr
[2] Sorbonne Universités, UPMC Univ Paris 06, UMR 7606, LIP6, Paris, France
{julien.sopena,fabrice.kordon}@lip6.fr
[3] CNRS, UMR 7606, LIP6, Paris, France
[4] Université Paris, Nanterre, France

Abstract. Over the last decade, parallel SAT solving has been widely
studied from both theoretical and practical aspects. There are now
numerous solvers that differ by parallelization strategies, programming
languages, concurrent programming, involved libraries, etc.

Hence, comparing the efficiency of the theoretical approaches is a challenging task. Moreover, the introduction of a new approach needs either
a deep understanding of the existing solvers, or to start from scratch the
implementation of a new tool.

We present PaInleSS: a framework to build parallel SAT solvers for
many-core environments. Thanks to its genericity and modularity, it provides the implementation of basics for parallel SAT solving like clause
exchanges, Portfolio and Divide-and-Conquer strategies. It also enables
users to easily create their own parallel solvers based on new strategies.
Our experiments show that our framework compares well with some of
the best state-of-the-art solvers.

Keywords: Parallel · Satisfiability · Clause sharing · Portfolio · Cube
and conquer

1 Introduction

Boolean satisfiability (SAT) has been used successfully in many contexts such
as planning decision [19], hardware and software verification [6], cryptology [27]
and computational biology [22], etc. This is due to the capability of modern SAT
solvers to solve complex problems involving millions of variables and billions of
clauses.

Most SAT solvers have long been sequential and based on the well-known
DPLL algorithm [8,9]. This initial algorithm has been dramatically enhanced by
introducing sophisticated heuristics and optimizations: decision heuristics [21,
29], clauses learning [25,32,35], aggressive cleaning [2], lazy data structures [29],
preprocessing [11,23,24], etc. The development of these enhancements has been
greatly simplified by the introduction of MiniSat [10], an extensible SAT solver
easing the integration of these heuristics in an efficient sequential solver.

© Springer International Publishing AG 2017
S. Gaspers and T. Walsh (Eds.): SAT 2017, LNCS 10491, pp. 233–250, 2017.
DOI: 10.1007/978-3-319-66263-3_15

The emergence of many-core machines opens new possibilities in this domain. Two classes of parallel techniques have been developed: competition-based (*a.k.a.,* Portfolio) and cooperation-based (*a.k.a.,* Divide-and-Conquer). In the Portfolio settings [14], many sequential SAT solvers compete for the solving of the whole problem. The first one to find a solution, or proving the problem to be unsatisfiable ends the computation. Divide-and-Conquer approaches use the guiding path method [34] to decompose, recursively and dynamically, the original problem in sub-problems that are solved separately by sequential solvers. In both approaches, sequential solvers can dynamically share learnt information. Many heuristics exist to improve this sharing by proposing trade-off between gains and overhead.

While the multiplication of strategies and heuristics provides perspectives for parallel SAT solving, it makes more complex development and evaluation of new proposals. Thus, any new contribution faces three main problems:

Problem 1: concurrent programming requires specific skills (synchronization, load balancing, data consistency, etc.). Hence, the theoretical efficiency of an heuristic may be annihilated by implementation choices.

Problem 2: most of the contributions mainly target a specific component in the solver while, to evaluate it, a complete one (either built from scratch or an enhancement of an existing one) must be available. This makes the implementation and evaluation of a contribution much harder.

Problem 3: an implementation, usually, only allows to test a single composition policy. Hence, it becomes hard to evaluate a new heuristic with different versions of the other mechanisms.

This paper presents PArallel INstantiabLE Sat Solver (`PaInleSS`)[1], a framework that simplifies the implementation and evaluation of new parallel SAT solvers for many-core environments. The components of `PaInleSS` can be instantiated independently to produce a new complete solver. The guiding principle is to separate the technical components dedicated to some specific aspect of concurrent programming, from the components implementing heuristics and optimizations embedded in a parallel SAT solver.

Our main contributions are the following:

– we propose a new modular and generic framework that can be used to implement new strategies with minimal effort and concurrent programming skills;
– we provide adaptors for some state-of-the-art sequential SAT solvers: glucose [2], `Lingeling` [5], `MiniSat` [10], and `MapleCOMSPS` [21].
– we show that it is easy to implemented strategies in `PaInleSS`, and provide some that are present in the classical solvers of the state-of-the-art: `glucose-syrup` [3], `Treengeling` [5], and `Hordesat` [4];
– we show the effectiveness of our modular design by instantiating, with a minimal effort, new original parallel SAT solver (by mixing strategies);

---

[1] painless.lrde.epita.fr.

– we evaluate our approach on the benchmark of the parallel track of the SAT Race 2015. We compare the performance of solvers instantiated using the framework with the original solvers. The results show that the genericity provided by PaInleSS does not impact the performances of the generated instances.

The rest of the paper is organized as follows: Sect. 2 introduces useful background to deal with sequential SAT solving. Section 3 is dedicated to parallel SAT solving. Section 4 shows the architecture of PaInleSS. Section 5 presents different solvers implemented using PaInleSS. Section 6 analyzes the results of our experiments and Sect. 7 concludes and gives some perspectives work.

2   About Sequential SAT Solving

In this section, after some preliminary definitions and notations, we introduce the most important features of modern sequential SAT solvers.

A *propositional variable* can have two possible values $\top$ (True) or $\bot$ (False). A *literal* l is a propositional variable (x) or its negation ($\neg x$). A *clause* ω is a finite disjunction of literals (noted $\omega = \bigvee_{i=1}^{k} \ell_i$). A clause with a single literal is called *unit clause*. A *conjunctive normal form (CNF) formula* φ is a finite conjunction of clauses (noted $\varphi = \bigwedge_{i=1}^{k} \omega_i$). For a given φ, the set of its variables is noted: V_φ. An assignment $\mathcal{A}$ of variables of φ, is a function $\mathcal{A} : V_\varphi \to \{\top, \bot\}$. $\mathcal{A}$ is total (complete) when all elements of V_φ have an image by $\mathcal{A}$, otherwise it is partial. For a given formula φ, and an assignment $\mathcal{A}$, a clause of φ is satisfied when it contains at least one literal evaluating to true, regarding $\mathcal{A}$. The formula φ is satisfied by $\mathcal{A}$ iff $\forall \omega \in \varphi, \omega$ is satisfied. φ is said to be SAT if there is at least one assignment that makes it satisfiable. It is defined as UNSAT otherwise.

Conflict Driven Clause Leaning. The majority of the complete state-of-the-art sequential SAT solvers are based on the Conflict Driven Clause Learning (CDCL) algorithm [25,32,35], that is an enhancement of the DPLL algorithm [8,9]. The main components of a CDCL are depicted in Algorithm 1.

At each step of the main loop, unitPropagation[2] (line 4) is applied on the formula. In case of conflict (line 5), two situations can be observed: the conflict is detected at decision level 0 ($dl == 0$), thus the formula is declared UNSAT (lines 6–7); otherwise, a new asserting clause is derived by the conflict analysis and the algorithm backjumps to the assertion level [25] (lines 8–10). If there is no conflict (lines 11–13), a new decision literal is chosen (heuristically) and the algorithm continues its progression (adding a new decision level: $dl \leftarrow dl + 1$). When all variables are assigned (line 3), the formula is said to be SAT.

---

[2] The unitPropagation function implements the Boolean Constraint Propagation (BCP) procedure that forces (in cascade) the values of the variables in asserting clauses [8].

```
1  function CDCL()
2  |  dl ← 0                                    // Current decision level
3  |  while not all variables are assigned do
4  |  |   conflict ← unitPropagation()
5  |  |   if conflict then
6  |  |   |   if dl = 0 then
7  |  |   |   |   return ⊥                       // φ is UNSAT
8  |  |   |   ω ← conflictAnalysis()
9  |  |   |   addLearntClause(ω)
10 |  |   |   dl ← backjump(ω)
11 |  |   else
12 |  |   |   assignDecisionLiteral()
13 |  |   |   dl ← dl + 1
14 |  return ⊤                                   // φ is SAT
```

Algorithm 1. CDCL algorithm.

The Learning Mechanism. The effectiveness of the CDCL lies in the *learning mechanism*. Each time a conflict is encountered, it is analyzed (`conflictAnalysis` function in Algorithm 1) in order to compute its reasons and to derive a *learnt clause*. While present in the system, this clause will avoid the same mistake to be made another time, and therefore allows faster deductions (conflicts/unit propagations).

Since the number of conflicts is very huge (in avg. 5000/s [2]), controlling the size of the database storing learnt clauses is a challenge. It can dramatically affect performance of the `unitPropagation` function. Many strategies and heuristics have been proposed to manage the cleaning of the stored clauses (*e.g.*, the Literal Block Distance (LBD) [2] measure).

With the two classical approaches used for parallel SAT solving: Portfolio and Divide-and-Conquer (see Sect. 3), multiple sequential solvers are used in parallel to solve the formula. With these paradigms sequential solvers can be seen as black boxes providing solving and clause sharing functionalities.

3   About Parallel SAT Solving

The arrival of many-core machines leads to new possibilities for SAT solving. Parallel SAT solving rely on two concepts: parallelization strategy and learnt clause exchanges. Two main parallelization methods have been developed: Portfolio and Divide-and-Conquer. We can also mention the hybrid approaches as alternatives, that are combinations of the first two techniques. With these parallelization strategies, it is possible to exchange learnt clauses, between the underling sequential solvers.

3.1 Parallelization Strategies

Portfolio. The Portfolio scheme has been introduced by [14], in `ManySat`. The main idea of this approach is to run sequential solvers working in parallel on the entire formula, in a competitive way. This strategy aims at increasing the probability of finding a solution using the *diversification* [12] (also known as swarming in others contexts) principle.

The diversification can only concern the used heuristics: several solvers (workers) with different heuristics are instantiated. They differ by their decision strategies, learning schemes, the used random seed, etc.

Another type of diversification, introduced in `HordeSat` [4], uses the *phase* of the variables: before starting the search each solver receives a special phase, acting as a soft division of the search space. Solvers are invited to visit a certain part of the search space but they can move out of this region during the search.

Another technique to ensure the diversification is the *block branching* [33]: each worker focuses on a particular subset (or block) of variables. Hence, the decision variables of a worker are chosen from the block it is in charge of.

Divide-and-Conquer. The Divide-and-Conquer approach is based on splitting the search space in disjoint parts. These parts are solved independently, in parallel, by different workers. As the parts are disjunct, if one of the partitions is proven to be SAT then the initial formula is SAT. The formula is UNSAT if all the partitions are UNSAT. The challenging points of the this method are: *dividing the search space* and *balancing jobs between workers*.

To divide the search space, the most used technique is based on the Shannon's decomposition, known as the *guiding path* [34]. The guiding path is a vector of literals (*a.k.a., cube*) that are assumed by the worker when solving the formula.

Choosing the best division variables is a hard problem requiring heuristics. If some parts are too easy this will lead to repeatedly divide the search space and ask for a new job (phenomenon known as *ping-pong effect*). As all the partitions do not require the same solving time, some workers may become idle and a mechanism for load balancing is needed. Each time a solver proves that its partition is UNSAT[3], it needs a new job. Another solver is chosen as target to divide its search space (*i.e.,* to extend its guiding path). The target will work on one of the new partition and the idle worker on the other one. This mechanism is often called *work stealing*.

Hybrid Approaches. As already presented, Portfolio and Divide-and-Conquer, are the two main explored approaches to parallelize SAT solving.

The Portfolio scheme is simple to implement, and uses the principle of diversification to increase the probability of solving the problem. However, since workers can overlap their search regions, the theoretical resulting speed-up is not as good as the one of the Divide-and-Conquer approach [17]. Surprisingly, while giving

---

[3] If the result is SAT the global solving ends.

a better theoretical speed-up, the Divide-and-Conquer approach suffers from the two challenging issues we mentioned: dividing the search space and balancing jobs between workers.

Emerging techniques, called *hybrid approaches*, propose to use simultaneously the two strategies, so that we benefit from the advantages of each, while trying to avoid their drawbacks.

A basic manner to mix the two approaches is to compose them. There are two possible strategies: Portfolio of Divide-and-Conquer (introduced by c-sat [30]), and Divide-and-Conquer of Portfolios (*e.g.*, ampharos [1] an adaptive Divide-and-Conquer that allows multiple workers on the same sub-part of the search space). Let us mention other more sophisticated ways to mix approaches like *scattering* [16,18] or *transition heuristics* based strategies [1,7,26,31].

3.2   Clauses Sharing

In all these parallelization paradigms, sharing information between workers is possible, the most important one being clauses learnt by each worker. Hence, the main questions are: *which clauses should be shared? And between which workers?* Indeed, sharing all clauses can have a bad impact on the overall behavior.

To answer the first question, many solvers rely on the standard measures, defined for sequential solvers (*i.e.*, activity, size, LBD): only clauses under a given threshold for these measures are shared. One simple way to get the threshold is to define it as constant it (*e.g.*, clauses up to size 8 are shared in ManySat [14]). More sophisticated approaches adapt thresholds dynamically in order to control the flow of shared clauses during the solving [4,13].

A simple solution to the second question, adopted in almost all parallel SAT solvers, is to share clauses between all workers. However, a finer (but more complex) solution is to let each worker choose its emitters [20].

As a conclusion of this section, we can say that parallel SAT solving is based on two distinct concepts. First, there exist numerous strategies to parallelize SAT solving by organizing the workers search. Secondly, with all these strategies is it possible to share clauses between the workers. This two concepts have been our intuition sources for the design of the architecture of PaInleSS.

4   Architecture of the Framework

There exist numerous strategies to parallelize SAT solving, their performances heavily relying on their implementation. The most difficult issues deal with concurrent programming. Languages and libraries provide abstractions to deal with this difficulties, and according to these abstractions developers have more or less control on mechanisms such as memory or threads management (*e.g.*, Java vs C++). This will affect directly the performance of the produced solver.

Therefore, it is difficult to compare the strategies without introducing technological bias. Indeed, it is difficult to integrate new strategies on top of existing

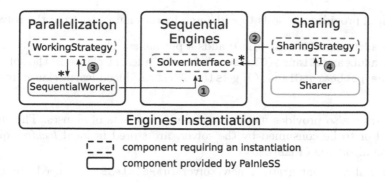

Fig. 1. Architecture of PaInleSS.

solvers, or to develop a new solver from scratch. Moreover, an implementation usually offers the possibility to modify a particular component, it is then difficult to test multiple combinations of components.

PaInleSS aims to be a solution to these problems. It is a generic, modular, and efficient framework, developed in C++11, allowing an easy implementation of parallel strategies. Taking black-boxed sequential solvers as input, it minimizes the effort to encode new parallelization and sharing strategies, thus enabling the implementation of complete SAT solvers at a reduced cost.

As mentioned earlier, a typical parallel SAT solver relies mainly on three core concepts: sequential engine(s), parallelization, and sharing. These last form the core of the PaInleSS architecture (see Fig. 1): the sequential engine is handled by the SolverInterface component. The parallelization is implemented by the WorkingStrategy and SequentialWorker components. Components Sharing-Strategy and Sharer are in charge of the sharing.

Sequential Engine. SolverInterface is an adapter for the basic functions expected from a sequential solver, it is divided in two subgroups: *solving* and *clauses export/import* (respectively represented by arrows 1 and 2 in Fig. 1). Subgroup 1 provides methods that interact with the solving process of the underling solver. The most important methods of this interface are:

- SatResult solve(int[*] cube): tries to solve the formula, with the given cube (that can be empty in case of Portfolio). This method returns SAT, UNSAT, or UNKNOWN.
- void setInterrupt(): stops the current search initiated using the solve method.
- void setPhase(int var, bool value): set the phase of variable var to value.
- void bumpVariableActivity(int var, int factor): bumps factor times the activity of variable var.
- void diversify(): adjusts internal parameters of the solver, to diversify its behaviour.

Subgroup 2 provides methods to add/fetch learnt clauses to/from the solver:

- `void addClause(Clause cls)`: adds a permanent clause to the solver.
- `void addLearntClause(Clause cls)`: adds a learnt clause to the solver.
- `Clause getLearntClause()`: gets the oldest produced learnt clause from the solver.

The interface also provides methods to manipulate sets of clauses. The clauses produced or to be consumed by the solver, are stored in local *lockfree* queues (based on algorithm of [28]).

Technically, to integrate a new solver in `PaInleSS`, one needs to create a new class inheriting from `SolverInterface` and implement the required methods (*i.e.,* wrapping the methods of the API offered by the underlying solver). The framework currently provides some basic adaptors for `Lingeling` [5], `glucose` [2], `Minisat` [10], and `MapleCOMSPS` [21].

Parallelization. Basic parallelization strategies, such as those introduced in Sect. 3, must be implemented easily. We also aim at creating new strategies and mixing them.

A tree-structured (of arbitrary depth) composition mechanism enables the mix of strategies: internal nodes represent parallelization strategies, and leaves solvers. As an example (see Fig. 2(a)), a Divide-and-Conquer of Portfolios is represented by a tree of depth 3: the root corresponds to the Divide-and-Conquer having children representing the Portfolios acting on several solvers (the leaves of the tree).

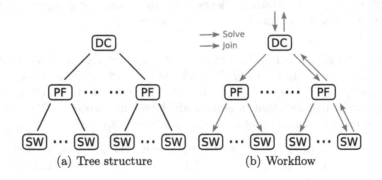

(a) Tree structure   (b) Workflow

Fig. 2. Example of a composed parallelization strategy.

`PaInleSS` implements nodes using the `WorkingStrategy` class, and leaves with the `SequentialWorker` class. This last is a subclass of `WorkingStrategy` that integrates an execution flow (a thread) operating the associated solver.

The overall solving workflow within this tree is user defined and guaranteed by the two main methods of the `WorkingStrategy` (arrows 3 in Fig. 1):

- void solve(int[*] cube): according to the strategy implemented, this method manages the organization of the search by giving orders to the children strategies.
- void join(SatResult res, int[*] model): used to notify the parent strategy of the solving end. If the result is SAT, model will contain an assignment that satisfies the sub-formula treated by this node.

It is worth noting that the workflow must start by a call to the root's solve method and eventually ends by a call to the root's join method. The propagation of solving orders from a parent to one of its child nodes, is done by a call to the solve method of this last. The results are propagated back from a child to its parent by a call to the join method of this last. The solving can not be effective without a call to the leaves' solve methods.

Back to the example of Fig. 2(a). Consider the execution represented in Fig. 2(b). The solving order starts by a call to the root's (DC node) solve method. It is relayed trough the tree structure to the leaves (SW nodes). Here, once its problem is found SAT by one of the SW, it propagates back the result to its PF node parent via a call to the join method. According to the strategy of the PF, the DC's join method is called and ends the global solving.

Hence, to develop its own parallelization strategy, the user should create one or more subclass of WorkingStrategy and to build the tree structure.

Sharing. In parallel SAT solving, we must pay a particular attention to the exchange of learnt clauses. Indeed, beside the theoretical aspects, a bad implementation of the sharing can dramatically impact the efficiency of the solver (*e.g.,* improper use of locks, synchronization problems). We now present how sharing is organized in PaInleSS.

When a solver learns a clause, it can share it according to a filtering policy such as the size or the LBD of the clause. To do so it puts the clause in a special buffer (buff_exp in Fig. 3). The sharing of the learnt clauses is realized by dedicated thread(s): Sharer(s). Each one is in charge of a set of producers and consumers (these are references to SolverInterface). Its behaviour reduces to

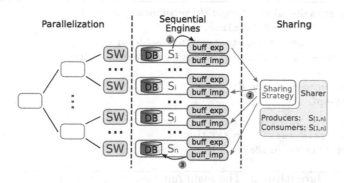

Fig. 3. Sharing mechanism implemented in PaInleSS.

a loop of sleeping and exchange phases. This last is done by calling the interface of SharingStrategy class (arrow 4 in Fig. 1). The main method of this class is the following:

- void doSharing(SolverInterface[*] producers, SolverInterface[*] consumers): according to the underlying strategy, this method gets clauses from the producers and add them to the consumers.

In the example of Fig. 3, the Sharer uses a given strategy, and all the solvers (S_i) are producers and consumers. The use of dedicated workflows (*i.e.,* threads) allows CPU greedy strategies to be run on a dedicated core, thus not interfering with the solving workers. Moreover, sharing phase can be done manipulating groups of clauses, allowing the use of more sophisticated heuristics. Finally, during its search a solver can get clauses from its import buffer (buff_imp in Fig. 3) to integrate them in its local database.

To define a particular sharing strategy the user only needs to provide a subclass of SharingStrategy. With our mechanism it is possible to have several groups of sharing each one manage by a Sharer. Moreover, solvers can be dynamically added/deleted from/to the producers and/or customers sets of a Sharer.

Engine Instantiation. To create a particular instance of PaInleSS, the user has to adapt the main function presented by Algorithm 2. The role of this function is to instantiate and bind all the components correctly. This actions are simplified by the use of parameters.

First, the concrete solver classes (inheriting from SolverInterface) are instantiated (line 2). Then the WorkingStrategy (including SequentialWorker) tree is implemented (line 3). This operation links SequentialWorker to their SolverInterface. Finally, the Sharer(s) and their concrete SharingStrategy(s) are created; the producers and consumers sets are initialized (line 4).

```
1  function main-PaInleSS (args: the program arguments)
2      solvers ← Create SolverInterface
3      root ← Create WorkingStrategy tree (solvers)
4      sharers ← Create SharingStrategy and Sharer (solvers)
5      root.solve()
6      while timeout or stop do
7          sleep(...)
8      print(root.getResult())
9      if root.getResult() == SAT then
10         print(root.getModel())
```

Algorithm 2. The main function of PaInleSS.

The solving starts by the call to the `solve` method of the root `WorkingStrategy` tree. The main thread will execute a loop, where it sleeps for an amount of time, and then checks if either the timeout has been reached or the solving ended (lines 6–7). It prints the final result (line 8), plus the model in case of a SAT instance (lines 9–10).

5   Implementing and Combining Existing Strategies

To validate the generic aspect of our approach, we selected three efficient state-of-the-art parallel SAT solvers: `glucose-syrup` [3], `Treengeling` [5], and `Hordesat` [4]. For each selected solver, we implemented a solver that mimics the original one using `PaInleSS`. To show the modularity of `PaInleSS`, we used the already developed components to instantiate two new original solvers that combine existing strategies.

Solver "à la Glucose-Syrup". The `glucose-syrup`[4] solver is the winner of the parallel track of the SAT Race 2015. It is a Portfolio based on the sequential solver `glucose` [2]. The sharing strategy exchanges all the exported clauses between all the workers. Beside, the workers have customized settings in order to diversify their search.

Hence, implementing a solver "à la `glucose-syrup`", namely `PaInleSS-breakglucose-syrup`, required the following components: `Glucose` an adaptor to use the `glucose` solver; `Portfolio` a simple `WorkingStrategy` that implements a Portfolio strategy; `SimpleSharing` a `SharingStrategy` that exchanges all the exported clauses from the producers to all the consumers with no filtering.

The implementation of `PaInleSS-glucose-syrup` required 355 lines of code (LoC) for the adaptor, 95 LoC for the Portfolio, and 44 LoC for the sharing strategy.

Solver "à la Treengeling". The `Treengeling`[5] solver is the winner of the parallel track of the SAT Competition 2016. It is based on the sequential engine `Lingeling` [5]. Its parallelization strategy is a variant of Divide-and-Conquer called Cube-and-Conquer [15]. The solving is organized in rounds. Some workers search for a given number of conflicts. When the limit is reached, some are selected to split their sub-spaces using a lookahead heuristic. The sharing is restricted to the exchange of unit clauses from a special worker. This last is also in charge of the solving of the whole formula during all the execution.

To implement a solver "à la `Treengeling`", namely `PaInleSS-treengeling`, we needed the following components: `Lingeling`, an adaptor of the sequential solver `Lingeling`; `CubeAndConquer` a `WorkingStrategy`, that implements a

---

[4] www.labri.fr/perso/lsimon/downloads/softwares/glucose-syrup.tgz.
[5] www.fmv.jku.at/lingeling/lingeling-bbc-9230380-160707.tar.gz.

Cube-and-Conquer [15]; `SimpleSharing` already used to define for the `glucose-syrup` like solver. In this case, the underlying sequential solvers are parametrized to export only unit clauses, and only the special worker is a producer.

For the `CubeAndConquer` we choose time to manage rounds because it allows, once one worker has encountered an UNSAT situation, to restart the worker with another guiding-path. In the original implementation, rounds are managed using numbers of conflicts, this makes the reuse of idle CPU much harder.

The implementation of `PaInleSS-treengeling` needed 377 LoC for the adaptor and 249 LoC for `CubeAndConquer`.

Solver "à la Hordesat". `Hordesat`[6] is a Portfolio-based solver with a modular design. `Hordesat` uses as sequential engine either `Minisat` [10] or `Lingeling`. It is a Portfolio where the sharing is realized by controlling the flow of exported clauses. Every second, 1500 literals (*i.e.*, sum of the size of the clauses) are exported from each sequential engine. Moreover, we used the `Lingeling` solver and the native diversification of `Plingeling` [5] (a Portfolio solver of `Lingeling`) combined to the random sparse diversification (presented as the best combination by [4]).

The solver "à la `Hordesat`", namely `PaInleSS-hordesat`, required the following components: `Lingeling` and `Portfolio` that have been implemented earlier; `HordesatSharing` a `SharingStrategy` that implements the `Hordesat` sharing strategy. This last required only 148 LoC.

Combining Existing Strategies. Based on the implemented solvers, we reused the obtained components to quickly build two new original solvers.

`PaInleSS-treengeling-hordesat`: it is a `PaInleSS-treengeling`-based solver that shares clauses using the strategy of `Hordesat`. The implementation of this solver reuses the `Lingeling`, `CubeAndConquer`, and `HordesatSharing` classes. To instantiate this solver we only needed a special parametrization. Beside, the modularity aspects, by this instantiation, we aimed to investigate the impact of a different sharing strategy on the overall performances of `PaInleSS-treengeling`.

`PaInleSS-treengeling-glucose`: it is a Portfolio solver that mixes Cube-and-Conquer of `Lingeling`, and a Portfolio of `Glucose` solvers. Here, `Glucose` workers export unit and *glue* clauses [2] (*i.e.*, clauses with LBD equals to 2) to the other solvers. This last solver reuses the following components: `Lingeling`, `Glucose`, `Portfolio`, `CubeAndConquer`, `SimpleSharing`. Only 15 LoC are required to build the parallelization strategy tree. By the instantiation of this solver, we aimed to study the effect of mixing some parallelisation strategies.

---

[6] baldur.iti.kit.edu/hordesat/files/hordesat.zip.

6   Numerical Results

This section presents the results of experiments we realized using the solvers described in Sect. 5: `PaInleSS-glucose-syrup`, `PaInleSS-treengeling`, `PaIn-leSS-hordesat`, `PaInleSS-treengeling-hordesat`, and `PaInleSS-treenge-ling-glucose`. The goal here is to show that the introduction of genericity does not add an overhead *w.r.t.* the original solvers.

All the experiments have been executed on a parallel machine with 40 processors Intel Xeon CPU E7- 2860 @ 2.27 GHz, and 500 Go of memory. We used the 100 instances of the parallel track of the SAT Race 2015[7]. All experiments have been conducted using the following parametrisations: each solver has been run once on each instance, with a time-out of 5000 s (as in the SAT Race). We limited the number of involved CPUs to 36.

Table 1. Results of the different solvers. The different columns represent: the number of UNSAT solved instances, SAT solved instances, total solved instances, and the cumulative time spent solving the instances solved by the two solvers.

Solver	UNSAT	SAT	Total	Cum. Time Inter.
glucose-syrup	30	41	71	15 h37
PaInleSS-glucose	**32**	**46**	**78**	**13 h18**
Treengeling	32	50	82	20 h55
PaInleSS-treengeling	32	50	82	**14 h12**
Hordesat	31	**44**	**75**	15 h05
PaInleSS-hordesat	31	43	74	**14 h19**

The number of solved instances per solver are reported in Table 1. Globally, these primary results show that our solvers compare well to the studied state-of-the-art solvers. We can deduce that the genericity offered by `PaInleSS` does not impact the global performances. Moreover, on instances solved by both, the original solver and our implementation, the cumulative solving time is in our favor (see column Cum. Time. Inter. in Table 1). A more detailed analysis is given for each solver in the rest of the section.

PaInleSS-glucose-syrup *vs.* glucose-syrup. Our implementation of the `glucose-syrup` parallelization strategy was able to solve 7 more instances compared to `glucose-syrup`. This concerns both SAT and UNSAT instances as shown in the scatter plot of Fig. 4(a) and, in the cactus plots of Figs. 5(a) and 6(b). This gain is due to our careful design of the sharing mechanism that is decentralized and uses lock-free buffers. Indeed in `glucose-syrup` a global buffer is used to exchange clauses, which requires import/export to use a unique lock, thus introducing a bottleneck. The absence of bottleneck in our implementation increases the parallel all over the execution, explaining our better performances.

---

[7] baldur.iti.kit.edu/sat-race-2015/downloads/sr15bench-hard.zip.

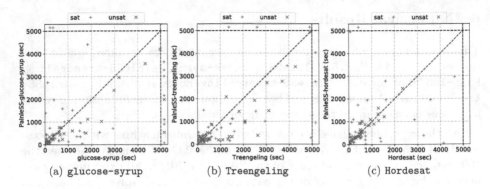

Fig. 4. Scatter plots of `PaInleSS`'s solvers against state-of-the-art ones.

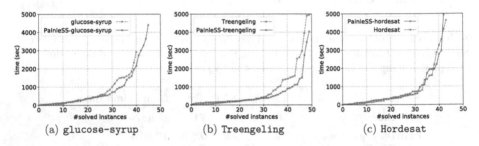

Fig. 5. Cactus plots of SAT instances of `PaInleSS`'s solvers against state-of-the-art ones.

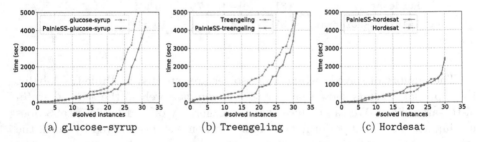

Fig. 6. Cactus plots of UNSAT instances of `PaInleSS`'s solvers against state-of-the-art ones.

PaInleSS-treengeling vs. Treengeling. Concerning `Treengeling`, our implementation has comparable results. Figure 4(b) shows that the average solving time of SAT instances is quite similar, while for the UNSAT instances, our implementation is in average faster. This is corroborated by the cactus plot depicted in Fig. 6(b). This speed up is due to our fine implementation of the Cube-and-Conquer strategy, thus increasing the real parallelism all over the execution and explaining our better performances on UNSAT instances.

PaInleSS-hordesat *vs.* Hordesat. Although Hordesat was able to solve 1 more instance than our tool, results are comparable. Moreover scatter plot of Fig. 4(c), and cactus plots of Figs. 5(c) and 6(c) exhibit quit similar results for the two tools. For instances solved by both tools, our tool was a beat faster and used almost 3000 seconds less as pointed out in Table 1. As the sharing strategy of Hordesat is mainly based on two parameters, namely the number of exchanged literals per round, and the sleeping time of sharer by round, we think that a finer tuning of this couple of parameters for our implementation could improve the performances of our tool.

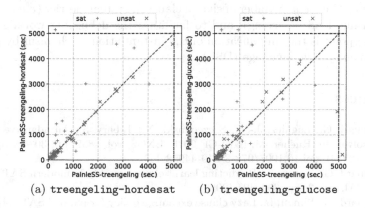

(a) `treengeling-hordesat`   (b) `treengeling-glucose`

Fig. 7. Scatter plots of the composed solvers against PaInleSS-treengeling.

Results of the Composed Solvers. PaInleSS-treengeling-hordesat solved 81 instances (49 SAT and 32 UNSAT), and PaInleSS-treengeling-glucose solved 81 instances (48 SAT and 33 UNSAT). The scatter plot of the two strategies (Fig. 7), show that these strategies are almost equivalent w.r.t. the original ones. These results allow us to conclude that the introduced strategies do not add any value to the original one.

7   Conclusion

Testing and implementing new strategies for parallel SAT solving has become a challenging issue. Any new contribution in the domain faces the following problems: concurrent programming requires specific skills, testing new strategies required a prohibitive development of a complete solver (either built from scratch or an enhancement of an existing one), an implementation often allows to test only a single composition policy and avoids the evaluation of a new heuristic with different versions of the other mechanisms.

To tackle these problems we proposed PaInleSS, a modular, generic and efficient framework for parallel SAT solving. We claimed that its modularity and genericity allow the implementation of basic strategies, as well as new onces and their combination with a minimal effort and concurrent programming skills.

We have proven our claims, first, by the implementation of strategies present in some state-of-the-art solvers: `glucose-syrup`, `Treengeling`, and `Hordesat`. Second, we reused the developed complements to derive, easily, new solvers that mix strategies. We also show that the instantiated solvers are as efficient as the original one (and even better), by conducting a set experiments using benchmarks of the SAT Race 2015.

As perspectives, we plan to adapt our framework for mutli-machine environments. We also would like to enhance `PaInleSS` with helpful tools to monitor algorithm metrics (*e.g.,* number of shared clauses), system metrics (*e.g.,* synchronization time, load balancing), and to facilitate the debugging work. Another interesting point is the simplification of the instantiation mechanism by providing a domain specific language (DSL).

References

1. Audemard, G., Lagniez, J.-M., Szczepanski, N., Tabary, S.: An adaptive parallel SAT solver. In: Rueher, M. (ed.) CP 2016. LNCS, vol. 9892, pp. 30–48. Springer, Cham (2016). doi:10.1007/978-3-319-44953-1_3
2. Audemard, G., Simon, L.: Predicting learnt clauses quality in modern SAT solvers. In: IJCAI, vol. 9, pp. 399–404 (2009)
3. Audemard, G., Simon, L.: Lazy clause exchange policy for parallel SAT solvers. In: Sinz, C., Egly, U. (eds.) SAT 2014. LNCS, vol. 8561, pp. 197–205. Springer, Cham (2014). doi:10.1007/978-3-319-09284-3_15
4. Balyo, T., Sanders, P., Sinz, C.: HordeSat: a massively parallel portfolio SAT Solver. In: Heule, M., Weaver, S. (eds.) SAT 2015. LNCS, vol. 9340, pp. 156–172. Springer, Cham (2015). doi:10.1007/978-3-319-24318-4_12
5. Biere, A.: Splatz, lingeling, plingeling, treengeling, yalsat entering the SAT competition 2016. SAT COMPETITION 2016, p. 44 (2016)
6. Biere, A., Cimatti, A., Clarke, E., Zhu, Y.: Symbolic model checking without BDDs. In: Cleaveland, W.R. (ed.) TACAS 1999. LNCS, vol. 1579, pp. 193–207. Springer, Heidelberg (1999). doi:10.1007/3-540-49059-0_14
7. Blochinger, W.: Towards robustness in parallel SAT solving. In: PARCO, pp. 301–308 (2005)
8. Davis, M., Logemann, G., Loveland, D.: A machine program for theorem-proving. Commun. ACM **5**(7), 394–397 (1962)
9. Davis, M., Putnam, H.: A computing procedure for quantification theory. J. ACM **7**(3), 201–215 (1960)
10. Eén, N., Sörensson, N.: An extensible SAT-solver. In: Giunchiglia, E., Tacchella, A. (eds.) SAT 2003. LNCS, vol. 2919, pp. 502–518. Springer, Heidelberg (2004). doi:10.1007/978-3-540-24605-3_37
11. Eén, N., Biere, A.: Effective preprocessing in SAT through variable and clause elimination. In: Bacchus, F., Walsh, T. (eds.) SAT 2005. LNCS, vol. 3569, pp. 61–75. Springer, Heidelberg (2005). doi:10.1007/11499107_5
12. Guo, L., Hamadi, Y., Jabbour, S., Sais, L.: Diversification and intensification in parallel SAT solving. In: International Conference on Principles and Practice of Constraint Programming, pp. 252–265. Springer (2010)

13. Hamadi, Y., Jabbour, S., Sais, J.: Control-based clause sharing in parallel SAT solving. In: Hamadi, Y., Monfroy, E., Saubion, F. (eds.) Auton. Search, pp. 245–267. Springer, Heidelberg (2011)
14. Hamadi, Y., Jabbour, S., Sais, L.: ManySAT: a parallel SAT solver. J. Satisfiability Boolean Model. Comput. **6**, 245–262 (2009)
15. Heule, M.J.H., Kullmann, O., Wieringa, S., Biere, A.: Cube and conquer: guiding CDCL SAT solvers by lookaheads. In: Eder, K., Lourenço, J., Shehory, O. (eds.) HVC 2011. LNCS, vol. 7261, pp. 50–65. Springer, Heidelberg (2012). doi:10.1007/978-3-642-34188-5_8
16. Hyvärinen, A.E.J., Junttila, T., Niemelä, I.: A distribution method for solving SAT in grids. In: Biere, A., Gomes, C.P. (eds.) SAT 2006. LNCS, vol. 4121, pp. 430–435. Springer, Heidelberg (2006). doi:10.1007/11814948_39
17. Hyvärinen, A.E.J., Junttila, T., Niemelä, I.: Partitioning search spaces of a randomized search. In: Serra, R., Cucchiara, R. (eds.) AI*IA 2009. LNCS, vol. 5883, pp. 243–252. Springer, Heidelberg (2009). doi:10.1007/978-3-642-10291-2_25
18. Hyvärinen, A.E.J., Manthey, N.: Designing scalable parallel SAT solvers. In: Cimatti, A., Sebastiani, R. (eds.) SAT 2012. LNCS, vol. 7317, pp. 214–227. Springer, Heidelberg (2012). doi:10.1007/978-3-642-31612-8_17
19. Kautz, H.A., Selman, B., et al.: Planning as satisfiability. In: ECAI, vol. 92, pp. 359–363 (1992)
20. Lazaar, N., Hamadi, Y., Jabbour, S., Sebag, M.: Cooperation control in Parallel SAT Solving: a Multi-armed Bandit Approach. Research Report RR-8070, INRIA, September 2012
21. Liang, J.H., Ganesh, V., Poupart, P., Czarnecki, K.: Learning rate based branching heuristic for SAT solvers. In: Creignou, N., Le Berre, D. (eds.) SAT 2016. LNCS, vol. 9710, pp. 123–140. Springer, Cham (2016). doi:10.1007/978-3-319-40970-2_9
22. Lynce, I., Marques-Silva, J.: SAT in bioinformatics: making the case with haplotype inference. In: Biere, A., Gomes, C.P. (eds.) SAT 2006. LNCS, vol. 4121, pp. 136–141. Springer, Heidelberg (2006). doi:10.1007/11814948_16
23. Manthey, N.: Coprocessor 2.0 – a flexible CNF simplifier. In: Cimatti, A., Sebastiani, R. (eds.) SAT 2012. LNCS, vol. 7317, pp. 436–441. Springer, Heidelberg (2012). doi:10.1007/978-3-642-31612-8_34
24. Manthey, N.: Coprocessor – a standalone SAT preprocessor. In: Tompits, H., Abreu, S., Oetsch, J., Pührer, J., Seipel, D., Umeda, M., Wolf, A. (eds.) INAP/WLP -2011. LNCS, vol. 7773, pp. 297–304. Springer, Heidelberg (2013). doi:10.1007/978-3-642-41524-1_18
25. Marques-Silva, J.P., Sakallah, K., et al.: GRASP: a search algorithm for propositional satisfiability. IEEE Trans. Comput. **48**(5), 506–521 (1999)
26. Martins, R., Manquinho, V., Lynce, I.: Improving search space splitting for parallel SAT solving. In: IEEE International Conference on Tools with Artificial Intelligence, vol. 1, pp. 336–343. IEEE (2010)
27. Massacci, F., Marraro, L.: Logical cryptanalysis as a SAT problem. J. Autom. Reasoning **24**(1), 165–203 (2000)
28. Michael, M.M., Scott, M.L.: Simple, fast, and practical non-blocking and blocking concurrent queue algorithms. In: Proceedings of the Fifteenth Annual ACM Symposium on Principles of Distributed Computing, pp. 267–275. ACM (1996)
29. Moskewicz, M.W., Madigan, C.F., Zhao, Y., Zhang, L., Malik, S.: Chaff: engineering an efficient SAT solver. In: 38th annual Design Automation Conference, pp. 530–535. ACM (2001)

30. Ohmura, K., Ueda, K.: c-sat: A parallel SAT solver for clusters. In: Kullmann, O. (ed.) SAT 2009. LNCS, vol. 5584, pp. 524–537. Springer, Heidelberg (2009). doi:10. 1007/978-3-642-02777-2_47

31. Schulz, S., Blochinger, W.: Cooperate and compete! a hybrid solving strategy for task-parallel SAT solving on peer-to-peer desktop grids. In: High Performance Computing and Simulation, pp. 314–323. IEEE (2010)

32. Silva, J.P.M., Sakallah, K.A.: GRASP–a new search algorithm for satisfiability. In: IEEE/ACM International Conference on Computer-Aided Design, pp. 220–227. IEEE (1997)

33. Sonobe, T., Inaba, M.: Portfolio with block branching for parallel SAT solvers. In: Nicosia, G., Pardalos, P. (eds.) LION 2013. LNCS, vol. 7997, pp. 247–252. Springer, Heidelberg (2013). doi:10.1007/978-3-642-44973-4_25

34. Zhang, H., Bonacina, M.P., Hsiang, J.: PSATO: a distributed propositional prover and its application to quasigroup problems. J. Symbolic Comput. **21**(4), 543–560 (1996)

35. Zhang, L., Madigan, C.F., Moskewicz, M.H., Malik, S.: Efficient conflict driven learning in a boolean satisfiability solver. In: IEEE/ACM International Conference on Computer-Aided Design, pp. 279–285. IEEE Press (2001)

A Propagation Rate Based Splitting Heuristic
for Divide-and-Conquer Solvers

Saeed Nejati[✉], Zack Newsham, Joseph Scott, Jia Hui Liang,
Catherine Gebotys, Pascal Poupart, and Vijay Ganesh

University of Waterloo, Waterloo, ON, Canada
{snejati,znewsham,j29scott,jliang,cgebotys,ppoupart,
vganesh}@uwaterloo.ca

Abstract. In this paper, we present a divide-and-conquer SAT solver, MapleAmpharos, that uses a novel *propagation-rate* (PR) based splitting heuristic. The key idea is that we rank variables based on the ratio of how many propagations they cause during the run of the worker conflict-driven clause-learning solvers to the number of times they are branched on, with the variable that causes the most propagations ranked first. The intuition here is that, in the context of divide-and-conquer solvers, it is most profitable to split on variables that maximize the propagation rate. Our implementation MapleAmpharos uses the AMPHAROS solver as its base. We performed extensive evaluation of MapleAmpharos against other competitive parallel solvers such as Treengeling, Plingeling, Parallel CryptoMiniSat5, and Glucose-Syrup. We show that on the SAT 2016 competition Application benchmark and a set of cryptographic instances, our solver MapleAmpharos is competitive with respect to these top parallel solvers. What is surprising that we obtain this result primarily by modifying the splitting heuristic.

1 Introduction

Over the last two decades, sequential Boolean SAT solvers have had a transformative impact on many areas of software engineering, security, and AI [8,10,25]. Parallel SAT algorithms constitute a natural next step in SAT solver research, and as a consequence there has been considerable amount of research in parallel solvers in recent years [21]. Unfortunately, developing practically efficient parallel SAT solvers has proven to be a much harder challenge than anticipated [12,13,27]. Nonetheless, there are a few solver architectures that have proven to be effective on industrial instances. The two most widely used architectures are portfolio-based and variants of divide-and-conquer approach.

Portfolio solvers [1,5,6,15] are based on the idea that a collection of solvers each using a different set of heuristics are likely to succeed in solving instances obtained from diverse applications. These portfolio solvers are further enhanced with clause sharing techniques. By contrast, divide-and-conquer techniques are based on the idea of dividing the search space of the input formula F and solving the resulting partitions using distinct processors. Partitions are defined as the

© Springer International Publishing AG 2017
S. Gaspers and T. Walsh (Eds.): SAT 2017, LNCS 10491, pp. 251–260, 2017.
DOI: 10.1007/978-3-319-66263-3_16

original formula F restricted with a set of assumptions P (i.e., $F \wedge P$). These sets of assumptions cover the whole search space and each pair of them has at least one clashing literal [11,16–18,26,29].

Different strategies for choosing assumptions to split the search space (known as splitting heuristics in parallel SAT solver parlance) and the relative speedup obtained by splitting some or all of the variables have been studied in detail in the literature [21]. It is well-known that the choice of splitting heuristic can have a huge impact on the performance of divide-and-conquer solvers. One successful approach is to use *look-ahead* to split the input instance into sub-problems, and solve these sub-problems using CDCL solvers. Such methods are generally referred to as cube-and-conquer solvers [16], where the term cube refers to a conjunction of literals. In their work, Heule et al. [16] split the input instances (cube phase) and solve the partitions (conquer phase) sequentially, one after another. A potential problem with this approach is that it might result in sub-problems of *unbalanced hardness* and might lead to a high solving time for a few sub-problems. By contrast, in concurrent-cube-and-conquer [29] these two phases are run concurrently.

In this paper, we propose a new propagation rate-based splitting heuristic to improve the performance of divide-and-conquer parallel SAT solvers. We implemented our technique as part of the AMPHAROS solver [2], and showed significant improvements vis-a-vis AMPHAROS on instances from the SAT 2016 Application benchmark. Our key hypothesis was that variables that are likely to maximize propagation are good candidates for splitting in the context of divide-and-conquer solvers because the resultant sub-problems are often simpler. An additional advantage of ranking splitting variables based on their *propensity to cause propagations* is that it can be cheaply computed using conflict-driven clause-learning (CDCL) solvers that are used as workers in most modern divide-and-conquer parallel SAT solvers.

Contributions. We present a new splitting heuristic based on propagation rate, where a formula is broken into two smaller sub-formulas by setting the highest propagating variable to True and False. We evaluate the improved solver against the top parallel solvers from the SAT 2016 competition on the Application benchmark and a benchmark of cryptographic instances obtained from encoding of preimage attacks on the SHA-1 cryptographic hash function. Our solver, called MapleAmpharos, outperforms the baseline AMPHAROS and is competitive against Glucose, parallel CryptoMiniSat5, Treengeling and Plingeling on the SAT 2016 Application benchmark. Additionally, MapleAmpharos has better solving time compared to all of the solvers on our crypto benchmark.

2   Background on Divide-and-Conquer Solvers

We assume that the reader is familiar with the sequential conflict-driven clause-learning (CDCL) solvers. For further details we refer the reader to the Handbook of Satisfiability [9].

All divide-and-conquer SAT solvers take as input a Boolean formula which is partitioned into many smaller sub-formulas that are subsequently solved by sequential CDCL solvers. Usually the architecture is of a master-slave type. The master maintains the partitioning using a tree, where each node is a variable and can have at most two children. Left branch represents setting parent variable to False and right branch is for setting it to True. It means that if the original formula is F, and variable x is used to split the formula, the two sub-formulas would be $F_1 = F \land \neg x$ and $F_2 = F \land x$. We refer to the variable x as the *splitting variable*. Each leaf in this tree is called a cube (conjunction of literals). The path from root to each of the leaves represents different assumption sets that will be conjuncted to the original formula to generate partitions. The slave processes are usually CDCL solvers and they solve the sub-formulas simplified against the respective cubes, and report back to the master the results of completely or partially solving these sub-formulas.

Depending on the input problem, there could be redundant work among the workers when the formula is split. When the workers are directed to work on the same problem (with different policies), it is called *intensification*, and when the search space of workers are less overlapping, we call it *diversification*. Maintaining a balance between these two usually leads to a better performance [14].

3   Propagation Rate-Based Splitting Heuristic

In this Section we describe our propagation rate based splitting heuristic, starting with a brief description of AMPHAROS that we use as our base solver [2]. We made our improvements in three steps: (1) We used MapleSAT [19] as the worker or backend solver. This change gave us a small improvement over the base AMPHAROS. We call this solver Ampharos-Maplesat in Table 1; (2) We then used a propagation-rate based splitting heuristic on top of Ampharos-Maplesat, and we call this solver MapleAmpharos-PR. This gave us the biggest boost; (3) MapleAmpharos: Here we combined 3 heuristics, namely, MapleSAT backend, PR-based splitting heuristics, and different restart policies at worker solvers of MapleAmpharos-PR. The last change gave us a small boost over the MapleAmpharos-PR.

3.1   The AMPHAROS Solver

AMPHAROS is a divide-and-conquer solver wherein each worker is a CDCL SAT solver. The input to each worker is the original formula together with assumptions corresponding to the path (from the root of the splitting tree to the leaf) assigned to the worker. The workers can switch from one leaf to another for the sake of load balancing and intensification/diversification. Each worker searches for a solution to the input formula until it reaches a predefined limit or upper bound on the number of conflicts. We call this the *conflict limit*. Once the conflict limit is reached, the worker declares that the cube[1] is hard and reports

---

[1] While the term cube refers to a conjunction of literals, we sometimes use this term to also refer to a sub-problem created by simplifying a formula with a cube.

the "best variable" for splitting the formula to the master. A variable is deemed "best" by a worker if it has the highest VSIDS activity over all the variables when the conflict limit is reached. The Master then uses a load balancing policy to decide whether to split the problem into two by creating False and True branches over the reported variable.

3.2   Propagation Rate Splitting Heuristic

As mentioned earlier, the key innovation in MapleAmpharos is a propagation rate-based splitting heuristic. Picking variables to split on such that the resultant sub-problems are collectively easier to solve plays a crucial role in the performance of divide-and-conquer solvers. Picking the optimum sequence of splitting variables such that the overall running time is minimized is in general an intractable optimization problem.

For our splitting heuristic, we use a dynamic metric inspired by the measures that look-ahead solvers compute as part of their "look-ahead policy". In a look-ahead solver, candidate variables for splitting are assigned values (True and False) one at a time, and the formula is simplified against this assigned variable. A measure proportional to the number of simplified clauses in the resultant formula is used to rank all the candidate variables in decreasing order, and the highest ranked variable is used as a split. However, look-ahead heuristics are computationally expensive, especially when the number of variables is large. Propagation rate-based splitting on the other hand is very cheap to compute.

In our solver MapleAmpharos when a worker reaches its conflict limit, it picks the variable that has caused the highest number of propagations per decision (the propagation rate) and reports it back to the Master node. More precisely, whenever a variable v is branched on, we sum up the number other variables propagated by that decision. The propagation rate of a variable v is computed as the ratio of the total number of propagations caused whenever v is chosen as a decision variable divided by the total number of times v is branched on during the run of the solver. Variables that have never been branched on during the search get a value of zero as their propagation rate.

When a worker solver stops working on a sub-problem due to it reaching the conflict limit, or proving the cube to be UNSAT, it could move to work on a completely different sub-problem which has a different set of assumptions. Even through this node switching we do not reset the propagation rate counters.

The computational overhead of our propagation rate heuristic is minimal, since all the worker solvers do is maintain counters for the number of propagations caused by each decision variable during their runs. An important feature of our heuristic is that the number of propagation per decision variable is deeply influenced by the branching heuristic used by the worker solver. Also, the search path taken by the worker solver determines the number of propagations per variable. For example, a variable v when set to the value True might cause lots of propagation, and when set to the value False may cause none at all. Despite these peculiarities, our results show that the picking splitting variables based on the propagation rate-based heuristic is competitive for Application and cryptographic instances.

3.3 Worker Diversification

Inspired by the idea of using different heuristics in a competitive solver setting [15], we experimented with the idea of using different restart policies in worker CDCL solvers. We configured one third of the workers to use Luby restarts [20], another third to use geometric restarts, and the last third to use MABR restarts. MABR is a Multi-Armed Bandit Restart policy [22], which adaptively switches between 4 different restart policies of linear, uniform, geometric and Luby. We note that while we get some benefit from worker diversification, the bulk of the improvement in the performance of MapleAmpharos over AMPHAROS and other solvers is due to the propagation rate splitting heuristic.

4   Experimental Results

In our experiments we compared MapleAmpharos against 5 other top-performing parallel SAT solvers over the SAT 2016 Application benchmark and a set of cryptographic instances obtained from encoding of SHA-1 preimage attacks as Boolean formulas.

4.1   Experimental Setup

We used the Application benchmark of the SAT competition 2016 which has 300 industrial instances obtained from diverse set of applications. Timeout for each instance was set at 2 h wall clock time. All jobs were run on 8 core Intel Xeon CPUs with 2.53 GHz and 8 GB RAM. We compared our solver MapleAmpharos against the top parallel solvers from the SAT 2016 competition, namely, Treengeling and Plingeling [7], CryptoMiniSat5 [28], Glucose-Syrup [4] and also baseline version of AMPHAROS solver [2]. Our parallel solver MapleAmpharos uses MapleSAT [19] as its worker CDCL solver.

4.2   Case Study 1: SAT 2016 Application Benchmark

Figure 1, shows the cactus plot comparing the performance of MapleAmpharos against the other top parallel SAT solvers we considered in our experiments. In this version of the MapleAmpharos solver we used the best version of worker diversification (which is a combination of Luby, Geometric and MAB-restart referred to Sect. 3.3). As can be seen from the cactus plot in Fig. 1 and the Table 1, MapleAmpharos outperforms the baseline AMPHAROS, and is competitive vis-a-vis Parallel CryptoMiniSat, Glucose-Syrup, Plingeling and Treengeling. However, MapleAmpharos performs the best compared to the other solvers when it comes to solving cryptographic instances.

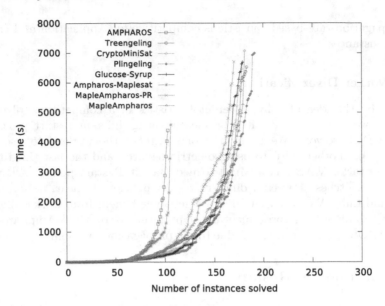

Fig. 1. Performance of MapleAmpharos vs. competing parallel solvers over the SAT 2016 Application benchmark

Table 1. Solving time details of MapleAmpharos and competing parallel solvers on SAT 2016 Application benchmark

	MapleAmpharos	MapleAmpharos-PR	Ampharos-Maplesat	AMPHAROS
Avg. time (s)	979.396	1035.73	310.94	392.933
# of solved	182	171	107	104
SAT	77	72	44	42
UNSAT	105	99	63	62
	CryptoMiniSat	Syrup	Plingeling	Treengeling
Avg. time (s)	942.894	898.767	965.167	969.467
# of solved	180	180	192	184
SAT	72	74	76	77
UNSAT	108	106	116	107

4.3   Case Study 2: Cryptographic Hash Instances

We also evaluated the performance of our solver against these parallel SAT solvers on instances that encode preimage attacks on the SHA-1 cryptographic hash function. These instances are known to be hard for CDCL solvers. The best solvers to-date can only invert at most 23 rounds automatically (out of maximum of 80 rounds in SHA-1) [22,23]. Our benchmark consists of instances corresponding to a SHA-1 function reduced to 21, 22 and 23 rounds, and for each number of rounds, we generate 20 different random hash targets. The solution to these instances are preimages that when hashed using SHA-1, generate

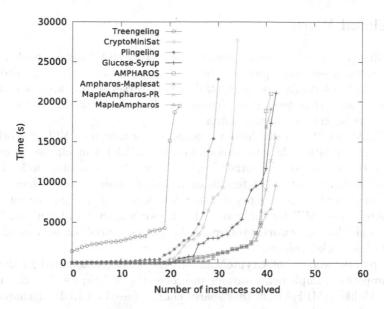

Fig. 2. Performance of MapleAmpharos vs. competing parallel solvers on SHA-1 instances

Table 2. Average solving time comparison on SHA-1 benchmark

	MapleAmpharos	MapleAmpharos-PR	Ampharos-Maplesat	AMPHAROS
Avg. time (s)	1048.53	1457.14	1518.76	1619.1
# of solved	43	43	42	42
	CryptoMiniSat	Syrup	Plingeling	Treengeling
Avg. time (s)	3056.31	2912.84	2668.48	4783.35
# of solved	35	43	31	23

the same hash targets. The instances were generated using the tool used for generating these type of instances in SAT competition [24]. The timeout for each instance was set to 8 hours. Figure 2 shows the performance comparison and Table 2 shows details of the average solving times on this benchmark. We compute the average for each solver only over the instances for which the resp. solvers finish. As can be seen from these results, MapleAmpharos performs the best compared to all of the other solvers. In particular, for the hardest instances in this benchmark (encoding of preimage attacks on 23 rounds of SHA-1), only Glucose-Syrup, AMPHAROS, and MapleAmpharos are able to invert some of the targets. Further, MapleAmpharos generally solves these SHA-1 instances much faster.

5   Related Work

Treengeling [7], one of the most competitive parallel SAT solvers to-date, uses a concurrent-cube-and-conquer architecture wherein a look-ahead procedure is used to split the formula, and sequential CDCL worker solvers are used to solve the sub-formulas thus generated. Treengeling is multi-threaded, and uses Lingeling as the backend sequential solver.

AMPHAROS [2] is a divide-and-conquer solver that uses VSIDS scoring as its splitting heuristic. The unit literals and low-LBD learnt clauses in each of the workers are shared with other workers through the master node. It also adaptively balances between intensification and diversification by changing the number of shared clauses between the workers as well as number of cubes. AMPHAROS uses MPI for communication between master and workers. It uses MiniSat and Glucose as worker solvers, and other sequential solvers can be easily retrofitted as worker solvers.

The parallel version of CryptoMiniSat5 [28] that participated in the SAT 2016 competition implements various in-processing techniques that can run in parallel. Unlike AMPHAROS that shares clauses based on LBD, it shares only unary and binary clauses.

Unlike AMPHAROS, Glucose-Syrup [3] uses a shared memory architecture. Additionally, it uses a lazy exchange policy for learnt clauses. They share clauses between cores when they are deemed to be useful locally rather than right after they are learnt. Furthermore, it implements a strategy for importing shared clauses into each of the workers. It checks whether the clauses are in the UNSAT proof of a cube and if not, they will be flagged as useless. This approach reduces the burden on propagation by reducing the sharing of not useful clauses.

Plingeling [7] is a portfolio solver which uses Lingeling as the CDCL solver. It launches workers with different configurations and shares clauses among the workers. The workers may also exchange facts derived using simplifications (e.g., the equivalence between variables) applied to their copy of the formula.

6   Conclusion

We present a propagation rate-based splitting heuristic for divide-and-conquer solvers, implemented on top of the AMPHAROS solver, that is competitive with respect to 5 other parallel SAT solvers on industrial and cryptographic instances. Many of these competing solvers were top performers in the SAT 2016 competition. Our key insight is that attempting to maximize propagations is a good strategy for splitting in divide-and-conquer solvers because the resultant sub-problems are significantly simplified and hence easier to solve. Finally, a crucial advantage of our approach is that the computational overhead associated with propagation rate-based splitting heuristic is minimal.

References

1. Audemard, G., Hoessen, B., Jabbour, S., Lagniez, J.-M., Piette, C.: Revisiting clause exchange in parallel SAT solving. In: Cimatti, A., Sebastiani, R. (eds.) SAT 2012. LNCS, vol. 7317, pp. 200–213. Springer, Heidelberg (2012). doi:10.1007/978-3-642-31612-8_16

2. Audemard, G., Lagniez, J.-M., Szczepanski, N., Tabary, S.: An adaptive parallel SAT solver. In: Rueher, M. (ed.) CP 2016. LNCS, vol. 9892, pp. 30–48. Springer, Cham (2016). doi:10.1007/978-3-319-44953-1_3

3. Audemard, G., Simon, L.: Lazy clause exchange policy for parallel SAT solvers. In: Sinz, C., Egly, U. (eds.) SAT 2014. LNCS, vol. 8561, pp. 197–205. Springer, Cham (2014). doi:10.1007/978-3-319-09284-3_15

4. Audemard, G., Simon, L.: Glucose and syrup in the sat 2016. SAT Compet. pp. 40–41 (2016)

5. Balyo, T., Sanders, P., Sinz, C.: HordeSat: A massively parallel portfolio SAT solver. In: Heule, M., Weaver, S. (eds.) SAT 2015. LNCS, vol. 9340, pp. 156–172. Springer, Cham (2015). doi:10.1007/978-3-319-24318-4_12

6. Biere, A.: Yet another local search solver and lingeling and friends entering the sat competition 2014. SAT Compet. 2014(2), 65 (2014)

7. Biere, A.: Splatz, lingeling, plingeling, treengeling, yalsat entering the sat competition 2016. SAT Compet. 2016, 44 (2016)

8. Biere, A., Cimatti, A., Clarke, E.M., Strichman, O., Zhu, Y.: Bounded model checking. Adv. Comput. 58, 117–148 (2003)

9. Biere, A., Heule, M., van Maaren, H.: Handbook of Satisfiability, vol. 185. IOS press, Amsterdam (2009)

10. Cadar, C., Ganesh, V., Pawlowski, P.M., Dill, D.L., Engler, D.R.: EXE: Automatically generating inputs of death. ACM Trans. Inf. Syst. Secur. 12(2), 10 (2008)

11. Chu, G., Stuckey, P.J., Harwood, A.: Pminisat: a parallelization of minisat 2.0. SAT race (2008)

12. Fujii, H., Fujimoto, N.: Gpu acceleration of bcp procedure for sat algorithms. In: Proceedings of the International Conference on Parallel and Distributed Processing Techniques and Applications (PDPTA), The Steering Committee of The World Congress in Computer Science, Computer Engineering and Applied Computing (WorldComp), p. 1 (2012)

13. Greenlaw, R., Hoover, H.J., Ruzzo, W.L.: Limits to Parallel Computation: P-Completeness Theory. Oxford University Press on Demand, New York (1995)

14. Guo, L., Hamadi, Y., Jabbour, S., Sais, L.: Diversification and intensification in parallel SAT solving. In: Cohen, D. (ed.) CP 2010. LNCS, vol. 6308, pp. 252–265. Springer, Heidelberg (2010). doi:10.1007/978-3-642-15396-9_22

15. Hamadi, Y., Jabbour, S., Sais, L.: Manysat: a parallel SAT solver. J. Satisf. Boolean Model. Comput. 6, 245–262 (2008)

16. Heule, M.J.H., Kullmann, O., Wieringa, S., Biere, A.: Cube and conquer: guiding CDCL SAT solvers by lookaheads. In: Eder, K., Lourenço, J., Shehory, O. (eds.) HVC 2011. LNCS, vol. 7261, pp. 50–65. Springer, Heidelberg (2012). doi:10.1007/978-3-642-34188-5_8

17. Hyvärinen, A.E.J., Junttila, T., Niemelä, I.: Partitioning SAT instances for distributed solving. In: Fermüller, C.G., Voronkov, A. (eds.) LPAR 2010. LNCS, vol. 6397, pp. 372–386. Springer, Heidelberg (2010). doi:10.1007/978-3-642-16242-8_27

18. Hyvärinen, A.E.J., Manthey, N.: Designing scalable parallel SAT solvers. In: Cimatti, A., Sebastiani, R. (eds.) SAT 2012. LNCS, vol. 7317, pp. 214–227. Springer, Heidelberg (2012). doi:10.1007/978-3-642-31612-8_17

19. Liang, J.H., Ganesh, V., Poupart, P., Czarnecki, K.: Learning rate based branching heuristic for SAT solvers. In: Creignou, N., Le Berre, D. (eds.) SAT 2016. LNCS, vol. 9710, pp. 123–140. Springer, Cham (2016). doi:10.1007/978-3-319-40970-2_9
20. Luby, M., Sinclair, A., Zuckerman, D.: Optimal speedup of Las Vegas algorithms. In: 1993 Proceedings of the 2nd Israel Symposium on the Theory and Computing Systems, pp. 128–133. IEEE (1993)
21. Manthey, N.: Towards next generation sequential and parallel SAT solvers. KI-Künstliche Intelligenz 30(3–4), 339–342 (2016)
22. Nejati, S., Liang, J.H., Ganesh, V., Gebotys, C., Czarnecki, K.: Adaptive restart and cegar-based solver for inverting cryptographic hash functions. arXiv preprint (2016). arxiv:1608.04720
23. Nossum, V.: SAT-based preimage attacks on SHA-1. Master's thesis (2012)
24. Nossum, V.: Instance generator for encoding preimage, second-preimage, and collision attacks on SHA-1. In: Proceedings of the SAT competition, pp. 119–120 (2013)
25. Rintanen, J.: Planning and SAT. Handb. Satisf. 185, 483–504 (2009)
26. Semenov, A., Zaikin, O.: Using Monte Carlo method for searching partitionings of hard variants of boolean satisfiability problem. In: Malyshkin, V. (ed.) PaCT 2015. LNCS, vol. 9251, pp. 222–230. Springer, Cham (2015). doi:10.1007/978-3-319-21909-7_21
27. Sohanghpurwala, A.A., Hassan, M.W., Athanas, P.: Hardware accelerated SAT solvers a survey. J. Parallel Distrib. Comput. 106, 170–184 (2016)
28. Soos, M.: The cryptominisat 5 set of solvers at SAT competition 2016. SAT Compet. 2016, 28 (2016)
29. Tak, P., Heule, M.J.H., Biere, A.: Concurrent cube-and-conquer. In: Cimatti, A., Sebastiani, R. (eds.) SAT 2012. LNCS, vol. 7317, pp. 475–476. Springer, Heidelberg (2012). doi:10.1007/978-3-642-31612-8_42

Quantified Boolean Formulas

Shortening QBF Proofs
with Dependency Schemes

Joshua Blinkhorn$^{(\boxtimes)}$ and Olaf Beyersdorff

School of Computing, University of Leeds, Leeds, UK
scjlb@leeds.ac.uk

Abstract. We provide the first proof complexity results for QBF dependency calculi. By showing that the reflexive resolution path dependency scheme admits exponentially shorter Q-resolution proofs on a known family of instances, we answer a question first posed by Slivovsky and Szeider in 2014 [30]. Further, we conceive a method of QBF solving in which dependency recomputation is utilised as a form of inprocessing. Formalising this notion, we introduce a new calculus in which a dependency scheme is applied dynamically. We demonstrate the further potential of this approach beyond that of the existing static system with an exponential separation.

1   Introduction

Proof complexity is the study of proof size in systems of formal logic. Since its beginnings the field has enjoyed strong connections to computational complexity [8,10] and bounded arithmetic [9,17], and has emerged in the past two decades as the primary means for the comparison of algorithms in automated reasoning.

Recent successes in that area, epitomised by progress in SAT solving, have motivated broader research into the efficient solution of computationally hard problems. Amongst them, the logic of quantified Boolean formulas (QBF) is an established field with a substantial volume of literature. QBF extends propositional logic with the addition of existential and universal quantification, and naturally accommodates more succinct encodings of problem instances. This gives rise to diverse applications in areas including conformant planning [11,23], verification [1], and ontologies [16].

It is fair to say that much of the early research into QBF solving [13,25,33], and later the proof complexity of associated theoretical models [4–6], was built upon existing techniques for SAT. For example, QCDCL [12] is a major paradigm in QBF solving based on conflict-driven clause learning (CDCL [28]), *the* dominant paradigm for SAT. By analogy, the fundamental theoretical model of QCDCL, the calculus Q-resolution (Q-Res [15]), is an extension of propositional resolution, the calculus that underpins CDCL. Given, however, that the decision problem for QBF is **PSPACE**-complete, it is perhaps unsurprising that the implementation of QCDCL presents novel obstacles for the practitioner, beyond those encountered at the level of propositional logic.

© Springer International Publishing AG 2017
S. Gaspers and T. Walsh (Eds.): SAT 2017, LNCS 10491, pp. 263–280, 2017.
DOI: 10.1007/978-3-319-66263-3_17

Arguably, the biggest challenge concerns the allowable order of variable assignments. In traditional QCDCL, the freedom to assign variables is limited according to a linear order imposed by the quantifier prefix. Whereas decision variables must be chosen carefully to ensure sound results, coercing the order of assignment to respect the prefix is frequently needlessly restrictive [19]. Moreover, limiting the choice adversely affects the impact of decision heuristics. In contrast, such heuristics play a major role in SAT solving [18,21,26,27], where variables may be assigned in an arbitrary order.

Dependency awareness, as implemented in the solver DepQBF [7], is a QBF-specific paradigm that attempts to maximise the impact of decision heuristics. By computing a *dependency scheme* before the search process begins, the linear order of the prefix is effectively supplanted by a partial order that better approximates the variable dependencies of the instance, granting the solver greater freedom regarding variable assignments. Use of the scheme is static; dependencies are computed only once and do not change during the search. Despite the additional computational cost incurred, empirical results demonstrate improved solving on many benchmark instances [19].

Dependency schemes themselves are tractable algorithms that identify dependency information by appeal to the syntactic form of an instance. From the plethora of schemes that have been proposed in the literature, two have emerged as principal objects of study. The *standard dependency scheme* ($\mathcal{D}^{\text{std}}$ [24]), a variant of which is used by DepQBF, was originally proposed in the context of backdoor sets. This scheme uses sequences of clauses connected by common existential variables to determine a dependency relation between the variables of an instance. The *reflexive resolution path dependency scheme* ($\mathcal{D}^{\text{rrs}}$ [31]) utilises the notion of a *resolution path*, a more refined type of connection introduced in [32].

A solid theoretical model for dependency awareness was only recently proposed in the form of the calculus $Q(\mathcal{D})$-Res [31], a parametrisation of Q-resolution by the dependency scheme $\mathcal{D}$. Whereas the body of work on $Q(\mathcal{D})$-Res and related systems has focused on soundness [2,22,31], authors of all three papers have cited open problems in proof complexity. Indeed, prior to this paper there were no proof-theoretic results to support any claims concerning the potential of dependency schemes in the practice of QBF solving.

In this work, not only do we provide the first such results, we also demonstrate the potential of dependency schemes to further reduce the size of proofs if they are applied *dynamically*. We summarise our contributions below.

1. The First Separations for QBF Dependency Calculi.

We use the well-known formulas of Kleine Büning et al. [15] to prove the first exponential separation for $Q(\mathcal{D})$-Res. We show that $\mathcal{D}^{\text{rrs}}$ can identify crucial independencies in these formulas, leading to short proofs in the system $Q(\mathcal{D}^{\text{rrs}})$-Res. In contrast, we show that $\mathcal{D}^{\text{std}}$ cannot identify any non-trivial independencies, allowing us to lift the exponential lower bound for Q-Res [3,15] to $Q(\mathcal{D}^{\text{std}})$-Res.

2. A Model of Dynamic Dependency Analysis. We propose the new calculus dyn-Q($\mathcal{D}$)-Res that models the dynamic application of a dependency scheme in Q-resolution. The system employs a so-called 'reference rule' that allows new axioms, called reference clauses, to be introduced into the proof. The key insight is that the application of an assignment to an instance formula may allow the dependency scheme to unlock new independencies. As such, the reference rule alludes to an explicit refutation of the formula under an appropriate restriction, and is analogous to the recomputation of dependencies at an arbitrary point of the QCDCL procedure. We prove that dyn-Q($\mathcal{D}$)-Res is sound whenever the dependency scheme $\mathcal{D}$ is fully exhibited.

3. Exponential Separation of Static and Dynamic Systems. Our final contribution demonstrates that the dynamic application of dependency schemes can shorten Q-resolution proofs even further, yielding an exponential improvement even over the static approach. Using a modification of the aforementioned formulas from [15], we prove that dyn-Q($\mathcal{D}^{rrs}$)-Res is exponentially stronger than Q($\mathcal{D}^{rrs}$)-Res.

2   Preliminaries

Quantified Boolean Formulas. In this paper, we consider *quantified Boolean formulas* (QBFs) in *prenex conjunctive normal form* (PCNF), typically denoted $\Phi = \mathcal{Q}.\phi$. A PCNF over Boolean variables $z_1, \ldots, z_n$ consists of a *quantifier prefix* $\mathcal{Q} = \mathcal{Q}_1 z_1 \cdots \mathcal{Q}_n z_n$, $\mathcal{Q}_i \in \{\exists, \forall\}$ for $i \in [n]$, in which all variables are quantified either existentially or universally, and a propositional conjunctive normal form (CNF) formula ϕ called the *matrix*. The prefix $\mathcal{Q}$ imposes a linear ordering $<_\Phi$ on the variables of Φ, such that $z_i <_\Phi z_j$ holds whenever $i < j$, in which case we say that z_j *is right of* z_i.

A literal is a variable or its negation, a clause is a disjunction of literals, and a CNF is a conjunction of clauses. Throughout, we refer to a clause as a set of literals and to a CNF as a set of clauses. We typically write x for existential variables, u for universals, and z for either. For a literal l, we write var$(l) = z$ iff $l = z$ or $l = \neg z$, for a clause C we write vars$(C) = \{\text{var}(l) \mid l \in C\}$, and for a PCNF Φ we write vars(Φ) for the variables in the prefix of Φ.

A (partial) assignment δ to the variables of Φ is represented as a set of literals, typically denoted $\{l_1, \ldots, l_k\}$, where literal z (resp. $\neg z$) represents the assignment $z \mapsto 1$ (resp. $z \mapsto 0$). The *restriction of* Φ by δ, denoted $\Phi[\delta]$, is obtained by removing from ϕ any clause containing a literal in δ, and removing the negated literals $\neg l_1, \ldots, \neg l_k$ from the remaining clauses, while the variables of δ and their associated quantifiers are removed from the prefix $\mathcal{Q}$. For assignments to single variables we may omit the braces; for example, we write $\Phi[l]$ for $\Phi[\{l\}]$.

QBF Resolution. *Resolution* is a well-studied refutational proof system for propositional CNF formulas with a single inference rule: the *resolvent* $C_1 \cup C_2$

may be derived from clauses $C_1 \cup \{x\}$ and $C_2 \cup \{\neg x\}$ (variable x is the *pivot*). Resolution is *refutationally* sound and complete: that is, the empty clause can be derived from a CNF iff it is unsatisfiable.

There exist a host of resolution-based QBF proof systems – see [3] for a detailed account. *Q-resolution* (Q-Res) introduced in [15] is the standard refutational calculus for PCNF. In addition to resolution over existential pivots, the calculus has a *universal reduction rule* which allows a clause C to be derived from $C \cup \{l\}$, provided var(l) is a universal variable right of all existentials in C. Tautologies are explicitly forbidden; one may not derive a clause containing both z and $\neg z$.

For a QBF resolution system P, a P *derivation* of a clause C from a PCNF Φ is a sequence $C_1, \ldots, C_m$ of clauses in which $C = C_m$, and each clause is either an axiom or is derived from previous clauses in the sequence using an inference rule. A *refutation* of Φ is a derivation of the empty clause from Φ.

A proof system P *p-simulates* a system Q (denoted Q $\leq_p$ P) if each Q-proof can be transformed in polynomial time into a P-proof of the same formula [10]. The systems P and Q are *p-equivalent* (denoted P $\equiv_p$ Q) if P $\leq_p$ Q and Q $\leq_p$ P.

QBF Models. Let $\Phi = Q_1 z_1 \cdots Q_n z_n . \phi$ be a PCNF over existential variables $V_\exists$ and universal variables $V_\forall$. A *model* f for Φ is a mapping from total assignments to $V_\forall$ to total assignments to $V_\exists$ that satisfies two conditions: (a) whenever α and α' agree on all universals left of a variable z_i, then $f(\alpha)$ and $f(\alpha')$ agree on all existential variables left of (and including) z_i; (b) for each α in the domain of f, $\alpha \cup f(\alpha)$ satisfies every clause $C \in \phi$ (that is, $C \cap (\alpha \cup f(\alpha)) \neq \emptyset$). A PCNF is true iff it has a model, otherwise it is false.

Following [25], a model can be depicted naturally as a tree, as shown in Fig. 1. For each α in the domain of f, the literals of the set $\alpha \cup f(\alpha)$ are written in prefix order on a unique path from the root of the tree to some leaf. As such, a model can be uniquely identified with a set of $2^{|V_\forall|}$ *paths*, each of which is one

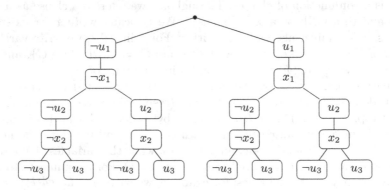

Fig. 1. Tree depiction of a model for the PCNF with prefix $\forall u_1 \exists x_1 \forall u_2 \exists x_2 \forall u_3$ and clauses $\{u_1, \neg x_1\}, \{\neg u_1, x_1\}$ and $\{\neg u_1, \neg u_2, x_2, \neg u_3\}$.

of the sets $\alpha \cup f(\alpha)$. This is a convenient interpretation (cf. [2]), and we adopt this approach for all technicalities concerning QBF models.

3   Static Dependency Awareness in Q-Resolution

In this section, we provide the necessary background for dependency schemes and their incorporation into Q-resolution. We recall the definitions of the standard [24] and reflexive resolution path [31] dependency schemes, and the definition of the dependency calculus $\mathsf{Q}(\mathcal{D})$-Res.

3.1   Overview of Dependency Schemes

For the duration of this work, we deal only with the (in)dependence of existential variables on universal variables[1]. This is a convenience afforded by the fact that we deal with *refutational* calculi, in which the (in)dependence of universals on existentials does not feature. We therefore take the opportunity to work with tighter (and in some cases considerably simpler) definitions than those referenced in the literature.

A dependency scheme is presented as a function mapping PCNFs to binary relations. The binary relations represent variable dependencies. For an arbitrary PCNF Φ, the *trivial dependency relation* captures the linear order of the quantifier prefix of Φ, and is given by $\mathcal{D}^{\mathrm{trv}}(\Phi) = \{(u, x) \in \mathrm{vars}_\forall(\Phi) \times \mathrm{vars}_\exists(\Phi) \mid u <_\Phi x\}$. Formally, a *dependency scheme* $\mathcal{D}$ is a mapping from the set of all PCNFs that satisfies $\mathcal{D}(\Phi) \subseteq \mathcal{D}^{\mathrm{trv}}(\Phi)$ for each PCNF Φ. The existence of a pair $(u, x) \in \mathcal{D}(\Phi)$ should be interpreted as 'existential x depends on universal u in Φ according to dependency scheme $\mathcal{D}$'. We say that $\mathcal{D}'$ is *at least as general* as $\mathcal{D}$ iff $\mathcal{D}'(\Phi) \subseteq \mathcal{D}(\Phi)$ for each PCNF Φ, and is *strictly more general* if the inclusion is strict for some PCNF.

All non-trivial dependency schemes that have appeared in the literature to date are based in some way or another on connections between clauses in the matrix. In the standard dependency scheme $\mathcal{D}^{\mathrm{std}}$, an existential x depends on a universal u whenever a clause containing variable x is connected to a clause containing variable u, whereby clauses are connected iff they share a common existential variable that is right of u. The absence of such a connection ensures that x is independent of u according to $\mathcal{D}^{\mathrm{std}}$.

Definition 1 (standard dependency scheme [24]). *Let $\Phi = Q.\phi$ be a PCNF. The pair $(u, x) \in \mathcal{D}^{\mathrm{trv}}(\Phi)$ is in $\mathcal{D}^{\mathrm{std}}(\Phi)$ iff there exists a sequence of clauses $C_1, \ldots, C_n \in \phi$ with $u \in \mathrm{vars}(C_1)$, $x \in \mathrm{vars}(C_n)$, such that, for each $i \in [n-1]$, $\mathrm{vars}(C_i) \cap \mathrm{vars}(C_{i+1})$ contains an existential variable right of u.*

Whereas connections in $\mathcal{D}^{\mathrm{std}}$ are based on common variables, the reflexive resolution path dependency scheme $\mathcal{D}^{\mathrm{rrs}}$ improves upon $\mathcal{D}^{\mathrm{std}}$ by taking polarity into account. The connecting existential variable must appear in opposite

---

[1] In practice, the dual notion of (in)dependence of universals on existentials is equally important.

polarities in the connected clauses, yielding a strictly more general scheme. As explained above, we present a simplified formulation of $\mathcal{D}^{\mathrm{rrs}}$ tailored to the current work.

Definition 2 (reflexive resolution path dependency scheme [31]). *Let $\Phi = \mathcal{Q}.\phi$ be a PCNF, and let $(u, x) \in \mathcal{D}^{\mathrm{trv}}(\Phi)$. Then $(u, x) \subset \mathcal{D}^{\mathrm{rrs}}(\Phi)$ iff there is a sequence of clauses $C_1, \ldots, C_n \in \phi$ and a sequence of existential literals $l_1, \ldots, l_{n-1}$ for which the following four conditions hold:*

(a) *$u \in C_1$ and $\neg u \in C_n$,*
(b) *$x = \mathrm{var}(l_i)$, for some $i \in [n-1]$,*
(c) *$u <_\Phi \mathrm{var}(l_i)$, $l_i \in C_i$ and $\neg l_i \in C_{i+1}$, for each $i \in [n-1]$,*
(d) *$\mathrm{var}(l_i) \neq \mathrm{var}(l_{i+1})$ for each $i \in [n-2]$.*

3.2   Dependency Schemes in Q-Resolution

The theoretical model for the use of dependency schemes in dependency-aware solving is captured by the calculus $\mathsf{Q}(\mathcal{D})$-Res, introduced in [31]. The main idea is to generalise Q-Res by replacing the implicit reference to the trivial dependency scheme with an explicit reference to a strictly more general scheme. Note that Q-Res allows a universal variable u to be reduced only if it is right of all existentials in the clause, or, equivalently, whenever all existentials in the clause are trivially independent of u. By contrast, in $\mathsf{Q}(\mathcal{D})$-Res u can be reduced whenever all existentials in the clause are $\mathcal{D}$-independent of u. We recall the rules of $\mathsf{Q}(\mathcal{D})$-Res in Fig. 2.

Soundness of the calculus $\mathsf{Q}(\mathcal{D})$-Res is not guaranteed, and hinges on the choice of the dependency scheme $\mathcal{D}$. Previous work has shown that the concept of *full*

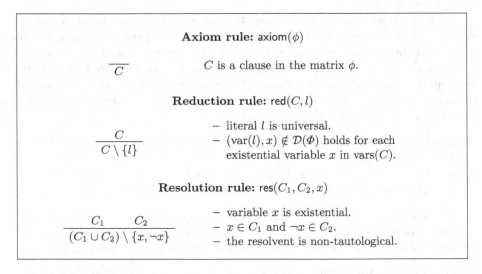

Axiom rule: $\mathsf{axiom}(\phi)$

$$\overline{C}$$

C is a clause in the matrix ϕ.

Reduction rule: $\mathsf{red}(C, l)$

$$\frac{C}{C \setminus \{l\}}$$

— literal l is universal.
— $(\mathrm{var}(l), x) \notin \mathcal{D}(\Phi)$ holds for each existential variable x in $\mathrm{vars}(C)$.

Resolution rule: $\mathsf{res}(C_1, C_2, x)$

$$\frac{C_1 \qquad C_2}{(C_1 \cup C_2) \setminus \{x, \neg x\}}$$

— variable x is existential.
— $x \in C_1$ and $\neg x \in C_2$.
— the resolvent is non-tautological.

Fig. 2. The rules of $\mathsf{Q}(\mathcal{D})$-Res [31]. $\mathcal{D}$ is a dependency scheme and $\Phi = \mathcal{Q}.\phi$ is a PCNF.

exhibition[2], which imposes a natural condition on $\mathcal{D}$, is sufficient to prove soundness in Q($\mathcal{D}$)-Res [29], and indeed in stronger dependency calculi for QBF [2]. Following [2], we say that a model f exhibits the independence of x on u iff, for each α in the domain of f, the assignment to x in $f(\alpha)$ remains unchanged when the assignment to u in α is flipped.

Definition 3 (full exhibition [2,29]). *A model f for a PCNF Φ is a $\mathcal{D}$-model iff, for each $(u, x) \in \mathcal{D}^{\mathrm{trv}}(\Phi) \setminus \mathcal{D}(\Phi)$, f exhibits the independence of x on u. A dependency scheme $\mathcal{D}$ is* fully exhibited *iff each true PCNF has a $\mathcal{D}$-model.*

Informally, full exhibition ensures that a true PCNF has a particular model in which existentials do not depend on the universals from which they are independent according to the dependency scheme. As in [29], we refer to such a model as a $\mathcal{D}$-model. In Sect. 5, we show that full exhibition remains sufficient for soundness when a dependency scheme is applied dynamically, as opposed to the static application offered in Q($\mathcal{D}$)-Res.

It should be clear that Q($\mathcal{D}$)-Res is simulated by Q($\mathcal{D}'$)-Res whenever $\mathcal{D}'$ is at least as general as $\mathcal{D}$. We conclude this section by noting the following trivial simulations for Q($\mathcal{D}$)-Res.

Proposition 4. Q-Res $\equiv_p$ Q($\mathcal{D}^{\mathrm{trv}}$)-Res $\leq_p$ Q($\mathcal{D}^{\mathrm{std}}$)-Res $\leq_p$ Q($\mathcal{D}^{\mathrm{rrs}}$)-Res.

4   Exponential Separation of Q($\mathcal{D}^{\mathrm{std}}$)-Res and Q($\mathcal{D}^{\mathrm{rrs}}$)-Res

In this section, we prove that Q($\mathcal{D}^{\mathrm{rrs}}$)-Res is exponentially stronger than Q($\mathcal{D}^{\mathrm{std}}$)-Res. Given that Q($\mathcal{D}^{\mathrm{std}}$)-Res p-simulates Q-Res (Proposition 4), we thereby separate Q($\mathcal{D}^{\mathrm{rrs}}$)-Res and Q-Res, thus answering the question initially posed by Slivovsky and Szeider in [30]. The separating formulas are a well-studied family of PCNFs, originally introduced in [15]. We recall the definition of this formula family, which is referred to as $\Psi(n)$ throughout this paper.

Definition 5. (formulas of Kleine Büning et al. [15]). *The formula family $\Psi(n) := \mathcal{Q}(n) . \psi(n)$ has prefixes $\mathcal{Q}(n) := \exists x_1 \exists y_1 \forall u_1 \cdots \exists x_n \exists y_n \forall u_n \exists t_1 \cdots \exists t_n$ and matrices $\psi(n)$ consisting of the clauses*

$$
\begin{aligned}
A &:= \{\neg x_1, \neg y_1\}, \\
B_i &:= \{x_i, u_i, \neg x_{i+1}, \neg y_{i+1}\} & B_i' &:= \{y_i, \neg u_i, \neg x_{i+1}, \neg y_{i+1}\} & i \in [n-1], \\
B_n &:= \{x_n, u_n, \neg t_1, \ldots, \neg t_n\} & B_n' &:= \{y_n, \neg u_n, \neg t_1, \ldots, \neg t_n\}, \\
C_i &:= \{u_i, t_i\} & C_i' &:= \{\neg u_i, t_i\} & i \in [n].
\end{aligned}
$$

We first show that the standard dependency scheme cannot identify any non-trivial independencies for $\Psi(n)$.

Proposition 6. *For each $n \in \mathbb{N}$, $\mathcal{D}^{\mathrm{std}}(\Psi(n)) = \mathcal{D}^{\mathrm{trv}}(\Psi(n))$.*

---

[2] The term 'full exhibition' was coined in [2]. The concept itself and the term '$\mathcal{D}$-model' originate from [29].

Proof. Let $n \in \mathbb{N}$ and let $i, j \in [n]$.

For $(u_i, t_j) \in \mathcal{D}^{\text{trv}}(\Psi(n))$, consider the sequence of clauses $B_i, \ldots, B_n$, and observe that $u_i \in \text{vars}(B_i)$ and $t_j \in \text{vars}(B_n)$. For each $k \in [i, n-1]$, the existential variable x_{k+1}, which is right of u_i, is in the set $\text{vars}(B_k) \cap \text{vars}(B_{k+1})$. Therefore $(u_i, t_j) \in \mathcal{D}^{\text{std}}(\Psi(n))$.

For each $(u_i, x_j) \in \mathcal{D}^{\text{trv}}(\Psi(n))$ with $i < j$, the fact that $(u_i, x_j) \in \mathcal{D}^{\text{std}}(\Psi(n))$ is shown similarly, using the sequence of clauses $B_i, \ldots, B_j$. For the final case $(u_i, y_j) \in \mathcal{D}^{\text{trv}}(\Psi(n))$ take the sequence $B'_i, \ldots, B'_j$.   □

The salient consequence of Proposition 6 is that every application of $\forall$-reduction in a $\mathsf{Q}(\mathcal{D}^{\text{std}})$-Res derivation from $\Psi(n)$ is also available in Q-Res. As a result, the Q-Res lower bound for $\Psi(n)$ lifts directly to $\mathsf{Q}(\mathcal{D}^{\text{std}})$-Res.

Theorem 7. *The QBFs $\Psi(n)$ require exponential-size $\mathsf{Q}(\mathcal{D}^{\text{std}})$-Res refutations.*

Proof. It is known that $\Psi(n)$ require exponential-size Q-Res refutations [3,15]. By Proposition 6, any $\mathsf{Q}(\mathcal{D}^{\text{std}})$-Res refutation of $\Psi(n)$ is a $\mathsf{Q}(\mathcal{D}^{\text{trv}})$-Res refutation of $\Psi(n)$. The result follows since Q-Res and $\mathsf{Q}(\mathcal{D}^{\text{trv}})$-Res are p-equivalent, by Proposition 4.   □

In contrast, the more general dependency scheme $\mathcal{D}^{\text{rrs}}$ can identify some crucial non-trivial independencies in $\Psi(n)$.

Proposition 8. *For each $n \in \mathbb{N}$ and for each $i, j \in [n]$, if $i \neq j$ then $(u_i, t_j) \notin \mathcal{D}^{\text{rrs}}(\Psi(n))$.*

Proof. Let $n \in \mathbb{N}$ and let $i, j \in [n]$ with $i \neq j$. Suppose that $D_1, \ldots, D_k \in \psi(n)$ and $l_1, \ldots, l_{k-1}$ are sequences of clauses and literals respectively, satisfying the four conditions of Definition 2 with respect to the pair $(u_i, t_j) \in \mathcal{D}^{\text{trv}}(\Psi(n))$. By condition (b), the literal sequence contains a literal in the variable t_j. Observe that, in the matrix $\psi(n)$, the positive literal t_j occurs only in the clauses $C_j = \{u_j, t_j\}$ and $C'_j = \{\neg u_j, t_j\}$. Hence, by condition (c), there is some clause D in the clause sequence such that $D = C_j$ or $D = C'_j$. Since t_j is the only existential literal in D, the clause must be an endpoint of the sequence by condition (d), and hence we must have $D = D_1$ or $D = D_k$. However, since $i \neq j$, this implies that either $u_i \notin D_1$ or $u_i \notin D_k$, contradicting condition (a).   □

According to Proposition 8, a $\mathsf{Q}(\mathcal{D}^{\text{rrs}})$-Res refutation of $\Psi(n)$ may contain $\forall$-reduction steps that are disallowed in Q-Res. For example, under $\mathcal{D}^{\text{rrs}}$ it is possible to remove literal u_n from the clause $\{x_n, u_n, \neg t_1, \ldots, \neg t_{n-1}\}$. As we demonstrate in the proof of the following theorem, it is precisely this step (which is unavailable in Q-Res due to the presence of existentials right of u) that permits the construction of $O(n)$-size $\mathsf{Q}(\mathcal{D}^{\text{rrs}})$-Res refutations.

Theorem 9. *The formulas $\Psi(n)$ have linear-size $\mathsf{Q}(\mathcal{D}^{\text{rrs}})$-Res refutations.*

Proof. A portion of a linear-size $\mathsf{Q}(\mathcal{D}^{\text{rrs}})$-Res refutation of $\Psi(n)$ is shown in Fig. 3. The clauses $\{x_{n-1}, u_{n-1}, \neg t_1, \ldots, \neg t_{n-1}\}$ and $\{y_{n-1}, \neg u_{n-1}, \neg t_1, \ldots, \neg t_{n-1}\}$ are derived in a constant number of steps, and the task is reduced to the refutation of $\Psi(n-1)$. The complete refutation is therefore linear in size.

According to Proposition 8, in a $\mathsf{Q}(\mathcal{D}^{\mathrm{rrs}})$-Res derivation from $\Psi(n)$ the variable u_i may be removed from a clause D provided that the existential variables in D that are right of u_i are contained in the set $\{t_1, \ldots, t_n\} \setminus \{t_i\}$. Such $\forall$-reduction steps, which would be disallowed in Q-Res, are marked with an asterisk $(*)$ in Fig. 3. □

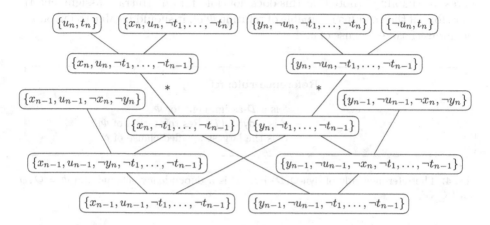

Fig. 3. Portion of a linear size $\mathsf{Q}(\mathcal{D}^{\mathrm{rrs}})$-Res refutation of $\Psi(n)$. The $\forall$-reduction steps marked with $*$ are forbidden in Q-Res, but are allowed in $\mathsf{Q}(\mathcal{D}^{\mathrm{rrs}})$-Res due to Proposition 8.

The following result is an immediate consequence of Theorems 7 and 9.

Theorem 10. $\mathsf{Q}(\mathcal{D}^{\mathrm{rrs}})$-Res *is exponentially stronger than* $\mathsf{Q}(\mathcal{D}^{\mathrm{std}})$-Res.

5   Modelling Dynamic Dependency Awareness

In this section, we introduce the dynamic dependency calculus dyn-$\mathsf{Q}(\mathcal{D})$-Res and prove that it is sound for a fully exhibited scheme $\mathcal{D}$.

5.1   Dynamic Dependencies in Q-Resolution

We first define a particular kind of assignment to the variables of a PNCF that, in a clear sense, 'respects' the dependency scheme $\mathcal{D}$.

Definition 11 ($\mathcal{D}$-assignment). *Let* $\mathcal{D}$ *be a dependency scheme and let* δ *be a partial assignment to the variables of a PCNF* Φ. *Then* δ *is a* $\mathcal{D}$-assignment *for* Φ *iff, whenever* δ *assigns an existential literal* l, *then* δ *assigns all universal variables in the set* $\{u \mid (u, \mathrm{var}(l)) \in \mathcal{D}(\Phi)\}$.

We also define the *largest falsified clause* of an assignment.

Definition 12 (largest falsified clause). *Let $\delta = l_1, \ldots, l_k$ be an assignment. The largest falsified clause of δ is $\{\neg l_1, \ldots, \neg l_k\}$.*

Definition of the Calculus. We define dyn-Q($\mathcal{D}$)-Res as the proof system that has the rules of Q($\mathcal{D}$)-Res in addition to the *reference rule* shown in Fig. 4. On an intuitive level, the reference rule is based on the following fact: Given a PCNF Φ and a fully exhibited dependency scheme $\mathcal{D}$, if Φ is false under restriction by a $\mathcal{D}$-assignment δ, then adding the largest falsified clause of δ to the matrix of Φ preserves satisfiability[3] (note that this does not hold for an arbitrary assignment δ). Therefore, if the calculus is capable of refuting $\Phi[\delta]$, it should be able to introduce the largest falsified clause of δ.

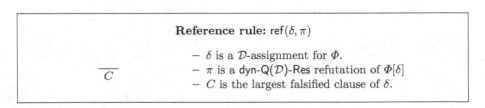

Fig. 4. The reference rule of dyn-Q($\mathcal{D}$)-Res. $\mathcal{D}$ is a dependency scheme and $\Phi = \mathcal{Q}.\phi$ is a PCNF.

We refer to a clause derived by application of the reference rule as a *reference clause*. As stated in the rule itself, a reference clause may only be introduced if an explicit refutation π of $\Phi[\delta]$ can be given. This feature allows the size of a dyn-Q($\mathcal{D}$)-Res derivation to be suitably defined. We refer to π as a *referenced refutation*.

The power of the reference rule lies in the fact that the dependency scheme $\mathcal{D}$ may identify (or *unlock*) new non-trivial independencies in the restricted formula, meaning that it may be easier to refute the restricted formula $\Phi[\delta]$ than to derive the reference clause from Φ directly. We note that the referenced refutation π, being a derivation from $\Phi[\delta]$, can make use of these newly unlocked independencies. In this way, the calculus models the recomputation of dependencies during the QCDCL search procedure. We elaborate on this point in Subsect. 5.3.

Reference Degree. The reference degree of a dyn-Q($\mathcal{D}$)-Res derivation is 0 iff it does not contain any reference clauses (i.e. it is a Q($\mathcal{D}$)-Res derivation). For all other derivations π, the reference degree is $d+1$, where d is the largest reference degree of a refutation referenced from π.

Proof Size. The size of a dyn-Q($\mathcal{D}$)-Res derivation π of reference degree 0 is the number of clauses in the proof. The size of a derivation π with non-zero reference degree is $a + b$, where a is the number of clauses in π and b is the sum of the sizes of refutations referenced from π.

---

[3] We prove this statement formally in Subsect. 5.2 (Lemma 14).

5.2   Soundness of dyn-Q($\mathcal{D}$)-Res

The task of proving that dyn-Q($\mathcal{D}$)-Res is sound for a fully exhibited dependency scheme $\mathcal{D}$ (Theorem 15) can essentially be reduced to proving that the reference clauses derived from a true PCNF are satisfied by a $\mathcal{D}$-model (Lemma 14).

In what follows, we find it convenient to introduce a notion of restriction for models. Let f be a model for a PCNF Φ, and let l be a literal with var$(l) \in$ vars(Φ). If l is universal, then $f[l]$ is obtained from f by removing all paths containing $\neg l$ and removing l from all remaining paths. If l is existential, then $f[l]$ is obtained from f by removing all occurrences of l and $\neg l$ from the paths of f.

It should be clear that $f[l]$ is a model for $\Phi[l]$ if l is universal. The same is also true for an existential literal l provided that it is unopposed in f, by which we mean that its negation $\neg l$ does not appear in any path in f. These facts are useful enough in the sequel to be the subject of the following proposition.

Proposition 13. *Let Φ be a PCNF, let f be a model for Φ and let l be a literal with* var$(l) \in$ vars(Φ). *Then $f[l]$ is a model for $\Phi[l]$ if either* (a) l *is universal, or* (b) l *is existential and unopposed in f.*

We extend the restriction of a model to an arbitrary assignment $\delta = \{l_1, \ldots, l_k\}$ in the natural way; that is, $f[\delta]$ is the result of the successive restriction of f by the literals in δ. It should be clear that the order of successive restrictions does not matter.

We proceed to prove that a $\mathcal{D}$-model of a PCNF Φ satisfies any reference clause derivable from it in dyn-Q($\mathcal{D}$)-Res.

Lemma 14. *Let $\mathcal{D}$ be a dependency scheme, let f be a $\mathcal{D}$-model for a PCNF Φ, and let δ be a $\mathcal{D}$-assignment for Φ. If $\Phi[\delta]$ is false, then f satisfies the largest falsified clause of δ.*

Proof. Let $\Phi = \mathcal{Q} . \phi$, and let C be the largest falsified clause of δ. We prove the contrapositive statement: if f does not satisfy C, then $\Phi[\delta]$ is true.

The idea of the proof is to restrict f by δ, obtaining a model for $\Phi[\delta]$. The simplest way to do this is to restrict first by the universal subassignment of δ, and then by the existential subassignment. To that end, let $\delta_\forall := \{l \in \delta \mid$ var(l) is universal$\}$ and define $\delta_\exists$ similarly.

By successive application of Proposition 13 (a), it follows that $f[\delta_\forall]$ is a model for $\Phi[\delta_\forall]$. We claim that every literal in $\delta_\exists$ is unopposed in $f[\delta_\forall]$. We will therefore prove the result since, by successive application of Proposition 13 (b), it follows that $(f[\delta_\forall])[\delta_\exists] = f[\delta]$ is a model for $(\Phi[\delta_\forall])[\delta_\exists] = \Phi[\delta]$.

It remains to prove that the literals in $\delta_\exists$ are indeed unopposed in $f[\delta_\forall]$. Suppose that f falsifies C. Then there is some path P in f that contains none of the literals in C. Since P contains a literal for every variable, it must therefore contain the negation of every literal in C. It follows, by definition of largest falsified clause (Definition 12), that $\delta \subseteq P$. Then, by definition of model restriction, there is some path $P' = P \setminus \delta_\forall$ in $f[\delta_\forall]$ with $\delta_\exists \subseteq P'$. The result follows since each existential variable in vars$(\delta_\exists)$ appears in $f[\delta_\forall]$ in a single polarity. To see this, let $x \in \text{vars}(\delta_\exists)$, and note that $\{u \mid (u, x) \in \mathcal{D}(\Phi)\} \subseteq \text{vars}(\delta_\forall)$, since δ is a $\mathcal{D}$-assignment. Hence, $f[\delta_\forall]$ exhibits the independence of x on all remaining universals, and x therefore occurs in a single polarity. $\square$

To prove that dyn-Q($\mathcal{D}$)-Res is sound, we must prove that one cannot derive the empty clause from any true PCNF. The proof is obtained by the addition of Lemma 14 to the literature's existing proof of soundness for Q($\mathcal{D}$)-Res.

Theorem 15. *The dynamic dependency calculus* dyn-Q($\mathcal{D}$)-Res *is sound if* $\mathcal{D}$ *is fully exhibited.*

Proof. In [29] it is shown that Q($\mathcal{D}$)-Res is sound if $\mathcal{D}$ is fully exhibited. The result may be proved by induction on derivation depth in the following way (for a detailed proof cf. [2]): Let $\mathcal{D}$ be a fully exhibited dependency scheme, let $\Phi := \mathcal{Q} \cdot \phi$ be a true PCNF and assume π is a Q($\mathcal{D}$)-Res refutation of Φ. Since $\mathcal{D}$ is fully exhibited, there exists a fully exhibiting model f for Φ (with respect to $\mathcal{D}$) that satisfies every matrix clause. Moreover, if f satisfies the antecedent clauses of any application of resolution or reduction, then f satisfies the consequent clause. We therefore reach a contradiction, since f satisfies the conclusion of π, the empty clause.

Now, if we instead let π be a dyn-Q($\mathcal{D}$)-Res refutation, the above method can be lifted provided that the fully exhibiting model f satisfies every clause introduced by application of the reference rule. In this way, we prove soundness by induction on the reference degree d of π.

The base case $d = 0$ is already established [29], since any dyn-Q($\mathcal{D}$)-Res refutation of degree 0 is a Q($\mathcal{D}$)-Res refutation. For the inductive step, let $d \geq 1$, and suppose that all dyn-Q($\mathcal{D}$)-Res refutations of reference degree less than d are sound. Further, let C be the first reference clause of π, introduced by application of ref(δ, π'). Since π' is a dyn-Q($\mathcal{D}$)-Res refutation of $\Phi[\delta]$ of reference degree at most $d-1$, $\Phi[\delta]$ is false by the inductive hypothesis. Since C is the largest falsified clause of δ, it is therefore satisfied by f, by Lemma 14. Successive application of the argument demonstrates that f satisfies every reference clause in π. $\square$

As it is known that $\mathcal{D}^{\text{rrs}}$ is fully exhibited [2], the fact that dyn-Q($\mathcal{D}^{\text{rrs}}$)-Res is sound is a corollary to Theorem 15. Since $\mathcal{D}^{\text{rrs}}$ is strictly more general than $\mathcal{D}^{\text{std}}$, every dyn-Q($\mathcal{D}^{\text{std}}$)-Res refutation is a dyn-Q($\mathcal{D}^{\text{rrs}}$)-Res refutation, hence dyn-Q($\mathcal{D}^{\text{std}}$)-Res is also sound.

Corollary 16. *The calculi* dyn-Q($\mathcal{D}^{\text{std}}$)-Res *and* dyn-Q($\mathcal{D}^{\text{rrs}}$)-Res *are both sound.*

5.3   Motivations for dyn-Q($\mathcal{D}$)-Res

We chose to define a $\mathcal{D}$-assignment in order to replicate the kind of assignment that is maintained by a QCDCL solver using a dependency scheme, whereby decision variables are assigned only after all others on which they depend. In line with our discussion in Sect. 3, we can relax the theoretical model so that only the (in)dependence of existentials on universals is considered, and hence universals may be assigned arbitrarily in a $\mathcal{D}$-assignment.

The motivation for dyn-Q($\mathcal{D}$)-Res is this observation: If, by recomputing dependencies, the solver is able to refute the formula under its current assignment δ, it should be able to learn the largest falsified clause of δ. In this way, the system shares similarities with 'Q-resolution with generalised axioms' [20]. Regarding proof complexity, a drawback of that calculus is that every false formula may be refuted in a single step. Our system resolves this difficulty, using the notion of referencing to accommodate a suitable definition of proof size.

In line with [20], we could have allowed assignments due to unit propagation and pure literal elimination in dyn-Q($\mathcal{D}$)-Res. This would allow additional existential literals to be included in a $\mathcal{D}$-assignment provided that they are valid assignments under Boolean constraint propagation. Doing so would result in a stronger version of dyn-Q($\mathcal{D}$)-Res, since such a modification extends the set of $\mathcal{D}$-assignments for any instance. However, we prefer to the present simpler system, since propagation is not necessary for the separation in the following section.

Soundness of the system with propagation can be proved by an extension of our argument in Lemma 14. This is because existential literals that become unit under restriction are always unopposed in the restricted model, and hence Proposition 13 still applies. Existential literals that become pure can be assigned unopposed throughout the restricted model without falsifying any clauses.

6   Static vs Dynamic Dependency Awareness in Q($\mathcal{D}$)-Res

In this section, we investigate the relative proof complexities of Q($\mathcal{D}$)-Res and dyn-Q($\mathcal{D}$)-Res. We prove an exponential separation when $\mathcal{D}$ is the reflexive resolution path dependency scheme. In contrast, the two systems are p-equivalent when $\mathcal{D}$ is the trivial dependency scheme. The latter result, while a perfectly natural conjecture, requires a non-trivial proof.

Theorem 17. Q-Res *and* dyn-Q($\mathcal{D}^{\mathrm{trv}}$)-Res *are p-equivalent proof systems.*

Proof (sketch). Since dyn-Q($\mathcal{D}^{\mathrm{trv}}$)-Res trivially p-simulates Q-Res (Proposition 4), we need only prove the reverse simulation. We prove by induction on reference degree that any dyn-Q($\mathcal{D}^{\mathrm{trv}}$)-Res derivation can be transformed into a Q-Res derivation of the same size, in time linear in the size of the original derivation. To that end, let π be a dyn-Q($\mathcal{D}^{\mathrm{trv}}$)-Res derivation of a clause C from a PCNF Φ of reference degree d.

If $d = 0$, then π is a Q-Res derivation, so the base case is established trivially. For the inductive step, let $d \geq 1$, and let R be a reference clause in π derived by application of rule ref(δ, π'). Note that the reference degree of π' is less than d, and hence, by the inductive hypothesis, π' can be transformed in linear time into a Q-Res refutation ρ of $\Phi[\delta]$ with $|\rho| = |\pi'|$. A $\mathcal{D}^{\mathrm{trv}}$-assignment assigns variables strictly in block order and assigns no variable before the preceding block is fully assigned. As a result, adding the literals in R to each clause of ρ cannot invalidate any $\forall$-reduction step, nor introduce a universal tautology. Moreover, doing so transforms ρ in linear time into a Q-Res derivation ρ' of R, with $|\rho'| = |\rho|$. Since every axiom clause in ρ' is subsumed by some clause in the matrix of Φ, ρ' can be transformed into a derivation of R from Φ simply by omitting any steps that are rendered unnecessary by the absence of a literal. This last transformation can clearly be carried out in linear time and does not increase the size of the derivation.

Successive application of this method to all the reference clauses in π yields a Q-Res derivation of C of size at most $|\pi|$. The complete transformation can be carried out in time linear in $|\pi|$. $\square$

The remainder of this section is devoted to the separation of dyn-Q($\mathcal{D}^{\mathrm{rrs}}$)-Res from Q($\mathcal{D}^{\mathrm{rrs}}$)-Res. The separating formulas are a modification of $\Psi(n)$, for which we make use of the following operation.

Definition 18 (clause-matrix product). *Let C be a clause and let ϕ be a CNF matrix. The clause-matrix product $C \otimes \phi$ is the CNF matrix with clauses $\{C \cup C' \mid C' \in \phi\}$.*

We modify $\Psi(n)$ by adding two fresh existential variables a and b, quantified at the very beginning and very end of the prefix, respectively. Taking two copies of the matrix $\psi(n)$, to each clause of the first copy we add literals a and b, and to each clause of the second we add literals $\neg a$ and $\neg b$. Finally, we add the clauses $\{a, \neg b\}$ and $\{\neg a, b\}$ so that the modified formulas are false.

Definition 19 (modification of the formulas of Kleine Büning et al.). *Let $\Psi(n) := \mathcal{Q}(n) . \psi(n)$ be the formulas of Kleine Büning et al. (as in Definition 5). We define the formula family*

$$\Xi(n) := \exists a \mathcal{Q}(n) \exists b . (\{a, b\} \otimes \psi(n)) \cup (\{\neg a, \neg b\} \otimes \psi(n)) \cup \{\{a, \neg b\}, \{\neg a, b\}\}.$$

The purpose of variable b is to introduce sufficiently many $\mathcal{D}^{\mathrm{rrs}}$ connections between clauses, such that $\mathcal{D}^{\mathrm{rrs}}$ can no longer identify any non-trivial independencies. This means that static application of $\mathcal{D}^{\mathrm{rrs}}$ cannot improve upon Q-Res. However, under either assignment to variable a, one copy of the matrix $\psi(n)$ vanishes, and the connections due to b disappear with it. The restricted formulas $\Xi(n)[a]$ and $\Xi(n)[\neg a]$ are sufficiently similar to $\Psi(n)$ to admit short Q($\mathcal{D}^{\mathrm{rrs}}$)-Res refutations. Hence, dynamic application of $\mathcal{D}^{\mathrm{rrs}}$ yields shorter proofs.

To prove the lower bound for the static calculus, we first show that $\mathcal{D}^{\mathrm{rrs}}(\Xi(n))$ $= \mathcal{D}^{\mathrm{trv}}(\Xi(n))$, from which it follows that any Q($\mathcal{D}^{\mathrm{rrs}}$)-Res refutation of $\Xi(n)$ is also a Q-Res refutation. We then show that any Q-Res refutation of $\Xi(n)$ contains an embedded refutation of $\Psi(n)$, which has size at least 2^n [3,15].

Theorem 20. *The QBFs $\Xi(n)$ require exponential-size Q($\mathcal{D}^{\mathrm{rrs}}$)-Res refutations.*

Proof. To see that $\mathcal{D}^{\mathrm{rrs}}$ does not identify any spurious existential dependencies for $\Xi(n)$ – or, equivalently, that $\mathcal{D}^{\mathrm{rrs}}(\Xi(n)) = \mathcal{D}^{\mathrm{trv}}(\Xi(n))$ – we must show that, for each pair $(v, z) \in \mathcal{D}^{\mathrm{trv}}(\Xi(n))$, there exists a sequence of k clauses and a sequence of $k - 1$ literals satisfying the four conditions of Definition 2.

Let $i, j \in [n]$. For $(u_i, b) \in \mathcal{D}^{\mathrm{trv}}(\Xi(n))$, the clauses $\{a, b\} \cup B_i, \{\neg a, \neg b\} \cup B_i'$ and the single literal b form suitable sequences. For $(u_i, t_j) \in \mathcal{D}^{\mathrm{trv}}(\Xi(n))$, the clauses $\{a, b\} \cup B_i, \{\neg a, \neg b\} \cup C_j, \{a, b\} \cup B_n, \{\neg a, \neg b\} \cup B_i'$ and the literals b, t_j, b are suitable. For $(u_i, x_j) \in \mathcal{D}^{\mathrm{trv}}(\Xi(n))$ (with $i < j$), the clauses $\{a, b\} \cup B_i, \{\neg a, \neg b\} \cup B_j, \{a, b\} \cup \{B_{j-1}\}, \{\neg a, \neg b\} \cup B_i'$ and the literals b, x_j, b are suitable, and the case for $(u_i, y_j) \in \mathcal{D}^{\mathrm{trv}}(\Xi(n))$ is similar.

Now, let π be a Q($\mathcal{D}^{\mathrm{rrs}}$)-Res refutation of $\Xi(n)$, and let α be the assignment $\{\neg a, \neg b\}$. Since α assigns only existential variables, $\pi[\alpha]$ is a refutation of $\Xi(n)[\alpha]$ that is no larger than π. Observe that $\Xi(n)[\alpha] = \Psi(n)$, hence the size of $\pi[\alpha]$ is at least 2^n [3, 15], and we must have $|\pi| \geq 2^n$.   □

The upper bound argument makes use of the construction of short refutations from the proof of Theorem 9. By referencing those refutations, dyn-Q($\mathcal{D}^{\mathrm{rrs}}$)-Res admits simple $O(n)$-size refutations of $\Xi(n)$.

Theorem 21. *The formulas $\Xi(n)$ have linear-size* dyn-Q($\mathcal{D}^{\mathrm{rrs}}$)-Res *refutations.*

Proof. We construct linear-size Q($\mathcal{D}^{\mathrm{rrs}}$)-Res refutations of $\Xi(n)[\neg a]$ and $\Xi(n)[a]$. Since a and $\neg a$ are $\mathcal{D}^{\mathrm{rrs}}$-assignments for $\Xi(n)$, in dyn-Q($\mathcal{D}^{\mathrm{rrs}}$)-Res one can introduce the unit clauses $\{a\}$ and $\{\neg a\}$ by application of the reference rule, from which the empty clause is derived by a single resolution step. As the two referenced refutations are of linear size, so is the complete refutation.

It remains to construct the referenced refutations of $\Xi(n)[\neg a]$ and $\Xi(n)[a]$. We describe the case for $\Xi(n)[\neg a]$ – the other case is similar.

Note that the formula $\Xi(n)[\neg a]$ may be obtained from $\Psi(n)$ by adding the literal b to every clause, and then adding the unit clause $\{\neg b\}$ to the matrix. We make two observations. First, since the negative literal $\neg b$ occurs only in a unit clause, such a modification of $\Psi(n)$ cannot introduce any new existential $\mathcal{D}^{\mathrm{rrs}}$ dependencies; no $\mathcal{D}^{\mathrm{rrs}}$ path can go through variable b. As a result, Proposition 8 lifts from $\Psi(n)$ to $\Xi(n)[\neg a]$; that is, $(u_i, t_j) \notin \mathcal{D}^{\mathrm{rrs}}(\Xi(n)[\neg a])$ for each $i, j \in [n]$ with $i \neq j$. Second, $\Psi(n)$ can be derived in $O(n)$ resolution steps from $\Xi(n)[\neg a]$ simply by resolving the unit clause $\{\neg b\}$ with every other clause (there are $O(n)$ clauses in $\Xi(n)[\neg a]$). It follows that Q($\mathcal{D}^{\mathrm{rrs}}$)-Res can refute $\Xi(n)[\neg a]$ in $O(n)$ steps by first deriving the clauses of $\Psi(n)$ and then replicating the refutation given in the proof of Theorem 9.   □

Our final result is immediate from Theorems 20 and 21.

Theorem 22. dyn-Q($\mathcal{D}^{\mathrm{rrs}}$)-Res *is exponentially stronger than* Q($\mathcal{D}^{\mathrm{rrs}}$)-Res.

7   Conclusions

We demonstrated that the use of dependency schemes in Q-resolution can yield exponentially shorter proofs, and thereby provided strong theoretical evidence supporting the notion that dependency schemes can be utilised for improved solving. We also demonstrated that the dynamic use of schemes has further potential, beyond that of the static approach in existing implementations.

We emphasize that, at the present time, there is no implementation for $\mathcal{D}^{\mathrm{rrs}}$, and that DepQBF uses (a refinement of) $\mathcal{D}^{\mathrm{std}}$. Since we do not separate Q-Res and $Q(\mathcal{D}^{\mathrm{std}})$-Res, our theoretical results are not *exactly* in line with experimental results. We believe, however, that our results should be viewed as sound motivation for further research into (dynamic) dependency-aware solving.

Finally, we suggest strongly that the results in this paper will lift to further QBF calculi, and most notably to expansion-based systems. We therefore highlight the potential for dependency schemes in expansion solving, and endorse the move in this direction mooted at the conclusion of [14].

References

1. Benedetti, M., Mangassarian, H.: QBF-based formal verification: experience and perspectives. J. Satisfiability Boolean Model. Comput. (JSAT) **5**(1–4), 133–191 (2008)
2. Beyersdorff, O., Blinkhorn, J.: Dependency schemes in QBF calculi: semantics and soundness. In: Rueher, M. (ed.) CP 2016. LNCS, vol. 9892, pp. 96–112. Springer, Cham (2016). doi:10.1007/978-3-319-44953-1_7
3. Beyersdorff, O., Chew, L., Janota, M.: Proof complexity of resolution-based QBF calculi. In: International Symposium on Theoretical Aspects of Computer Science (STACS). Leibniz International Proceedings in Informatics (LIPIcs), vol. 30, pp. 76–89 (2015)
4. Beyersdorff, O., Chew, L., Mahajan, M., Shukla, A.: Feasible interpolation for QBF resolution calculi. In: Halldórsson, M.M., Iwama, K., Kobayashi, N., Speckmann, B. (eds.) ICALP 2015. LNCS, vol. 9134, pp. 180–192. Springer, Heidelberg (2015). doi:10.1007/978-3-662-47672-7_15
5. Beyersdorff, O., Chew, L., Mahajan, M., Shukla, A.: Are short proofs narrow? QBF resolution is not simple. In: Symposium on Theoretical Aspects of Computer Science (STACS), pp. 15:1–15:14 (2016)
6. Beyersdorff, O., Chew, L., Sreenivasaiah, K.: A game characterisation of tree-like Q-resolution size. In: Dediu, A.-H., Formenti, E., Martín-Vide, C., Truthe, B. (eds.) LATA 2015. LNCS, vol. 8977, pp. 486–498. Springer, Cham (2015). doi:10.1007/978-3-319-15579-1_38
7. Lonsing, F., Biere, A.: Integrating dependency schemes in search-based QBF solvers. In: Strichman, O., Szeider, S. (eds.) SAT 2010. LNCS, vol. 6175, pp. 158–171. Springer, Heidelberg (2010). doi:10.1007/978-3-642-14186-7_14
8. Buss, S.R.: Towards NP-P via proof complexity and search. Ann. Pure Appl. Logic **163**(7), 906–917 (2012)
9. Cook, S.A., Nguyen, P.: Logical Foundations of Proof Complexity. Cambridge University Press, Cambridge (2010)
10. Cook, S.A., Reckhow, R.A.: The relative efficiency of propositional proof systems. J. Symbolic Logic **44**(1), 36–50 (1979)
11. Egly, U., Kronegger, M., Lonsing, F., Pfandler, A.: Conformant planning as a case study of incremental QBF solving. In: Artificial Intelligence and Symbolic Computation (AISC 2014), pp. 120–131 (2014)
12. Giunchiglia, E., Marin, P., Narizzano, M.: Reasoning with quantified boolean formulas. In: Handbook of Satisfiability, pp. 761–780. IOS Press (2009)
13. Giunchiglia, E., Narizzano, M., Tacchella, A.: Clause/term resolution and learning in the evaluation of quantified boolean formulas. J. Artif. Intell. Res. (JAIR) **26**, 371–416 (2006)

14. Janota, M., Klieber, W., Marques-Silva, J., Clarke, E.M.: Solving QBF with counterexample guided refinement. J. Artif. Intell. **234**, 1–25 (2016)
15. Kleine Büning, H., Karpinski, M., Flögel, A.: Resolution for quantified boolean formulas. Inf. Comput. **117**(1), 12–18 (1995)
16. Kontchakov, R., Pulina, L., Sattler, U., Schneider, T., Selmer, P., Wolter, F., Zakharyaschev, M.: Minimal module extraction from DL-lite ontologies using QBF solvers. In: International Joint Conference on Artificial Intelligence (IJCAI), pp. 836–841. AAAI Press (2009)
17. Krajíček, J.: Bounded Arithmetic, Propositional Logic, and Complexity Theory, Encyclopedia of Mathematics and Its Applications, vol. 60. Cambridge University Press, Cambridge (1995)
18. Liang, J.H., Ganesh, V., Zulkoski, E., Zaman, A., Czarnecki, K.: Understanding VSIDS branching heuristics in conflict-driven clause-learning SAT solvers. In: Piterman, N. (ed.) HVC 2015. LNCS, vol. 9434, pp. 225–241. Springer, Cham (2015). doi:10.1007/978-3-319-26287-1_14
19. Lonsing, F.: Dependency schemes and search-based QBF solving: theory and practice. Ph.D. thesis, Johannes Kepler University (2012)
20. Lonsing, F., Egly, U., Seidl, M.: Q-resolution with generalized axioms. In: Creignou, N., Le Berre, D. (eds.) SAT 2016. LNCS, vol. 9710, pp. 435–452. Springer, Cham (2016). doi:10.1007/978-3-319-40970-2_27
21. Moskewicz, M.W., Madigan, C.F., Zhao, Y., Zhang, L., Malik, S.: Chaff: engineering an efficient SAT solver. In: Proceedings of Design Automation Conference (DAC), pp. 530–535 (2001)
22. Peitl, T., Slivovsky, F., Szeider, S.: Long distance Q-resolution with dependency schemes. In: Creignou, N., Le Berre, D. (eds.) SAT 2016. LNCS, vol. 9710, pp. 500–518. Springer, Cham (2016). doi:10.1007/978-3-319-40970-2_31
23. Rintanen, J.: Asymptotically optimal encodings of conformant planning in QBF. In: National Conference on Artificial Intelligence (AAAI), pp. 1045–1050. AAAI Press (2007)
24. Samer, M., Szeider, S.: Backdoor sets of quantified boolean formulas. J. Autom. Reasoning **42**(1), 77–97 (2009)
25. Samulowitz, H., Bacchus, F.: Using SAT in QBF. In: Beek, P. (ed.) CP 2005. LNCS, vol. 3709, pp. 578–592. Springer, Heidelberg (2005). doi:10.1007/11564751_43
26. Shacham, O., Zarpas, E.: Tuning the VSIDS decision heuristic for bounded model checking. In: International Workshop on Microprocessor Test and Verification (MTV), p. 75 (2003)
27. Silva, J.P.M.: The impact of branching heuristics in propositional satisfiability algorithms. In: Portugese Conference on Progress in Artificial Intelligence (EPIA), pp. 62–74 (1999)
28. Silva, J.P.M., Sakallah, K.A.: GRASP - a new search algorithm for satisfiability. In: International Conference on Computer-Aided Design (ICCAD), pp. 220–227 (1996)
29. Slivovsky, F.: Structure in #SAT and QBF. Ph.D. thesis, Vienna University of Technology (2015)
30. Slivovsky, F., Szeider, S.: Variable dependencies and Q-resolution. In: Sinz, C., Egly, U. (eds.) SAT 2014. LNCS, vol. 8561, pp. 269–284. Springer, Cham (2014). doi:10.1007/978-3-319-09284-3_21
31. Slivovsky, F., Szeider, S.: Soundness of Q-resolution with dependency schemes. TCS **612**, 83–101 (2016)

32. Van Gelder, A.: Variable independence and resolution paths for quantified boolean formulas. In: Lee, J. (ed.) CP 2011. LNCS, vol. 6876, pp. 789–803. Springer, Heidelberg (2011). doi:10.1007/978-3-642-23786-7_59
33. Zhang, L., Malik, S.: Conflict driven learning in a quantified boolean satisfiability solver. In: International Conference on Computer-aided Design (ICCAD), pp. 442–449 (2002)

A Little Blocked Literal Goes a Long Way

Benjamin Kiesl[1]([✉]), Marijn J.H. Heule[2], and Martina Seidl[3]

[1] Institute of Information Systems, Vienna University of Technology, Austria
kiesl@kr.tuwien.ac.at
[2] Department of Computer Science, The University of Texas at Austin, USA
[3] Institute for Formal Models and Verification, JKU Linz, Austria

Abstract. Q-resolution is a generalization of propositional resolution that provides the theoretical foundation for search-based solvers of quantified Boolean formulas (QBFs). Recently, it has been shown that an extension of Q-resolution, called long-distance resolution, is remarkably powerful both in theory and in practice. However, it was unknown how long-distance resolution is related to QRAT, a proof system introduced for certifying the correctness of QBF-preprocessing techniques. We show that QRAT polynomially simulates long-distance resolution. Two simple rules of QRAT are crucial for our simulation—*blocked-literal addition* and *blocked-literal elimination*. Based on the simulation, we implemented a tool that transforms long-distance-resolution proofs into QRAT proofs. In a case study, we compare long-distance-resolution proofs of the well-known Kleine Büning formulas with corresponding QRAT proofs.

1 Introduction

Quantified Boolean formulas (QBF) [19] extend propositional formulas with existential and universal quantifiers over the propositional variables. These quantifiers lead to increased expressiveness, which makes QBF attractive for reasoning problems in areas such as formal verification and artificial intelligence [3].

To obtain a better understanding of the strengths and limitations of different QBF-solving approaches, their underlying proof systems have been extensively analyzed, providing a comprehensive proof-complexity landscape for QBF [4,6,9,10,16]. Two kinds of proof systems have received particular attention: instantiation-based proof systems [5,6], which provide the foundation for expansion-based solvers like RAReQS [18], and resolution-based proof systems [1,2,7,16,17,20,23,24,26], which provide the foundation for search-based solvers like DepQBF [22]. Apart from these, also sequent systems have been studied [8,10]. There is, however, another practically useful proof system—quite different from the aforementioned ones—whose place in the complexity landscape was still unclear: the QRAT proof system [15].

The QRAT proof system is a generalization of DRAT [25] (the de-facto standard for proofs in practical SAT solving) that has its strengths when it comes

---

This work has been supported by the Austrian Science Fund (FWF) under projects W1255-N23 and S11408-N23, and by the National Science Foundation (NSF) under grant number CCF-1618574.

S. Gaspers and T. Walsh (Eds.): SAT 2017, LNCS 10491, pp. 281–297, 2017.
DOI: 10.1007/978-3-319-66263-3_18

to preprocessing: Many QBF solvers use preprocessing techniques to simplify a QBF before they actually evaluate its truth. With the QRAT system, it is possible to certify the correctness of virtually all preprocessing simplifications performed by state-of-the-art QBF solvers and preprocessors. Additionally, there exist efficient tools for checking the correctness of QRAT proofs as well as for extracting winning strategies (so-called *Skolem functions*) from QRAT proofs of satisfiability [15].

It can be easily seen that QRAT simulates the basic Q-resolution calculus [20] that allows only resolution upon existential variables. Likewise, it simulates the calculus QU-Res [24], which extends Q-resolution by allowing resolution upon universal variables. So far, however, it was unclear how QRAT is related to the long-distance-resolution calculus [1,26]—a calculus that is particularly popular because it allows for short proofs both in theory and in practice [11].

In this paper, we prove that QRAT can polynomially simulate the long-distance-resolution calculus. For our simulation, we need only Q-resolution and universal reduction together with blocked-literal elimination and blocked-literal addition using fresh variables [14,21]. These four rules are allowed in QRAT. To illustrate the power of blocked literals, we present handcrafted QRAT proofs of the formulas commonly used to display the strength of long-distance resolution—the well-known *Kleine Büning formulas* [20]. Our proofs are slightly smaller than the long-distance resolution proofs of these formulas described by Egly et al. [11].

To put our simulation into practice, we implemented a tool that transforms long-distance-resolution proofs into QRAT proofs. With this tool it is now possible to obtain QRAT proofs that certify the correctness of both the preprocessing and the actual solving, even when using a QBF solver based on long-distance resolution. We used our tool to transform long-distance-resolution proofs of the Kleine Büning formulas into QRAT proofs. We compare the resulting proofs with the handcrafted QRAT proofs as well as with the original proofs. Rounding off the picture, we locate QRAT in the proof-complexity landscape of resolution-based proof systems and discuss open questions.

2   Preliminaries

In the following, we introduce the background required to understand the rest of the paper. A *literal* is either a variable x (a *positive literal*) or the negation $\bar{x}$ of a variable x (a *negative literal*). The complement $\bar{l}$ of a literal l is defined as $\bar{x}$ if $l = x$ and as x if $l = \bar{x}$. A *clause* is a disjunction of literals. A (*propositional*) *formula* in *conjunctive normal form* (CNF) is a conjunction of clauses. A clause can be seen as a set of literals and a formula can be seen as a set of clauses.

A *quantifier prefix* has the form $Q_1 X_1 \ldots Q_q X_q$, where all the X_i are mutually disjoint sets of variables, $Q_i \in \{\forall, \exists\}$, and $Q_i \neq Q_{i+1}$. A *quantified Boolean formula* (QBF) ϕ in *prenex conjunctive normal form* (PCNF) is of the form $\Pi.\psi$ where Π is a quantifier prefix and ψ, called the *matrix* of ϕ, is a propositional formula in CNF. The quantifier $Q(\Pi, l)$ of a literal l is Q_i if $var(l) \in X_i$. Let $Q(\Pi, l) = Q_i$ and $Q(\Pi, k) = Q_j$, then $l \leq_\Pi k$ if $i \leq j$, and $l <_\Pi k$ if $i < j$.

Using the truth constants 1 (*true*) and 0 (*false*), a QBF $\forall x \Pi.\psi$ is false iff at least one of $\Pi.\psi[x/1]$ and $\Pi.\psi[x/0]$ is false where $\Pi.\psi[x/t]$ is obtained from $\Pi.\psi$

by replacing all occurrences of x in ψ by t and removing x from Π. Respectively, a QBF $\exists x \Pi.\psi$ is false iff both $\Pi.\psi[x/1]$ and $\Pi.\psi[x/0]$ are false. If the matrix ψ of ϕ contains the empty clause (denoted by $\bot$) after eliminating the truth constants according to standard rules, then ϕ is false. If ψ is empty, ϕ is true.

2.1 Resolution-Based Calculi

In resolution-based calculi, a proof P of a QBF $\Pi.\psi = \Pi.C_1 \wedge \cdots \wedge C_m$ is a sequence $C_{m+1}, \ldots, C_n$ of clauses with $C_n = \bot$ and for every C_i ($m+1 \le i \le n$), it holds that C_i has been derived from clauses in ψ or from earlier clauses in P (i.e., from clauses with index strictly smaller than i) by applications of either the $\forall$-*red* rule (also called *universal reduction*) or instantiations of the *resolution* rule which are defined as follows:

$$\frac{C \vee x}{C} \text{ ($\forall$-red)} \qquad \frac{C \vee l \qquad D \vee \bar{l}}{C \vee D} \text{ (resolution)}$$

The rule $\forall$-red is only applicable if the literal x is universal and if for every existential literal $l \in C$, it holds that $l <_{\Pi} x$. In the resolution rule, the resolvent $C \vee D$ is derived from its two antecedent clauses. We assume that no clause in ψ contains complementary literals, otherwise the $\forall$-red rule is unsound.

The most basic resolution-based calculus for QBF is the *Q-resolution calculus* (Q-Res) [20]. It uses the resolution rule *Q-res* which requires that (1) l is existential and (2) C does not contain a literal x such that $\bar{x} \in D$. In contrast, the *long-distance-resolution calculus* (LQ-Res) [1,26] uses a less restrictive variant of the resolution rule, called *LQ-res*, which requires that (1) l is existential and (2) for every literal $x \in C$ such that $\bar{x} \in D$, it holds that x is universal and $l <_{\Pi} x$. Note that every Q-res step is also an LQ-res step. In the rest of the paper, we refer to resolution steps as LQ-res steps only if they are not Q-res steps, otherwise we refer to them as Q-res steps. Note that in the literature a complementary pair $x, \bar{x}$ is also represented by a so-called *merged literal* x^*.

Example 1. Consider the QBF $\phi = \exists a \forall x \exists b \exists c.(\bar{a} \vee \bar{x} \vee c) \wedge (\bar{x} \vee b \vee \bar{c}) \wedge (a \vee x \vee b) \wedge (\bar{b})$. The following is a long-distance-resolution proof of ϕ: $\bar{a} \vee \bar{x} \vee b$, $x \vee \bar{x} \vee b$, $x \vee \bar{x}$, x, $\bot$. We explain this proof in more detail later (also see Fig. 1).

2.2 The QRAT Proof System Light

In this paper, we do not need the power of the full QRAT proof system [15]. We therefore introduce only a very restricted version of QRAT that is sufficient for the simulation of the long-distance-resolution calculus.

One of the main concepts in this variant of QRAT is the concept of a *blocked literal*. For the definition of blocked literals, we first have to introduce so-called *outer resolvents*. Given two clauses $C \vee x, D \vee \bar{x}$ of a QBF $\Pi.\psi$, the outer resolvent $C \vee x \bowtie_{\Pi}^{x} D \vee \bar{x}$ of $C \vee x$ with $D \vee \bar{x}$ upon x is the clause consisting of all literals in C together with those literals of D that occur outer to x, i.e., the outer resolvent is the clause $C \cup \{l \mid l \in D \text{ and } l \le_{\Pi} x\}$. We can now define blocked literals:

Definition 1. *A universal literal x is* blocked *in a clause $C \vee x$ w.r.t. a QBF $\Pi.\psi$ if, for every clause $D \vee \bar{x} \in \psi \setminus \{C \vee x\}$, the outer resolvent $C \vee x \bowtie_{\Pi}^{x} D \vee \bar{x}$ contains a pair of complementary literals.*

Example 2. Let $\phi = \exists a \forall x, y \exists b.(a \vee x \vee y) \wedge (\bar{a} \vee \bar{x} \vee b) \wedge (\bar{y} \vee \bar{x} \vee b)$. The literal x is blocked in $a \vee x \vee y$ w.r.t. ϕ: There are two outer resolvents of $a \vee x \vee y$ upon x, namely $a \vee y \vee \bar{a}$, obtained by resolving with $\bar{a} \vee \bar{x} \vee b$, and $a \vee y \vee \bar{y}$, obtained by resolving with $\bar{y} \vee \bar{x} \vee b$. Both contain a pair of complementary literals.   □

If a literal is blocked in a clause, its removal is called *blocked-literal elimination* (BLE) [14]. If, after adding a literal to a clause, the literal is blocked in that clause, then this addition is called *blocked-literal addition* (BLA). Both BLE and BLA do not change the truth value of a formula.

In our restricted variant of QRAT, a *derivation* for a QBF ϕ is a sequence $M_1, \ldots, M_n$ of proof steps. Starting with $\phi_0 = \phi$, every M_i modifies ϕ_{i-1} in one of the following four ways, which results in a new formula ϕ_i: (1) It adds to ϕ_{i-1} a clause that is derived from two clauses in ϕ_{i-1} via a resolution step. (2) It adds to ϕ_{i-1} a clause C that is obtained from a clause $C \vee x \in \phi_{i-1}$ by a $\forall$-red step, with the additional restriction that C does not contain $\bar{x}$. (3) It adds a blocked literal to a clause in ϕ_{i-1}. (4) It removes a blocked literal from a clause in ϕ_{i-1}.

A QRAT derivation $M_1, \ldots, M_n$ therefore gradually derives new formulas $\phi_1, \ldots, \phi_n$ from the starting formula ϕ. If the final formula ϕ_n contains the empty clause $\bot$, then the derivation is a *(refutation) proof* of ϕ. Note that the $\forall$-red rule in QRAT is more restricted than the $\forall$-red rule from the resolution-based calculi, making it sound also when clauses contain complementary literals.

To simplify the presentation, we do not specify how the modification steps M_i are represented syntactically. We also do not include clause deletion. Note that certain proof steps can modify the quantifier prefix by introducing new or removing existing variables. Note also that Q-resolution proofs do not contain complementary literals, so they can be simply rewritten into QRAT proofs using only Q-res and $\forall$-red steps. Finally, we want to highlight that for our simulation, we do not need the unrestricted resolution rule; the Q-res rule suffices.

3   Illustration of the Simulation

We start by illustrating on an example how our restricted variant of QRAT can simulate the long-distance-resolution calculus. As already mentioned, the $\forall$-red rule used in QRAT is more restricted than the one in the long-distance-resolution calculus because it does not allow us to remove a literal x from a clause that contains $\bar{x}$. This means that once we derive a clause that contains both a literal x and its complement $\bar{x}$, we cannot simply get rid of the two literals by using the $\forall$-red rule. We therefore want to avoid the derivation of clauses with complementary literals entirely. Now, the only way the long-distance-resolution calculus can derive such clauses is via resolution (LQ-res) steps. So to avoid the complementary literals, we eliminate them already before performing the resolution steps. We demonstrate this on an example:

$$\frac{a \vee x \vee b \quad \dfrac{\dfrac{\bar{x} \vee b \vee \bar{c} \quad \bar{a} \vee \bar{x} \vee c}{\bar{a} \vee \bar{x} \vee b} \text{(Q-res)}}{} }{x \vee \bar{x} \vee b} \text{(LQ-res)}$$

$$\frac{x \vee \bar{x} \vee b \qquad\qquad \bar{b}}{\dfrac{x \vee \bar{x}}{\dfrac{x}{\bot} \text{(\textforall-red)}} \text{(\textforall-red)}} \text{(Q-res)}$$

Fig. 1. LQ-res proof of QBF $\phi = \exists a \forall x \exists b \exists c.(\bar{a} \vee \bar{x} \vee c) \wedge (\bar{x} \vee b \vee \bar{c}) \wedge (a \vee x \vee b) \wedge (\bar{b})$.

Example 3. Consider the QBF $\phi = \exists a \forall x \exists b \exists c.(\bar{a} \vee \bar{x} \vee c) \wedge (\bar{x} \vee b \vee \bar{c}) \wedge (a \vee x \vee b) \wedge (\bar{b})$ from Example 1. To increase readability, we illustrate its long-distance-resolution proof as a proof tree in Fig. 1. To simulate this proof with QRAT, we first add the resolvent $\bar{a} \vee \bar{x} \vee b$ to ϕ via a Q-res step to obtain the new formula ϕ'. Now we cannot simply perform the next derivation step (the LQ-res step) because the resulting resolvent $x \vee \bar{x} \vee b$ would contain complementary literals. To deal with this, we try to eliminate x from the clause $a \vee x \vee b$. This is where the addition and elimination of blocked literals come into play.

We cannot yet eliminate x from ϕ' because x is not blocked in $a \vee x \vee b$ with respect to ϕ': For x to be blocked, all outer resolvents of $a \vee x \vee b$ upon x must contain complementary literals. The clauses that can be resolved with $a \vee x \vee b$ are $\bar{a} \vee \bar{x} \vee c$, $\bar{a} \vee \bar{x} \vee b$, and $\bar{x} \vee b \vee \bar{c}$. While the outer resolvents with the former two clauses contain the complementary literals a and $\bar{a}$, the outer resolvent $a \vee b$, obtained by resolving with $\bar{x} \vee b \vee \bar{c}$, does not contain complementary literals.

Now we use a feature of QRAT to make x blocked in $a \vee x \vee b$: We add a new literal x' (which goes to the same quantifier block as x) to $a \vee x \vee b$ to turn it into $a \vee x' \vee x \vee b$. The addition of x' is clearly a blocked-literal addition as there are no outer resolvents of $a \vee x' \vee x \vee b$ upon x'. Likewise, we add the complement $\bar{x}'$ of x' to $\bar{x} \vee b \vee \bar{c}$ to turn it into $\bar{x}' \vee \bar{x} \vee b \vee \bar{c}$. Again this is a blocked-literal addition since $a \vee x' \vee x \vee b$ (which is the only clause containing the complement x' of $\bar{x}'$) contains x while $\bar{x}' \vee \bar{x} \vee b \vee \bar{c}$ contains $\bar{x}$.

Now the complementary pair $x', \bar{x}'$ is contained in the outer resolvent of $a \vee x' \vee x \vee b$ with $\bar{x}' \vee \bar{x} \vee b \vee \bar{c}$ upon x. Thus, the literal x becomes blocked in $a \vee x' \vee x \vee b$ and so we can remove it to obtain $a \vee x' \vee b$. We have thus replaced x in $a \vee x \vee b$ by x' and now we can resolve $a \vee x' \vee b$ with $\bar{a} \vee \bar{x} \vee b$ upon a to obtain the resolvent $x' \vee \bar{x} \vee b$ (instead of $x \vee \bar{x} \vee b$ as in the original proof). Finally, we resolve $x' \vee \bar{x} \vee b$ with $\bar{b}$ to obtain $x' \vee \bar{x}$ from which we derive the empty clause $\bot$ via $\forall$-red steps.   □

To summarize, we start by adding clauses of a given long-distance-resolution proof to our formula until we bump into an LQ-res step. To avoid complementary literals in the resolvent of the LQ-res step, we then use blocked-literal addition and blocked-literal elimination to replace these literals. After this, we can derive a resolvent without complementary literals and move on until we encounter the next LQ-res step, which we again eliminate. We repeat this procedure until the whole long-distance-resolution proof is turned into a QRAT proof.

Note that the modification of existing clauses has an impact on later derivations. For instance, by replacing $a \lor x \lor b$ in the above example with $a \lor x' \lor b$, we not only affected the immediate resolvent $x \lor \bar{x} \lor b$, which we turned into $x' \lor \bar{x} \lor b$, but also the later resolvent $x \lor \bar{x}$, which became $x' \lor \bar{x}$. We therefore have to show that these modifications are harmless in the sense that they do not lead to an invalid proof. We do so in the next section, where we define our simulation in detail before proving that it indeed produces a valid QRAT proof.

4   Simulation

We first describe our simulation procedure on a high level before we specify the details and prove its correctness. As we have seen, given a long-distance-resolution proof, we can use QRAT to derive all clauses up to the first LQ-res step. The crucial part of the simulation is then the elimination of complementary literals from this LQ-res step, which might involve the modification of several clauses via the addition and elimination of blocked literals.

Let $\phi = \Pi.C_1 \land \cdots \land C_m$ be a QBF and $P = C_{m+1}, \ldots, C_r, \ldots, C_n$ be a long-distance-resolution proof of ϕ where C_r is the first clause derived via an LQ-res step. If there is no such C_r, the proof can be directly translated to QRAT. Otherwise, in a first step, our procedure produces a QRAT derivation that adds all the clauses $C_{m+1}, \ldots, C_{r-1}$ to ϕ by using Q-res and $\forall$-red steps. It then uses blocked-literal addition and blocked-literal elimination to avoid complementary literals in the resolvent C_r, which it thereby turns into a different resolvent C_r'. After this, it adds C_r' to ϕ via a Q-res step. The result is a QRAT derivation of a formula ϕ' from ϕ. We explain this first step in Sect. 4.1.

In a second step, the procedure first removes all the clauses $C_{m+1}, \ldots, C_r$ from P since they—or their modified variants—are now all contained in ϕ'. As several clauses have been modified via blocked-literal addition and blocked-literal elimination in the first step, it then propagates these modifications through the remaining part of P. This turns P into a long-distance resolution proof P' of ϕ'. We explain this second step in Sect. 4.2.

By repeating these two steps for every LQ-res step, we finally obtain a QRAT proof of ϕ. Thus, we have to show that after the above two steps (i.e., after one iteration of our procedure), ϕ' is obtained by a valid QRAT derivation and the proof P' is a valid long-distance-resolution proof of ϕ' that is shorter than P. The correctness of the simulation follows then simply by induction.

To simplify the presentation, we assume that the long-distance resolvent C_r contains only one pair of complementary literals, i.e., $C_r = C \lor D \lor x \lor \bar{x}$ was derived from two clauses $C \lor l \lor x$ and $D \lor \bar{l} \lor \bar{x}$ where C does not contain a literal k such that $\bar{k}$ is contained in D. Although this assumption leads to a loss of generality, we show later that our argument can be easily extended to the more general case where C and D are allowed to contain multiple pairs of complementary literals.

4.1   QRAT Derivation of the Formula ϕ'

Below we describe the QRAT derivation of ϕ' from ϕ. Initially, $\phi' = \phi$.

1. Add the clauses $C_{m+1}, \dots C_{r-1}$ to ϕ' via Q-res and $\forall$-red steps.
2. Consider the LQ-res step that derived $C_r = C \vee D \vee x \vee \bar{x}$ from two clauses $C \vee l \vee x$ and $D \vee \bar{l} \vee \bar{x}$:

$$\frac{C \vee l \vee x \qquad D \vee \bar{l} \vee \bar{x}}{C \vee D \vee x \vee \bar{x}} \text{ (LQ-res)}$$

   Towards making x blocked in $C \vee l \vee x$, add a new literal x' (that goes to the same quantifier block as x) to $C \vee l \vee x$ to turn it into $C \vee l \vee x' \vee x$.
3. Add $\bar{x}'$ to each clause $C_i \in \phi'$ for which (1) $\bar{x} \in C_i$, and (2) the outer resolvent of $C \vee l \vee x' \vee x$ and C_i upon x is not a tautology.
4. Now x is a blocked literal in $C \vee l \vee x' \vee x$. Eliminate it to obtain $C \vee l \vee x'$.
5. Add the clause $C \vee D \vee x' \vee \bar{x}$ to ϕ' by performing a Q-res step of $C \vee l \vee x'$ and $D \vee \bar{l} \vee \bar{x}$ upon l.

To see that this results in a valid QRAT derivation, observe the following: In step 2, the addition of x' is a blocked-literal addition, since $\bar{x}'$ is not contained in any of the clauses. In step 3, for every C_i with $\bar{x} \in C_i$, the addition of $\bar{x}'$ is a blocked-literal addition as only $C \vee l \vee x' \vee x$ can be resolved with C_i upon $\bar{x}'$ and the corresponding outer resolvent contains x and $\bar{x}$. Note that instead of eliminating x from $C \vee l \vee x$, we could have also eliminated $\bar{x}$ from $D \vee \bar{l} \vee \bar{x}$. It remains to modify the long-distance-resolution proof P of ϕ so that it becomes a valid proof of ϕ'.

4.2   Modification of the Long-Distance-Resolution Proof

We next turn the proof $P = C_{m+1}, \dots, C_r, \dots, C_n$ of ϕ into a proof P' of ϕ'. First, we remove the clauses $C_{m+1}, \dots, C_r$ from P since ϕ' already contains variants $C'_{m+1}, \dots, C'_r$ of these clauses. Second, since we have modified the clauses in ϕ', we have to propagate these modifications through the remaining proof.

   Assume, for instance, that in P the clause C_{r+1} has been obtained by resolving a clause C_i with a clause C_j. Both C_i and C_j might have been affected by blocked-literal additions so that they are now different clauses $C'_i, C'_j \in \phi'$. To account for these modifications of C_i and C_j, we replace C_{r+1} in P by the resolvent of C'_i and C'_j. Moreover, in cases where P removes x from a clause via a $\forall$-red step, we now also remove x'. Analogously for $\bar{x}'$ and $\bar{x}$.

   To formalize these modifications, we first assign to every clause C_i with $1 \leq i \leq r$ its corresponding clause of ϕ' as follows:

$$C'_i = \begin{cases} C_i \cup \{\bar{x}'\} & \text{if } \bar{x} \in C_i \text{ and the outer resolvent of } C \vee l \vee x \vee x' \\ & \text{and } C_i \text{ upon } x \text{ is not a tautology;} \\ (C_i \setminus \{x\}) \cup \{x'\} & \text{if } C_i = C_r \text{ or } C_i = C \vee l \vee x; \\ C_i & \text{otherwise.} \end{cases}$$

Note that, by construction, $C_i' \in \phi'$ for $1 \le i \le r$. For every i such that $r < i \le n$, we step-by-step, starting with $i = r+1$, define C_i' based on the derivation rule that was used for deriving C_i in P. We distinguish between clauses derived by resolution steps and clauses derived by $\forall$-red steps:

CASE 1: C_i has been derived via a resolution step of two clauses $C_j = C \vee l$ and $C_k = D \vee \bar{l}$ upon l, i.e., $C_i = C \vee D$. We define $C_i' = C_j' \setminus \{l\} \vee C_k' \setminus \{\bar{l}\}$.

CASE 2: C_i has been derived from a clause C_j via a $\forall$-red step. If the $\forall$-red step removes a literal l with $var(l) \neq var(x)$, we define $C_i' = C_j' \setminus \{l\}$. If it removes x, we define $C_i' = C_j' \setminus \{x, x'\}$, and if it removes $\bar{x}$, we define $C_i' = C_j' \setminus \{\bar{x}, \bar{x}'\}$.

Note that $\forall$-red steps of x and $\bar{x}$ in P' might remove two literals at once. Although such $\forall$-red steps do not constitute valid derivation steps in a strict sense, this is not a serious problem: These steps can be easily rewritten into two distinct $\forall$-red steps since x and x' are in the same quantifier block. For instance, the left step below can be rewritten into the two steps on the right:

$$\frac{C \vee x \vee x'}{C} \text{ ($\forall$-red)} \qquad\qquad \frac{\dfrac{C \vee x \vee x'}{C \vee x} \text{ ($\forall$-red)}}{C} \text{ ($\forall$-red)}$$

Next, we show that the resulting proof P' is—apart from the minor detail just mentioned—a valid long-distance-resolution proof of ϕ'.

4.3 Correctness of the Simulation

To prove the correctness of our simulation, we first introduce a lemma that guarantees that the modified long-distance-resolution proof P' has a similar structure as the original proof P:

Lemma 1. *Let $\phi' = \Pi'.C_1' \wedge \cdots \wedge C_r'$ and $P' = C_{r+1}', \ldots, C_n'$ be obtained from $\phi = \Pi.C_1 \wedge \cdots \wedge C_m$ and $P = C_{m+1}, \ldots, C_r, \ldots, C_n$ as defined above. Then, for every clause C_i' with $1 \le i \le n$, the following holds: (1) If x' or x is in C_i', then $x \in C_i$. (2) If $\bar{x}'$ or $\bar{x}$ is in C_i', then $\bar{x} \in C_i$. (3) C_i' agrees with C_i on all literals whose variables are different from x and x', i.e., $C_i' \setminus \{x, \bar{x}, x', \bar{x}'\} = C_i \setminus \{x, \bar{x}\}$.*

Proof. By induction on i.

BASE CASE $(i \le r)$: The claim holds by the definition of C_i'.

INDUCTION STEP $(r < i)$: Consider the clause C_i in P that corresponds to C_i'. We proceed by a case distinction based on how C_i was derived in P.

CASE 1: C_i is a resolvent $C_j \setminus \{l\} \vee C_k \setminus \{\bar{l}\}$ of two clauses C_j, C_k. In this case, $C_i' = C_j' \setminus \{l\} \vee C_k' \setminus \{\bar{l}\}$. By the induction hypothesis, the statement holds for C_j' and C_k'. Now, if C_i' contains x' or x, then at least one of C_j' and C_k' must contain x' or x and thus one of C_j and C_k must contain x, hence $x \in C_i$. Analogously, if C_i' contains $\bar{x}'$ or $\bar{x}$, then C_i contains $\bar{x}$. Now, C_j' agrees with C_j on all literals whose variables are different from x and x', and the same holds for C'_k and C_k. Thus, C'_i agrees with C_i on all literals whose variables are different from x and x'.

CASE 2: C_i has been derived from a clause C_j via a $\forall$-red step, i.e., $C_i = C_j \setminus \{y\}$ for some universal literal y. By the induction hypothesis, the statement holds for C'_j. If $var(y) \neq var(x')$, then $C'_i = C'_j \setminus \{y\}$ and thus the claim holds. If $y = x$, then $C'_i = C'_j \setminus \{x, x'\}$ and thus the claim holds too. The case where $y = \bar{x}$ is analogous to the case where $y = x$. $\square$

We can now show that the proof P', produced by our simulation procedure, is a valid long-distance-resolution proof of ϕ':

Theorem 2. *Let $\phi' = \Pi'.C'_1 \wedge \cdots \wedge C'_r$ and $P' = C'_{r+1}, \ldots, C'_n$ be obtained from $\phi = \Pi.C_1 \wedge \cdots \wedge C_m$ and $P = C_{m+1}, \ldots, C_r, \ldots, C_n$ by our procedure. Then, P' is a valid long-distance-resolution proof of ϕ'.*

Proof. We have to show that every clause C'_i in P' has been derived from clauses in $C'_1, \ldots, C'_{i-1}$ via a valid application of a derivation rule and that $C'_n = \bot$. To show that every clause in P' has been derived via a valid application of a derivation rule, let C'_i be a clause in P'. We proceed by a case distinction based on the rule via which its counterpart C_i has been derived in P:

CASE 1: C_i has been derived from two clauses C_j, C_k via a Q-res step or an LQ-res step upon some existential literal l. In this case, $C'_i = C'_j \setminus \{l\} \vee C'_k \setminus \{\bar{l}\}$. We have to show that $l \in C'_j$, $\bar{l} \in C'_k$, and for every literal $l' \in C'_j$ such that $l' \neq l$ and $\bar{l'} \in C'_k$, it holds that l' is universal and $l <_{\Pi'} l'$. By Lemma 1, C'_j agrees with C_j on all literals whose variables are different from the universal literals x and x'. Likewise for C'_k and C_k. Therefore, $l \in C'_j$ and $\bar{l} \in C'_k$.

Now, assume C'_j contains a literal l' such that $l' \neq l$ and $\bar{l'} \in C'_k$. If the variable of l' is different from x and x', then it must be the case that l' is universal and $l <_{\Pi'} l'$, for otherwise the derivation of C_i in P were not valid. Assume thus that the variable of l' is either x or x'. If l' is either x or x', then Lemma 1 implies that C_j contains x and also, since $\bar{l'} \in C'_k$, that C_k contains $\bar{x}$. Therefore, it holds that $l <_{\Pi'} x$ (since otherwise the derivation of C_i in P were not valid) and since x' and x are in the same quantifier block, it also holds that $l <_{\Pi'} x'$, hence $l <_{\Pi'} l'$. The case where l' is $\bar{x}$ or $\bar{x'}$ is symmetric.

CASE 2: C_i has been derived from a clause C_j via a $\forall$-red step, that is, by removing a universal literal y such that for every existential literal $l' \in C_j$, it holds that $l' <_{\Pi} y$. If $var(y) \neq x$, then $C'_j = C'_i \setminus \{y\}$ and since, by Lemma 1, C'_i coincides with C_i on all existential variables, it holds for every existential literal $l' \in C'_i$ that $l' <_{\Pi'} y$. If $var(y) = x$, then C'_j is of the form $C'_i \setminus \{x, x'\}$ or $C'_i \setminus \{\bar{x}, \bar{x'}\}$. Now, x and x' are in the same quantifier block and thus, with the same argument as for $var(y) = x$, it holds for every existential literal $l' \in C'_j$ that $l' <_{\Pi'} y$.

Finally, to see that $C'_n = \bot$, observe the following: By Lemma 1, since x and $\bar{x}$ are not in C_n, it follows that x' and $\bar{x'}$ are not in C'_n. Moreover, again by Lemma 1, C_n and C'_n agree on all other literals. Therefore, $C'_n = C_n = \bot$. $\square$

We can also show that our simulation does not introduce new LQ-res steps. Hence, if a long-distance-resolution proof contains n LQ-res steps, our simulation terminates after at most n iterations (the proof is omitted due to space reasons):

Theorem 3. *Let P' be obtained from $\phi = \Pi.\psi$ and P by our procedure. Then, P' contains fewer LQ-res steps than P.*

4.4 Clashes of Several Universal Literals

Until now, we assumed that LQ-res steps involve only one pair of complementary universal literals. When multiple such pairs are involved, the procedure changes only slightly: Instead of eliminating only a single literal from one of the clauses that are involved in the LQ-res step, we now eliminate several of them. If we start with the outermost one and gradually move inwards, we ensure that at most one blocked literal is added per clause. We illustrate this on an example. Consider the QBF $\phi = \exists a\exists b\forall x\exists c\forall y\exists d.(b\lor x\lor c\lor y\lor d)\land(a\lor\bar{x}\lor c)\land(\bar{a}\lor\bar{b}\lor\bar{y}\lor d)$ and the following derivations in a long-distance-resolution proof:

$$\cfrac{b\lor x\lor c\lor y\lor d \qquad \cfrac{a\lor\bar{x}\lor c \qquad \bar{a}\lor\bar{b}\lor\bar{y}\lor d}{\bar{b}\lor\bar{x}\lor c\lor\bar{y}\lor d}\text{(Q-res)}}{x\lor\bar{x}\lor c\lor y\lor\bar{y}\lor d}\text{(LQ-res)}$$

In the LQ-res step, there are two pairs of complementary universal literals, namely $x,\bar{x}$ and $y,\bar{y}$. We therefore try to get rid of both x and y in the left antecedent $L = b\lor x\lor c\lor y\lor d$ of the LQ-res step. As in the case where only one literal is removed, we start by deriving in QRAT all clauses that occur before the LQ-res step. In this case, we add $\bar{b}\lor\bar{x}\lor c\lor\bar{y}\lor d$ to ϕ via a Q-res step and denote the resulting formula by ϕ'.

Now we want to remove x from L via blocked-literal elimination. In order for x to be blocked in ϕ', all outer resolvents of L upon x have to be tautologies. The formula ϕ' contains two clauses that can be resolved with L upon x, namely $\bar{b}\lor\bar{x}\lor c\lor\bar{y}\lor d$ and $a\lor\bar{x}\lor c$. As the first clause contains $\bar{b}$ and L contains b, the corresponding outer resolvent upon x contains $b,\bar{b}$. But there are no complementary literals in the outer resolvent $a\lor b$ with the second clause. We therefore add a fresh literal x' to L and add its complement $\bar{x}'$ to $a\lor\bar{x}\lor c$ to obtain $\phi' = \exists a\exists b\forall x\forall x'\exists c\forall y\exists d.(b\lor x\lor x'\lor c\lor y\lor d)\land(a\lor\bar{x}\lor\bar{x}'\lor c)\land(\bar{a}\lor\bar{b}\lor\bar{y}\lor d)\land(\bar{b}\lor\bar{x}\lor c\lor\bar{y}\lor d)$.

We can now remove the blocked literal x from $(b\lor x\lor x'\lor c\lor y\lor d)$ to obtain $L' = b\lor x'\lor c\lor y\lor d$. If we now resolved L' with $\bar{b}\lor\bar{x}\lor c\lor\bar{y}\lor d$, we would get the following LQ-res step:

$$\cfrac{b\lor x'\lor c\lor y\lor d \qquad \bar{b}\lor\bar{x}\lor c\lor\bar{y}\lor d}{x'\lor\bar{x}\lor c\lor y\lor\bar{y}\lor d}\text{(LQ-res)}$$

Since there is still a clash of y and $\bar{y}$, we need to get rid of y in L'. We are lucky because we do not need to perform any blocked-literal additions: The only clauses in ϕ' that contain $\bar{y}$ are $\bar{a}\lor\bar{b}\lor\bar{y}\lor d$ and $\bar{b}\lor\bar{x}\lor c\lor\bar{y}\lor d$, and the outer resolvents of L' with both of them contain complementary literals. We can thus remove y from L' and use a Q-res step to add the resulting resolvent to ϕ':

$$\frac{b \vee x' \vee c \vee d \qquad \bar{b} \vee \bar{x} \vee c \vee \bar{y} \vee d}{x' \vee \bar{x} \vee c \vee \bar{y} \vee d} \text{ (Q-res)}$$

Similarly to the case where we only eliminated one literal, we then propagate the corresponding changes through the rest of the proof to turn it into a valid long-distance resolution proof of ϕ'.

5   Complexity of the Simulation

After showing how a long-distance-resolution proof can be translated into a QRAT proof, we still have to prove that the size (the number of derivation steps) of the resulting QRAT proof is polynomial w.r.t. the size of the original proof and the formula. We have seen that the long-distance-resolution proof and the QRAT proof correspond one-to-one on resolution steps and $\forall$-red steps. Therefore, we only need to estimate the number of blocked-literal addition and blocked-literal elimination steps to obtain an upper bound on the size of the QRAT proof.

Consider a long-distance-resolution proof $C_{m+1}, \ldots, C_r, \ldots, C_n$ of a QBF $\Pi.C_1 \wedge \cdots \wedge C_m$, where C_r is the first clause that is derived via an LQ-res step:

$$\frac{C \vee l \vee x_1 \vee \cdots \vee x_k \qquad D \vee \bar{l} \vee \bar{x}_1 \vee \cdots \vee \bar{x}_k}{C_r = C \vee D \vee x_1 \vee \bar{x}_1 \vee \cdots \vee x_k \vee \bar{x}_k} \text{ (LQ-res)}$$

We can make the following observation: To remove all the literals $x_1, \ldots, x_k$ from $C \vee l \vee x_1 \vee \cdots \vee x_k$ via blocked-literal elimination, we have to add at most one new literal of the form $\bar{x}'_i$ to every clause $C_1, \ldots, C_{r-1}$ if we start by eliminating the outermost universal literal x_1 and step-by-step work ourselves towards the innermost literal x_k. The reason this works is as follows:

Assume we have added the literal x'_1 to $C \vee l \vee x_1 \vee \cdots \vee x_k$ and the corresponding literal $\bar{x}'_1$ to another clause $C_i = C'_i \vee \bar{x}_1$ to obtain complementary literals in the outer resolvent of the resulting clauses $C \vee l \vee x_1 \vee x'_1 \vee \cdots \vee x_k$ and $C' \vee \bar{x}_1 \vee \bar{x}'_1$ upon x_1. Then, the outer resolvent of $C \vee l \vee x_1 \vee x'_1 \vee \cdots \vee x_k$ with $C' \vee \bar{x}_1 \vee \bar{x}'_1$ upon a literal x_j that is inner to x_1 (i.e., $x_1 <_\Pi x_j$) contains the complementary pair $x'_1, \bar{x}'_1$, so we have to add no further literals to $C' \vee \bar{x}_1 \vee \bar{x}'_1$.

Hence, the number of blocked-literal additions for literals of the form $\bar{x}'_i$ is bounded by the number of clauses, that is, by n. Moreover, for every addition of a literal $\bar{x}'_i$ to some clause, there is at most one addition of the corresponding literal x'_i. Therefore, there are at most $2n$ blocked-literal additions per LQ-res step. Now, for every addition of a literal x'_i, there is exactly one elimination of the corresponding literal x_i. Thus, overall there are at most $3n$ blocked-literal additions and eliminations for every LQ-res step. Since the number of LQ-res steps is bounded by the number of clauses in the proof, the size of the QRAT derivation is at most $3n^2$. It follows that whenever a QBF has a long-distance-resolution proof of polynomial size, it also has a polynomial-size QRAT proof:

Theorem 4. *The QRAT proof system polynomially simulates the long-distance-resolution calculus.*

6   Evaluation

We now know that QRAT can polynomially simulate long-distance resolution. But what does it mean in practice? Can we have short QRAT proofs for formulas that have short long-distance-resolution proofs? To answer this question at least partly, we consider the formulas well-known for having short long-distance-resolution proofs while only having long Q-resolution proofs—the Kleine Büning formulas [20]. A Kleine Büning formula of size n, in short $KBKF_n$, has the prefix $\exists a_0, a_1, b_1 \forall x_1 \exists a_2, b_2 \forall x_2 \ldots \exists a_n, b_n \forall x_n \exists c_1, c_2, \ldots, c_n$ and the following clauses:

$I\ :\ \bar{a}_0$	$I'\ :\ a_0 \vee \bar{a}_1 \vee \bar{b}_1$
$A_i\ :\ a_i \vee \bar{x}_i \vee \bar{a}_{i+1} \vee \bar{b}_{i+1}$	$B_i\ :\ b_i \vee x_i \vee \bar{a}_{i+1} \vee \bar{b}_{i+1}$ for $i \in \{1..n-1\}$
$C\ :\ a_n \vee \bar{x}_n \vee \bar{c}_1 \vee \cdots \vee \bar{c}_n$	$C'\ :\ b_n \vee x_n \vee \bar{c}_1 \vee \cdots \vee \bar{c}_n$
$X_i\ :\ \bar{x}_i \vee c_i$	$X'_i\ :\ x_i \vee c_i$ for $i \in \{1..n\}$

We can reduce a formula $KBKF_n$ to a formula $KBKF_{n-1}$ by using only Q-res, blocked-literal elimination, and clause-deletion steps[1] (no $\forall$-red steps or resolution upon universal literals). To do so, we use the clauses A_n, B_n, C, C', X_n, and X'_n of $KBKF_n$ to construct the clauses C and C' of $KBKF_{n-1}$. The required 12 steps are shown below. The last two clauses (11 and 12) respectively correspond to the clauses C and C' of $KBKF_{n-1}$.

1.	$a_n \vee \bar{x}_n \vee \bar{c}_1 \vee \cdots \vee \bar{c}_{n-1}$	(Q-res of C and X_n)
2.	$b_n \vee x_n \vee \bar{c}_1 \vee \cdots \vee \bar{c}_{n-1}$	(Q-res of C' and X'_n)
3.		(delete C, C', X_n, X'_n)
4.	$a_{n-1} \vee \bar{x}_{n-1} \vee \bar{b}_n \vee \bar{x}_n \vee \bar{c}_1 \vee \cdots \vee \bar{c}_{n-1}$	(Q-res of 1 and A_{n-1})
5.	$b_{n-1} \vee x_{n-1} \vee \bar{a}_n \vee x_n \vee \bar{c}_1 \vee \cdots \vee \bar{c}_{n-1}$	(Q-res of 2 and B_{n-1})
6.	$a_{n-1} \vee \bar{x}_{n-1} \vee \bar{b}_n \vee \bar{c}_1 \vee \cdots \vee \bar{c}_{n-1}$	(BLE of $\bar{x}_n$ from 4)
7.	$b_{n-1} \vee x_{n-1} \vee \bar{a}_n \vee \bar{c}_1 \vee \cdots \vee \bar{c}_{n-1}$	(BLE of x_n from 5)
8.	$a_{n-1} \vee \bar{x}_{n-1} \vee x_n \vee \bar{c}_1 \vee \cdots \vee \bar{c}_{n-1}$	(Q-res of 6 and B_{n-1})
9.	$b_{n-1} \vee x_{n-1} \vee \bar{x}_n \vee \bar{c}_1 \vee \cdots \vee \bar{c}_{n-1}$	(Q-res of 7 and A_{n-1})
10.		(delete 4, 5, 6, 7, A_{n-1}, B_{n-1})
11.	$a_{n-1} \vee \bar{x}_{n-1} \vee \bar{c}_1 \vee \cdots \vee \bar{c}_{n-1}$	(BLE of x_n from 8)
12.	$b_{n-1} \vee x_{n-1} \vee \bar{c}_1 \vee \cdots \vee \bar{c}_{n-1}$	(BLE of $\bar{x}_n$ from 9)

Table 1 shows the sizes of the Kleine Büning formulas as well as of the corresponding long-distance-resolution proofs (in the QRP format) and QRAT proofs. The latter are obtained by the construction mentioned in this section. The size of both types of proofs is linear in the size of the formula. Although QRAT proofs use about twice as many proof steps (including deletion steps), the file size of QRAT proofs is smaller. The explanation for this is that long-distance-resolution proofs increase the length of clauses, while QRAT proofs decreases their length.

Short proofs of the $KBKF$ formulas can also be obtained by using resolution upon universal variables as in the calculus QU-Res [24]. There is, however, a variant of the $KBKF$ formulas, called $KBKF_n-qu$ [2], which has only exponential proofs in the QU-Res calculus. A $KBKF_n-qu$ formula is obtained from $KBKF_n$

---

[1] Clause deletion was not used in the simulation, but is allowed in the QRAT system.

Table 1. The size of Kleine Büning formulas in the number of variables (#var) and clauses (#cls). Additionally, the size of their long-distance-resolution proofs (in the QRP format) in the number of Q-res steps (#Q), LQ-res steps (#L), ∀-red steps (#∀), and the file size in KB (ignoring the part that represents the formula). On the right, the number of Q-res (#Q), BLE (#B), and deletion (#D) steps as well as the file size for the manual QRAT proofs.

	input		LD proofs (QRP)				QRAT proofs			
formula	#var	#cls	#Q	#L	#∀	file size	#Q	#B	#D	file size
$KBKF_{10}$	41	42	41	18	38	6	57	38	92	6
$KBKF_{50}$	201	202	201	98	198	138	297	198	492	112
$KBKF_{100}$	401	402	401	198	398	573	597	398	992	421
$KBKF_{200}$	801	802	801	398	798	2321	1197	798	1992	1627
$KBKF_{500}$	2001	2002	2001	998	1998	16 259	2997	1998	4992	11 890

by adding a universal literal y_i (occurring in the same quantifier block as x_i) to every clause in $KBKF_n$ that contains x_i, and a literal $\bar{y}_i$ to every clause in $KBKF_n$. For these formulas, blocked-literal elimination can remove all the y_i and $\bar{y}_i$ literals, which reduces a $KBKF_n-qu$ formula to a $KBKF_n$ formula that can then be efficiently proved using resolution upon universal literals.

In addition to the handcrafted QRAT proofs, we implemented a tool (called `ld2qrat`) that, based on our simulation, transforms long-distance-resolution proofs into QRAT proofs. We used `ld2qrat` to transform the long-distance-resolution proofs of the $KBKF_n$ formulas (by Egly et al. [11]) into QRAT proofs and validated the correctness of these proofs with the proof checker `QRAT-trim`. In the plain mode, `ld2qrat` closely follows our simulation. Additionally, it features two optimizations: (1) Given an LQ-res step upon l with the antecedents $C \vee l \vee x$ and $D \vee \bar{l} \vee \bar{x}$, if one of x or $\bar{x}$ is already a blocked literal, it is removed with blocked-literal elimination. This avoids the introduction of new variables. (2) Clauses are deleted as soon as they are not needed later in the proof anymore.

Table 2 shows properties of the QRAT proofs produced by `ld2qrat` from the long-distance-resolution proofs of the $KBKF$ formulas. On the left are the sizes of proofs obtained without the clause-deletion optimization. On the right are the sizes of proofs with this optimization. A (least squares) regression analysis confirms that the length (number of steps) of the QRAT proofs without deletion is quadratically related to the length of the corresponding long-distance-resolution proofs: The function $f(x) = 0.22x^2 - 4.48x + 54.58$ (where x is the length of the long-distance-resolution proof and $f(x)$ is the length of the QRAT proof) fits the data from the above tables perfectly (the error term R^2 of the regression is 1).

7 QRAT in the Complexity Landscape

After the analysis of QRAT in theory and practice, we now locate it in the proof-complexity landscape of resolution-based calculi for QBF, which is shown in Fig. 2. Besides the long-distance-resolution calculus LQ-Res, another well-known proof system is the calculus QU-Res [24], which extends the basic Q-resolution

Table 2. Comparison of the QRAT proofs obtained by applying ld2qrat to long-distance-resolution proofs (in the QRP format) of the Kleine Büning formulas. The file size is given in KB and the time for translating the proof (time) is given in seconds.

| | QRP to QRAT w/o deletion | | | | QRP to QRAT w/ deletion | | | |
formula	#var	#step	file size	time	#var	#step	file size	time
$KBKF_{10}$	59	1690	103	0.07	59	448	26	0.01
$KBKF_{50}$	299	52170	18774	0.45	299	6288	2227	0.12
$KBKF_{100}$	599	214270	154299	3.77	599	22588	16192	0.86
$KBKF_{200}$	1199	868470	1309559	30.70	1199	85188	126375	7.95
$KBKF_{500}$	2999	5471070	23622369	497.32	2999	512988	2229195	124.10

calculus (Q-Res) by allowing resolution upon universals literals if the resulting resolvent does not contain complementary literals. As QRAT also allows resolution upon universal literals, it simulates QU-Res. Balabanov et al. [2] showed the incomparability between LQ-Res and QU-Res by exponential separations. It thus follows that QRAT is strictly stronger than both LQ-Res and QU-Res.

Another system that is stronger than both LQ-Res and QU-Res is the calculus LQU$^+$-Res [2], which extends LQ-Res by allowing (long-distance) resolution upon universals literals. We know that either QRAT is strictly stronger than LQU$^+$-Res or the two systems are incomparable: On purely existentially-quantified formulas, LQU$^+$-Res boils down to ordinary propositional resolution (without complementary literals in resolvents) whereas the QRAT system boils down to the RAT system [25]. As the RAT system is strictly stronger than resolution—there exist polynomial-size RAT proofs of the well-known pigeon hole formulas [13] while resolution proofs of these formulas are necessarily exponential in size [12]—LQU$^+$-Res cannot simulate QRAT.

On the other hand, QRAT might be able to simulate LQU$^+$-Res, but not with our simulation of the long-distance-resolution calculus, because the simulation cannot convert all LQU$^+$-Res proofs into QRAT proofs. To see this, consider the QBF $\exists a \forall x \forall y \exists b.(a \vee x \vee b) \wedge (\bar{a} \vee \bar{x} \vee b) \wedge (x \vee \bar{b}) \wedge (\bar{x} \vee \bar{y} \vee \bar{b})$ with the following LQU$^+$-Res proof [2]: $x \vee \bar{x} \vee b$, $\bar{y} \vee \bar{b}$, $x \vee \bar{x} \vee \bar{y}$, $x \vee \bar{x}$, x, $\bot$. The proof can be illustrated as follows:

$$\frac{\dfrac{a \vee x \vee b \quad \bar{a} \vee \bar{x} \vee b}{x \vee \bar{x} \vee b}\text{(LQ-res)} \quad \dfrac{x \vee \bar{b} \quad \bar{x} \vee \bar{y} \vee \bar{b}}{\dfrac{\bar{y} \vee \bar{b}}{}}\text{(QU-res)}}{\dfrac{\dfrac{x \vee \bar{x} \vee \bar{y}}{\dfrac{x \vee \bar{x}}{\dfrac{x}{\bot}\text{(∀-red)}}\text{(∀-red)}}\text{(∀-red)}}{}}\text{(Q-res)}$$

In our simulation, we first replace the literal x in $a \vee x \vee b$ by x' before resolving the resulting clause $a \vee x' \vee b$ with $\bar{a} \vee \bar{x} \vee b$. The replacement of x by x' also leads to the addition of $\bar{x}'$ to $\bar{x} \vee \bar{y} \vee \bar{b}$. If we now perform the universal resolution step of $x \vee \bar{b}$ with $\bar{x} \vee \bar{x}' \vee \bar{y} \vee \bar{b}$, then we obtain the following partial proof:

$$\frac{a \vee x' \vee b \quad \bar{a} \vee \bar{x} \vee b}{x' \vee \bar{x} \vee b}\text{(Q-res)} \qquad \frac{x \vee \bar{b} \quad \bar{x} \vee \bar{x}' \vee \bar{y} \vee \bar{b}}{\bar{x}' \vee \bar{y} \vee \bar{b}}\text{(QU-res)}$$

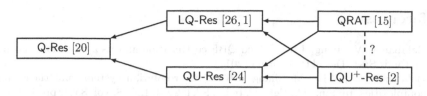

Fig. 2. Complexity landscape including QRAT. A directed edge from a proof system A to a proof system B means that A is strictly stronger than B.

The Q-res step upon b is now impossible because x' is in $x' \vee \bar{x} \vee b$ and $\bar{x}'$ is in $\bar{x}' \vee \bar{y} \vee \bar{b}$. We also cannot eliminate x' from $x' \vee \bar{x} \vee b$ via blocked-literal elimination: This would require us to add a new literal x'' to $x' \vee \bar{x} \vee b$ and to add $\bar{x}''$ to $\bar{x}' \vee \bar{y} \vee \bar{b}$ leading to the new pair $x'', \bar{x}''$ of complementary literals.

Our key result, Lemma 1, does not hold anymore when allowing resolution over universal literals. Lemma 1 guarantees that whenever a new literal $\bar{x}'$ is in a proof clause C_i' of the modified long-distance-resolution proof, then $\bar{x}$ was contained in the corresponding clause C_i in the original proof. The above example shows that resolution over universal literals destroys this property: Although $\bar{x}'$ is contained in the clause $\bar{x}' \vee \bar{y} \vee \bar{b}$, the literal x is not contained in the corresponding clause $y \vee \bar{y} \vee b$ of the original proof because we resolved it away.

8 Conclusion

We showed that the QRAT proof system polynomially simulates long-distance resolution. In our simulation, we used only a small subset of the QRAT rules: Q-resolution, universal reduction, blocked-literal addition, and blocked-literal elimination. Based on our simulation, we implemented a tool that transforms long-distance-resolution proofs into QRAT proofs. The tool allows to merge a QRAT derivation produced by a QBF-preprocessor with a long-distance-resolution proof produced by a search-based solver. The correctness of the resulting QRAT proof can then be checked with a proof checker such as QRAT-trim [15]. We evaluated the tool on long-distance-resolution proofs of the well-known Kleine Büning formulas and manually constructed QRAT proofs of these formulas that are smaller than their long-distance counterparts.

We further noted that our simulation breaks down if the long-distance-resolution calculus is extended by resolution upon universal literals, as in the calculus LQU$^+$-Res. Investigating the exact relationship between LQU$^+$-Res and QRAT therefore remains open for future work. Another open question is whether blocked-literal elimination can be polynomially simulated in LQU$^+$-Res. We also do not know whether it is possible to simulate long-distance resolution with only Q-resolution, universal reduction, clause deletion, and blocked-literal elimination (but no blocked-literal addition). Finally, what is still unclear is how QRAT relates to instantiation-based proof systems and sequent proof systems. Answers to these questions will shed more light on the proof-complexity landscape of QBF.

References

1. Balabanov, V., Jiang, J.R.: Unified QBF certification and its applications. Formal Methods Syst. Des. **41**(1), 45–65 (2012)
2. Balabanov, V., Widl, M., Jiang, J.R.: QBF resolution systems and their proof complexities. In: Sinz, C., Egly, U. (eds.) SAT 2014. LNCS, vol. 8561, pp. 154–169. Springer, Cham (2014). doi:10.1007/978-3-319-09284-3_12
3. Benedetti, M., Mangassarian, H.: QBF-based formal verification: experience and perspectives. J. Satisf. Boolean Model. Comput. (JSAT) **5**(1–4), 133–191 (2008)
4. Beyersdorff, O., Bonacina, I., Chew, L.: Lower bounds: from circuits to QBF proof systems. In: Proceedings of the 2016 ACM Conference on Innovations in Theoretical Computer Science (ITCS 2016), pp. 249–260. ACM (2016)
5. Beyersdorff, O., Chew, L., Janota, M.: On unification of QBF resolution-based calculi. In: Csuhaj-Varjú, E., Dietzfelbinger, M., Ésik, Z. (eds.) MFCS 2014. LNCS, vol. 8635, pp. 81–93. Springer, Heidelberg (2014). doi:10.1007/978-3-662-44465-8_8
6. Beyersdorff, O., Chew, L., Janota, M.: Proof complexity of resolution-based QBF calculi. In: Proceedings of the 32nd Internation Symposium on Theoretical Aspects of Computer Science (STACS 2015). LIPIcs, vol. 30, pp. 76–89. Schloss Dagstuhl - Leibniz-Zentrum für Informatik (2015)
7. Beyersdorff, O., Chew, L., Mahajan, M., Shukla, A.: Are short proofs narrow? QBF resolution is not simple. In: Proceedings of the 33rd Symposium on Theoretical Aspects of Computer Science (STACS 2016). LIPIcs, vol. 47, pp. 15:1–15:14. Schloss Dagstuhl - Leibniz-Zentrum fuer Informatik (2016)
8. Beyersdorff, O., Pich, J.: Understanding Gentzen and Frege systems for QBF. In: Proceedings of the 31st Annual ACM/IEEE Symposium on Logic in Computer Science (LICS 2016), pp. 146–155. ACM (2016)
9. Chen, H.: Proof complexity modulo the polynomial hierarchy: understanding alternation as a source of hardness. In: Proceedings of the 43rd International Colloquium on Automata, Languages, and Programming (ICALP 2016). LIPIcs, vol. 55, pp. 94:1–94:14. Schloss Dagstuhl - Leibniz-Zentrum fuer Informatik (2016)
10. Egly, U.: On stronger calculi for QBFs. In: Creignou, N., Le Berre, D. (eds.) SAT 2016. LNCS, vol. 9710, pp. 419–434. Springer, Cham (2016). doi:10.1007/978-3-319-40970-2_26
11. Egly, U., Lonsing, F., Widl, M.: Long-distance resolution: proof generation and strategy extraction in search-based QBF solving. In: McMillan, K., Middeldorp, A., Voronkov, A. (eds.) LPAR-19. LNCS, vol. 8312, pp. 291–308. Springer, Heidelberg (2013). doi:10.1007/978-3-642-45221-5_21
12. Haken, A.: The intractability of resolution. Theor. Comput. Sci. **39**, 297–308 (1985)
13. Heule, M.J.H., Hunt Jr., W.A., Wetzler, N.D.: Expressing symmetry breaking in DRAT proofs. In: Felty, A.P., Middeldorp, A. (eds.) CADE 2015. LNCS, vol. 9195, pp. 591–606. Springer, Cham (2015). doi:10.1007/978-3-319-21401-6_40
14. Heule, M.J.H., Seidl, M., Biere, A.: Blocked literals are universal. In: Havelund, K., Holzmann, G., Joshi, R. (eds.) NFM 2015. LNCS, vol. 9058, pp. 436–442. Springer, Cham (2015). doi:10.1007/978-3-319-17524-9_33
15. Heule, M.J.H., Seidl, M., Biere, A.: Solution validation and extraction for QBF preprocessing. J. Autom. Reason. **58**(1), 97–125 (2016)
16. Janota, M.: On Q-resolution and CDCL QBF solving. In: Creignou, N., Le Berre, D. (eds.) SAT 2016. LNCS, vol. 9710, pp. 402–418. Springer, Cham (2016). doi:10.1007/978-3-319-40970-2_25

17. Janota, M., Grigore, R., Marques-Silva, J.P.: On QBF proofs and preprocessing. In: McMillan, K., Middeldorp, A., Voronkov, A. (eds.) LPAR-19. LNCS, vol. 8312, pp. 473–489. Springer, Heidelberg (2013). doi:10.1007/978-3-642-45221-5_32
18. Janota, M., Klieber, W., Marques-Silva, J.P., Clarke, E.M.: Solving QBF with counterexample guided refinement. Artif. Intell. **234**, 1–25 (2016)
19. Kleine Büning, H., Bubeck, U.: Theory of quantified Boolean formulas. In: Handbook of Satisfiability, pp. 735–760. IOS Press (2009)
20. Kleine Büning, H., Karpinski, M., Flögel, A.: Resolution for quantified Boolean formulas. Inf. Comput. **117**(1), 12–18 (1995)
21. Kullmann, O.: On a generalization of extended resolution. Discrete Appl. Math. **96–97**, 149–176 (1999)
22. Lonsing, F., Egly, U.: DepQBF 6.0: a search-based QBF solver beyond traditional QCDCL. CoRR abs/1702.08256 (2017)
23. Slivovsky, F., Szeider, S.: Variable dependencies and Q-resolution. In: Sinz, C., Egly, U. (eds.) SAT 2014. LNCS, vol. 8561, pp. 269–284. Springer, Cham (2014). doi:10.1007/978-3-319-09284-3_21
24. Van Gelder, A.: Contributions to the theory of practical quantified Boolean formula solving. In: Milano, M. (ed.) CP 2012. LNCS, pp. 647–663. Springer, Heidelberg (2012). doi:10.1007/978-3-642-33558-7_47
25. Wetzler, N.D., Heule, M.J.H., Hunt Jr., W.A.: DRAT-trim: efficient checking and trimming using expressive clausal proofs. In: Sinz, C., Egly, U. (eds.) SAT 2014. LNCS, vol. 8561, pp. 422–429. Springer, Cham (2014). doi:10.1007/978-3-319-09284-3_31
26. Zhang, L., Malik, S.: Conflict driven learning in a quantified Boolean satisfiability solver. In: Proceedings of the 2002 IEEE/ACM International Conference on Computer-Aided Design (ICCAD 2002), pp. 442–449. ACM/IEEE Computer Society (2002)

Dependency Learning for QBF

Tomáš Peitl, Friedrich Slivovsky$^{(\boxtimes)}$, and Stefan Szeider

Algorithms and Complexity Group, TU Wien, Vienna, Austria
{peitl,fslivovsky,sz}@ac.tuwien.ac.at

Abstract. Quantified Boolean Formulas (QBFs) can be used to succinctly encode problems from domains such as formal verification, planning, and synthesis. One of the main approaches to QBF solving is Quantified Conflict Driven Clause Learning (QCDCL). By default, QCDCL
assigns variables in the order of their appearance in the quantifier prefix
so as to account for dependencies among variables. Dependency schemes
can be used to relax this restriction and exploit independence among
variables in certain cases, but only at the cost of nontrivial interferences
with the proof system underlying QCDCL. We propose a new technique
for exploiting variable independence within QCDCL that allows solvers
to learn variable dependencies on the fly. The resulting version of QCDCL
enjoys improved propagation and increased flexibility in choosing variables for branching while retaining ordinary (long-distance) Q-resolution
as its underlying proof system. In experiments on standard benchmark
sets, an implementation of this algorithm shows performance comparable
to state-of-the-art QBF solvers.

1 Introduction

Conflict Driven Clause Learning (CDCL) represents the state of the art in
propositional satisfiability (SAT) solving (see, e.g., [23]). Modern CDCL solvers
are able to handle input formulas with thousands of variables and millions of
clauses [22]. Their remarkable performance has led to the adoption of SAT solving in electronic design automation (for a survey, see [33]), it has turned algorithms relying on SAT oracles into viable tools for solving hard problems (see,
e.g., [24]), and it has even helped resolve open questions in combinatorics [12].

Encouraged by this success, researchers are turning to an even harder problem: the satisfiability problem of Quantified Boolean Formulas (QSAT). Quantified Boolean Formulas (QBFs) enrich propositional formulas with universal
and existential quantification over truth values and offer much more succinct
encodings of problems from domains such as planning and synthesis [9]. This
expressive power comes at a price: QSAT is complete for the complexity class
PSPACE and thus believed to be significantly harder than SAT.

Quantified CDCL [7,34] is a natural generalization of CDCL and one of two
dominant algorithmic paradigms in QSAT solving (the other being approaches
broadly based on quantifier expansion [3,14,15,21,27,29,31]). While the performance of QCDCL solvers has much improved over the past years, they have so

© Springer International Publishing AG 2017
S. Gaspers and T. Walsh (Eds.): SAT 2017, LNCS 10491, pp. 298–313, 2017.
DOI: 10.1007/978-3-319-66263-3_19

far failed to replicate the success of CDCL in the domain of SAT. One of the main obstacles in lifting CDCL to QSAT is that the alternation of existential and universal quantifiers in the quantifier prefix of a QBF (we consider formulas in prenex normal form) introduces dependencies among variables that have to be respected by the order of variable assignments. The resulting constraints not only reduce the empirical effectiveness of branching heuristics but impose severe theoretical limits on the power of QCDCL [13]. By default, QCDCL only considers variables from the outermost quantifier block with unassigned variables for branching. In the worst case, this forces solvers into a fixed branching order. Among several techniques that have been introduced to relax this restriction, *dependency schemes* are arguably the most general. Given a QBF, a dependency scheme efficiently computes an overapproximation of its variable dependencies—that is, the result contains every pair of variables for which there is a "real" dependency, but it may contain "spurious" dependencies. Lonsing and Biere [5] introduced a generalization of QCDCL that uses dependency schemes to relax constraints on the order of variable assignments and implemented this algorithm in the solver DepQBF.

The use of dependency schemes within DepQBF often leads to performance improvements, but it has several drawbacks. First, it changes the proof system underlying constraint learning, and proving soundness of the resulting algorithm is nontrivial even for a simple version of QCDCL and common dependency schemes [25,30]. The continuous addition of solver features makes QCDCL a moving target, and the integration of dependency schemes with any new technique usually requires a new soundness proof. Second, even if soundness of the resulting proof system can be established, efficient (linear-time) strategy extraction from proofs—a common requirement for applications—does not follow. Third, and perhaps most importantly, the syntactic criteria for identifying dependencies used by common dependency schemes (such as the standard dependency scheme or the resolution-path dependency scheme) are fairly coarse, so that the set of dependencies computed by such schemes frequently coincides with the "trivial" dependencies implicit in the quantifier prefix (see Table 3 in Sect. 5).

In this paper, we describe a new approach to exploiting variable independence in QCDCL solvers we call *dependency learning*. The idea is that the solver maintains a set D of dependencies that is used in propagation and choosing variables for branching just like in QCDCL with dependency schemes: a clause is considered unit under the current assignment if it contains a single existential variable that does not depend, according to D, on any universal variable remaining in the clause; a variable is eligible for branching if it does not depend, according to D, on any variable that is currently unassigned (cf. [5]). But instead of initializing D using a dependency scheme, dependencies are added on the fly as needed. Initially, the set D is empty, so every clause containing a single existential variable is considered unit and variables can be assigned in any order. When propagation runs into a conflict, the solver attempts to derive a new clause by long-distance Q-resolution [1,8]. Because propagation implicitly performs universal reduction relative to D but Q-resolution applies universal reduction according to the prefix order, the solver may be unable to generate a learned clause. If this happens, a set of variable dependencies can be identified as the reason for this failure and added to D, preventing this situation from occurring in the future. The resulting version of QCDCL potentially improves on the flexibility provided by dependency schemes but retains long-distance Q-resolution as its underlying proof system and therefore supports linear-time strategy extraction [2].

To explore the effectiveness of this technique, we implemented *Qute*, a QCDCL solver that supports dependency learning. In experiments with benchmark instances from the 2016 QBF evaluation, Qute is competitive with state-of-the-art QBF solvers on formulas in prenex conjunctive normal form (PCNF). For formulas represented as quantified circuits in the QCIR format, Qute solves more instances than any other available solver. Additional experiments show that the number of dependencies learned by Qute on PCNF instances preprocessed by Bloqqer is typically only a fraction of those identified by the standard dependency scheme and even the (reflexive) resolution-path dependency scheme, and that dependency learning allows QCDCL to deal with formulas that are provably hard to solve for vanilla QCDCL [13].

The remainder of this paper is organized as follows. In Sect. 2, we cover basic definitions and notation. In Sect. 3, we introduce QCDCL and (long-distance) Q-resolution, its underlying proof system. In Sect. 4, we describe how to modify QCDCL to support dependency learning, and prove that the resulting algorithm is sound and terminating. In Sect. 5, we provide an experimental evaluation of Qute. In Sect. 6, we conclude with a discussion of our results and future research directions.

2   Preliminaries

We consider QBFs in Prenex Conjunctive Normal Form (PCNF), i.e., formulas $\Phi = \mathcal{Q}.\varphi$ consisting of a (quantifier) prefix $\mathcal{Q}$ and a propositional CNF formula φ, called the *matrix* of Φ. The *prefix* is a sequence $\mathcal{Q} = Q_1 x_1 \ldots Q_n x_n$, where $Q_i \in \{\forall, \exists\}$ is a universal or existential quantifier and the x_i are variables. We write $x_i \prec_\Phi x_j$ if $1 \le i < j \le n$ and $Q_i \ne Q_j$, dropping the subscript if the formula Φ is understood. A *CNF formula* is a finite conjunction $C_1 \wedge \cdots \wedge C_m$ of clauses, a *clause* is a finite disjunction $(\ell_1 \vee \cdots \vee \ell_k)$ of literals, and a *literal* is a variable x or a negated variable $\neg x$. Dually, a *DNF formula* is a finite disjunction of $T_1 \vee \cdots \vee T_k$ of terms, and a *term* is a finite conjunction $(\ell_1 \wedge \cdots \wedge \ell_k)$ of literals. Whenever convenient, we consider clauses and terms as sets of literals, CNF formulas as sets of clauses, and DNF formulas as sets of terms. We assume that PCNF formulas are *closed*, so that every variable occurring in the matrix appears in the prefix, and that each variable appearing in the prefix occurs in the matrix. We write $var(x) = var(\neg x) = x$ to denote the variable associated with a literal and let $var(C) = \{ var(\ell) : \ell \in C \}$ if C is a clause, $var(\varphi) = \bigcup_{C \in \varphi} var(C)$ if φ is a CNF formula, and $var(\Phi) = var(\varphi)$ if $\Phi = \mathcal{Q}.\varphi$ is a PCNF formula.

An *assignment* is a sequence $\sigma = (\ell_1, \ldots, \ell_k)$ of literals such that $var(\ell_i) \ne var(\ell_j)$ for $1 \le i < j \le n$. If S is a clause or term, we write $S[\sigma]$ for the the result of applying σ to S. For a clause C, we define $C[\sigma] = \top$ if $\ell_i \in C$ for some $1 \leq i \leq k$, and $C[\sigma] = C \setminus \{\overline{\ell_1}, \ldots, \overline{\ell_k}\}$ otherwise, where $\overline{x} = \neg x$ and $\overline{\neg x} = x$. For a term T, we let $T[\sigma] = \bot$ if $\overline{\ell_i} \in T$ for some $1 \leq i \leq k$, and $T[\sigma] = T \setminus \{\ell_1, \ldots, \ell_k\}$ otherwise. An assignment σ *falsifies* a clause C if $C[\sigma] = \emptyset$; it *satisfies* a term T if $T[\sigma] = \emptyset$. For CNF formulas φ, we let $\varphi[\sigma] = \{ C[\sigma] : C \in \varphi, C[\sigma] \neq \top \}$, and for PCNF formulas $\Phi = Q.\varphi$, we let $\Phi[\sigma] = Q'.\varphi[\sigma]$, where Q' is obtained by deleting variables from Q not occurring in $\varphi[\sigma]$.

The semantics of a PCNF formula Φ are defined as follows. If Φ does not contain any variables then Φ is true if its matrix is empty and false if its matrix contains the empty clause $\emptyset$. Otherwise, let $\Phi = QxQ.\varphi$. If $Q = \exists$ then Φ is true if $\Phi[(x)]$ is true or $\Phi[(\neg x)]$ is true, and if $Q = \forall$ then Φ is true if both $\Phi[(x)]$ and $\Phi[(\neg x)]$ are true.

3 QCDCL and Q-Resolution

We briefly review QCDCL and Q-resolution [17], its underlying proof system. More specifically, we consider *long-distance Q-resolution*, a version of Q-resolution that admits the derivation of tautological clauses in certain cases. Although this proof system was already used in early QCDCL solvers [34], the formal definition shown in Fig. 1 was given only recently [1]. A dual proof system called *(long-distance) Q-consensus*, which operates on terms instead of clauses, is obtained by swapping the roles of existential and universal variables (the analogue of universal reduction for terms is called *existential reduction*).

$$\frac{C_1 \vee e \qquad \neg e \vee C_2}{C_1 \vee C_2} \text{ (resolution)}$$

The *resolution* rule allows the derivation of $C_1 \vee C_2$ from clauses $C_1 \vee e$ and $\neg e \vee C_2$, provided that the *pivot* variable e is existential and that $e \prec var(\ell_u)$ for each universal literal $\ell_u \in C_1$ such that $\overline{\ell_u} \in C_2$. The clause $C_1 \vee C_2$ is called the resolvent of $C_1 \vee e$ and $\neg e \vee C_2$.

$$\frac{C}{C \setminus \{u, \neg u\}} \text{ (universal reduction)}$$

The *universal reduction* rule admits the deletion of a universal variable u from a clause C under the condition that $e \prec u$ for each existential variable e in C.

Fig. 1. Long-distance Q-resolution.

A (long-distance) Q-resolution *derivation* from a PCNF formula Φ is a sequence of clauses such that each clause appears in the matrix of Φ or can be derived from clauses appearing earlier in the sequence using resolution or universal reduction. A derivation of the empty clause is called a *refutation*, and one can show that a PCNF formula is false, if, and only if, it has a long-distance Q-resolution refutation [1]. Dually, a PCNF formula is true, if, and only if, the empty term can be derived from a DNF representation of its matrix by Q-consensus.

Starting from an input PCNF formula, QCDCL generates ("learns") *constraints*—clauses and terms—until it produces an empty constraint. Every clause learned by QCDCL can be derived from the input formula by Q-resolution, and every term learned by QCDCL can be derived by Q-consensus [8,10]. Accordingly, the solver outputs TRUE if the empty term is learned, and FALSE if the empty clause is learned.

One can think of QCDCL solving as proceeding in rounds. Along with a set of clauses and terms, the solver maintains an assignment σ. During each round, this assignment is extended by quantified Boolean constraint propagation (QBCP) and—possibly—branching.

Quantified Boolean constraint propagation consists in the exhaustive application of universal and existential reduction in combination with unit assignments.[1] More specifically, QBCP reports a clause C as falsified if $C[\sigma] \neq \top$ and universal reduction can be applied to $C[\sigma]$ to obtain the empty clause. Dually, a term T is considered satisfied if $T[\sigma] \neq \bot$ and $T[\sigma]$ can be reduced to the empty term. A clause C is *unit* under σ if $C[\sigma] \neq \top$ and universal reduction yields the clause $C' = (\ell)$, for some existential literal ℓ such that $var(\ell)$ is unassigned. Dually, a term T is *unit* under σ if $T[\sigma] \neq \bot$ and existential reduction can be applied to obtain a term $T' = (\ell)$ containing a single universal literal ℓ. If $C = (\ell)$ is a unit clause, then the assignment σ has to be extended by ℓ in order not to falsify C, and if $T = (\ell)$ is a unit term, then σ has to be extended by $\overline{\ell}$ in order not to satisfy T. If several clauses or terms are unit under σ, QBCP nondeterministically picks one and extends the assignment accordingly. This is repeated until a constraint is empty or no unit constraints remain.

If QBCP does not lead to an empty constraint, the assignment σ is extended by *branching*. That is, the solver chooses an unassigned variable x such that every variable y with $y \prec x$ is assigned, and extends the assignment σ by x or $\neg x$.

The resulting assignment can be partitioned into so-called *decision levels*. The decision level of an assigment σ is the number of literals in σ that were assigned by branching. The decision level of a literal ℓ in σ is the decision level of the prefix of σ that ends with ℓ. Note that each decision level greater than 0 can be associated with a unique variable assigned by branching.

Eventually, the assignment maintained by QCDCL must falsify a clause or satisfy a term. When this happens (this is called a *conflict*), the solver proceeds to *conflict analysis* to derive a learned constraint C. Initially, C is the falsified clause (satisfied term), called the *conflict clause (term)*. The solver finds the existential (universal) literal in C that was assigned last by QBCP, and the antecedent clause (term) R responsible for this assignment. A new constraint is derived by resolving C and R and applying universal (existential) reduction.

---

[1] We do not consider the pure literal rule as part of QBCP.

This is done repeatedly until the resulting constraint C is *asserting*. A clause (term) S is asserting if there is a unique existential (universal) literal $\ell \in S$ with maximum decision level among literals in S, the corresponding decision variable is existential (universal), and every universal (existential) variable $y \in var(S)$ such that $y \prec var(\ell)$ is assigned at a lower decision level (an asserting constraint becomes unit after backtracking). Once an asserting constraint has been found, it is added to the solver's set of constraints. Finally, QCDCL *backtracks*, undoing variable assignments until reaching a decision level computed from the learned constraint.

4   QCDCL with Dependency Learning

We now describe how to modify QCDCL to support dependency learning. First, the solver maintains a set $D \subseteq \{(x, y) : x \prec y\}$ of variable dependencies. Second, both QBCP and the decision rule must be modified as follows:

- QBCP() uses universal and existential reduction relative to D. Universal reduction relative to D removes each universal variable u from a clause C such that there is no existential variable $e \in var(C)$ with $(u, e) \in D$ (existential reduction relative to D is defined dually).

Algorithm 1. QCDCL with Dependency Learning

```
 1: procedure SOLVE( )
 2:     D = ∅
 3:     while TRUE do
 4:         conflict = QBCP()
 5:         if conflict == NONE then
 6:             DECIDE()
 7:         else
 8:             constraint, btlevel = ANALYZECONFLICT(conflict)
 9:             if constraint != NONE then
10:                 if ISEMPTY(constraint) then
11:                     if ISTERM(constraint) then
12:                         return TRUE
13:                     else
14:                         return FALSE
15:                     end if
16:                 else
17:                     ADDLEARNEDCONSTRAINT(constraint)
18:                 end if
19:             end if
20:             BACKTRACK(btlevel)
21:         end if
22:     end while
23: end procedure
```

- DECIDE() may assign any variable y such that there is no unassigned variable x with $(x, y) \in D$ (note that $(x, y) \in D$ implies $x \prec y$).

This is how DepQBF uses the dependency relation D computed by a dependency scheme in propagation and decisions [5]. Unlike DepQBF, QCDCL with dependency learning does *not* use the generalized reduction rules during conflict analysis (RESOLVE and REDUCE in lines 7 and 8 refer to resolution and reduction as defined in Fig. 1). As a consequence, the algorithm cannot always construct a learned constraint during conflict analysis (see the example below). Such cases are dealt with in lines 9 through 12 of ANALYZECONFLICT (Algorithm 2):

- EXISTSRESOLVENT(*constraint*, *reason*, *pivot*) determines whether the resolvent of *constraint* and *reason* exists.
- If this is not the case, there has to be a variable v (universal for clauses, existential for terms) satisfying the following condition: $v \prec pivot$ and there exists a literal $\ell \in constraint$ with $var(\ell) = v$ and $\bar{\ell} \in reason$. The set of such variables is computed by ILLEGALMERGES. For each such variable, a new dependency is added to D. No learned constraint is returned by conflict analysis, and the backtrack level (*btlevel*) is set so as to cancel the decision level at which *pivot* was assigned.

The criteria for a constraint to be asserting must also be slightly adapted: a clause (term) S is asserting with respect to D if there is a unique existential (universal) literal $\ell \in S$ with maximum decision level among literals in S, the corresponding decision variable is existential (universal), and every universal (existential) variable $y \in var(S)$ such that $(y, var(\ell)) \in D$ is assigned (again, this corresponds to the definition of asserting constraints used in DepQBF [19, p. 119]). Finally, in the main QCDCL loop, we have to implement a case distinction to account for the fact that conflict analysis may not return a constraint (line 9).

Algorithm 2. Conflict Analysis with Dependency Learning

1: **procedure** ANALYZECONFLICT(*conflict*)
2:     *constraint* = GETCONFLICTCONSTRAINT(*conflict*)
3:     **while** NOT ASSERTING(*constraint*) **do**
4:         *pivot* = GETPIVOT(*constraint*)
5:         *reason* = GETANTECEDENT(*pivot*)
6:         **if** EXISTSRESOLVENT(*constraint*, *reason*, *pivot*) **then**
7:             *constraint* = RESOLVE(*constraint*, *reason*, *pivot*)
8:             *constraint* = REDUCE(*constraint*)
9:         **else**
10:            *illegal_merges* = ILLEGALMERGES(*constraint*, *reason*, *pivot*)
11:            $D = D \cup \{ (v, pivot) : v \in illegal\_merges \}$
12:            **return** NONE, DECISIONLEVEL(*pivot*)
13:        **end if**
14:    **end while**
15:    *btlevel* = GETBACKTRACKLEVEL(*constraint*)
16:    **return** *constraint*, *btlevel*
17: **end procedure**

4.1 Examples

We now illustrate the two possible outcomes of conflict analysis with simple examples. First, take the QBF

$$\Phi = \forall u \exists e.(u \vee e).$$

Starting from an empty set D of dependencies, QCDCL with dependency learning initially assumes that e is independent of u. By applying universal reduction relative to D to the clause $(u \vee e)$, one derives the unit clause (e). Accordingly, the solver appends e to its current assignment and finds the matrix satisfied. Since (e) is a term in the DNF representation of Φ's matrix and (e) can be reduced to the empty term by existential reduction, QCDCL learns the empty term and correctly reports that Φ is true. Now consider

$$\Psi = \forall u \exists e.(u \vee e) \wedge (\neg u \vee \neg e).$$

Again, QCDCL with dependency learning starting with empty D considers the first clause unit and appends e to its assignment. Propagating this assignment to the second clause results in a conflict, as $(\neg u \vee \neg e)[e \mapsto 1] = (\neg u)$, which simplifies to the empty clause by universal reduction. During conflict analysis, the solver attempts to construct a learned clause by resolving the conflict clause $(\neg u \vee \neg e)$ with the clause $(u \vee e)$ responsible for propagating e. But since $u \prec e$ is universal and appears negated in the first and unnegated in the second clause, these two clauses do not have a resolvent in long-distance Q-resolution, and the solver is unable to learn a clause. Instead, it adds the dependency (u, e) to D and backtracks until e is unassigned. Now that D contains the dependency (u, e), the universal variable u can no longer be reduced from a clause that contains e, so neither clause is unit. Moreover, the solver cannot branch on e while u is unassigned. It is easy to see that from this point on, the solver behaves just like ordinary QCDCL on this example.

4.2 Soundness and Termination

Soundness of QCDCL with dependency learning is an immediate consequence of the following observation.

Observation 1. Every constraint learned by QCDCL with dependency learning can be derived from the input formula by long-distance Q-resolution or Q-consensus.

To prove termination, we argue that the algorithm learns a new constraint or a new dependency after each conflict. Just as in QCDCL, every learned constraint is asserting, so learning does not introduce duplicate constraints.

Observation 2. QCDCL with dependency learning never learns a constraint already present in the database.

The only additional argument required to prove termination is one that tells us that the algorithm cannot indefinitely "learn" the same dependencies.

Lemma 1. *If QCDCL with dependency learning does not learn a constraint during conflict analysis, it learns a new dependency.*

Proof. To simplify the presentation, we are only going to consider clause learning (the proof for term learning is analogous). We first establish an invariant of intermediate clauses derived during conflict analysis: they are empty under the partial assignment obtained by backtracking to the last literal in the assignment that falsifies an existential literal in the clause. Formally, let C be a clause and let $\sigma = (\ell_1, \ldots, \ell_k)$ be an assignment. We define $last_C(\sigma) = \max(\{ i \in [k] : \overline{\ell_i} \in C, var(\ell_i) \in var_\exists \} \cup \{0\})$ and let $\sigma_C = (\ell_1, \ldots, \ell_{last_C(\sigma)})$. In particular, if $last_C(\sigma) = 0$ then σ_C is empty.

We now prove the following claim: if σ is an assignment that falsifies a clause, then, for every intermediate clause C constructed during conflict analysis, $C[\sigma_C]$ is empty after universal reduction. The proof is by induction on the number of resolution steps in the derivation of C. If C is the conflict clause then $C[\sigma]$ reduces to the empty clause. That means $C[\sigma_C]$ can only contain universal literals and can also be reduced to the empty clause by universal reduction. Suppose C is the result of resolving clauses C' and R and applying universal reduction, where C' is an intermediate clause derived during conflict analysis and R is a clause that triggered unit propagation. The induction hypothesis tells us that $C'[\sigma_{C'}]$ reduces to the empty clause. Since the pivot literal ℓ is chosen to be the last existential literal falsified in C', we must have $\sigma_{C'} = (\ell_1, \ldots, \ell_k)$ where $\ell_k = \overline{\ell}$. Let $\tau = (\ell_1, \ldots, \ell_{k-1})$. We must have $C'[\tau] = U' \cup \{\ell\}$, where U' is a purely universal clause. Because R is responsible for propagating $\overline{\ell}$, we further must have $R[\tau] = U'' \cup \{\overline{\ell}\}$, where U'' again is a purely universal clause. It follows that their resolvent $C[\tau] = (C' \setminus \{\ell\})[\tau] \cup (R \setminus \{\overline{\ell}\})[\tau] = U' \cup U''$ is a purely universal clause. Since τ is a prefix of σ, it follows that $C[\sigma_C]$ is a purely universal clause as well and therefore empty after universal reduction. This proves the claim.

We proceed to prove the lemma. If the algorithm does not learn a clause during conflict analysis, this must be due to a failed attempt at resolving an intermediate clause C with a clause R responsible for unit propagation. That is, if e is the existential pivot variable, there must be a universal variable $u \prec e$ such that $u \in var(C) \cap var(R)$ and $\{u, \neg u\} \subseteq C \cup R$. Towards a contradiction, suppose that $(u, e) \in D$. Let σ denote the assignment that caused the conflict and assume without loss of generality that $\{u, e\} \subseteq R$ and $\{\neg u, \neg e\} \subseteq C$. Since R caused propagation of e but $(u, e) \in D$, the variable u must have been assigned before e and $\neg u \in \sigma$. As the pivot $\neg e$ is the last existential literal falsified in C, it follows that $\neg u \in \sigma_C$. Because $\neg u \in C$, this implies that the assignment σ_C satisfies C, in contradiction with the claim proved above.

The number of dependencies and constraints is bounded by a function of the number n of variables, and QCDCL runs into a conflict at least every n variable assignments, so Observation 2 and Lemma 1 imply termination.

Theorem 1. *QCDCL with dependency learning is sound and terminating.*

5   Experiments

To see whether dependency learning works in practice, we implemented a QCDCL solver that supports this technique named Qute.[2] We evaluated the performance of Qute in several experiments. First, we measured the number of instances solved by Qute on benchmark sets from the 2016 QBF evaluation [26]. We compare these numbers with those of the best performing publicly available solvers for each benchmark set. In a second experiment, we computed the dependency sets given by the standard dependency scheme [4, 28] and the reflexive resolution-path dependency scheme [30, 32] for preprocessed instances, and compared their sizes to the number of dependencies learned by Qute. Finally, we revisit an instance family which is known to be hard to solve for QCDCL [13] and show they pose no challenge to Qute. For our experiments, we used a cluster with Intel Xeon E5649 processors at 2.53 GHz running 64-bit Linux.

5.1   Solved Instances for QBF Evaluation 2016 Benchmark Sets

In our first experiment, we used the prenex non-CNF (QCIR [16]) benchmark set from the 2016 QBF evaluation consisting of 890 formulas. Time and memory limits were set to 10 min and 4 GB, respectively. The results are summarized in Table 1 and Fig. 2. Qute is able to solve signficantly more instances within the timeout than the other solvers, and this appears to be in large part due to dependency learning: when dependency learning is deactivated, the number of solved instances drops significantly.

Table 1. Instances from the 2016 QBF evaluation prenex non-CNF (QCIR) benchmark set solved within 10 min.

Solver	Total	Sat	Unsat
Qute+dl	581	274	307
GhostQ	524	228	296
QuAbS	515	225	290
Qute	494	228	266
RAReQS	403	174	229

For our second experiment, we used the prenex CNF (PCNF) benchmark set from the 2016 QBF evaluation consisting of 825 instances. Time and memory limits were again set to 10 min and 4 GB. We performed this experiment twice: with and without preprocessing using bloqqer [6]. In order not to introduce variance in overall runtime through preprocessing, each instance was preprocessed only once and solvers were run on the preprocessed instances with a timeout corresponding to the overall timeout minus the time spent on preprocessing.

---

[2] http://github.com/perebor/qute.

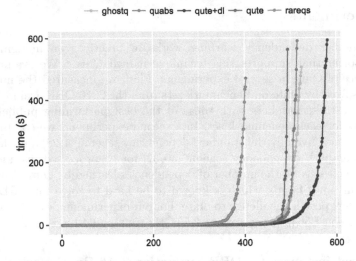

Fig. 2. Solved instances from the 2016 QBF evaluation prenex non-CNF (QCIR) benchmark set (x-axis) sorted by runtime (y-axis).

Since Qute shows good performance on QCIR instances, we included configurations that perform partial circuit reconstruction using $qcir\text{-}conv$[3] and then solve the resulting QCIR instance.

The results obtained without using bloqqer are shown on the left hand side of Table 2. When not using qcir-conv, Qute solves more instances with dependency learning than without dependency learning. Curiously, the opposite is the case when qcir-conv is used: in this case, Qute solves 3 more instances when dependency learning is turned off. Overall, we see that circuit reconstruction

Table 2. Instances from the QBF evaluation 2016 prenex CNF (PCNF) benchmark set solved within 10 min without preprocessing (left) and with preprocessing using bloqqer (right).

solver	total	sat	unsat
GhostQ	584	297	287
Qute+qcir-conv	538	283	255
Qute+dl+qcir-conv	535	277	258
DepQBF	451	200	251
Qute+dl	434	190	244
Qute	416	191	225
CAQE	358	167	191
RAReQS	335	128	207

solver	total	sat	unsat
RAReQS	615	299	316
DepQBF	585	294	291
CAQE	577	288	289
GhostQ	563	289	274
Qute+dl+qcir-conv	556	276	280
Qute+qcir-conv	541	266	275
Qute+dl	519	252	267
Qute	510	242	268

---

[3] http://www.cs.cmu.edu/~wklieber/qcir-conv.py.

(also used internally by GhostQ [18]) substantially increases the performance of Qute.

The results including preprocessing with bloqqer are shown on the right hand side of Table 2. With the exception of GhostQ, all solvers and configurations solve more instances when paired with bloqqer. However, the increase is less significant for Qute compared to other solvers, in particular in combination with qcir-conv. Notably, dependency learning increased the number of instances solved by Qute regardless of whether qcir-conv was used.

5.2   Learned Dependencies Compared to Dependency Relations

To get an idea of how well QCDCL with dependency learning is able to exploit independence, we compared the number of dependencies learned by Qute with the number of standard and resolution-path dependencies for instances from the PCNF benchmark set after preprocessing with bloqqer. We only considered instances with at least one quantifier alternation after preprocessing. Qute was run with a 10 min timeout (excluding preprocessing). If an instance was not solved we used the number of dependencies learned within that time limit.[4]

Summary statistics are shown in Table 3. On average, the standard dependency scheme does not provide a significant improvement over trivial dependencies. The reflexive resolution-path dependency scheme does better, but the high median shows that the set of trivial dependencies it can identify as spurious is still small in many cases. The fraction of learned dependencies is much smaller than either dependency relation on average, and the median fraction of trivial dependencies learned is even below 1%.

This indicates that proof search in QCDCL with dependency learning is less constrained than in QCDCL with either dependency scheme: since QCDCL is allowed to branch on a variable x only if every variable that x depends on has already been assigned, decision heuristics are likely to have a larger pool of variables to choose from if fewer dependencies are present.

Table 3. Learned dependencies, standard dependencies, and reflexive resolution-path dependencies for instances preprocessed by bloqqer, as a fraction of trivial dependencies.

Dependencies	Mean	Median	Variance
Learned	0.082	0.008	0.030
Standard	0.938	1.000	0.030
Resolution-path	0.660	0.942	0.172

---

[4] We cannot rule out that, for unsolved instances, Qute would have to learn a larger fraction of trivial dependencies before terminating. However, the solver tends to learn most dependencies at the beginning of a run, with the fraction of learned trivial dependencies quickly converging to a value that does not increase much until termination.

5.3 Dependency Learning on Hard Instances for QCDCL

For our third experiment, we chose a family of instances CR_n recently used to show that ordinary QCDCL does not simulate tree-like Q-resolution [13]. Since the hardness of these formulas is tied to QCDCL not propagating across quantifier levels, they represent natural test cases for QCDCL with dependency learning. We recorded the number of backtracks required to solve CR_n by Qute with and without dependency learning, for $n \in \{1, ..., 50\}$. As a reference, we used DepQBF.[5] For this experiment, we kept the memory limit of 4 GB but increased the timeout to one hour. The results are summarized in Fig. 3. As one would expect, Qute without dependency learning and DepQBF were only able to solve instances up to $n = 7$ and $n = 8$, respectively. Furthermore, it is evident from the plot that the number of backtracks grows exponentially with n for both solvers. By contrast, Qute with dependency learning was able to solve all instances within the timeout.

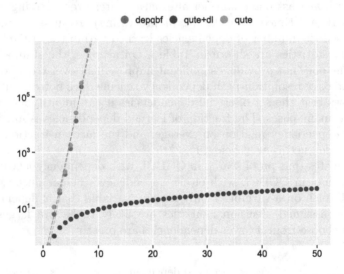

Fig. 3. Backtracks for instances CR_n based on the completion principle [13], as a function of n.

6 Discussion

In our experiments, Qute performed much better when presented with non-CNF input. In particular, dependency learning was most effective on the prenex non-CNF (QCIR) benchmark set, accounting for a 15% increase in the number of solved instances. Even for PCNF formulas, the best configuration(s)

---

[5] For sake of comparing with Qute in prefix mode, we disabled features recently added to DepQBF such as dynamic quantified blocked clause elimination [20] and oracle calls to the expansion-based solver Nenofex.

used tools for partially recovering circuit structure from CNF. This is consistent with the fact that Qute did not benefit from preprocessing nearly as much as other solvers, since preprocessing is known to adversely affect circuit reconstruction [11]. Whether this bias towards non-CNF representations is inherent to QCDCL with dependency learning or an artifact of other design choices implemented in our solver remains to be seen.

Dependency learning has several advantages over the use of dependency schemes within QCDCL: by retaining long-distance Q-resolution as its underlying proof system, the resulting algorithm is amenable to a simple correctness proof and supports linear-time strategy extraction. Moreover, our experiments indicate that proof search is less constrained with dependency learning, since typically only a small fraction of the dependencies computed by known dependency schemes has to be learned.

Sometimes, this additional freedom can be detrimental to performance, and a significant proportion of the overall runtime has to be spent on learning dependencies that are not spurious. To deal with such cases, we hope to find some middle ground between our current "tabula rasa" implementation of dependency learning and approaches that include too many spurious dependencies, by introducing a (small) set of dependencies that steer proof search in the right direction. For instance, Qute uses Tseitin conversion to obtain a set of initial clauses and terms from non-CNF (QCIR) instances. We found that assigning a Tseitin variable before a variable used in its definition often results in learning a dependency, so that it pays off to simply include dependencies of a Tseitin variable on the variables used in its definition from the start. For similar reasons, users may want to initialize D with pairs of variables that they know are dependent by construction. We hope to address this question by designing heuristics for "seeding" dependencies in a smart way as part of future work.

Acknowledgments. The authors thank Florian Lonsing for helpful discussions related to QCDCL. This research was kindly supported by FWF grants P27721 and W1255-N23.

References

1. Balabanov, V., Jiang, J.R.: Unified QBF certification and its applications. Formal Methods Syst. Des. **41**(1), 45–65 (2012)
2. Balabanov, V., Jiang, J.R., Janota, M., Widl, M.: Efficient extraction of QBF (counter)models from long-distance resolution proofs. In: Bonet, B., Koenig, S. (eds.) Proceedings of the Twenty-Ninth AAAI Conference on Artificial Intelligence, Austin, Texas, USA, 25–30 January 2015, pp. 3694–3701. AAAI Press (2015)
3. Biere, A.: Resolve and expand. In: Hoos, H.H., Mitchell, D.G. (eds.) SAT 2004. LNCS, vol. 3542, pp. 59–70. Springer, Heidelberg (2005). doi:10.1007/11527695_5
4. Lonsing, F., Biere, A.: A compact representation for syntactic dependencies in QBFs. In: Kullmann, O. (ed.) SAT 2009. LNCS, vol. 5584, pp. 398–411. Springer, Heidelberg (2009). doi:10.1007/978-3-642-02777-2_37

5. Lonsing, F., Biere, A.: Integrating dependency schemes in search-based QBF solvers. In: Strichman, O., Szeider, S. (eds.) SAT 2010. LNCS, vol. 6175, pp. 158–171. Springer, Heidelberg (2010). doi:10.1007/978-3-642-14186-7_14

6. Biere, A., Lonsing, F., Seidl, M.: Blocked clause elimination for QBF. In: Bjørner, N., Sofronie-Stokkermans, V. (eds.) CADE 2011. LNCS (LNAI), vol. 6803, pp. 101–115. Springer, Heidelberg (2011). doi:10.1007/978-3-642-22438-6_10

7. Cadoli, M., Schaerf, M., Giovanardi, A., Giovanardi, M.: An algorithm to evaluate Quantified Boolean Formulae and its experimental evaluation. J. Autom. Reason. **28**(2), 101–142 (2002)

8. Egly, U., Lonsing, F., Widl, M.: Long-distance resolution: proof generation and strategy extraction in search-based QBF solving. In: McMillan, K., Middeldorp, A., Voronkov, A. (eds.) LPAR 2013. LNCS, vol. 8312, pp. 291–308. Springer, Heidelberg (2013). doi:10.1007/978-3-642-45221-5_21

9. Faymonville, P., Finkbeiner, B., Rabe, M.N., Tentrup, L.: Encodings of bounded synthesis. In: Legay, A., Margaria, T. (eds.) TACAS 2017. LNCS, vol. 10205, pp. 354–370. Springer, Heidelberg (2017). doi:10.1007/978-3-662-54577-5_20

10. Giunchiglia, E., Narizzano, M., Tacchella, A.: Clause/term resolution and learning in the evaluation of Quantified Boolean Formulas. J. Artif. Intell. Res. **26**, 371–416 (2006)

11. Goultiaeva, A., Bacchus, F.: Recovering and utilizing partial duality in QBF. In: Järvisalo, M., Van Gelder, A. (eds.) SAT 2013. LNCS, vol. 7962, pp. 83–99. Springer, Heidelberg (2013). doi:10.1007/978-3-642-39071-5_8

12. Heule, M.J.H., Kullmann, O., Marek, V.W.: Solving and verifying the boolean pythagorean triples problem via cube-and-conquer. In: Creignou, N., Le Berre, D. (eds.) SAT 2016. LNCS, vol. 9710, pp. 228–245. Springer, Cham (2016). doi:10.1007/978-3-319-40970-2_15

13. Janota, M.: On Q-resolution and CDCL QBF solving. In: Creignou, N., Le Berre, D. (eds.) SAT 2016. LNCS, vol. 9710, pp. 402–418. Springer, Cham (2016). doi:10.1007/978-3-319-40970-2_25

14. Janota, M., Klieber, W., Marques-Silva, J., Clarke, E.: Solving QBF with counterexample guided refinement. In: Cimatti, A., Sebastiani, R. (eds.) SAT 2012. LNCS, vol. 7317, pp. 114–128. Springer, Heidelberg (2012). doi:10.1007/978-3-642-31612-8_10

15. Janota, M., Marques-Silva, J.: Solving QBF by clause selection. In: Yang, Q., Wooldridge, M. (eds.) Proceedings of the Twenty-Fourth International Joint Conference on Artificial Intelligence, IJCAI 2015, pp. 325–331. AAAI Press (2015)

16. Jordan, C., Klieber, W., Seidl, M.: Non-cnf QBF solving with QCIR. In: Darwiche, A. (ed.) Beyond NP, Papers from the 2016 AAAI Workshop. AAAI Workshops, vol. WS-16-05. AAAI Press (2016)

17. Kleine Büning, H., Karpinski, M., Flögel, A.: Resolution for quantified Boolean formulas. Inf. Comput. **117**(1), 12–18 (1995)

18. Klieber, W., Sapra, S., Gao, S., Clarke, E.: A non-prenex, non-clausal QBF solver with game-state learning. In: Strichman, O., Szeider, S. (eds.) SAT 2010. LNCS, vol. 6175, pp. 128–142. Springer, Heidelberg (2010). doi:10.1007/978-3-642-14186-7_12

19. Lonsing, F., Schemes, D., Solving, Search-Based QBF: Theory and Practice. PhD thesis, Johannes Kepler University, Linz, Austria, April 2012

20. Lonsing, F., Bacchus, F., Biere, A., Egly, U., Seidl, M.: Enhancing search-based QBF solving by dynamic blocked clause elimination. In: Davis, M., Fehnker, A., McIver, A., Voronkov, A. (eds.) LPAR 2015. LNCS, vol. 9450, pp. 418–433. Springer, Heidelberg (2015). doi:10.1007/978-3-662-48899-7_29

21. Lonsing, F., Biere, A.: Nenofex: expanding NNF for QBF solving. In: Kleine Büning, H., Zhao, X. (eds.) SAT 2008. LNCS, vol. 4996, pp. 196–210. Springer, Heidelberg (2008). doi:10.1007/978-3-540-79719-7_19

22. Malik, S., Zhang, L.: Boolean satisfiability from theoretical hardness to practical success. Commun. ACM **52**(8), 76–82 (2009)

23. Marques-Silva, J.P., Lynce, I., Malik, S.: Conflict-driven clause learning sat solvers. In: Biere, A., Heule, M., van Maaren, H., Walsh, T. (eds.) Handbook of Satisfiability, pp. 131–153. IOS Press (2009)

24. Meel, K.S., Vardi, M.Y., Chakraborty, S., Fremont, D.J., Seshia, S.A., Fried, D., Ivrii, A., Malik, S.: Constrained sampling and counting: universal hashing meets SAT solving. In Darwiche, A. (ed.) Beyond NP, Papers from the 2016 AAAI Workshop, Phoenix, Arizona, USA. AAAI Workshops, February 12, 2016, vol. WS-16-05. AAAI Press (2016)

25. Peitl, T., Slivovsky, F., Szeider, S.: Long distance Q-resolution with dependency schemes. In: Creignou, N., Le Berre, D. (eds.) SAT 2016. LNCS, vol. 9710, pp. 500–518. Springer, Cham (2016). doi:10.1007/978-3-319-40970-2_31

26. Pulina, L.: The ninth QBF solvers evaluation - preliminary report. In: Lonsing, F., Seidl, M. (eds.) Proceedings of the 4th International Workshop on Quantified Boolean Formulas (QBF 2016). CEUR Workshop Proceedings, vol. 1719, pp. 1–13. CEUR-WS.org (2016)

27. Rabe, M.N., Tentrup, L.: CAQE: a certifying QBF solver. In: Kaivola, R., Wahl, T. (eds.) Formal Methods in Computer-Aided Design - FMCAD 2015, pp. 136–143. IEEE Computer Soc. (2015)

28. Samer, M., Szeider, S.: Backdoor sets of quantified Boolean formulas. J. Autom. Reason. **42**(1), 77–97 (2009)

29. Scholl, C., Pigorsch, F.: The QBF solver AIGSolve. In: Lonsing, F., Seidl, M. (eds.) Proceedings of the 4th International Workshop on Quantified Boolean Formulas (QBF 2016). CEUR Workshop Proceedings, vol. 1719, pp. 55–62. CEUR-WS.org (2016)

30. Slivovsky, F., Szeider, S.: Soundness of Q-resolution with dependency schemes. Theoret. Comput. Sci. **612**, 83–101 (2016)

31. Tentrup, L.: Non-prenex QBF solving using abstraction. In: Creignou, N., Le Berre, D. (eds.) SAT 2016. LNCS, vol. 9710, pp. 393–401. Springer, Cham (2016). doi:10.1007/978-3-319-40970-2_24

32. Gelder, A.: Variable independence and resolution paths for quantified boolean formulas. In: Lee, J. (ed.) CP 2011. LNCS, vol. 6876, pp. 789–803. Springer, Heidelberg (2011). doi:10.1007/978-3-642-23786-7_59

33. Vizel, Y., Weissenbacher, G., Malik, S.: Boolean satisfiability solvers and their applications in model checking. Proc. IEEE **103**(11), 2021–2035 (2015)

34. Zhang, L., Malik, S.: Conflict driven learning in a quantified Boolean satisfiability solver. In: Pileggi, L.T., Kuehlmann, A. (eds.) Proceedings of the 2002 IEEE/ACM International Conference on Computer-Aided Design, ICCAD 2002, San Jose, California, USA, 10–14 November 2002, pp. 442–449. ACM/IEEE Computer Society (2002)

A Resolution-Style Proof System for DQBF

Markus N. Rabe[✉]

University of California, Berkeley, Berkeley, USA
rabe@berkeley.edu

Abstract. This paper presents a sound and complete proof system for
Dependency Quantified Boolean Formulas (DQBF) using resolution, universal reduction, and a new proof rule that we call fork extension. This
opens new avenues for the development of efficient algorithms for DQBF.

1 Introduction

Dependency quantified Boolean formulas (DQBF) extend quantified Boolean formulas (QBF) by *Henkin quantifiers* $\exists x : Y.\ \varphi$, which bind a variable x and also
specify a *dependency set* Y of universal variables that x may depend on [17,22].
Henkin quantifiers allow us to succinctly express the existence of functions satisfying quantified constraints. For example, we can formulate the existence of a
function $add : \mathbb{B}^{32} \times \mathbb{B}^{32} \to \mathbb{B}^{32}$ implementing the axioms of addition in 32-bit
arithmetic: $\forall x.\ add(x,0) = x \wedge \forall x,y.\ add(x,y+1) = add(x,y)+1$, where $x+1$
stands for the increment-by-one circuit encoded as constraints. The formula can
be rewritten into DQBF syntax without much overhead, as we discuss in Sect. 3.

DQBF enables elegant encodings for applications such as bounded reactive
synthesis [9] and partial equivalence checking of circuits [14]. Its elegance and
expressiveness make DQBF a potential candidate logic to serve as the interface between algorithms and applications in synthesis, verification, and artificial
intelligence that require impractically large encodings in other logics such as
QBF and propositional Boolean logic (SAT). Recently, there has been a surge
of interest in practical algorithms for DQBF [11–15,32], but their performance
is still unsatisfactory [8]. We argue that this is due to the lack of suitable proof
systems.

Resolution is one of the fundamental proof rules for first-order logic [7]. It
says that given two clauses $(x \vee a_1 \vee \cdots \vee a_n)$ and $(\neg x \vee b_1 \vee \cdots \vee b_m)$ we can
infer the clause $(a_1 \vee \cdots \vee a_n \vee b_1 \vee \cdots \vee b_m)$, which we call the *resolvent*. The
best known application of resolution is the conflict analysis step of the conflict-
driven clause learning (CDCL) algorithm, which is the basis for modern SAT
solvers [26].

While resolution is sufficient to prove or disprove any propositional Boolean
formula, we need a second proof rule for QBF. The *universal reduction* rule says
that we can delete a literal of a universal variable from a clause, if no variable in
the clause may depend on it, i.e. no existential variable in this clause is bound in
the scope of the universal variable. Resolution and universal reduction form the

© Springer International Publishing AG 2017
S. Gaspers and T. Walsh (Eds.): SAT 2017, LNCS 10491, pp. 314–325, 2017.
DOI: 10.1007/978-3-319-66263-3_20

Q-resolution proof system, which plays a similar role for QBF as resolution plays for SAT [6]. Some of the recently popular algorithms for QBF, QCDCL [20,33], CEGAR [18,19,24,27], and Incremental Determinization [23], can be phrased as variants of Q-Resolution [28].

While resolution is complete for propositional Boolean logic, and Q-resolution is complete for QBF [6], resolution and Q-resolution were shown to be incomplete for DQBF [1,2]. The only known complete proof systems for DQBF resort to expansion or (similarly) annotating literals with partial instantiations of the universal variables [2]. While expansion is widely used in preprocessors for QBF [4,32], using expansion as a solving technique is less popular due to its often excessive memory requirements. Instead, most of the popular solvers for SAT and QBF rely on resolution in their reasoning. A resolution-style proof system for DQBF might therefore help to lift efficient algorithms to DQBF.

In this paper, we present a resolution-style proof system for DQBF, called *Fork-Resolution*. It uses resolution, universal reduction, and, to achieve completeness, introduces a new proof rule called *fork extension*. Fork extension is derived from the extension rule from extended resolution [29] and says that, given a clause $(a_1 \vee \cdots \vee a_n \vee b_1 \vee \cdots \vee b_m)$, we can infer the clauses $(x \vee a_1 \vee \cdots \vee a_n)$ and $(\neg x \vee b_1 \vee \cdots \vee b_m)$, where x is a fresh variable. The key insight in this paper is that it is sufficient to apply extension to a specific type of clauses that we call *information forks*.[1]

An information fork is a clause that contains variables with incomparable dependency sets. Consider a DQBF with a single clause $\exists x_1 : \{y_1\}.\ \exists x_2 : \{y_2\}.\ (x_1 \vee x_2)$. To satisfy the clause with a Skolem function, we have to decide in which cases the clause is satisfied by x_1 and in which cases x_2 is responsible for satisfying the clause. As x_1 and x_2 have disjoint dependency sets, they cannot depend on the value of the other variable. So the only way to make sure the clause is always satisfied is to require that one of x_1 and x_2 is *always* responsible for satisfying the clause. In other words, there must be a constant that indicates which side of the clause is responsible for satisfying the clause. With the fork extension rule, we introduce a variable x that represents this constant. We split the clause into two clauses $(x_1 \vee x)$ and $(\neg x \vee x_2)$ and introduce x with the empty dependency set $(\exists x : \emptyset)$. In general, the decision of which side of the clause has to satisfy the clause must be decided based only on the information that is available to both literals, i.e. the intersection of their dependency sets.

This paper is structured as follows. We introduce basic notation in Sect. 2 and provide an encoding of functions in DQBF in Sect. 3. In Sect. 4, we introduce the Fork-Resolution proof system. Section 5 demonstrates that Fork-Resolution can disprove a formula for which Q-resolution is incomplete. The proofs of soundness and completeness can be found in Sect. 6. We briefly discuss how Fork-Resolution is separated exponentially from other proof systems for DQBF in Sect. 7. Section 8 discusses related work.

---

[1] The concept is loosely connected to information forks in reactive synthesis of distributed systems [10].

2   Dependency Quantified Boolean Formulas

Dependency quantified Boolean formulas (DQBFs) over a set of variables V are generated by the following grammar with starting symbol ψ:

$$\psi := \exists v : \{v, \ldots, v\}. \psi \mid \varphi$$
$$\varphi := 0 \mid 1 \mid v \mid \neg\varphi \mid \varphi \vee \varphi \mid \varphi \wedge \varphi,$$

where $v \in V$, and $\{v, \ldots, v\}$ stands for a finite subset of V. Quantifiers in DQBF $\exists x : \{y_1, \ldots, y_n\}$ specify a single *existential variable* x and also a finite set of *universal variables* $\{y_1, \ldots, y_n\}$ that x may depend on. We call $\{y_1, \ldots, y_n\}$ also the *dependency set* of x, denoted $dep(x)$. The variables that occur in dependency sets are the *universal variables*. All variables that are not universal are called *existential variables*. We say that a quantifier $\exists x : Y. \varphi$ *binds* the existential variable x in its subformula φ. To simplify the discussion, we assume that existential variables are bound at most once and that universal variables are never bound. A DQBF is *closed* if it only contains universal variables and bound existential variables.

In the following we define the semantics of DQBF for which we assume that the reader is familiar with the natural semantics of propositional Boolean formulas. An *assignment* $\mathbf{y}$ to a set of variables Y is a function $\mathbf{y} : Y \to \mathbb{B}$ that maps each variable $y \in Y$ to either 1 or 0. Given a set of variables Y, we denote the set of assignments to Y with 2^Y. Given a propositional formula φ over variables X and an assignment $\mathbf{x}'$ for $X' \subseteq X$, we define $\varphi(\mathbf{x}')$ to be the formula obtained by replacing the variables X' by their truth value in $\mathbf{x}'$. By $\varphi(\mathbf{x}', \mathbf{x}'')$ we denote the replacement by multiple assignments for disjoint sets $X', X'' \subseteq X$. A *Skolem function* f_x maps assignments to $dep(x)$ to assignments to x, interpreted as assignments to x. We define the truth of a DQBF φ with existential variables $X = \{x_1, \ldots, x_n\}$ and universal variables Y as the existence of Skolem functions $f_X = \{f_{x_1}, \ldots, f_{x_n}\}$, such that $\varphi(\mathbf{y}, f_X(\mathbf{y}))$ is satisfiable for every assignment $\mathbf{y}$ to Y.

A *literal* l is either a variable $x \in X$, or its negation $\neg x$. We use $\bar{l}$ to denote the *complement* operation that gives the literal with the negated value of l and we define $var(l)$ to be the variable that the literal contains. Given a set of literals $\{l_1, \ldots, l_n\}$, their disjunction $(l_1 \vee \ldots \vee l_n)$ is called a *clause*. As usual in the community, we use set notation for the manipulation of clauses. A clause C is called *tautological* if it contains a literal l and its complement $\bar{l}$. We use $vars(C)$ to denote the variables occurring in C.

A propositional formula is in conjunctive normal form (CNF), if it is a conjunction of clauses. A closed DQBF is in CNF if its propositional subformula is in CNF. Every DQBF φ can be transformed into a closed DQBF in CNF with size $O(|\varphi|)$, preserving satisfiability. In the rest of this work we assume DQBFs to be in CNF.

We extend the notation of dependency sets to negated variables (and thus literals), $dep(\neg v) = dep(v)$. Applied to clauses, $dep(C)$ denotes the union of the universal variables and the dependencies of the existential variables in C.

3   On Quantification over Functions

Our interest in DQBF is rooted in the fact that we can encode existential quantification over functions. In the following, we present a transformation that allows us to eliminate functions from a formula. For the (first-order) theory of equality and uninterpreted functions, there is a similar transformation, called *Ackermannization* [5]. Unlike Ackermannization, the transformation below is linear in the size of the formula.

Consider a formula $\varphi = \exists f : \mathbb{B}^N \to \mathbb{B} . \psi$. We can turn that into an equivalent formula φ' that does not use f. Let f have k function applications $f(e_{1,1}, \ldots, e_{1,N}), \ldots, f(e_{k,1}, \ldots, e_{k,N})$ in ψ, where $e_{i,j}$ are terms. We introduce fresh variables $f_1, \ldots, f_k$ and dependency sets $X_1, \ldots, X_k$, each $X_i = \{x_{i,1}, \ldots, x_{i,N}\}$ consisting of N fresh variables, for the function applications. We then define φ' as

$$\exists f_1 : X_1. \ \ldots \ \exists f_k : X_k.$$
$$\bigwedge_{i=2}^{k} \left(\left(\bigwedge_{j=1}^{N} x_{1,j} = x_{i,j} \right) \implies f_1 = f_i \right) \wedge$$
$$\left[\left(\bigwedge_{i=1}^{k} \bigwedge_{j=1}^{N} e_{i,j} = x_{i,j} \right) \implies \psi[\{f(e_{i,1}, \ldots, e_{i,N})/f_i \mid 1 \leq i \leq k\}] \right]$$

The conjunct $\bigwedge_{i=2}^{k} \left(\left(\bigwedge_{j=1}^{N} x_{1,j} = x_{i,j} \right) \implies f_1 = f_i \right)$ requires that variables $f_1, \ldots, f_k$ represent the same function. If f_1 and f_i have different values for some "input" x^*, we set $x_1 = x_i = x^*$ and see that $f_1 = f_i$ is violated. Transitivity of equality allows us to avoid encoding the pairwise equality explicitly.

The second conjunct of the formula states that if $f_1, \ldots, f_k$ represent the outputs of this function for the k function applications, the formula where we replace the function applications by the f_i has to hold.

The constraints can be transformed into a CNF of linear size with the Tseitin transformation [29].

If φ contains multiple functions they can be replaced independently, by first introducing new variables such that no argument of a function contains a function application. Then replacing one function does not affect the function applications of other functions. So, the linear increase in size does not multiply over multiple applications of the function elimination and we only get a linear increase in the size of the formula overall.

4   The Fork-Resolution Proof System

The Fork-Resolution proof system consists of the following three rules:

Resolution [7]:
$$\frac{C_1 \cup \{l\} \qquad C_2 \cup \{\bar{l}\}}{C_1 \cup C_2} \quad \text{(Res)}$$

We call the clause $(C_1 \setminus \{l\}) \cup (C_2 \setminus \{\bar{l}\})$ that is obtained by the resolution rule the *resolvent* of C_1 and C_2 and we call $var(l)$ the *pivot*. The resolution rule is only applied when the resolvent is not a tautology, i.e. does not contain two complementary literals.

Universal Reduction [6]:

$$\frac{C \quad var(l) \notin dep(C\setminus\{l\}) \quad \bar{l} \notin C \quad var(l) \text{ is universal}}{C\setminus\{l\}} \text{(} \forall \text{Red)}$$

Fork Extension:

$$\frac{C_1 \cup C_2 \quad dep(C_1) \not\subseteq dep(C_2) \quad dep(C_1) \not\supseteq dep(C_2) \quad x \text{ is fresh}}{\exists x : dep(C_1) \cap dep(C_2). \; C_1 \cup \{x\} \; \wedge \; C_2 \cup \{\neg x\}} \text{(FEx)}$$

The fork extension rule introduces a new quantified variable x to split a clause into two parts. The dependency set of x is defined as the intersection of $dep(C_1)$ and $dep(C_2)$. We only apply fork extension to clauses $C_1 \cup C_2$ that consist of two parts have *incomparable dependency sets* ($dep(C_1) \not\subseteq dep(C_2)$ and $dep(C_1) \not\supseteq dep(C_2)$). We call such clauses *information forks*.

The idea is that for each assignment to the universal variables one of the two parts C_1 and C_2 is "responsible" for satisfying the original clause $C_1 \cup C_2$. The variables in C_1 and C_2, however, may have different dependency sets, and so they must coordinate their responsibility only based on the information that is common to them.

5 Example

We demonstrate the proof system along an example. We use a shortened version of the example used to show incompleteness of Q-resolution for DQBF [1]. The formula states $\exists y_1 : \{x_1\}. \; \exists y_2 : \{x_2\}. \; (x_1 \wedge x_2) \leftrightarrow (y_1 = y_2)$. The formula states that the existential variables y_1 and y_2 have to be equal iff x_1 and x_2 are true. But y_1 and y_2 can each only see one of x_1 and x_2 and thus cannot coordinate to satisfy the constraint. That is, the formula is false. The propositional part of the formula can be represented in CNF as follows:

$$y_1 \vee y_2 \vee x_1 \tag{1}$$

$$\neg y_1 \vee \neg y_2 \vee x_1 \tag{2}$$

$$y_1 \vee y_2 \vee x_2 \tag{3}$$

$$\neg y_1 \vee \neg y_2 \vee x_2 \tag{4}$$

$$y_1 \vee \neg y_2 \vee \neg x_1 \vee \neg x_2 \tag{5}$$

$$\neg y_1 \vee y_2 \vee \neg x_1 \vee \neg x_2 \tag{6}$$

Despite the formula being false, we can see with a little effort that all resolvents of the clauses above are tautologies. This demonstrates that Q-resolution for DQBF is incomplete [1].

The Fork-Resolution proof system can disprove the formula as we show in the following. Variables y_1 and y_2 are leaf-existentials, and all six clauses are information forks. Applying FEx to clauses (1)–(6) introduces variables t_1 to t_6 with empty dependencies ($\exists t_1 : \emptyset. \ldots \exists t_6 : \emptyset.$) and yields the following clauses:

$$t_1 \vee y_1 \vee x_1 \quad (1a) \qquad \neg t_1 \vee y_2 \quad (1b)$$
$$t_2 \vee \neg y_1 \vee x_1 \quad (2a) \qquad \neg t_2 \vee \neg y_2 \quad (2b)$$
$$t_3 \vee y_1 \quad (3a) \qquad \neg t_3 \vee y_2 \vee x_2 \quad (3b)$$
$$t_4 \vee \neg y_1 \quad (4a) \qquad \neg t_4 \vee \neg y_2 \vee x_2 \quad (4b)$$
$$t_5 \vee y_1 \vee \neg x_1 \quad (5a) \qquad \neg t_5 \vee \neg y_2 \vee \neg x_2 \quad (5b)$$
$$t_6 \vee \neg y_1 \vee \neg x_1 \quad (6a) \qquad \neg t_6 \vee y_2 \vee \neg x_2 \quad (6b)$$

Next, we derive six resolvents as listed below. Their names indicate the pair of clauses they originate from. (We drop universal variables by universal reduction.)

$$t_1 \vee t_4 \quad (1a4a) \qquad \neg t_1 \vee \neg t_5 \quad (1b5b)$$
$$t_3 \vee t_6 \quad (3a6a) \qquad \neg t_2 \vee \neg t_6 \quad (2b6b)$$
$$t_4 \vee t_5 \quad (4a5a) \qquad \neg t_3 \vee \neg t_4 \quad (3b4b)$$

Resolving clauses $(4a5a)$ and $(1b5b)$ gives us $(\neg t_1 \vee t_4)$, which we resolve with $(1a4a)$ to get t_4. Similarly, we resolve clauses $(3a6a)$ and $(2b6b)$ to get $(\neg t_2 \vee t_3)$, which we resolve with $(t_2 \vee t_3)$ to get t_3. Resolving t_3 and t_4 with $(3b4b)$ derives the empty clause, which shows that the formula is false.

6   Proofs of Soundness and Completeness

The central insight of this paper is that when we eliminate information forks from a DQBF, we can eliminate existential variables using resolution like in QBF. Eliminating variables may introduce new information forks, so in general we have to alternate fork extension and resolution.

An important technique we are going to use is that variables will be defined only in terms of the variables that they share clauses with. The following definitions help us to zoom in on the neighborhood of a variable.

Definition 1 (Projection of assignments). Given two sets of variables X and X' with $X' \subseteq X$ and an assignment $\mathbf{x}$ to X. We call an assignment $\mathbf{x}'$ to X' the *projection* of $\mathbf{x}$ to X', if $\mathbf{x}'(x) = \mathbf{x}(x)$ for all $x \in X'$. We denote the projection of $\mathbf{x}$ to X' with $\mathbf{x}|_{X'}$.

Definition 2 (Projection of Skolem functions). Let φ be a true DQBF in PCNF, let C be a clause in φ, and let $f : 2^Y \rightarrow 2^X$ be a Skolem function. We call a function $f_C : 2^{dep(C)} \rightarrow 2^{vars(C)}$ the *projection* of f to C, if for all assignments $\mathbf{y}$ to Y and variables x in C it holds $f(\mathbf{y})(x) = f_C(\mathbf{d})(x)$, where $\mathbf{d}$ is the projection $\mathbf{d}$ of $\mathbf{y}$ to $dep(C)$. We denote the projection of f to C with $f|_C$.

In the following lemma we recall the soundness of resolution and universal reduction and we prove the soundness of the fork extension rule.

Lemma 1. *The fork-resolution proof system for DQBF is sound.*

Proof. We show soundness for each proof rule individually.

Res. The resolution rule is as usual, with the exception that it is possible to apply resolution to tautological clauses, in which case the resolvent is subsumed by of $C_1 \cup \{l\}$ or $C_2 \cup \{\bar{l}\}$. For all other clauses soundness follows from the soundness of QU-resolution for DQBF [2,30].

∀Red. (Similarly in [1]). Let l be a literal of a universal variable x and let C be a clause with $l \in C$ and $\bar{l} \notin C$ and let $x \notin dep(C)$. If the DQBF is true, there is a function f_C mapping assignments $\boldsymbol{d}$ to $dep(C)$ to values $f_C(\boldsymbol{d})$ for the existential variables in C, such that clause C is satisfied. In particular, C is satisfied for each assignment to the universals setting l to 0 because $\bar{l} \notin C$. Since f_C is independent of x, we know that after removing l from C, the clause is still satisfied.

FEx. Let $\mathcal{Q}.\varphi$ be a true DQBF in PCNF with quantifier prefix $\mathcal{Q}$, universal variables Y, and existential variables X, let $C_1 \cup C_2$ be a clause in φ, and let x be a fresh variable. We show that $\mathcal{Q}.\exists x : dep(C_1) \cap dep(C_2).\ \varphi \wedge C_1 \cup \{x\} \wedge C_2 \cup \{\neg x\}$ is true by constructing a Skolem function $f_x : dep(C_1) \cap dep(C_2) \rightarrow 2^{\{x\}}$ for x that together with f_C, satisfies the (new) constraints.

Consider an arbitrary Skolem function f for $\mathcal{Q}.\varphi$ and let $\mathbf{y}$ be an assignment to Y. We fix $f_x(\mathbf{y}|_{dep(x)})$ to be 1, if C_1 is not satisfied by $\mathbf{y}$ or $f(\mathbf{y})$, and we fix $f_x(\mathbf{y}|_{dep(x)})$ to be 0, if C_2 is not satisfied by $\mathbf{y}$ or $f(\mathbf{y})$. In case both C_1 and C_2 are satisfied, we (arbitrarily) fix x to be 1. The case that neither C_1 and C_2 are satisfied contradicts the fact that f is a Skolem function.

Let us assume that for the given $\mathbf{y}$ and Skolem function f one part of the clause is violated, which assume w.l.o.g. to be C_1. To prove the correctness of f_x we have to show that there cannot exist a second assignment $\mathbf{y}'$ to Y that agrees with $\mathbf{y}$ on the variables $dep(C_1) \cap dep(C_2)$ but violates C_2 instead. Assuming that there is such an assignment $\mathbf{y}'$, we can construct an assignment $\mathbf{y}''$ to Y that agrees with $\mathbf{y}$ on *all but* the variables of $dep(C_2)$, and agrees with $\mathbf{y}'$ on the variables of $dep(C_2)$. Since $\mathbf{y}$ and $\mathbf{y}''$ agree on $dep(C_1) \cap dep(C_2)$, assignment $\mathbf{y}''$ violates both C_1 and C_2, which contradicts the satisfaction of $C_1 \cup C_2$. The soundness of the FEx rule follows by way of contradiction.   □

The Lemmas 2 and 4 show that after fork extension and variable elimination the clauses to which they were applied become obsolete and can be removed. Even though the proof system does not allow us to remove clauses this will be an important tool to understand which clauses we do not have to worry about any more.

Lemma 2. *Let $\mathcal{Q}.\varphi \wedge C_1 \cup C_2$ be a DQBF in PCNF with quantifier prefix $\mathcal{Q}$ and let x be a fresh variable. Then $\mathcal{Q}.\varphi \wedge C_1 \cup C_2$ and $\mathcal{Q}.\exists y : dep(C_1) \cap dep(C_2).\ \varphi \wedge C_1 \cup \{x\} \wedge C_2 \cup \{\neg x\}$ are equivalent.*

Proof. Soundness of the FEx rule was proven in Lemma 1. The other direction follows from the soundness of resolution and the soundness of removing constraints and unused variables.   □

Variable elimination by resolution is a well known technique for SAT and QBF [3,6]. Given a propositional formula φ in CNF over variables X we can eliminate a variable $x \in X$ by adding all resolvents for pivot x and removing all clauses with literals of x. We denote the formula obtained through variable elimination $\mathsf{elim}(x, \varphi)$.

Proposition 1 (Variable elimination [3,6]). *Let φ be a propositional formula in CNF over variables X and a variable $x \in X$. Then for all assignments $\mathbf{x}$ to $X \backslash \{x\}$ we have that $\mathsf{elim}(x, \varphi)(\mathbf{x})$ if, and only if, $\exists x. \varphi(\mathbf{x})$.*

To lift variable elimination to DQBF we need a stronger property of resolution. Instead of expressing the notion of equivalence between φ and $\mathsf{elim}(x, \varphi)$ via existential quantification, we provide a function with minimal dependencies resolving the quantification.

Lemma 3 (Variable elimination with dependencies). *Let φ be a propositional formula in CNF over variables X and a variable $x \in X$ and let $X' \subseteq X$ be the set of variables occurring together with x in some clause in φ but not x itself. Then there is a function f_x mapping assignments to X' to assignments to x such that for all assignments $\mathbf{x}$ to $X \backslash \{x\}$ we have that $\mathsf{elim}(x, \varphi)(\mathbf{x})$ if, and only if, $\varphi(\mathbf{x}, f_x(\mathbf{x}|_{X'}))$.*

Proof. Let φ' be the clauses that contain a literal of x. For every assignment $\mathbf{x}'$ to X', we define $f_x(\mathbf{x}')$ as the value that satisfies φ', if one exists. If there is no such value, we arbitrarily fix $f_x(\mathbf{x}')$ to be 1.

"$\Longleftarrow$": Whenever φ is true, also $\mathsf{elim}(x, \varphi)$ must be true, because of the soundness of the elimination rule and because removing clauses only makes it easier to satisfy the formula.

"$\Longrightarrow$": Let $\mathbf{x}$ be an assignment to $X \backslash \{x\}$ such that $\varphi(\mathbf{x}, f_x(\mathbf{x}|_{X'}))$ is false. If a clause that does not contain x is violated, then also $\mathsf{elim}(x, \varphi)(\mathbf{x})$ is false, as it contains the same clause. Otherwise, a clause C containing x is violated. By choice of f_x we know that x is chosen to be 1, but also for value 0 not all constraints could be satisfied. Thus there must be a clause C' that is violated if we changed the assignment of x. The resolvent of C and C' is violated for $\mathbf{x}$. $\square$

Next we show that leaf-existentials in DQBFs without information forks can be eliminated through resolution. We call an existential variable x in a DQBF φ a *leaf-existential* if all existentials x' in φ have a smaller or incomparable dependency set than x ($dep(x) \not\subseteq dep(x')$).

Lemma 4 (Similarly in [31], Theorem 4). *Let $\psi = \exists x_1 : Y_1 \ldots \exists x_n : Y_n. \varphi$ be a DQBF without information forks. Further let x_1 be a leaf-existential. Then there is an equivalent DQBF $\psi' = \exists x_2 : Y_2 \ldots \exists x_n : Y_n. \mathsf{elim}(x_1, \varphi)$.*

Proof. The soundness of resolution immediately gives us that $\psi \implies \psi'$. We prove $\psi' \implies \psi$ in the following:

Let φ'' be the conjunction of all resolvents with pivot x_1. Let ψ' be true and let $f_{x_2}, \ldots, f_{x_k}$ be the Skolem functions proving ψ' to be true. Further, let X' be the variables occurring in φ'. By Lemma 3 we obtain a function f_{x_1} for x_1 mapping the assignments to $X'\backslash\{x_1\}$ to $\mathbb{B}$ such that $\varphi'[x_1/f_{x_1}(Y')]$ and φ'' are equivalent. We can transform f_{x_1} to a Skolem function for x_1 with domain Y_1 by inlining the definitions for the other Skolem functions, as $dep(X'\backslash\{x_1\}) \subseteq dep(x_1)$ due to the lack of information-forks and x_1 being a leaf-existential. □

Theorem 1 shows how the lemmas can be put into action. We give a sound and complete algorithm that just applies the proof rules of Fork Resolution. Like algorithms for SAT and QBF that just rely on variable elimination, this algorithm is only a theoretical possibility and will likely not scale to interesting applications. Yet, more efficient algorithms may be possible using the same proof rules, just as CDCL is based on the resolution proof system.

Theorem 1. *A DQBF is false if, and only if, we can derive an empty clause in the Fork-Resolution proof system.*

Proof. We provide an round-based algorithm based on the proof rules. Each round consists of two steps. The first step is to eliminate all information forks using Lemma 2. The second step is to eliminate a leaf-existential with a dependency set of maximal size using Lemma 4. The proof system does not allow us to remove clauses as suggested in Lemmas 2 and 4, but additional clauses can never impede the derivation of the empty clause.

The termination guarantee for the (Q-)resolution proof systems for SAT and QBF can be easily established, since variable elimination reduces the number of variables in the formula. However, Fork-Resolution introduces variables with the fork extension rule, so its termination guarantee is based on the dependency sets that occur in the formula: Each round reduces either the number of existentials with a maximally sized dependency set or the size of the maximally sized dependency set. The lexicographic ordering of the two provides the termination relation. □

7 Separation from Other Proof Systems

Our definition of resolution admits universal variables as pivots and thus includes QU-resolution [30]. This already gives us the result that Fork-Resolution has exponentially smaller proofs for some formulas than D-IR-calc and ∀Exp-Res [2]. For the other direction consider a QBF, where the FEx rule cannot be applied and Fork-Resolution coincides with QU-resolution. IR-calc and ∀Exp-Res are exponentially more succinct on some formulas than QU-resolution [2].

The question whether without resolution on universal variables Fork-Resolution proofs can be exponentially more succinct than proofs by D-IR-calc is open.

8 Related Work

Previous (sound and complete) proof systems for DQBF rely on universal expansion or instantiation [2], which is also the basis for DQBF solvers, such as iDQ [13]. Earlier DQBF solvers relied on a search procedure generalizing DPLL [12], on generating refutation proofs of increasing size [11], or on expansion in symbolic datastructures [15].

In QBF, *clausal abstraction* is a technique to split clauses with variables in order to separate the quantifiers [18,24]. Clausal abstraction is applied when the dependency sets of two parts of a clause are different but ordered, while FEx is applied when they are incomparable.

The Bernays-Schönfinkel class of first-order logic, also called the effectively propositional fragment (EPR), and DQBF are related in that they share the same complexity class NEXPTIME [21]. Translations between the two logics are known. Like iDQ, EPR solvers rely on instantiation and a number of proof rules that are quite different from SAT solvers and QBF solvers. A recent attempt to run an EPR solver in the QBF competition suggested that EPR is not competitive on the type of problems in the QBF libraries [16]. Hence, by lifting the resolution-based solver technologies from SAT and QBF to DQBF, Fork-Resolution may enable solving new classes of problems.

9 Conclusion

The beauty of DQBF lies it its unified representation of propositional variables, quantified variables, and functions. It offers incredible succinctness but still presents a big challenge for the development of practical algorithms. In a recent work solvers for SAT, QBF, and DQBF were compared on encodings of the same problem [9]. The experiment showed that current DQBF solvers on their compact encodings fall far behind QBF solvers on the longer QBF encoding, and even SAT solver on the much longer propositional encoding of the same problem. This suggests that the current solving technologies for DQBF are *unable to leverage the succinctness* of DQBF encodings. Arguably, instantiation and expansion—the currently used techniques for solving DQBF—are aimed at making a DQBF more QBF-like. So it is maybe not that surprising that directly encoding the problem in QBF outperforms the DQBF approach.

In this paper, we presented an alternative approach to DQBF. The Fork-Resolution proof system is sound and complete and lifts resolution to *reasoning about functions* without instantiating their inputs. This opens new paths in the development of practical algorithms for DQBF. It remains to be shown that efficient algorithms can be built based on Fork-Resolution.

Acknowledgements. The author expresses his gratitude to Armin Biere, Benjamin Caulfield, Daniel J. Fremont, Martina Seidl, Sanjit A. Seshia, Martin Suda, Leander Tentrup, and Ralf Wimmer for supportive comments and detailed discussions on this work.

This work was supported in part by NSF grants CCF-1139138, CNS-1528108, and CNS-1646208, and by TerraSwarm, one of six centers of STARnet, a Semiconductor Research Corporation program sponsored by MARCO and DARPA.

References

1. Balabanov, V., Chiang, H.J.K., Jiang, J.H.R.: Henkin quantifiers and Boolean formulae: a certification perspective of DQBF. Theor. Comput. Sci. **523**, 86–100 (2014)
2. Beyersdorff, O., Chew, L., Schmidt, R.A., Suda, M.: Lifting QBF resolution calculi to DQBF. In: Creignou, N., Le Berre, D. (eds.) SAT 2016. LNCS, vol. 9710, pp. 490–499. Springer, Cham (2016). doi:10.1007/978-3-319-40970-2_30
3. Biere, A.: Resolve and expand. In: Hoos, H.H., Mitchell, D.G. (eds.) SAT 2004. LNCS, vol. 3542, pp. 59–70. Springer, Heidelberg (2005). doi:10.1007/11527695_5
4. Biere, A., Lonsing, F., Seidl, M.: Blocked clause elimination for QBF. In: Bjørner, N., Sofronie-Stokkermans, V. (eds.) CADE 2011. LNCS (LNAI), vol. 6803, pp. 101–115. Springer, Heidelberg (2011). doi:10.1007/978-3-642-22438-6_10
5. Bruttomesso, R., Cimatti, A., Franzén, A., Griggio, A., Santuari, A., Sebastiani, R.: To Ackermann-ize or not to Ackermann-ize? On efficiently handling uninterpreted function symbols in $SMT(\mathcal{EUF} \cup \mathcal{T})$. In: Hermann, M., Voronkov, A. (eds.) LPAR 2006. LNCS, vol. 4246, pp. 557–571. Springer, Heidelberg (2006). doi:10.1007/11916277_38
6. Buning, H.K., Karpinski, M., Flogel, A.: Resolution for quantified boolean formulas. Inf. Comput. **117**(1), 12–18 (1995)
7. Davis, M., Putnam, H.: A computing procedure for quantification theory. J. ACM (JACM) **7**(3), 201–215 (1960)
8. Faymonville, P., Finkbeiner, B., Rabe, M.N., Tentrup, L.: 3 encodings of reactive synthesis. In: Proceedings of QUANTIFY, pp. 20–22 (2015)
9. Faymonville, P., Finkbeiner, B., Rabe, M.N., Tentrup, L.: Encodings of bounded synthesis. In: Legay, A., Margaria, T. (eds.) TACAS 2017. LNCS, vol. 10205, pp. 354–370. Springer, Heidelberg (2017). doi:10.1007/978-3-662-54577-5_20
10. Finkbeiner, B., Schewe, S.: Uniform distributed synthesis. In: Proceedings of LICS, Washington, DC, USA, pp. 321–330. IEEE Computer Society (2005)
11. Finkbeiner, B., Tentrup, L.: Fast DQBF refutation. In: Sinz, C., Egly, U. (eds.) SAT 2014. LNCS, vol. 8561, pp. 243–251. Springer, Cham (2014). doi:10.1007/978-3-319-09284-3_19
12. Fröhlich, A., Kovásznai, G., Biere, A.: A DPLL algorithm for solving DQBF. In: Proceedings of Pragmatics of SAT 2012 (2012)
13. Fröhlich, A., Kovásznai, G., Biere, A., Veith, H.: iDQ: instantiation-based DQBF solving. In: Proceedings of Pragmatics of SAT, pp. 103–116 (2014)
14. Gitina, K., Reimer, S., Sauer, M., Wimmer, R., Scholl, C., Becker, B.: Equivalence checking of partial designs using dependency quantified Boolean formulae. In: Proceedings of ICCD, pp. 396–403, October 2013
15. Gitina, K., Wimmer, R., Reimer, S., Sauer, M., Scholl, C., Becker, B.: Solving DQBF through quantifier elimination. In: Proceedings of DATE (2015)
16. Giunchiglia, E., Narizzano, M., Pulina, L., Tacchella, A.: Quantified Boolean formulas satisfiability library (QBFLIB) (2005). www.qbflib.org
17. Henkin, L.: Some remarks on infinitely long formulas. J. Symb. Logic **30**, 167–183 (1961)

18. Janota, M., Marques-Silva, J.: Solving QBF by clause selection. In: Proceedings of IJCAI, pp. 325–331. AAAI Press (2015)
19. Janota, M., Marques-Silva, J.: Abstraction-based algorithm for 2QBF. In: Sakallah, K.A., Simon, L. (eds.) SAT 2011. LNCS, vol. 6695, pp. 230–244. Springer, Heidelberg (2011). doi:10.1007/978-3-642-21581-0_19
20. Lonsing, F., Biere, A.: DepQBF: a dependency-aware QBF solver. JSAT **7**(2–3), 71–76 (2010)
21. Peterson, G., Reif, J., Azhar, S.: Lower bounds for multiplayer noncooperative games of incomplete information. Comput. Math. Appl. **41**(7), 957–992 (2001)
22. Peterson, G.L., Reif, J.H.: Multiple-person alternation. In: Proceedings of FOCS, pp. 348–363. IEEE (1979)
23. Rabe, M.N., Seshia, S.A.: Incremental determinization. In: Creignou, N., Le Berre, D. (eds.) SAT 2016. LNCS, vol. 9710, pp. 375–392. Springer, Cham (2016). doi:10.1007/978-3-319-40970-2_23
24. Rabe, M.N., Leander Tentrup, C.: A certifying QBF solver. In: Proceedings of FMCAD, pp. 136–143 (2015)
25. Siekmann, J., Wrightson, G.: Automation of Reasoning: 2: Classical Papers on Computational Logic 1967–1970. Springer, Heidelberg (1983). doi:10.1007/978-3-642-81955-1
26. Silva, J.P.M., Sakallah, K.A.: GRASP - a new search algorithm for satisfiability. In: Proceedings of CAD, pp. 220–227. IEEE (1997)
27. Tentrup, L.: Non-prenex QBF solving using abstraction. In: Creignou, N., Le Berre, D. (eds.) SAT 2016. LNCS, vol. 9710, pp. 393–401. Springer, Cham (2016). doi:10.1007/978-3-319-40970-2_24
28. Tentrup, L.: On expansion and resolution in CEGAR based QBF solving. In: Majumdar, R., Kunčak, V. (eds.) CAV 2017. LNCS, vol. 10427, pp. 475–494. Springer, Cham (2017)
29. Tseitin, G.S.: On the complexity of derivation in propositional calculus. Stud. Constr. Math. Math. Logic **2**, 115–125 (1968). Reprinted in [2]: 10–13
30. Gelder, A.: Contributions to the theory of practical quantified Boolean formula solving. In: Milano, M. (ed.) CP 2012. LNCS, pp. 647–663. Springer, Heidelberg (2012). doi:10.1007/978-3-642-33558-7_47
31. Wimmer, R., Gitina, K., Nist, J., Scholl, C., Becker, B.: Preprocessing for DQBF. In: Heule, M., Weaver, S. (eds.) SAT 2015. LNCS, vol. 9340, pp. 173–190. Springer, Cham (2015). doi:10.1007/978-3-319-24318-4_13
32. Wimmer, R., Reimer, S., Marin, P., Becker, B.: HQSpre-an effective preprocessor for QBF and DQBF. In: Proceedings of TACAS (2017)
33. Zhang, L., Malik, S.: Conflict driven learning in a quantified Boolean satisfiability solver. In: Proceedings of ICCAD, pp. 442–449, November 2002

From DQBF to QBF by Dependency Elimination

Ralf Wimmer[1]([⊠]), Andreas Karrenbauer[2], Ruben Becker[2], Christoph Scholl[1],
and Bernd Becker[1]

[1] Albert-Ludwigs-Universität Freiburg, Freiburg im Breisgau, Germany
{wimmer,scholl,becker}@informatik.uni-freiburg.de
[2] MPI for Informatics, Saarland Informatics Campus, Saarbrücken, Germany
{karrenba,ruben}@mpi-inf.mpg.de

Abstract. In this paper, we propose the elimination of dependencies
to convert a given dependency quantified Boolean formula (DQBF) to
an equisatisfiable QBF. We show how to select a set of dependencies to
eliminate such that we arrive at a smallest equisatisfiable QBF in terms
of existential variables that is achievable using dependency elimination.
This approach is improved by taking so-called don't-care dependencies
into account, which result from the application of dependency schemes
to the formula and can be added to or removed from the formula at
no cost. We have implemented this new method in the state-of-the-art
DQBF solver HQS. Experiments show that dependency elimination is
clearly superior to the previous method using variable elimination.

1 Introduction

Dependency quantified Boolean formulas (DQBFs) have received considerable
attention in research during the last years. They are a generalization of ordinary quantified Boolean formulas (QBFs). While the latter have the restriction that every existential variable depends on all universal variables in whose
scope it is, DQBFs allow arbitrary dependencies, which are explicitly specified
in the formula. This makes DQBFs more expensive to solve than QBFs – for
DQBF the decision problem is NEXPTIME-complete, for QBF 'only' PSPACE-
complete. However, there are practically relevant applications that require the
higher expressiveness of DQBF for a natural and tremendously more compact
modeling. Among them is the analysis of multi-player games with incomplete
information [22], the synthesis of safe controllers [4] and of certain classes of LTL
properties [8], and the verification of incomplete combinational and sequential
circuits [12,27,35].

Driven by the needs of the applications mentioned above, research on DQBF
solving has not only led to fundamental theoretical results on DQBF [1,3], but
also to first solvers like IDQ and HQS [10,11,13,32].

This work was partly supported by the German Research Council (DFG) as part
of the project "Solving Dependency Quantified Boolean Formulas" and by the
Max Planck Center for Visual Computing and Communication (www.mpc-vcc.org).
Ruben Becker is a member of the Saarbrücken Graduate School of Computer Science.

S. Gaspers and T. Walsh (Eds.): SAT 2017, LNCS 10491, pp. 326–343, 2017.
DOI: 10.1007/978-3-319-66263-3_21

While ıDQ uses instantiation-based solving, i.e., it reduces deciding a DQBF to deciding a series of SAT problems which correspond to *partial* universal expansions, HQS [13] uses the elimination of universal variables to turn the DQBF at hand into an equisatisfiable QBF, which can be solved by an arbitrary QBF solver. The basic method is complemented by several preprocessing techniques for DQBF [9,32,33] and the application of dependency schemes [34] for manipulating the dependency sets of the DQBF formula without changing its truth value.

In this paper we improve on the state-of-the-art solver HQS by making the following contributions:

(1) We introduce a **novel technique called** '*dependency elimination*' for transforming a DQBF into an equisatisfiable QBF. While [13] uses a minimal number of universal expansions for turning a DQBF $\forall x_1 \ldots \forall x_n \exists y_1(D_{y_1}) \ldots \exists y_m(D_{y_m}) : \phi$ into an equisatisfiable QBF with linearly ordered dependency sets, dependency elimination is able not only to remove universal variables x_i *completely* from the formula, but also to remove universal variables x_i from *single* dependency sets D_{y_j}, i.e., it works with a finer granularity. Dependency elimination is used with the goal of producing fewer copies of existential variables in the final QBF.

(2) We provide a **method for selecting an optimal elimination set**. The main ingredients of this method are:

(a) Dependencies can be represented in a natural way by a bipartite tournament graph, also called the dependency graph. Determination of an optimal elimination set then corresponds to *breaking cycles* in this graph *by flipping a cost-minimal set of edges*. A (non-linear) cost function takes into account the number of existential variables after eliminating a set of dependencies.

(b) An exact and efficient solution for the optimization problem in (a) is presented. It is based on *integer linear programming with dynamically added constraints* similar to the so-called cutting plane approach [36].

(c) The efficiency of the optimal elimination set computation is significantly increased by integrating *symmetry reduction*. Symmetry reduction is based on the observation that in typical applications the number of different dependency sets is rather small. We prove that optimal solutions based on symmetry-reduced graphs are optimal solutions for the original graphs as well. Based on research on dependency schemes [34] we consider in our optimization also dependencies which can be removed 'free of charge' without dependency elimination, since their removal does not change the truth value of the DQBF.

(d) Furthermore, we prove that the problem of finding an optimal elimination set for DQBFs with 'don't-care dependencies' is NP-complete. Note that there are related problems in the literature like 'Minimal Feedback Arc Set' (FAS) for bipartite tournament graphs. FAS for bipartite tournament graphs has been shown to be NP-complete as well in [14], but it differs from our problem in two aspects: We are only allowed to flip a *subset* of all edges (only the edges representing dependencies of existential variables on universal ones) and our cost function does not simply count the number of flipped edges, but it is more complicated and non-linear.

(3) We perform an **extensive experimental evaluation**, proving that the computation time for selecting an optimal elimination set is typically negligible, the number of variable copies produced by the optimal dependency elimination is much smaller compared to full universal variable elimination in many cases, and – overall – that the performance of the solver HQS could be improved to a great extent by the novel approach.

The paper is structured as follows: In the next section we introduce the necessary foundations. Section 3 presents dependency elimination and our procedure which selects an appropriate set of dependencies to eliminate. In Sect. 4 we experimentally evaluate this novel method. Finally, in Sect. 5 we draw conclusions and point out future work.

2   Foundations

For a finite set V of Boolean variables, $\mathcal{A}(V)$ denotes the set of variable assignments of V, i.e., $\mathcal{A}(V) = \{\nu : V \rightarrow \mathbb{B}\}$ with $\mathbb{B} = \{0,1\}$. Given quantifier-free Boolean formulas ϕ and κ over V and a Boolean variable $v \in V$, $\phi[\kappa/v]$ denotes the Boolean formula which results from ϕ by replacing all occurrences of v simultaneously by κ (simultaneous replacement is necessary when κ contains the replaced variable v).

Dependency quantified Boolean formulas are obtained by prefixing Boolean formulas with so-called Henkin quantifiers [15].

Definition 1 (Syntax of DQBF). *Let* $V = \{x_1, \ldots, x_n, y_1, \ldots, y_m\}$ *be a finite set of Boolean variables. A* dependency quantified Boolean formula (DQBF) ψ *over* V *has the form* $\psi := \forall x_1 \ldots \forall x_n \exists y_1(D_{y_1}) \ldots \exists y_m(D_{y_m}) : \phi$, *where* $D_{y_i} \subseteq \{x_1, \ldots, x_n\}$ *is the* dependency set *of* y_i *for* $i = 1, \ldots, m$, *and* ϕ *is a quantifier-free Boolean formula over* V, *called the* matrix *of* ψ.

We denote the existential variables of a DQBF ψ with $V_\psi^\exists = \{y_1, \ldots, y_m\}$ and its universal variables by $V_\psi^\forall = \{x_1, \ldots, x_n\}$. As the order of the variables in the quantifier prefix Q does not matter, we can regard it as a set: For instance, $Q \backslash \{v\}$ with a variable $v \in V$ is the prefix which results from Q by removing the variable v together with its quantifier (as well as its dependency set in case v is existential, and all its occurrences in dependency sets if it is universal).

The semantics of a DQBF is typically defined in terms of so-called Skolem functions.

Definition 2 (Semantics of DQBF). *Let* ψ *be a DQBF as above. It is satisfied if there are functions* $s_y : \mathcal{A}(D_y) \rightarrow \mathbb{B}$ *for* $y \in V_\psi^\exists$ *such that replacing each existential variable* y *by (a Boolean expression for)* s_y *turns* ϕ *into a tautology. The functions* $(s_y)_{y \in V_\psi^\exists}$ *are called* Skolem functions *for* ψ.

Deciding whether a given DQBF is satisfied is NEXPTIME-complete [22].

Definition 3 (Equisatisfiability of DQBFs). *Two DQBFs* ψ *and* ψ' *are* equisatisfiable *iff they are either both satisfied or both not satisfied.*

The elimination of universal variables in solvers like HQS [13] is done by *universal expansion* and leads to an equisatisfiable DQBF [1,5,6,12]:

Definition 4 (Universal Expansion). *For a DQBF* $\psi = \forall x_1 \ldots \forall x_n \exists y_1$ $(D_{y_1}) \ldots \exists y_m (D_{y_m}) : \varphi$ *with* $Z_{x_i} = \{ y_j \in V_\psi^\exists \mid x_i \in D_{y_j} \}$, *the universal expansion of variable* $x_i \in V_\psi^\forall$ *is defined by*

$$(Q \backslash \{x_i\}) \cup \{ \exists y_j' (D_{y_j} \backslash \{x_i\}) \mid y_j \in Z_{x_i} \} : \varphi[1/x_i] \wedge \varphi[0/x_i][y_j'/y_j \text{ for all } y_j \in Z_{x_i}].$$

An important special case of DQBFs is known as *quantified Boolean formulas*. They exhibit a linearly ordered quantifier prefix, where each existential variable y depends on all universal variables in whose scope it is:

Definition 5 (Syntax of QBF, Equivalent QBFs). *Let* $V = \{x_1, \ldots, x_n, y_1,$ $\ldots, y_m\}$ *be a finite, non-empty set of Boolean variables,* $X_1, \ldots, X_k \subseteq$ $\{x_1, \ldots, x_n\}$ *a partition of* $\{x_1, \ldots, x_n\}$ *such that* $X_i \neq \emptyset$ *for* $i = 2, \ldots, k$, *and* $Y_1, \ldots, Y_k \subseteq \{y_1, \ldots, y_m\}$ *a partition of* $\{y_1, \ldots, y_m\}$ *such that* $Y_i \neq \emptyset$ *for* $i = 1, \ldots, k - 1$. *Additionally let* ϕ *be a quantifier-free Boolean formula over* V.

A quantified Boolean formula (QBF) Ψ *(in prenex form) is given by*

$$\Psi := \forall X_1 \exists Y_1 \forall X_2 \exists Y_2 \ldots \forall X_k \exists Y_k : \phi.$$

The QBF Ψ *is equivalent to the DQBF* $\psi := \forall x_1 \ldots \forall x_n \exists y_1 (D_{y_1}) \ldots \exists y_m (D_{y_m}) :$ ϕ, *if* $D_{y_i} = \bigcup_{\ell=1}^{L} X_\ell$ *such that* L *is the unique index with* $y_i \in Y_L$. *In this case we say that the DQBF* ψ *'can be written as a QBF* Ψ*' or the DQBF* ψ *'has an equivalent QBF prefix'.*

Lemma 1 ([13]). *A DQBF* ψ *has an equivalent QBF prefix if* $D_y \subseteq D_{y'}$ *or* $D_{y'} \subseteq D_y$ *holds for all* $y, y' \in V_\psi^\exists$.

QBFs can be solved more efficiently than general DQBFs. For QBF, the decision problem is "only" PSPACE-complete [21], and rather efficient solvers for QBF are available like DepQBF [19,20], AIGSolve [23,24], Qesto [17], RAReQS [16], to name just a few. Therefore the goal is to manipulate the DQBF at hand – preserving the truth value – in a way such that the resulting formula has an equivalent QBF prefix and can be solved by any available QBF solver.

3   Dependency Elimination

The DQBF solver HQS [13] uses universal expansion to turn the DQBF at hand into an equisatisfiable QBF. It determines a smallest possible set of universal variables whose elimination yields a QBF. This is done by solving a MAXSAT problem. Using universal expansion has the drawback that it copies *all* variables which depend on the eliminated universal variable.

Example 1. The DQBF $\forall x_1 \forall x_2 \exists y_1(x_1) \exists y_2(x_2) \exists y_3(x_1, x_2) \ldots \exists y_n(x_1, x_2) : \varphi$ does not have an equivalent QBF prefix. Therefore the expansion of either x_1 or x_2 is necessary. When x_1 is eliminated, $y_1, y_3, \ldots, y_n$ are doubled, creating $n - 1$ additional existential variables. The elimination of x_2 creates copies of $y_2, y_3, \ldots, y_n$. However, only the dependencies of y_1 on x_1 and of y_2 on x_2 are responsible for the formula not being a QBF.

Therefore we propose an alternative operation that we can use to obtain an equisatisfiable DQBF with an equivalent QBF prefix, namely dependency elimination, which allows to remove single dependencies from a formula.

Theorem 1 (Dependency Elimination). *Assume ψ is a DQBF as in Definition 1 and, w.l.o.g., $x_1 \in D_{y_1}$. Then ψ is equisatisfiable to:*

$$\psi' := \forall x_1 \ldots \forall x_n \; \exists y_1^0(D_{y_1} \backslash \{x_1\}) \; \exists y_1^1(D_{y_1} \backslash \{x_1\}) \; \exists y_2(D_{y_2}) \ldots \exists y_m(D_{y_m}) :$$
$$\phi\Big[\big((\neg x_1 \wedge y_1^0) \vee (x_1 \wedge y_1^1) \big) / y_1 \Big].$$

Proof. Assume ψ is satisfiable with Skolem functions s_{y_i} for y_i ($1 \le i \le m$). We have $s_{y_1} = (\neg x_1 \wedge s_{y_1 |x_1=0}) \vee (x_1 \wedge s_{y_1 |x_1=1})$ for the negative cofactor $s_{y_1 |x_1=0}$ w.r.t. x_1 and the positive cofactor $s_{y_1 |x_1=1}$ w.r.t. x_1. Then ψ' is satisfiable, too, with Skolem functions s_{y_i} for y_i ($2 \le i \le m$), $s_{y_1 |x_1=0}$ for y_1^0 and $s_{y_1 |x_1=1}$ for y_1^1. Conversely, if ψ' is satisfiable with Skolem functions s_{y_i} for y_i ($2 \le i \le m$), $s_{y_1^0}$ for y_1^0 and $s_{y_1^1}$ for y_1^1, then ψ is satisfiable with Skolem function $s_{y_1} = (x_1 \wedge s_{y_1^1}) \vee (\neg x_1 \wedge s_{y_1^0})$ for y_1 and s_{y_i} for y_i ($2 \le i \le m$). $\square$

Example 2. Consider again the formula from Example 1. If we eliminate the dependency of y_1 on x_1 we obtain the formula

$$\forall x_1 \forall x_2 \exists y_1^0(\emptyset) \exists y_1^1(\emptyset) \exists y_2(x_2) \exists y_3(x_1, x_2) \ldots \exists y_n(x_1, x_2) : \varphi[(\neg x_1 \wedge y_1^0) \vee (x_1 \wedge y_1^1)/y_1].$$

This formula can be written as the QBF

$$\exists y_1^0 \exists y_1^1 \forall x_2 \exists y_2 \forall x_1 \exists y_3 \ldots \exists y_n : \varphi[(\neg x_1 \wedge y_1^0) \vee (x_1 \wedge y_1^1)/y_1].$$

Instead of creating $n-1$ additional existential variables as in Example 1, we only had to double y_1 in order to obtain an equisatisfiable QBF.

The main question that we have to answer is which dependencies should be eliminated in order to obtain an equisatisfiable QBF. If we eliminate n dependencies of an existential variable y, we have to create $2^n - 1$ new copies of y. Therefore it is typically not feasible to eliminate all dependencies, but we have to take care to find a set of dependencies which requires the fewest variable copies and still turns the formula into a QBF.

3.1 Selecting Dependencies to Eliminate

In order to facilitate the selection of dependencies to eliminate we make use of the following dependency graph:

Definition 6 (Dependency Graph). *Let ψ be a DQBF as above. The* dependency graph $G_\psi = (V_\psi, E_\psi)$ *is a directed graph with the set $V_\psi = V$ of variables as nodes and edges*

$$E_\psi = \left\{(x,y) \in V_\psi^\forall \times V_\psi^\exists \mid x \in D_y\right\} \cup \left\{(y,x) \in V_\psi^\exists \times V_\psi^\forall \mid x \notin D_y\right\}.$$

G_ψ is a so-called *bipartite tournament graph* [2,7,14]: The nodes can be partitioned into two disjoint sets according to their quantifier and there are only edges that connect variables with different quantifiers – this is the bipartiteness property. We also write $G_\psi = (V_\psi^\forall, V_\psi^\exists, E_\psi)$ to make the two disjoint node sets apparent. Additionally, for each pair $(x,y) \in V_\psi^\forall \times V_\psi^\exists$ there is either an edge from x to y or vice-versa – this property is referred to by the term 'tournament'.

Theorem 2. *Let ψ be a DQBF and G_ψ its dependency graph. The graph G_ψ is acyclic iff ψ has an equivalent QBF prefix.*

Proof. Assume that ψ has an equivalent QBF prefix. The left-to-right order of this QBF prefix defines a total order $\prec$ on V with $D_y = \{x \in V_\psi^\forall \mid x \prec y\}$ for all $y \in V_\psi^\exists$. Then for all edges $(x,y) \in E_\psi$ we have $x \prec y$, and $y \prec x$ holds for all edges $(y,x) \in E_\psi$. That means all edges point to larger elements w.r.t. $\prec$. Therefore G_ψ is acyclic.

Now assume that G_ψ is acyclic. Then we can find a topological order for G_ψ, i.e., there exists a total order $\prec$ on the nodes of G_ψ such that we have: If $(v_1, v_2) \in E_\psi$ then $v_1 \prec v_2$. Now choose the QBF prefix from left to right according to the total order $\prec$. If $x \in D_y$, then $(x,y) \in E_\psi$, $x \prec y$ and x is to the left of y; if $x \notin D_y$, then $(y,x) \in E_\psi$, $y \prec x$ and x is to the right of y. Thus we have found an equivalent QBF prefix.   □

Eliminating a dependency essentially corresponds to flipping the direction of an edge $(x,y) \in E_\psi \cap (V_\psi^\forall \times V_\psi^\exists)$ from a universal to an existential variable. The cost of copying existential variables will be taken into account by choosing an appropriate cost for flipping sets of edges. This cost will count the number of existential variables after eliminating a set of dependencies. Our goal is to find a cost-minimal set of edges such that flipping those edges makes the dependency graph acyclic.

In the following, we will first determine a set $R \subseteq E_\psi \cap (V_\psi^\forall \times V_\psi^\exists)$ of edges whose *deletion* makes $G_{\text{del}}^R := (V_\psi^\forall, V_\psi^\exists, E_\psi \backslash R)$ acyclic. However, if R is such a set of edges, then we can turn it into a set $R' \subseteq R$ such that *flipping* the edges in R' yields an acyclic graph: Let $\prec$ be a topological order of G_{del}^R's nodes. Then we set $R' := \left\{(x,y) \in R \mid x \not\prec y\right\}$, i.e., we flip only those edges of R which point backward according to $\prec$. Then $\prec$ is also a topological order of $G_{\text{flip}}^{R'} := (V_\psi^\forall, V_\psi^\exists, (E_\psi \backslash R') \cup \{(y, x) \mid (x, y) \in R'\})$. Of course, $\prec$ is a topological order of $G_{\text{del}}^{R'} := (V_\psi^\forall, V_\psi^\exists, E_\psi \backslash R')$ as well and $G_{\text{del}}^{R'}$ is acyclic. Thus, if we choose a *minimal* set $R^{\min} \subseteq E_\psi \cap (V_\psi^\forall \times V_\psi^\exists)$ such that $G_{\text{del}}^{R^{\min}}$ is acyclic, then $G_{\text{flip}}^{R^{\min}}$ is acyclic as well. So we can restrict our attention to *removing* edges from $E_\psi \cap (V_\psi^\forall \times V_\psi^\exists)$ for turning G_ψ into an acyclic graph, as long as we remove a *minimal* set of edges – although in our application (turning a DQBF into a QBF by elimination of dependencies) we are only able to *flip* edges from $V_\psi^\forall$ to $V_\psi^\exists$.

We define an elimination set as follows.

Definition 7 (Elimination set). *A set $R \subseteq E_\psi \cap (V_\psi^\forall \times V_\psi^\exists)$ is an elimination set for the bipartite tournament graph $G_\psi = (V_\psi^\forall, V_\psi^\exists, E_\psi)$ if $G_{\text{del}}^R = (V_\psi^\forall, V_\psi^\exists, E_\psi \backslash R)$ is acyclic. An elimination set R is* minimal *if $R \backslash \{e\}$ is not an elimination set for every $e \in R$.*

Let R be a minimal elimination set. For $y \in V_\psi^\exists$, we set $R_y = \{x \in V_\psi^\forall \mid (x, y) \in R\}$. The cost of R is then given by $\text{cost}(R) := \sum_{y \in V_\psi^\exists} 2^{|R_y|}$. The cost of R corresponds to the number of existential variables in the formula after the dependencies in R have been eliminated. Hence, our goal is to determine an elimination set of minimal cost.

3.2 Symmetry Reduction

Before we look into the optimization problem of computing an elimination set of minimal cost, we consider reducing the size of the dependency graph by exploiting symmetries: The existential and universal variables are partitioned according to the dependency sets. We define an equivalence relation $\sim$ by:

$$y_i \sim y_j \quad \Leftrightarrow \quad D_{y_i} = D_{y_j}$$
$$x_i \sim x_j \quad \Leftrightarrow \quad \{y_\ell \in V_\psi^\exists \mid x_i \in D_{y_\ell}\} = \{y_\ell \in V_\psi^\exists \mid x_j \in D_{y_\ell}\}.$$

The dependency graph modulo $\sim$ is based on the equivalence classes $[v]_\sim$ of $\sim$ and is defined by $G_{\widetilde{\psi}} = (V_{\widetilde{\psi}}, E_{\widetilde{\psi}})$ where

$$V_{\widetilde{\psi}} = \{[v]_\sim \mid v \in V\} \text{ and}$$
$$E_{\widetilde{\psi}} = \{([x_i]_\sim, [y_j]_\sim) \mid x_i \in D_{y_j}\} \mathbin{\dot\cup} \{([y_j]_\sim, [x_i]_\sim) \mid x_i \notin D_{y_j}\}.$$

By definition, the resulting graph $G_{\widetilde{\psi}}$ is a bipartite tournament graph again. It is well defined: If $([x]_\sim, [y]_\sim) \in E_{\widetilde{\psi}}$ holds for some $x \in V_\psi^\forall$ and $y \in V_\psi^\exists$, then $x' \in D_{y'}$ holds for all $x' \in [x]_\sim$ and all $y' \in [y]_\sim$. If $([y]_\sim, [x]_\sim) \in E_{\widetilde{\psi}}$ holds, then $x' \notin D_{y'}$ for all $x' \in [x]_\sim$ and all $y' \in [y]_\sim$.

A further reduction can be obtained by the following observation: Incident edges of sources (i.e., nodes without incoming edges) and sinks (i.e., nodes without outgoing edges) never need to be flipped, since they cannot occur on a cycle. They can be removed from the graph, together with their incident edges. This can be repeated until the graph does not change anymore. The result is again a bipartite tournament graph.

If $R^\sim$ is the set of edges to be eliminated, transferring the cost function to the reduced graph yields:

$$\text{cost}^\sim(R^\sim) = \sum_{[y]_\sim \in V_\psi^\sim} |[y]_\sim| \cdot 2^{\sum_{([x]_\sim, [y]_\sim) \in R^\sim} |[x]_\sim|}.$$

The intuition behind the definition of $\text{cost}^\sim(R^\sim)$ is that eliminating $([x]_\sim, [y]_\sim)$ in $G_\psi^\sim$ 'means' eliminating all edges (x', y') in G_ψ with $x' \in [x]_\sim$ and $y' \in [y]_\sim$.

The following Theorem 3 justifies the application of symmetry reduction to the dependency graph: A cost-optimal elimination set for the reduced graph induces a cost-optimal elimination set for the original graph.

Theorem 3. *(a) If $R^\sim$ is a minimal elimination set for $G_\psi^\sim$, then $R = \{(x, y) \in (V_\psi^\forall \times V_\psi^\exists) \cap E_\psi \mid ([x]_\sim, [y]_\sim) \in R^\sim\}$ is a minimal elimination set for G_ψ such that $\text{cost}(R) = \text{cost}^\sim(R^\sim)$.*

(b) If R is a minimal elimination set for G_ψ, then the set $R^\sim := \{([x]_\sim, [y]_\sim) \mid (x, y) \in R\}$ is a minimal elimination set for $G_\psi^\sim$ such that $\text{cost}(R) = \text{cost}^\sim(R^\sim)$.

Before we prove Theorem 3, we show the following Lemma 2.

Lemma 2. *Let R be a minimal elimination set for G_ψ. Then for all $(x, y) \in V_\psi^\forall \times V_\psi^\exists$, we have $(x, y) \in R$ iff $[x]_\sim \times [y]_\sim \subseteq R$.*

Proof (Lemma 2). Let $x_1 \to y_1 \to x_2 \to \cdots \to y_k \to x_1$ be a cycle of G_ψ. Assume that for all $1 \le i \le k$ $[x_i]_\sim \times [y_i]_\sim \subseteq R$ does *not* hold. Then for all $1 \le i \le k$ there are (x_i', y_i') with $x_i' \sim x_i$ and $y_i' \sim y_i$ such that $(x_i', y_i') \notin R$. According to the definition of the relation $\sim$, G_ψ contains the cycle $x_1' \to y_1' \to x_2' \to \cdots \to y_k' \to x_1'$ which is not broken by R. This contradicts our assumption that R is an elimination set.

We conclude that for each cycle $x_1 \to y_1 \to x_2 \to \cdots \to y_k \to x_1$ we have $[x_i]_\sim \times [y_i]_\sim \subseteq R$ for some $1 \le i \le k$. All other (x_j, y_j) with $[x_j]_\sim \times [y_j]_\sim \not\subseteq R$ are not needed to break cycles in G_ψ and are thus not included in R due to minimality of R. $\qquad\square$

Proof (Theorem 3). Proof of part (a): Let $x_1 \to y_1 \to x_2 \to \cdots \to y_k \to x_1$ be an arbitrary cycle in G_ψ (if there is no cycle in G_ψ, then it trivially follows that R is an elimination set). By definition of $G_\psi^\sim$, $[x_1]_\sim \to [y_1]_\sim \to [x_2]_\sim \to \cdots \to [y_k]_\sim \to [x_1]_\sim$ is a cycle in $G_\psi^\sim$. Since $R^\sim$ is an elimination set for $G_\psi^\sim$, $([x_i]_\sim, [y_i]_\sim) \in R^\sim$ for some $i \in \{1, \ldots, k\}$ and by definition of R we have

$(x_i, y_i) \in R$. Therefore R is an elimination set. That the values of the cost functions coincide is easy to see.

We still have to show that R is minimal. Assume the contrary, i.e., there is $(x_1, y_1) \in R$ such that $R \setminus \{(x_1, y_1)\}$ is still an elimination set. By construction, $([x_1]_\sim, [y_1]_\sim) \in R^\sim$. Due to minimality of $R^\sim$, there has to be a cycle in $G_{\widetilde{\psi}}^\sim$ containing $([x_1]_\sim, [y_1]_\sim)$. Let $[x_1]_\sim \to [y_1]_\sim \to [x_2]_\sim \to \cdots \to [y_k]_\sim \to [x_1]_\sim$ be such a cycle without node repetitions (i.e., a simple cycle). By definition of $G_{\widetilde{\psi}}^\sim$, $x_1 \to y_1 \to x_2 \to \cdots \to y_k \to x_1$ is a cycle in G_ψ and $(x_i, y_i) \in R$ for some $i \in \{2, \ldots, k\}$, since $R \setminus \{(x_1, y_1)\}$ is an elimination set. By definition of R, $([x_i]_\sim, [y_i]_\sim) \in R^\sim$ and $([x_i]_\sim, [y_i]_\sim) \neq ([x_1]_\sim, [y_1]_\sim)$, since the cycle is simple. Therefore, all simple cycles broken by $([x_1]_\sim, [y_1]_\sim)$ are also broken by another edge from $R^\sim$. That means, $R^\sim$ is not minimal, contradicting our assumption.

The proof of part (b) immediately follows from Lemma 2.   □

3.3   An Optimization Approach

The Underlying Optimization Problem. Our goal in the above described problem is to determine an elimination set $R^\sim$ of minimal cost $\mathrm{cost}^\sim(R^\sim)$. We can determine such an elimination set by selecting one edge from each simple cycle in the dependency graph. A cycle is called simple if it does not contain any subcycle. This can be formulated as an optimization problem in a given arbitrary bipartite tournament graph G with disjoint node sets X, Y and node weights $\omega : X \cup Y \to \mathbb{R}_{>0}$. (Remember that in our application the nodes $X \cup Y$ represent equivalence classes $[v]_\sim$. Their weights correspond to the cardinality of $[v]_\sim$.) We introduce a decision variable $d_{(x,y)} \in \{0, 1\}$ for each edge $(x, y) \in E_{XY}$, where $E_{XY} = E \cap (X \times Y)$ with the interpretation that $d_{(x,y)} = 1$ indicates that (x, y) belongs to the elimination set. Let C denote the set of all simple cycles $c = x_1 \to y_1 \to x_2 \to \cdots \to y_k \to x_1$ such that $x_i \in X$, $y_i \in Y$ for all $i = 1, \ldots, k$ and $x_i \neq x_j$, $y_i \neq y_j$ for $i \neq j$. Moreover, for a simple cycle $c \in C$, let us denote with $F(c) = E(c) \cap (X \times Y)$, the set of all arcs in c that are directed from a node in X to a node in Y. The optimization problem can then be formulated as

$$\text{minimize} \quad \sum_{y \in Y} \omega(y) \cdot 2^{\sum_{x \in \mathrm{pre}(y)} \omega(x) \cdot d_{(x,y)}} \tag{1a}$$

$$\text{such that} \quad \sum_{(x,y) \in F(c)} d_{(x,y)} \geq 1 \qquad \forall c \in C \tag{1b}$$

$$d_{(x,y)} \in \{0, 1\} \qquad \forall (x, y) \in E_{XY} \tag{1c}$$

Two challenges make solving this optimization problem difficult: First, the objective function is non-linear as it is the sum of exponential functions. Second, the number of cycles in the dependency graph may be prohibitively large.

Solving the Optimization Problem. In order to bring the optimization problem into a more convenient form, we first rewrite it by introducing variables $f_y \in \mathbb{Z}$ and $z_y \in \mathbb{Z}$ for every $y \in Y$:

$$\text{minimize} \quad \sum_{y \in Y} \omega(y) \cdot z_y \tag{2a}$$

$$\text{such that} \quad \sum_{(x,y) \in F(c)} d_{(x,y)} \geq 1 \qquad \forall c \in C \tag{2b}$$

$$\sum_{x \in \text{pre}(y)} \omega(x) \cdot d_{(x,y)} = f_y \qquad \forall y \in Y \tag{2c}$$

$$z_y \geq 2^{f_y} \qquad \forall y \in Y \tag{2d}$$

$$d_{(x,y)} \in \{0, 1\} \qquad \forall (x,y) \in E_{XY} \tag{2e}$$

After this transformation, (2d) is the only non-linear constraint and the objective function is linear. We will handle these non-linear constraints dynamically as follows. We first solve the optimization problem consisting only of (2b), (2c), and (2e). These constraints form an integer linear program (ILP). Solving this ILP yields a solution $\bar{d}, \bar{f}, \bar{z}$ that are not necessarily feasible for the complete optimization problem (2). Thus, we check whether, for some $y \in Y$, the constraint $\bar{z}_y \geq 2^{\bar{f}_y}$ is violated. In this case, we add a linear inequality (lazy constraint approach) that cuts off this infeasible solution, but none of the feasible points. Such an inequality is, e.g., the constraint that z_y lies on or above the tangent to the function 2^{f_y} in the current value $\bar{f}_y$ of f_y. This tangent is described by $t(f_y) = 2^{\bar{f}_y} \cdot (1 + (f_y - \bar{f}_y) \cdot \ln 2)$. Thus, we could add the inequality $z_y \geq 2^{\bar{f}_y} \cdot (1 + (f_y - \bar{f}_y) \cdot \ln 2)$. However, such inequalities are not rational and their closure yields non-integral extreme points. Instead, we can take the secants through two adjacent extreme points of the convex hull of the integer points satisfying (2d) (feasible solutions are integer). The secants through $\bar{f}_y$ and $\bar{f}_y + 1$ and through $\bar{f}_y - 1$ and $\bar{f}_y$ yield the constraints

$$z_y \geq 2^{\bar{f}_y}(1 - \bar{f}_y + f_y) \quad \text{and} \quad z_y \geq 2^{\bar{f}_y - 1}(2 - \bar{f}_y + f_y).$$

Taken together, the two secant constraints are tighter than the tangent constraint and moreover their description contains only integer coefficients. This is why the secant constraints are preferable (see Fig. 1).

To further increase efficiency, we also relax the cycle constraints (2b) by only adding constraints for C_4, the set of all 4-cycles, first.[1] The longer cycle constraints are handled dynamically as well: If we obtain a solution, we check by depth-first search whether the induced graph is acyclic. If it is not, we add (2b) for the found cycle.[2] The described approach leads to Algorithm 1. Adding

---

[1] Note that each cyclic bipartite tournament graph has a cycle of length 4.

[2] The approach of dynamically or lazily adding constraints is similar to the cutting plane approach [36] and is used as one of the main ingredients for efficiently solving many NP-hard problems for which only a description with exponentially many constraints is at hand, as for example the traveling salesman problem.

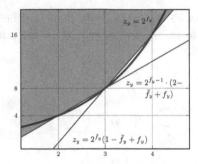

Fig. 1. The two secants on the function $z_y = 2^{f_y}$ at $\bar{f}_y = 2$. The shaded area denotes the feasible region defined by the two inequalities corresponding to the two secants.

separation constraints dynamically is typically supported by ILP solvers like Gurobi using call-back functions.

3.4 Don't-Care Dependencies

During preprocessing, often dependencies can be identified, which are only pseudo-dependencies (also called don't-care dependencies), i.e., an existential variable y contains a universal variable x in its dependency set, but it can be shown that removing the dependency does not change the satisfiability of the formula. Don't-care dependencies correspond to edges in the dependency graph which can be flipped without any costs.

Definition 8 (DQBF with Don't-Care Dependencies). *Let $\psi = \forall x_1 \ldots \forall x_n \exists y_1(D_{y_1}) \ldots \exists y_m(D_{y_m}) : \phi$ be a DQBF as before and $S_{y_i} \subseteq D_{y_i}$ for $i = 1, \ldots, m$. The sets S_{y_i} are called* don't-care sets *of ψ if ψ is equisatisfiable to*

$$\forall x_1 \ldots \forall x_n \exists y_1(D_{y_1} \setminus S_{y_1}) \ldots \exists y_m(D_{y_m} \setminus S_{y_m}) : \phi.$$

Detecting Don't-Care Dependencies. Using the same proof idea as described in [26] for QBF, we can show that deciding whether a dependency is a don't-care dependency has the same complexity as deciding DQBF itself. Therefore one usually resorts to efficient approximations for computing don't-care dependencies.

Lemma 3. *Let $\psi = \forall x_1 \ldots \forall x_n \exists y_1(D_{y_1}) \ldots \exists y_m(D_{y_m}) : \phi$ be a DQBF. Deciding whether $x \in D_y$ for $x \in V_\psi^\forall$ and $y \in V_\psi^\exists$ is a don't-care dependency is NEXPTIME-complete.*

Don't-care dependencies can be detected using so-called dependency schemes, which provide over-approximations of the actually dependent variables, see [25, 28–30] for QBF and [32, 34] for DQBF. Dependency schemes are based on efficient syntactic criteria, and by over-approximating the dependent variables they under-approximate the sets of don't-care dependencies.

Algorithm 1. SOLVEEXACT$(G = (X, Y, E), \omega)$

$C \leftarrow C_4,\ P \leftarrow \emptyset$
while True **do**
    Determine $(\bar{d}, \bar{f}, \bar{z})$ as the optimal solution of

$$\text{minimize} \quad \sum_{y \in Y} \omega(y) \cdot z_y$$

$$\text{such that} \quad \sum_{(x,y) \in F(c)} d_{(x,y)} \geq 1 \qquad\qquad \forall c \in C$$

$$\sum_{x \in \text{pre}(y)} \omega(x) \cdot d_{(x,y)} = f_y \qquad\qquad \forall y \in Y$$

$$\begin{aligned} z_y &\geq 2^{\bar{f}_y}(1 - \bar{f}_y + f_y) \\ z_y &\geq 2^{\bar{f}_y - 1}(2 - \bar{f}_y + f_y) \end{aligned} \qquad \forall (y, \bar{f}_y) \in P$$

$$d_{(x,y)} \in \{0, 1\} \qquad\qquad \forall (x, y) \in E_{XY}$$

    constr_added $\leftarrow$ **false**
    for $y \in Y$ **do**
        if $\bar{z}_y < 2^{\bar{f}_y}$ **then**
            $P \leftarrow P \cup \{(y, \bar{f}_y)\}$, constr_added $\leftarrow$ **true**
    if *cycle c exists in* $G' := (X, Y, E \setminus \{(x, y) : d_{(x,y)} = 1\})$ **then**
        $C \leftarrow C \cup c$, constr_added $\leftarrow$ **true**
    if *not* constr_added **then**
        return $(\bar{d}, \bar{f}, \bar{z})$

Exploiting Don't-Care Dependencies. To exploit don't-care dependencies, we first have to refine the symmetry reduction to take not only the dependency sets, but also the don't-care dependencies into account. This yields the following refined equivalence relation $\approx \subseteq V \times V$:

$$\begin{aligned} y_i \approx y_j \quad &\Leftrightarrow \quad D_{y_i} = D_{y_j} \wedge S_{y_i} = S_{y_j}, \\ x_i \approx x_j \quad &\Leftrightarrow \quad \{y \in V_\psi^\exists \mid x_i \in D_y\} = \{y \in V_\psi^\exists \mid x_j \in D_y\} \\ &\qquad\quad \wedge \{y \in V_\psi^\exists \mid x_i \in S_y\} = \{y \in V_\psi^\exists \mid x_j \in S_y\}. \end{aligned}$$

The graph $G_\psi^\approx$, resulting from G_ψ by merging equivalent nodes, is again a bipartite tournament graph.

Let $G = (X, Y, E)$ be the bipartite graph corresponding to a DQBF ψ with don't-care sets. We define the weight function $\omega : X \cup Y \rightarrow \mathbb{R}$ as follows: $\omega(v) = 1$ for all $v \in X \cup Y$ if the graph was not reduced, otherwise $\omega(v) = |[v]_\approx|$ if we have applied symmetry reduction using $\approx$. Let $DC = \{(x, y) \mid x \in S_y\}$ be the set of all don't-care dependencies and let $R \subseteq E_{XY}$ be an elimination set. Then we just do not count the cost for eliminating don't-care dependencies, i.e.,

$$\text{cost}(R) = \sum_{y \in Y} \omega(y) \cdot 2^{\sum_{(x,y) \in (R \setminus DC)} \omega(x)}.$$

The don't-care sets can easily be taken into account in Algorithm 1: Before applying Algorithm 1 we delete all edges in DC (i.e., all edges corresponding to don't-care dependencies) from the dependency graph, since those eliminations are free of charge, and apply Algorithm 1 without any other change. This means that we implicitly start with $R = DC$ and then add to the elimination set all $(x, y) \in V_\psi^\forall \times V_\psi^\exists$ with $\bar{d}_{(x,y)} = 1$ in the solution returned by Algorithm 1. The resulting elimination set R is not necessarily minimal and thus eliminating all those dependencies (which corresponds to flipping all edges in R) does not necessarily make the dependency graph acyclic. However, we can find an appropriate subset $R' \subseteq R$ such that flipping all edges from R' makes the dependency graph acyclic using the method from Sect. 3.1. (Another option would be to remove elements of DC from R one after the other as long as the resulting set R remains an elimination set, finally arriving at a minimal elimination set.)

Finally, we provide a complexity result for computing cost-minimal elimination sets in the presence of don't-care dependencies.

Lemma 4. *Given a bipartite tournament graph $G_\psi = ((X, Y), E_\psi)$, a set of don't-care dependencies $S_\psi \subseteq E_\psi$, and an integer $c \geq 0$, deciding whether an elimination set R with $\text{cost}(R) \leq c$ exists is NP-complete.*

This lemma can be proven by a reduction from vertex cover [18]. It is easy to see that Lemma 4 holds for symmetry-reduced graphs as well.

Since there is a one-to-one correspondence between dependency graphs and DQBF prefixes, we can conclude:

Theorem 4. *Given a DQBF with don't-care dependencies and an integer $c \geq 0$, deciding whether there is an elimination set R with $\text{cost}(R) \leq c$ is NP-complete.*

4   Experimental Evaluation

We have extended the DQBF solver HQS [13] to support dependency elimination on its internal formula representation as an And-Inverter Graph (AIG). To determine an optimal elimination set, we use a Python script which is called by HQS and which in turn calls the MILP solver Gurobi 7.0.2 to solve the optimization problem as in Algorithm 1. We use our preprocessor HQSPRE [33] to simplify the instances before the actual solution process starts. Since the benchmarks used for evaluation were generated from incomplete circuits and controller synthesis problems, we run HQSPRE in its gate-preserving mode. Additionally, we apply the reflexive quadrangle resolution path dependency scheme [34] to identify don't-care dependencies. As the last step of preprocessing, we use syntactic gate detection to reconstruct the underlying circuit structure; this removes variables which have been introduced artificially to obtain a formula in conjunctive normal form and leads to more compact AIGs.

All experiments were run on one core of an Intel Xeon CPU E5-2450 (8 cores) running at 2.10 GHz clock frequency and having 32 GB of main memory. Ubuntu 16.04 in 64 bit mode was used as the operating system. We aborted each experiment which took more than 3600 s of CPU time or more than 10 GB of memory. We used the same 4811 benchmark instances as in [13,31,32,34]. They mainly encompass partial equivalence checking problems [12,27] for combinational circuits, controller synthesis problems for sequential circuits and safety properties [4].

For the comparison of variable and dependency elimination, we switched off the UNSAT filtering procedure, which is based on QBF abstractions [9]; it affects both solution procedures in exactly the same way. Additionally we skipped those instances which were solved by the preprocessor or which the preprocessor could turn into QBFs. This led to a benchmark set of 3618 instances.

We solved all instances with the original version of HQS [13] using variable elimination and by dependency elimination for a cost-minimal elimination set as described in Sect. 3. In the latter case, if the elimination set contains, for some universal variable $x \in V_\psi^\forall$, all dependencies x is involved in, then we call universal expansion for x instead as it has the same effect as first eliminating the dependencies and then expanding x, but is slightly faster. Otherwise we eliminate the selected dependencies according to Theorem 1. We distinguished instances for which only universal expansion needed to be applied from those which also required the elimination of dependencies.

For 3233 out of 3618 instances, the optimal elimination set removed universal variables from all dependency sets they were involved in, i.e., only universal expansion was used. Those instances do not profit from dependency elimination in terms of copied existential variables. From those instances, variable elimination as in [13] was able to solve 2429 instances, and the novel dependency elimination 2411 instances. The reason for the small difference of 18 instances is that both methods do not necessarily expand the same variables. Since the selection procedures only consider the formula's prefix and not the structure of the matrix, there is no guarantee that an elimination set is found which leads to small formulas during subsequent solution of the resulting QBF. This can also be observed in Fig. 2 where we compare dependency and variable elimination. The left plot compares the computation times, the right one the number of existential variables in the resulting QBF.

For the remaining 385 instances, dependency elimination has an advantage over variable elimination. It can yield an equisatisfiable QBF with fewer existential variables. Accordingly, using dependency elimination, we could solve 325 instances, while variable elimination succeeded only for 177. All instances which could be solved using variable elimination were also solved using dependency elimination. A more detailed comparison of the two methods on these 385 instances is shown in Fig. 3. We can distinguish two subsets of instances: There are some for which the difference between variable and dependency elimination

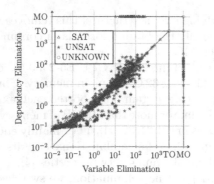
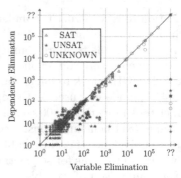

Fig. 2. Comparing the total computation times (in seconds) (left) and the number of existential variables in the resulting QBF (right) after variable and dependency elimination on the instances for which variable elimination *is* optimal.

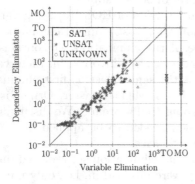
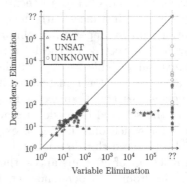

Fig. 3. Comparing the total computation times (in seconds) (left) and the number of existential variables in the resulting QBF (right) after variable and dependency elimination on the instances for which variable elimination is *not* optimal.

is small. Here the computation times are similar. Note that the number of existential variables in the QBF can even be slightly smaller for variable elimination because during expansion often unit and pure variables are detected which are immediately replaced by appropriate constants [13]. In contrast, dependency elimination replaces an existential variable by the representation of a multiplexer. This is an efficient local operation on AIGs, but typically does not allow to detect unit and pure variables. They are found later during the QBF solution process.

For the other subset of instances, dependency elimination is by orders of magnitude superior compared to variable elimination. While the latter runs exceed the memory limit for most of the instances, dependency elimination was able to solve them in little time and with much less memory consumption.

The computation times for selecting an optimal elimination set are negligible in most cases and are always below 10 s for our benchmark set. One reason for the small computation times is that most benchmarks contain only a small number of *different* dependency sets and thus our symmetry reduction from Sect. 3.2 works nicely; typically the reduced graphs consisted of 10–50 nodes only; the largest one had 54 nodes.

In our current (preliminary) implementation the effect of don't-care dependencies is negligible. The reason for this lies (a) in a restricted preprocessing that is tailored to our backend QBF solver AIGSolve and (b) in the structure of most of our benchmarks. The dependency schemes find a considerable amount of don't-care dependencies on our set of benchmarks [34], but only after intensive preprocessing including operations like blocked clause elimination. Since AIG-Solve profits from structure extraction from a CNF, we omit such preprocessing steps that destroy the structure. If we used a different backend QBF solver which does not rely on structure extraction, then we could use the full power of preprocessing, obtain many more don't-care dependencies, and profit much more from don't-care dependencies than in the current scenario.

In summary, for instances where variable elimination is already optimal, dependency elimination yields similar results. However, dependency elimination can yield equisatisfiable QBF that are smaller by orders of magnitude and allows to solve more instances in less time when variable elimination is not optimal. Therefore, dependency elimination is clearly superior to variable elimination.

5   Conclusion

We have presented a novel method to turn a DQBF into an equisatisfiable QBF. This is done by eliminating an appropriate set of dependencies from the formula, which requires to create copies of the involved existential variables. To determine an optimal elimination set that requires the fewest variable copies, we formulate this problem as a constraint system with non-linear objective function. This is solved using an MILP solver by handling the non-linearities by separation. Experiments show that dependency elimination allows to solve more instances with less memory consumption compared to variable elimination. Future research will try to integrate the structure of the formula into the selection process (which is currently only based on the quantifier prefix).

References

1. Balabanov, V., Chiang, H.K., Jiang, J.R.: Henkin quantifiers and Boolean formulae: a certification perspective of DQBF. Theor. Comput. Sci. **523**, 86–100 (2014)
2. Beineke, L.W., Little, C.H.C.: Cycles in bipartite tournaments. J. Comb. Theor. Ser. B **32**(2), 140–145 (1982)
3. Beyersdorff, O., Chew, L., Schmidt, R.A., Suda, M.: Lifting QBF resolution calculi to DQBF. In: Creignou, N., Le Berre, D. (eds.) SAT 2016. LNCS, vol. 9710, pp. 490–499. Springer, Cham (2016). doi:10.1007/978-3-319-40970-2_30
4. Bloem, R., Könighofer, R., Seidl, M.: SAT-based synthesis methods for safety specs. In: McMillan, K.L., Rival, X. (eds.) VMCAI 2014. LNCS, vol. 8318, pp. 1–20. Springer, Heidelberg (2014). doi:10.1007/978-3-642-54013-4_1

5. Bubeck, U.: Model-based transformations for quantified Boolean formulas. Ph.D. thesis, University of Paderborn (2010)
6. Bubeck, U., Büning, H.K.: Dependency quantified horn formulas: models and complexity. In: Biere, A., Gomes, C.P. (eds.) SAT 2006. LNCS, vol. 4121, pp. 198–211. Springer, Heidelberg (2006). doi:10.1007/11814948_21
7. Cai, M., Deng, X., Zang, W.: A min-max theorem on feedback vertex sets. Math. Oper. Res. **27**(2), 361–371 (2002)
8. Chatterjee, K., Henzinger, T.A., Otop, J., Pavlogiannis, A.: Distributed synthesis for LTL fragments. In: FMCAD 2013, pp. 18–25. IEEE, October 2013
9. Finkbeiner, B., Tentrup, L.: Fast DQBF refutation. In: Sinz, C., Egly, U. (eds.) SAT 2014. LNCS, vol. 8561, pp. 243–251. Springer, Cham (2014). doi:10.1007/978-3-319-09284-3_19
10. Fröhlich, A., Kovásznai, G., Biere, A.: A DPLL algorithm for solving DQBF. In: International Workshop on Pragmatics of SAT (POS), Trento, Italy (2012)
11. Fröhlich, A., Kovásznai, G., Biere, A., Veith, H.: iDQ: instantiation-based DQBF solving. In: Le Berre, D. (ed.) International Workshop on Pragmatics of SAT (POS 2014), Vienna, Austria. EPiC Series, vol. 27, pp. 103–116. EasyChair, July 2014
12. Gitina, K., Reimer, S., Sauer, M., Wimmer, R., Scholl, C., Becker, B.: Equivalence checking of partial designs using dependency quantified Boolean formulae. In: ICCD 2013, Asheville, NC, USA, pp. 396–403. IEEE CS, October 2013
13. Gitina, K., Wimmer, R., Reimer, S., Sauer, M., Scholl, C., Becker, B.: Solving DQBF through quantifier elimination. In: DATE 2015, Grenoble, France. IEEE, March 2015
14. Guo, J., Hüffner, F., Moser, H.: Feedback arc set in bipartite tournaments is NP-complete. Inf. Process. Lett. **102**(2–3), 62–65 (2007)
15. Henkin, L.: Some remarks on infinitely long formulas. In: Infinitistic Methods: Proceedings of the 1959 Symposium on Foundations of Mathematics, Warsaw, pp. 167–183. Panstwowe Wydawnictwo Naukowe, Panstwowe, September 1961
16. Janota, M., Klieber, W., Marques-Silva, J., Clarke, E.: Solving QBF with counterexample guided refinement. In: Cimatti, A., Sebastiani, R. (eds.) SAT 2012. LNCS, vol. 7317, pp. 114–128. Springer, Heidelberg (2012). doi:10.1007/978-3-642-31612-8_10
17. Janota, M., Marques-Silva, J.: Solving QBF by clause selection. In: Yang, Q., Wooldridge, M. (eds.) IJCAI 2015, Buenos Aires, Argentina, pp. 325–331. AAAI Press (2015). http://ijcai.org/Abstract/15/052
18. Karp, R.M.: Reducibility among combinatorial problems. In: Miller, R.E., Thatcher, J.W. (eds.) Proceedings of the Symposium on the Complexity of Computer Computations. The IBM Research Symposia Series, pp. 85–103. Plenum Press, New York, IBM Thomas J. Watson Research Center, Yorktown Heights (1972)
19. Lonsing, F., Biere, A.: DepQBF: a dependency-aware QBF solver. J. Satisf. Boolean Model. Comput. **7**(2–3), 71–76 (2010)
20. Lonsing, F., Egly, U.: Incremental QBF solving by DepQBF. In: Hong, H., Yap, C. (eds.) ICMS 2014. LNCS, vol. 8592, pp. 307–314. Springer, Heidelberg (2014). doi:10.1007/978-3-662-44199-2_48
21. Meyer, A.R., Stockmeyer, L.J.: Word problems requiring exponential time: preliminary report. In: STOC, pp. 1–9. ACM Press (1973)
22. Peterson, G., Reif, J., Azhar, S.: Lower bounds for multiplayer non-cooperative games of incomplete information. Comput. Math. Appl. **41**(7–8), 957–992 (2001)
23. Pigorsch, F., Scholl, C.: Exploiting structure in an AIG based QBF solver. In: DATE 2009, Nice, France, pp. 1596–1601. IEEE, April 2009

24. Pigorsch, F., Scholl, C.: An AIG-based QBF-solver using SAT for preprocessing. In: Sapatnekar, S.S. (ed.) DAC 2010, Anaheim, CA, USA, pp. 170–175. ACM Press, July 2010
25. Samer, M.: Variable dependencies of quantified CSPs. In: Cervesato, I., Veith, H., Voronkov, A. (eds.) LPAR 2008. LNCS, vol. 5330, pp. 512–527. Springer, Heidelberg (2008). doi:10.1007/978-3-540-89439-1_49
26. Samer, M., Szeider, S.: Backdoor sets of quantified Boolean formulas. J. Autom. Reason. **42**(1), 77–97 (2009)
27. Scholl, C., Becker, B.: Checking equivalence for partial implementations. In: DAC 2001, Las Vegas, NV, USA, pp. 238–243. ACM Press, June 2001
28. Slivovsky, F., Szeider, S.: Computing resolution-path dependencies in linear time. In: Cimatti, A., Sebastiani, R. (eds.) SAT 2012. LNCS, vol. 7317, pp. 58–71. Springer, Heidelberg (2012). doi:10.1007/978-3-642-31612-8_6
29. Slivovsky, F., Szeider, S.: Quantifier reordering for QBF. J. Autom. Reason. **56**, 459–477 (2015)
30. Gelder, A.: Variable independence and resolution paths for quantified Boolean formulas. In: Lee, J. (ed.) CP 2011. LNCS, vol. 6876, pp. 789–803. Springer, Heidelberg (2011). doi:10.1007/978-3-642-23786-7_59
31. Wimmer, K., Wimmer, R., Scholl, C., Becker, B.: Skolem functions for DQBF. In: Artho, C., Legay, A., Peled, D. (eds.) ATVA 2016. LNCS, vol. 9938, pp. 395–411. Springer, Cham (2016). doi:10.1007/978-3-319-46520-3_25
32. Wimmer, R., Gitina, K., Nist, J., Scholl, C., Becker, B.: Preprocessing for DQBF. In: Heule, M., Weaver, S. (eds.) SAT 2015. LNCS, vol. 9340, pp. 173–190. Springer, Cham (2015). doi:10.1007/978-3-319-24318-4_13
33. Wimmer, R., Reimer, S., Marin, P., Becker, B.: HQSpre – an effective preprocessor for QBF and DQBF. In: Legay, A., Margaria, T. (eds.) TACAS 2017. LNCS, vol. 10205, pp. 373–390. Springer, Heidelberg (2017). doi:10.1007/978-3-662-54577-5_21
34. Wimmer, R., Scholl, C., Wimmer, K., Becker, B.: Dependency schemes for DQBF. In: Creignou, N., Le Berre, D. (eds.) SAT 2016. LNCS, vol. 9710, pp. 473–489. Springer, Cham (2016). doi:10.1007/978-3-319-40970-2_29
35. Wimmer, R., Wimmer, K., Scholl, C., Becker, B.: Analysis of incomplete circuits using dependency quantified Boolean formulas. In: International Workshop on Logic and Synthesis (IWLS) (2016)
36. Wolsey, L.A.: Integer Programming. Wiley-Interscience, New York (1998)

Satisfiability Modulo Theories

Theory Refinement for Program Verification

Antti E.J. Hyvärinen[1]($\boxtimes$), Sepideh Asadi[1]($\boxtimes$), Karine Even-Mendoza[2]($\boxtimes$),
Grigory Fedyukovich[3]($\boxtimes$), Hana Chockler[2]($\boxtimes$), and Natasha Sharygina[1]($\boxtimes$)

[1] Università della Svizzera italiana, Lugano, Switzerland
{antti.hyvaerinen,sepideh.asadi,natasha.sharygina}@usi.ch
[2] King's College London, London, UK
{karine.even_mendoza,hana.chockler}@kcl.ac.uk
[3] University of Washington, Seattle, USA
grigory@cs.washington.edu

Abstract. Recent progress in automated formal verification is to a large degree due to the development of constraint languages that are sufficiently light-weight for reasoning but still expressive enough to prove properties of programs. Satisfiability modulo theories (SMT) solvers implement efficient decision procedures, but offer little direct support for adapting the constraint language to the task at hand. *Theory refinement* is a new approach that modularly adjusts the modeling precision based on the properties being verified through the use of combination of theories. We implement the approach using an augmented version of the theory of bit-vectors and uninterpreted functions capable of directly injecting non-clausal refinements to the inherent Boolean structure of SMT. In our comparison to a state-of-the-art model checker, our prototype implementation is in general competitive, being several orders of magnitudes faster on some instances that are challenging for flattening, while computing models that are significantly more succinct.

1 Introduction

The satisfiability modulo theories (SMT) [14] reasoning framework is currently one of the most successful approaches to verifying software in a scalable way. The approach is based on modeling the software and its specifications in propositional logic, while expressing domain-specific knowledge with first-order theories connected to the logic through equalities. Once a satisfying assignment is found for the propositional model, its consistency is queried as equalities from the theory solvers, which, in case of inconsistency, provide an explanation as a propositional clause. Successful verification of software relies on finding a model that is expressive enough to capture software behavior relevant to correctness, while sufficiently high-level to prevent reasoning from becoming prohibitively expensive. Since in general more precise theories are both more expensive computationally and potentially distracting for the automatic reasoning, finding such a balance is a non-trivial task.

We introduce *theory refinement*, a counter-example-guided abstraction refinement (CEGAR) [11,12] approach for modeling software modularly using theories

© Springer International Publishing AG 2017
S. Gaspers and T. Walsh (Eds.): SAT 2017, LNCS 10491, pp. 347–363, 2017.
DOI: 10.1007/978-3-319-66263-3_22

that are partially ordered with respect to their precision. Our main contribution is the process of gradually encoding a program using the most precise theory only for a critical subset of all program statements, while keeping lower precision for the rest of the statements. The critical subset of theories is identified based on counter-examples, and theories of different precision are bound to each other through special identities. We study several automatic heuristics for guiding the encoding and provide also a manual encoding option. We apply theory refinement on verification of safety properties of software through bounded model checking. However, we believe that the technique is applicable in most verification techniques where higher level information is available on the problem structure. This includes model checking [5] and upgrade checking [15], k-induction [23], the IC3 algorithm [6], and generation of inductive invariants [16]. We show that the modular composition of the theories preferring lower precision can be used to both obtain speed-up in solving and identifying statements whose precise semantics do not affect the program safety, providing the model checker with cleaner proofs.

Many SMT solvers use over-approximation through theories as a means of speeding up solving. For instance [8,9,17] organizes the theory solvers into layers that solve problems represented in QF_BV. The query is first given to fast and less precise theory solvers, and only passed on to the exact solver if previous layers fail to show unsatisfiability. In contrast to low-level SMT solving, this work studies how to automatically identify statements whose exact semantics can be ignored in model-checking. This shift of view point has several advantages: (i) the approach can be used both to obtain speed-up in solving, and as a means for synthesis and finding fix-points for transition relations; (ii) the guidance from the source code allows the use of more powerful heuristics for choosing which statements should remain abstract; and (iii) the refinement takes place on the level of the program, not at the level of the theory query, an approach potentially more natural from the point of view of the semantics of the program.

We present theory refinement with two new theories called *uninterpreted functions for programs* (UFP) and *bit vectors for programs* (BVP) that are based on the theories of quantifier-free uninterpreted functions with equality (QF_UF), and bit vectors (QF_BV), respectively. The two theories were chosen since they represent two natural extremes in precision and are commonly used in the layered solver approach (see, e.g., [17]). In addition to the functionality of QF_UF, UFP provides interpretations for constants, conversion of abstract values to concrete values, and commutativity for uninterpreted functions when applicable. The key difference in BVP compared to QF_BV is that BVP is capable of directly injecting non-clausal refinements, modeling the program statements bit-precisely, to the inherent Boolean structure maintained in the SMT solver.

We implemented theory refinement on the SMT solver OPENSMT [19] and the bounded model checker HIFROG [3] supporting a subset of the C language. We report promising results both with respect to speed and the amount of refined program statements on both instances from a software verification competition and our own regression test suite. We demonstrate that the approach has a potential of several orders of magnitude of improvement over the approach based solely on flattened bit-vectors, as implemented in the state-of-the-art tool CBMC and in our own tool. The implementation and the benchmarks are available at [1].

Related Work. Solving bit-vector problems with layers of theory solvers is introduced in [9] and further developed in [17]. While we work directly on software verification instead of bit-vectors, our approach is related, as we also use hierarchy of solvers combined with rewriting techniques. However, we work explicitly on the modeling language by automatically adjusting the precision to be different in different parts of the problem, and adding additional constraints that seams these parts together. In [8] a CEGAR based approach is used for solving problems involving arrays by transforming an abstract representation into clauses. We differ from this approach in that we integrate the system on the theory solver level, employing in the experiments the congruence closure algorithm together with a propositional solver. To the best of our knowledge, no existing approach uses this level of granularity in the modeling. Furthermore, we use counter-examples that are checked against the bit-precise implementation, and this way can avoid refinement of program parts that would need to be refined in approaches based on layered theory solvers.

Exploiting simultaneously several theories for one verification goal is not new. For example, [16] presents a system for synthesizing safe bit-precise inductive invariants for software. Compared to our work, the refinement direction is inverted: the software is first flattened, and in case of a time-out, converted to a domain-specific theory. Furthermore, we integrate seamlessly the theories UFP and BVP into an SMT solver whereas [16] considers real arithmetics.

Uninterpreted functions have been used together with the bit-precise encoding for verifying the equivalence of Verilog designs in [7,18]. The approach uses machine learning to identify sub-components that can likely be abstracted. In contrast, our emphasis is on software verification and integration to the SMT solver. A related approach [22] constructs test cases for scientific software by computing difference constraints from non-linear mathematical functions. This approach can be viewed as a special case of the framework we present in this paper; the formulas we derive can also be used for generating test cases, although this is not the focus of this paper. Similarly, [10] combines linear real arithmetic and equality of uninterpreted functions (QF_UF) for the SMT encoding of the program. The algorithm initially uses QF_UF to abstract non-linear operators, and then uses the monotonicity and the multiplication checks to identify spurious counterexample thus avoiding simulation and code execution. Both checks might result in a refinement formula, which is added then to the current SMT encoding. Unlike ours, their approach cannot be applied as such for bit-precise reasoning. In [3] we report early, very positive results on using the combination of EUF, LRA, and propositional flattening for encoding model checking problems. The current work which explores the possibilities in much more depth and rigor is motivated by this early result.

Another program-based refinement approach was proposed in [20], where compositional program is approximated with a program-specific theory of transition systems. Our approach is orthogonal to this, as we are able to handle programs in a more general way through the eventual flattening, while the theory of transition systems could likely be integrated as an additional theory.

2 Preliminaries

Let P be a loop-free program represented as a transition system, and t a *safety property*, that is, a logical formula over the variables of P. We are interested in determining whether all reachable states of P satisfy t. Given a program P and a safety property t, the task of a model checker is to find a counter-example, that is, an execution of P that does not satisfy t, or prove the absence of counter-examples on P. In the bounded, symbolic model checking approach followed in this paper the model checker encodes P into a logical formula, conjoins it with the negation of t, and checks the satisfiability of the encoding using an SMT solver. If the encoding is unsatisfiable, the program is safe, and we say that t holds in P. Otherwise, the satisfying assignment the SMT solver found is used to build a counter-example.

A *sort* is a set of constants. For example the Boolean sort $\mathbb{B} = \{\top, \bot\}$ consists of the Boolean constants, true and false. Given a set of sorts $\{T_0, \ldots, T_n\}$, a *function op* : $T_1 \times \ldots \times T_n \to T_0$ maps a (possibly empty) sequence of constants $v_1, \ldots, v_n$ such that $v_i \in T_i$ to a *return value* $v_0 \in T_0$. Functions mapping empty sequences are *variables*, and a *term* is either a constant, a variable, or an application of a function $op(t_1, \ldots, t_n)$ where t_i are, recursively, terms with a return value in the sort T_i. In most cases in this paper we use the usual infix notation together with parentheses to express the well-known arithmetic and logical functions.

3 Combination of Theories in Theory Refinement

This section fixes a notation for describing instances of the safety problem using SMT, and provides two communicating theories for solving the safety problem. The goal of the presentation is to clarify how the modeling works in the SMT framework, placing particular emphasis to the use of symbols and their semantic.

In modeling programs we consider sets of quantifier-free symbolic statements of the form $x = t$, where x is a variable, and t is a term. This form essentially corresponds to the Single static assignment (SSA) form [13] for loop-free programs. The symbolic statements are defined over a sort of bounded integers Sz and a Boolean sort $Sb = \{\top_l, \bot_l\}$; we distinguish between these sorts and, for instance, the sorts of integers $\mathbb{Z}$ and Booleans $\mathbb{B}$ to clarify the difference between this *symbolic encoding* (hence the S) and the representation used by an SMT solver. Table 1 lists the non-variable functions we consider in our encoding. Note that unlike some programming languages, including C and C++, we do not allow the encodings to interpret terms from Sz as terms from Sb or vice versa.

Table 1. The functions used in the encoding we consider. Note that unsigned and signed sum coincide.

Functions		Descriptions
Logical functions		
&&, ||	$Sb \times Sb \to Sb$	Logical and, or
!	$Sb \to Sb$	Logical not
Non-logical functions		
+, $*_u$, $*_s$, $/_u$, $/_s$	$Sz \times Sz \to Sz$	Sum, unsigned and signed product and division
$\%_u$, $\%_s$	$Sz \times Sz \to Sz$	Unsigned and signed remainder
$\ll$, $\gg_a$, $\gg_l$	$Sz \times Sz \to Sz$	left shift, arithmetic and logical right shift
&, |, ^	$Sz \times Sz \to Sz$	Bitwise and, or, exclusive or
$\sim$:	$Sz \to Sz$	bitwise complement
$\leq_s$, $\leq_u$, $<_s$, $<_u$, $\geq_s$, $\geq_u$, $>_s$, $>_u$	$Sz \times Sz \to Sb$	Signed and unsigned less than or equal to and greater than or equal to

We distinguish between the functions defined over the sort Sb and those defined over Sz, calling the former logical functions and the latter non-logical functions. The control-flow structures, such as if-then-elses, are encoded using the functions !, ||, and &&. For the purpose of this presentation we assume that the encodings do not contain arrays and pointers.[1] Figure 1 (*left*) shows an example sequence of statements that we will use as a running example in the discussion of this section.

3.1 Bit Vectors for Programs

Our theory of bit vectors for programs (BVP) has a single sort BVz^{bw} containing the integers representable in $bw \in \mathbb{N}$ bits. When the bit-width of the sort is clear from the context we simply write BVz for the sort. Each BVP term t of sort BVz^{bw} is associated with the bits $t_1, \ldots, t_{bw}$ which are variables from the sort $\mathbb{B}$. The bits t_1 and t_{bw} are called, respectively, the *least significant bit* and the *most significant bit* of t.

The BVP theory has two special constants 1^b and 0^b. For the constant 0^b, $0_i^b = \bot$, $1 \leq i \leq bw$. For the constant 1^b, $1_1^b = \top$ and $1_i^b = \bot$ for $2 \leq i \leq bw$. The equality of BVP is $=_{BVz}: BVz \times BVz \to BVz$. The interpretation of the equality is that if $x =_{BVz} y$ holds, then the value of the equality term is 1^b and otherwise 0^b. Finally, BVP has the functions defined in Table 1 with all sorts

---

[1] We do support these in our implementation, but their results are treated nondeterministically, that is, as unbound variables from Sz.

$$\left(c^b =_{BVz} \left((a^b \%_u 2^b) + (b^b \%_u 2^b) \right) \%_u 2^b \right)_1 \land$$

$c = \left((a \%_u 2) + (b \%_u 2) \right) \%_u 2$ $\quad \left((c')^b =_{BVz} (a^b + b^b) \%_u 2^b \right)_1 \land$

$c' = (a+b) \%_u 2$

$d = f *_u e *_u c$ $\quad\quad\quad\quad\quad \left(d^u = f^u *_u e^u *_u c^u \right) \land$

$d' = e *_u f *_u c'$ $\quad\quad\quad\quad \left((d')^u = e^u *_u f^u *_u (c')^u \right) \land$

$$\left(c^u = (c')^u \right) \leftrightarrow \left((c_1^b \leftrightarrow (c')_1^b) \land \ldots \land (c_{bw}^b \leftrightarrow (c')_{bw}^b) \right)$$

Fig. 1. (*Left*) a sequence of statements and (*right*) the corresponding encoding in combined UFP and BVP (to be described in Sect. 3.3). On the left all the variables are of sort Sz, and e and f are unbound.

replaced by the sort BVz. For a term t, the Boolean functions determining the bits t_i are computed through propositional flattening (see, e.g., [21]).

We encode a sequence of statements $P = \{x_1 = t_1, \ldots, x_n = t_n\}$ in BVP as follows. Each statement $x_i = t_i$ is converted to $|x_i|^b =_{BVz} |t_i|^b$, where the operator $| \cdot |^b$ is defined for a symbolic term t recursively:

$$|t|^b \stackrel{\text{def}}{=} \begin{cases} x^b & \text{if } t \doteq x \text{ is a variable or a constant} \\ |x|^b \bowtie |y|^b & \text{if } t \doteq x \bowtie y \text{ where } \bowtie \text{ is a binary function,} \\ \circ|x|^b & \text{if } t \doteq \circ x \text{ where } \circ \text{ is a unary function} \end{cases} \quad (1)$$

where $a \doteq b$ denotes that the term a matches the form of b. Conjunction of the least significant bits of encoded statements in P defines its BVP-encoding $[P]^b$:

$$[P]^b \stackrel{\text{def}}{=} (|x_1|^b =_{BVz} |t_1|^b)_1 \land \ldots \land (|x_n|^b =_{BVz} |t_n|^b)_1 \quad (2)$$

We say that a safety property t holds in program P if and only if $[P]^b \land \neg[t]_1^b$ is unsatisfiable. Based on the definition we can see that the symbolic encoding in Fig. 1 satisfies the safety property $(d = d')$ due to properties of modular arithmetics. The BVP encoding is often inefficient due to the quadratic growth of the formula with respect to bw. However, in many cases, the bit-precise encoding of statements (e.g., $*_u$ in Fig. 1) are irrelevant to the safety property, and can therefore be over-approximated. This motivates the use of less precise but more efficiently solvable encodings such as those based on uninterpreted functions.

3.2   Uninterpreted Functions for Programs

The logic UFP (Uninterpreted Functions for Programs) is the standard logic of quantifier-free uninterpreted functions having the Boolean sort $\mathbb{B}$, the standard Boolean functions $op : \mathbb{B} \times \ldots \times \mathbb{B} \to \mathbb{B}$ where op is an operator such as $\lor, \land$, and $\neg$, and an unbounded number of variables. In addition the logic is augmented with

- a sort $UFPn$ of real or integer numbers;
- the functions listed in Table 1 treated as uninterpreted functions with the sorts $UFPn$ and $\mathbb{B}$ instead of Sz and Sb respectively;
- commutativity of the functions $+$, $*_u$, $*_s$, $\&$, and $|$; and
- the concept of constants beyond the Boolean $\top$ and $\bot$.

As usual, UFP also contains the equality function $=_S: T \times T \to \mathbb{B}$ for all sorts T. As in the symbolic encoding, also in UFP we differentiate between two types of functions: those with a return sort $\mathbb{B}$, and those with a return sort $UFPn$.

Given a sequence of statements $P = \{x_1 = t_1, \ldots, x_n = t_n\}$, we denote its encoding in UFP by $[P]^u \overset{\text{def}}{=} ([x_1]^u =_{T_1} [t_1]^u) \wedge \ldots \wedge ([x_n]^u =_{T_n} [t_n]^u)$, where T_i is either $UFPn$ or $\mathbb{B}$ depending on the related sort. The encoding operator $[\cdot]^u$ is defined as follows for a term t:

$$[t]^u \overset{\text{def}}{=} \begin{cases} x^u & \text{if } t \doteq x \text{ is a variable or a constant} \\ [x]^u \wedge [y]^u & \text{if } t \doteq x \,\&\&\, y \\ [x]^u \vee [y]^u & \text{if } t \doteq x \,||\, y \\ \neg[x]^u & \text{if } t \doteq \,!\, x \\ [x]^u \bowtie [y]^u & \text{if } t \doteq x \bowtie y \text{ where } \bowtie \text{ is a non-logical function.} \end{cases} \tag{3}$$

We distinguish between the notions of program safety in UFP and in BVP. In particular, we say that a safety property t holds in program P in UFP if and only if $[P]^u \wedge \neg[t]^u$ is unsatisfiable.

The program in Fig. 1 is safe with respect to the safety property $!\,(c = c')\,||\,(d = d')$ in UFP and therefore also in BVP. However, it is not safe in UFP with respect to the safety property $d = d'$ that is safe in BVP. For checking safety of programs in UFP we use a theory solver implementing a congruence closure algorithm [14] that is modified to support constants and commutativity. The modifications are described in more detail in Sect. 5.1.

In our recent experiments [3] we showed that safety of many programs can be established by interpreting the arithmetic functions as uninterpreted functions. In the next subsection we describe how the UFP logic and the BVP logic can be combined.

3.3 Combination of UFP and BVP

We present the theory refinement approach using a seamless integration of the UFP and BVP encoding, and therefore require a form of theory combination. However, unlike in conventional theory combination on bit vectors (see, e.g., [17]), we do not need to consider bit-vectors as theories, but instead they are embedded directly to the Boolean structure of the SMT solver. The two theories UFP and BVP are combined using a *binding formula* defined as follows.

Definition 1. *Given a symbolic statement t, let $[t]^u$ and $[t]^b$ be its UFP and BVP-encodings respectively. If both $[t]^u$ and $[t]^b$ appear together in a formula,* *we say that t is* bound. *Let B be the set of all bound statements. The* binding formula *for B (denoted F_B) is defined as*

$$F_B \stackrel{\text{def}}{=} \bigwedge_{t,t' \in B} ([t]^u = [t']^u) \leftrightarrow (([t]^b_1 \leftrightarrow [t']^b_1) \wedge \ldots \wedge ([t]^b_{bw} \leftrightarrow [t']^b_{bw})) \qquad (4)$$

Intuitively, the combination of the theories UFP and BVP with F_B allow us to express an over-approximation of the symbolic encoding of a program. This is stated more formally in the following theorem.

Theorem 1. *Let P be a program. Then $[P]^b \wedge F_B \models [P]^u$.*

Proof (sketch). By simulation of executions in BVP: if there exist values $v^b_1, \ldots, v^b_n$ for the variables $x^b_1, \ldots, x^b_n$ in a term $[a = t]^b$ then the same values $v^u_1, \ldots, v^u_n$ satisfy the corresponding equality $[a]^u = [t]^u$. □

Figure 2 shows the combined UFP and BVP encoding schematically. The symbolic encoding of a program is partitioned by the model checker into three parts: the UFP encoding, the BVP encoding, and the binding formula F_B. The conjunction of these is solved by the SMT solver. Figure 1 *(right)* describes a combination encoding of UFP and BVP together with the necessary binding formula for the running example.

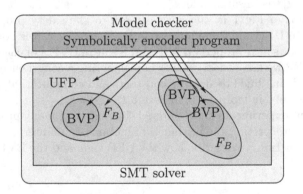

Fig. 2. A symbolic encoding of a program and the corresponding SMT formula. In the schematic example most of the program is encoded using UFP, while certain critical parts are encoded in BVP and made to communicate with the UFP encoding using the binding formula F_B.

4   Counterexample-Guided Theory Refinement

This section provides an algorithm for verifying safety of programs by gradually refining the *precision* ρ of the symbolic encoding from UFP to BVP in parts where satisfying truth assignments show that it is necessary for soundness. Algorithm 1 describes the high-level idea. The algorithm takes as input

---

Algorithm 1. The Counterexample-Guided Theory Refinement Algorithm

input : $P = \{(x_1 = t_1), \ldots, (x_n = t_n)\}$: a program, and t: a safety property
output: $\langle$**Safe**, $\bot\rangle$ or $\langle$**Unsafe**, $CE^b\rangle$

1   For all $1 \leq i \leq n$ initialize $\rho[x_i = t_i] \leftarrow [x_i = t_i]^u$
2   $\rho[t] \leftarrow [t]^u$
3   $F_B \leftarrow \top$
4   **while true do**
5      $Query \leftarrow \rho[x_1 = t_1] \wedge \ldots \wedge \rho[x_n = t_n] \wedge \neg\rho[t] \wedge F_B$
6      $\langle result, CE\rangle \leftarrow$ checkSAT($Query$)
7   **if** *result* is **UnSAT then**
8   **return** $\langle$**Safe**, $\bot\rangle$
9   **end**
10     $CE^b \leftarrow$ getValues(CE)
11   **foreach** $s \in P \cup \{t\}$ *s.t.* $\rho[s] \not\models [s]^b$ **do**
12        $\langle result, \_\rangle \leftarrow$ checkSAT($[s]^b \wedge CE^b$)
13   **if** *result* is **UnSAT then**
14           $\rho[s] \leftarrow$ refine^s($\rho[s]$)
15           $F_B \leftarrow$ computeBinding(ρ)
16   **break**
17   **end**
18   **end**
19   **if** *No s was refined at line 14* **then**
20   **return** $\langle$**Unsafe**, $CE^b\rangle$
21   **end**
22 **end**

---

a symbolically encoded problem P and a safety property t, and returns either
Safe, if t holds in P, or **Unsafe** with a bit-precise counter-example if t does
not hold in P. During the execution the algorithm picks statements $s \in P \cup \{t\}$
and refines their approximations in ρ until $\rho[s]$ is equivalent to $[s]^b$. Based on ρ,
the algorithm constructs the binding formula F_B sufficient to connect the UFP
and BVP terms.

The safety of the program is tested at lines 5–9 using the current precision
ρ and the binding formula. If the check succeeds, the algorithm terminates at
line 9. Otherwise, a satisfying truth assignment is extracted at line 10 and then
used to refine ρ at lines 11–18.

The need for refinement is checked for every statement s with a precision $\rho[s]$
not equivalent to $[s]^b$. If the truth assignment CE^b is inconsistent with $[s]^b$ then
$\rho[s]$ is refined to block the truth assignment. If at least one such replacement
happens in the current iteration, the execution proceeds to line 5. In practice it
is a good idea to refine several statements based on a single counter-example, as
discussed in Sect. 6. If no refinement is done, the truth assignment corresponds
to a counter-example and the algorithm terminates at line 20.

The algorithm uses four sub-procedures checkSAT, getValues, refine^s, and
computeBinding. checkSAT(F) determines the satisfiability of a formula F,

getValues(CE) computes a BVP encoding of CE through substituting the abstract values from UFP with concrete BVP values. refine$^s(F)$ refines the statement s with respect to the previous precision F, and computeBinding(ρ) computes the binding formula using Definition 1. Below we give a definition for the refine procedure, while the other procedures will be discussed in more detail in Sect. 5.3.

Definition 2. *The procedure* refine$^s(F)$ *returns an iterative refinement of the statement s of the symbolic encoding with respect to F, such that (i)* refine$^s(F) \models F$, *and (ii)* refine^s *has a fix-point that is equivalent to $[s]^b$ and reachable in a finite number of applications of* refine^s.

While in the implementation discussed in Sect. 5 we use refine$^s(F) = [s]^b \wedge [s]^u$, we want to point out the possibility of using interpolation-based methods (see, e.g., [4]) for the refinement.

Theorem 2. *Algorithm 1 terminates in a finite number of steps.*

Proof. Assume that Algorithm 1 does not terminate. Then there is a term in $P \cup \{t\}$ that can be refined an unbounded number of times before the fix-point equivalent to $[s]^b$ is reached, which contradicts Definition 2.   □

Theorem 3. *Algorithm 1 returns* **Unsafe** *if and only if the symbolic encoding P has an execution violating the safety property t.*

Proof. The algorithm maintains the invariants

$$\begin{aligned} &\text{Inv1 } [x_1 = t_1]^b \wedge \ldots \wedge [x_n = t_n]^b \models \rho[x_1 = t_1] \wedge \ldots \wedge \rho[x_n = t_n] \\ &\text{Inv2 } [t]^b \models \rho[t] \end{aligned} \tag{5}$$

at line 14 by Definition 2 and Theorem 1. Assume that the algorithm returns **Unsafe** but there is no execution violating the safety property t. Then there is a truth assignment σ such that $\rho[x_1 = t_1] \wedge \ldots \wedge \rho[x_n = t_n] \wedge F_B$ is true and $\rho[t]$ is false. The truth assignment σ must also satisfy $[x_1 = t_1]^b \wedge \ldots \wedge [x_n = t_n]^b$. By Inv2, if $\rho[t]$ is false also $[t]^b$ is false, hence contradicting the unsafety of (P,t). Now assume the algorithm returns **Safe** but there is an execution of P violating t. Then there is a truth assignment satisfying $[P]^b \wedge \neg[t]^b$. Since by Theorem 1 both $[P]^b \wedge F_B \models \rho[x_1 = t_1] \wedge \ldots \wedge \rho[x_n = t_n]$ and $\neg[t]^b \wedge F_B \models \neg\rho[t]$, also the query on line 5 is satisfiable, contradicting the assumption.   □

5   Implementation of Theory Refinement Algorithm

This section describes the prototype implementation of the theory refinement algorithm. The algorithm is implemented on the SMT solver OpenSMT [19] and the bounded model checker HiFrog [3]. The overview of implementation including the three main components and interactions between them is depicted in Fig. 3.

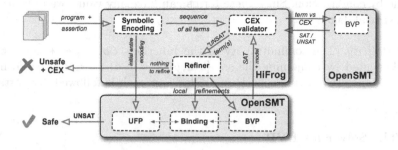

Fig. 3. The SMT-based model checking framework implementing a theory refinement approach used in the experiments.

5.1   The Solver for UFP

The UFP theory solver is based on the co-operation between a congruence closure algorithm, which maintains sets of equivalence classes and inequalities between the classes, and a SAT solver, which enforces a propositional structure describing the relations between the equalities. We refer the reader to [14] for the full description of the *egraph* algorithm that the UFP solver bases on.

Constants. The original egraph algorithm does not support constants other than the Boolean $\top$ and $\bot$, but constants play often an important role in our benchmarks. The egraph algorithm can represent an inequality between two terms t_1, t_2 by asserting explicitly the inequality $t_1 \neq t_2$ over these terms. This representation grows quadratically in the number of constants and therefore is not scalable. We adopt a different strategy for representing the inequalities between constants. An equivalence class in the egraph algorithm is represented by a linked list binding together the terms in the same class. Each class is represented by a canonical term from the linked list. In the original algorithm of [14], when two equivalence classes a and b are joined, the canonical term of the new class $a \cup b$ is the representative of whichever class a or b contains more terms. This is done to allow efficient joining and splitting in the backtracking search driven by the SMT solver. In our implementation the representative of a class a is always a constant if a contains a constant. The implicit inequality between constants is then implemented by a check that the respective equivalence classes are not both represented by a constant term. This approach fits naturally into the egraph algorithm and explanation generation. In the experiments we observed no noticeable slowdown compared to the original approach.

Values. Algorithm 1 requires concrete values from the UFP theory to construct a counter-example candidate. In general the values for UFP are obtained by assigning a running number for each equivalence class that the egraph algorithm maintains. However, there are two special cases for the values. First, if the equivalence class contains a constant, the value is that of the constant. Second, a preprocessing step in the SMT solver removes terms that only appear on clauses that are true by construction. Since these terms can have any value, we indicate this with a special flag.

Commutativity. The commutativity of the functions $Co = \{+, *_u, *_s, \&, |\}$ is implemented by conjoining the set $\{\circ(a, b) \leftrightarrow \circ(b, a) \mid \circ \in Co, \circ(a, b) \text{ in } P\}$ to the instance $[P]^u$ being solved. A similar approach is followed, for instance, in [10].

5.2   The Solver for BVP

The BVP theory is solved through propositional flattening [21]. The solver supports the operations listed in Table 1, and allows the use of arbitrary bit-widths.[2] Based on an extensive testing the implementation is robust, but still prototypical in the sense that we implement no sophisticated pre-processing techniques that are available in many other bit-vector solvers (see, e.g., [9]).

Unlike many other SMT solvers (see, e.g., [17]), we do not implement the bit-vector solver as a separate SAT solver working on the flattening and driven by the main SAT solver. Instead, we flatten the problem directly to the main SAT solver. This has several advantages: we avoid the overhead of duplicate solver instantiation, and we enable the solver to potentially learn much more intricate relationships between the flattened formula and the formula in UFP. However, an in-depth analysis of the implications of this design is beyond the scope of this paper.

5.3   Theory Refinement in Model Checking

We integrated Algorithm 1 into the bounded model checker HiFrog for C programs. HiFrog obtains first the symbolic encoding of the program P and a safety property t through a sequence of pre-processing steps, builds then the UFP formula, and finally gradually transforms parts of the UFP formula into BVP based on truth assignments until the safety is determined. We follow the approach where safety properties are expressed as assertions in the C code. The architecture is depicted in Fig. 3. HiFrog maintains two SMT solvers during the execution and which are represented by the checkSAT calls in Algorithm 1: the *main solver* for checking the satisfiability query constructed at line 5 (shown on the bottom of Fig. 3) and the *refinement solver* for checking the spuriousness of each counter-example at line 10 (shown on the right of Fig. 3). This choice was taken so that the expensive calls on the main solver would not be slowed down by unnecessary clauses at the refinement solver.

The counter-examples are flattened to propositional logic through the call to getValues by mapping the values in UFP to a unique bit-vector constant of the given bit width bw. At this stage of the development we ignore the case where the UFP solver gives more equivalence classes than what is representable in bw bits, since this limitation did not affect our results.

---

[2] The shift operations $\ll, \gg_a, \gg_l$ assume a bit-width that is a power of two.

The binding formula (see Definition 1) is updated whenever a statement $x = t$ is refined. This is done by first constructing the BVP formulas $[x]^b$ and $[t]^b$, and then adding the missing equalities to F_B with the call to `computeBinding`.

6   Experimental Results

We evaluated the theory-refinement mode of HiFrog on C programs mostly coming from the software model checking competition (SV-COMP). The benchmarks were split into the *safe* (128 instances) and *unsafe* (30 instances) sets, indicating whether the bad behavior is reachable or not. Among safe instances, 17 require refinements.

For benchmarking we used Ubuntu 14.04 Linux system with two Intel Xeon E5620 CPUs clocked at 2.40 GHz and 12 GB memory limit per process using a timeout of 300 s CPU time. The model checker was compiled with the GNU C++ compiler and the O3 optimization level. The complete experimental results, the source code, and a virtual machine are all available at [1].

Figure 4 shows the verification results on safe properties. We compared (Fig. 4, *left*) the HiFrog's theory-refinement mode against CBMC version 5.7, the winner of the software model checking competition falsification track in 2017.[3]

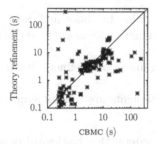 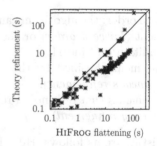

Fig. 4. Timings of CBMC (*left*) and HiFrog's flattening (*right*) against HiFrog's theory refinement for the safe instances.

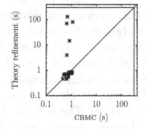 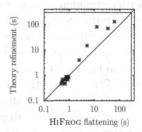

Fig. 5. Timings of CBMC (*left*) and HiFrog's flattening (*right*) against HiFrog's theory refinement for the unsafe instances.

---

[3] OpenSMT2:   https://scm.ti-edu.ch/repogit/opensmt2.git,   git   ID: 99c960e4c; HiFrog (including CBMC that shares the CProver framework [2] with HiFrog): https://scm.ti-edu.ch/repogit/hifrog, git ID b35956f2c.

In 101 cases, HiFROG was either as fast or faster than CBMC, sometimes by orders of magnitude. Furthermore, HiFROG's theory refinement mode is compared against HiFROG's propositional flattening (Fig. 4, *right*), hence ensuring that the only difference in the solvers is in how the symbolic encoding is presented to the SMT solver. In 115 cases, the theory refinement was either as fast or faster than flattening in determining safety, providing a more convincing evidence that the theory refinement approach works well in practice.

The verification results of unsafe benchmarks are shown in Fig. 5. In five cases, bug detection by HiFROG was slower than the one by CBMC since HiFROG required iterative refining of all the expressions to confirm the validity of the counter-example. However, in the remaining cases, HiFROG was comparable to CBMC.

6.1 Experiments on Refinement Heuristic

Algorithm 1 does not address which exact statement should be refined based on a counter-example on Line 11 in case there are several possibilities. However this selection affects the run time of the model checking and is therefore of practical interest. We consider the following three features while building a refinement heuristic:

- Traversal order: the algorithm can proceed either by choosing from P the first statement (*forward order*) or the last statement (*backward order*) satisfying the condition on Line 11.
- All statements falsified by the counter-example are refined simultaneously (*simultaneous refinement*).
- All statements that depend on refined statements are refined simultaneously (*dependency refinement*).

The heuristics are as follows: H0 – Forward order; H1 – Backward order; H2 – Forward order with simultaneous refinement; H3 – Backward order with simultaneous refinement; H4 – Forward order with dependency refinement; H5 – Backward order with dependency refinement; H6 – Forward order with simultaneous and dependency refinement; and H7 – Backward order with simultaneous and dependency refinement. Based on the experimentation, the fastest solver on average results from using Forward order with dependency refinement. This is the heuristic we use in the results on Figs. 4 and 5. We briefly report on the results of the heuristics in Table 2 over the 17 instances of our total benchmark set where statements were refined. This benchmark set contains three crafted instances and the rest from the *bitvector* category of SV-COMP. The row labeled *#solved* reports how many instances the heuristic could solve before the timeout, *#ref* reports how many statements in total had to be refined over the set, and *time* reports the total run time. As a reference the table also reports results on the heuristic *Min* that requires no run time and computes a minimum set of refinements required to prove the property.

Table 2. Comparison of the heuristics against Min on instances requiring refinement.

	H0	H1	H2	H3	H4	H5	H6	H7	*Min*
#solved	**17**	16	**17**	**17**	**17**	**17**	**17**	**17**	17
#ref	660	2218	1250	1250	**533**	2266	1442	1831	162
time (s)	538	223	257	317	**123**	166	147	158	46.2

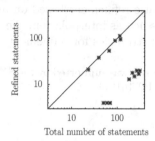

Fig. 6. The number of refined statements using the *Min* heuristic with respect to the total number of statements.

Finally, in Fig. 6 we show the reduction in the number of refined statements when using the *Min* heuristic on the 17 instances. As expected, the performance of the heuristic depends on the instance, but when effective, dramatically reduces the amount of flattened statements.

While the results are still preliminary mostly due to the prototype nature of the tools we are developing, we believe that they make a very strong point for the potential of the theory refinement approach in software model checking.

7   Conclusions and Future Work

We presented a new approach for abstraction refinement in software verification with SMT solvers. Our approach introduces iterative *theory refinement* and supports solving of formulas of combined theories in the SMT solver, where the binding to the theory is maintained by a series of identities in the original formula. Our main contribution is the gradual encoding process that uses the most precise theory only for a subset of all program statements, while handling the rest of the statements by using the less precise theories. This subset of the statements could either be identified by checking spurious counter-examples or simply specified by the user. Our framework can be extended by sets of theories with a partial order of refinement defined among them. In this paper, we demonstrated the framework on the UFP theory with the partial refinement to the BVP theory. We implemented this framework in the OpenSMT [19] solver and the model checker HiFrog [3].

We study different refinement strategies and compare them against a strategy computed off-line, as well as with the encoding into propositional logic, known as flattening or bit-blasting. Improvement is seen both in the running time and in the size of the resulting formula, demonstrating that the spurious counterexamples are usually eliminated by refining a small number of statements in the formula.

In future we plan to progress in several directions. We will study theory refinement with arithmetic theories and arrays, defining a partial order among theories based on the level of abstraction/refinement that they provide. We will further improve the automatic refinement based on an analysis of the counterexamples using approaches such as interpolation. We also plan to develop more sophisticated heuristics and strategies for refinement.

Acknowledgements. This work was supported by the SNF grants 163001 and 166288 and the SNF fellowship P2T1P2_161971.

References

1. http://verify.inf.usi.ch/hifrog/theoref
2. http://www.cprover.org/
3. Alt, L., Asadi, S., Chockler, H., Even Mendoza, K., Fedyukovich, G., Hyvärinen, A.E.J., Sharygina, N.: HiFrog: SMT-based function summarization for software verification. In: Legay, A., Margaria, T. (eds.) TACAS 2017. LNCS, vol. 10206, pp. 207–213. Springer, Heidelberg (2017). doi:10.1007/978-3-662-54580-5_12
4. Alt, L., Fedyukovich, G., Hyvärinen, A.E.J., Sharygina, N.: A proof-sensitive approach for small propositional interpolants. In: Gurfinkel, A., Seshia, S.A. (eds.) VSTTE 2015. LNCS, vol. 9593, pp. 1–18. Springer, Cham (2016). doi:10.1007/978-3-319-29613-5_1
5. Biere, A., Cimatti, A., Clarke, E., Zhu, Y.: Symbolic model checking without BDDs. In: Cleaveland, W.R. (ed.) TACAS 1999. LNCS, vol. 1579, pp. 193–207. Springer, Heidelberg (1999). doi:10.1007/3-540-49059-0_14
6. Bradley, A.R.: SAT-based model checking without unrolling. In: Jhala, R., Schmidt, D. (eds.) VMCAI 2011. LNCS, vol. 6538, pp. 70–87. Springer, Heidelberg (2011). doi:10.1007/978-3-642-18275-4_7
7. Brady, B.A., Bryant, R.E., Seshia, S.A.: Learning conditional abstractions. In: Proceedings of FMCAD 2011, pp. 116–124. FMCAD Inc. (2011)
8. Brummayer, R., Biere, A.: Lemmas on demand for the extensional theory of arrays. J. Satisfiability Boolean Model. Comput. **6**, 165–201 (2009)
9. Bruttomesso, R., et al.: A lazy and layered SMT($\mathcal{BV}$) solver for hard industrial verification problems. In: Damm, W., Hermanns, H. (eds.) CAV 2007. LNCS, vol. 4590, pp. 547–560. Springer, Heidelberg (2007). doi:10.1007/978-3-540-73368-3_54
10. Cimatti, A., Griggio, A., Irfan, A., Roveri, M., Sebastiani, R.: Invariant checking of NRA transition systems via incremental reduction to LRA with EUF. In: Legay, A., Margaria, T. (eds.) TACAS 2017. LNCS, vol. 10205, pp. 58–75. Springer, Heidelberg (2017). doi:10.1007/978-3-662-54577-5_4
11. Clarke, E.M., Grumberg, O., Jha, S., Lu, Y., Veith, H.: Counterexample-guided abstraction refinement. In: Emerson, E.A., Sistla, A.P. (eds.) CAV 2000. LNCS, vol. 1855, pp. 154–169. Springer, Heidelberg (2000). doi:10.1007/10722167_15
12. Clarke, E.M., Grumberg, O., Jha, S., Lu, Y., Veith, H.: Counterexample-guided abstraction refinement for symbolic model checking. J. ACM **50**(5), 752–794 (2003)

13. Cytron, R., Ferrante, J., Rosen, B., Wegman, M., Zadeck, F.: An efficient method of computing static single assignment form. In: Proceedings of POPL 1989, pp. 25–35. ACM (1989)
14. Detlefs, D., Nelson, G., Saxe, J.B.: Simplify: a theorem prover for program checking. J. ACM **52**(3), 365–473 (2005)
15. Fedyukovich, G., Sery, O., Sharygina, N.: eVolCheck: incremental upgrade checker for C. In: Piterman, N., Smolka, S.A. (eds.) TACAS 2013. LNCS, vol. 7795, pp. 292–307. Springer, Heidelberg (2013). doi:10.1007/978-3-642-36742-7_21
16. Gurfinkel, A., Belov, A., Marques-Silva, J.: Synthesizing safe bit-precise invariants. In: Ábrahám, E., Havelund, K. (eds.) TACAS 2014. LNCS, vol. 8413, pp. 93–108. Springer, Heidelberg (2014). doi:10.1007/978-3-642-54862-8_7
17. Hadarean, L., Bansal, K., Jovanović, D., Barrett, C., Tinelli, C.: A tale of two solvers: eager and lazy approaches to bit-vectors. In: Biere, A., Bloem, R. (eds.) CAV 2014. LNCS, vol. 8559, pp. 680–695. Springer, Cham (2014). doi:10.1007/978-3-319-08867-9_45
18. Ho, Y.S., Chauhan, P., Roy, P., Mishchenko, A., Brayton, R.: Efficient uninterpreted function abstraction and refinement for word-level model checking. In: Proceedings of FMCAD 2016, pp. 65–72. ACM (2016)
19. Hyvärinen, A.E.J., Marescotti, M., Alt, L., Sharygina, N.: OpenSMT2: an SMT solver for multi-core and cloud computing. In: Creignou, N., Le Berre, D. (eds.) SAT 2016. LNCS, vol. 9710, pp. 547–553. Springer, Cham (2016). doi:10.1007/978-3-319-40970-2_35
20. Katz, G., Barrett, C., Harel, D.: Theory-aided model checking of concurrent transition systems. In: Proceedings of FMCAD 2015, pp. 81–88. IEEE (2015)
21. Kroening, D., Strichman, O.: Decision Procedures - An Algorithmic Point of View. Texts in Theoretical Computer Science. An EATCS Series, 2nd edn. Springer, Heidelberg (2016)
22. Kutsuna, T., Ishii, Y., Yamamoto, A.: Abstraction and refinement of mathematical functions toward SMT-based test-case generation. Int. J. Softw. Tools Technol. Transf. **18**(1), 109–120 (2016)
23. McMillan, K.L.: An interpolating theorem prover. Theor. Comput. Sci. **345**(1), 101–121 (2005)

On Simplification of Formulas
with Unconstrained Variables and Quantifiers

Martin Jonáš[✉] and Jan Strejček

Masaryk University, Brno, Czech Republic
{xjonas,strejcek}@fi.muni.cz

Abstract. Preprocessing of the input formula is an essential part of all modern SMT solvers. An important preprocessing step is formula simplification. This paper elaborates on simplification of quantifier-free formulas containing unconstrained terms, i.e. terms that can have arbitrary values independently on the rest of the formula. We extend the idea in two directions. First, we introduce partially constrained terms and show some simplification rules employing this notion. Second, we show that unconstrained terms can be used also for simplification of formulas with quantifiers. Moreover, both these extensions can be merged in order to simplify partially constrained terms in formulas with quantifiers. We experimentally evaluate the proposed simplifications on formulas in the bit-vector theory.

1 Introduction

For most of the modern SMT solvers, preprocessing of the input formula is a crucial step for the efficiency of the solver. Therefore, modern SMT solvers employ hundreds of rewrite rules in order to simplify the input formula [10]. The aim of most of the simplifications is to reduce the size of the input formula and to replace expensive operations by easier ones. One class of these simplification rules focuses on formulas containing unconstrained variables. An unconstrained variable is a variable that occurs only once in the formula and therefore can be set to any suitable value without affecting the rest of the formula. For example, the formula $x + (5 * y + z) = y * z$ can be rewritten to an equisatisfiable formula $u = y * z$ because, regardless of the values of y and z, the term $x + (5 * y + z)$ can be evaluated to any value of u by choosing a suitable value of x. Such terms, which can be set to an arbitrary value by a well-suited choice of values of unconstrained variables, are called *unconstrained terms*. The principle of simplifications of unconstrained terms is recalled in more detail in Sect. 3. This simplification technique was proposed by Bruttomesso [8] and Brummayer [7], who independently observed that industrial benchmarks often contain non-trivial amount of unconstrained variables. For example, consider SMT queries coming from symbolic execution of a program, where a query is satisfiable if and only if the symbolically executed program path is feasible. There are basically two

---

The research was supported by Czech Science Foundation, grant GBP202/12/G061.

S. Gaspers and T. Walsh (Eds.): SAT 2017, LNCS 10491, pp. 364–379, 2017.
DOI: 10.1007/978-3-319-66263-3_23

sources of unconstrained variables in such queries. One source is input variables: such a variable is unconstrained in all queries corresponding to the symbolic execution of a path that reads the input variable at most once. The second source is program variables that are assigned on an executed path, but not read yet. For instance, the execution of an assignment y := x + 5 leads to a conjunct $y = x + 5$ in the path condition query, where y does not appear anywhere else in the query (unless it is read) and thus it is unconstrained. Such situations are especially frequent when analyzing Static Single Assignment (SSA) code such as LLVM, which uses many program variables.

1.1 Contribution and Structure of the Paper

In this paper, we extend the notion of unconstrained terms in several ways:

- In some cases, the definition of unconstrained term is too restrictive by allowing only terms that can evaluate to every possible value by a suitable choice of values of unconstrained variables. For example, Bruttomesso and Brummayer describe the simplification rule that replaces the bit-vector term $c \cdot x$ by a fresh variable y, if x is an unconstrained variable and c is an odd constant. However, if c is even, the simplification is no longer possible. We describe a less restrictive simplification using *partially constrained terms*, which for example allows replacing the term $6 \cdot x$ by the term $2 \cdot y$; although these two terms can not evaluate to all possible values, they can evaluate to precisely the same set of values.

  Partially constrained terms are studied in Sect. 4. This section also shows that several *ad-hoc* simplification rules introduced by Bruttomesso can be seen as instances of simplification of partially constrained terms. Our definition of partially constrained terms allows construction of more similar rules.
- Previously, the simplifications of unconstrained terms were described only on quantifier-free formulas. In Sect. 5, we formalize the conditions under which a simplification of unconstrained terms can be performed on *quantified* formulas.
- Sect. 6 combines techniques from the two preceding sections and describes simplification of *partially constrained terms in quantified formulas*. Furthermore, the resulting technique is combined with quantifier-specific simplification rules to allow more efficient and straightforward applications.

Sect. 7 experimentally evaluates the influence of proposed simplifications on performance of state-of-the-art SMT solvers Z3 [9], Boolector [13], and Q3B [12] on quantified bit-vector formulas arising in software and hardware verification.

We emphasize that the presented approach is not tied to any particular theory. We use the bit-vector theory in many examples and in evaluation as its functions tend to produce unconstrained terms when at least one argument is an unconstrained.

2   Preliminaries

This section briefly recalls the *theory of fixed sized bit-vectors* (*BV* or *bit-vector theory* for short). It is a multi-sorted first-order theory with infinitely many sorts corresponding to bit-vectors of various lengths. The BV theory uses only three predicates, namely *equality* (=), *unsigned inequality* of binary-encoded natural numbers ($\leq_u$), and *signed inequality* of integers in two's complement representation ($\leq_s$). The theory also contains various functions, namely *addition* (+), *multiplication* (·), *unsigned division* (÷), bit-wise *and* (bvand), bit-wise *or* (bvor), bit-wise *exclusive or* (bvxor), *left-shift* ($\ll$), *right-shift* ($\gg$), *concatenation* (concat), and *extraction* of n bits starting from position p (extract_p^n). The signature of BV theory also contains constants $c^{[n]}$ for each bit-width $n > 0$ and a number $0 \leq c \leq 2^n - 1$. Additionally, as in SMT-LIB [1] and in Hadarean [11], we suppose a distinguished sort *Boolean* and instead of treating formulas and terms differently, formulas are merely the terms of sort *Boolean*. This sort is similar to bit-vectors of length 1, but *Boolean* uses standard logic operators ($\wedge, \vee, \neg$) and not the bit-vector ones. If a bit-width of a constant or a variable is not specified, we suppose that it is equal to 32. The precise description of the multi-sorted logic can be found for example in Barrett et al. [3]. For a precise description of the syntax and semantics of the BV theory, we refer the reader to Hadarean [11].

For a valuation μ that assigns to each variable a value in its domain, $[\![\_]\!]_\mu$ denotes the evaluation function, which assigns to each formula φ the value obtained by substituting free variables in φ by values given by μ and evaluating all functions, predicates, logic operators etc. A formula φ is *satisfiable* if $[\![\varphi]\!]_\mu = \top$ for some valuation μ; it is *unsatisfiable* otherwise. Formulas φ and ψ are *equivalent* if they have the same set of free variables and for each valuation μ of these free variables, the equality $[\![\varphi]\!]_\mu = [\![\psi]\!]_\mu$ holds. Formulas φ and ψ are *equisatisfiable*, if either both are satisfiable, or both are unsatisfiable.

If φ is a formula and t, s are terms of the same sort, we use $\varphi[t \leftarrow s]$ to denote the formula φ with every occurrence of the term t replaced by the term s. In particular, if x is a variable, $\varphi[x \leftarrow t]$ is the result of substituting the variable x by the term t. Further, $vars(\varphi)$ denotes the set of free variables in φ. Finally, a variable $v \in vars(\varphi)$ is called *unconstrained in* φ, if it occurs only once in the formula φ and it is called *constrained* otherwise.

If convenient, we work with functions as with sets of pairs. For example, the union of functions $f : A \rightarrow B$ and $g : C \rightarrow D$ where $A \cap C = \emptyset$ is a function $f \cup g : (A \cup C) \rightarrow (B \cup D)$. Similarly, $\{(a, b)\}$ is a function from the set $\{a\}$ to the set $\{b\}$. This function is also denoted as $\{a \mapsto b\}$.

3   Unconstrained Terms in Quantifier-Free Formulas

This section formalizes known simplifications of quantifier-free formulas containing *unconstrained terms*. Intuitively, a term t is unconstrained in the formula φ if for every assignment to the variables occurring in the term, every possible value of the sort of the term t can be obtained by changing values of only variables that are unconstrained in φ. The idea of simplification is to replace a nontrivial unconstrained term by a fresh variable, which leads to a smaller equisatisfiable formula. For example, consider the formula $(x + 3y = 0 \ \wedge \ y > 0)$ in the theory of integers. The formula contains one unconstrained variable x. The term $x + 3y$ is unconstrained as it can attain any integer value, regardless of the value of y. If we replace the term $x + 3y$ by a fresh variable v, we get the equisatisfiable formula $(v = 0 \ \wedge \ y > 0)$. Alternatively, one can realize that the whole term $x + 3y = 0$ is unconstrained and thus it can be replaced by a fresh *Boolean* variable w. In this way, we get an equisatisfiable formula $(w \ \wedge \ y > 0)$. In both cases, the variable y of the simplified formula become unconstrained and the formula can be further simplified.

To formalize the simplification principle, we define when a term is *unconstrained due to* a set of variables U, which means that a term can evaluate to an arbitrary value by changing only values of variables in U. Further, we define when a term is unconstrained in a formula φ, which means that it is unconstrained due to a set of variables that are unconstrained in φ.

Definition 1. *Let t be a term and $U \subseteq vars(t)$ be a set of variables. We say that the term t is* unconstrained due to U *if, for each valuation μ of variables in $(vars(t) \setminus U)$ and every value b of the same sort as the term t, there exists a valuation ν of variables in U such that $[\![t]\!]_{\mu \cup \nu} = b$.*

Example 1. In the bit-vector theory, the following terms are unconstrained due to $\{x\}$ for any term t' not containing x:

- $x + t'$ and $t' + x$,
- $c^{[n]} \cdot x$ and $x \cdot c^{[n]}$ if c is an odd constant,
- $\mathsf{bvnot}(x)$, $\mathsf{bvxor}(x, t')$ and $\mathsf{bvxor}(t', x)$,
- $x <_u c^{[n]}$ if $c \neq 0$,
- $c^{[n]} <_u x$ if $c \neq 2^n - 1$,
- $x = t'$ and $x \neq t'$.

Note that the last two terms are unconstrained due to $\{x\}$ because each sort of the bit-vector theory contains at least two elements. Further, the terms $x \cdot y$, $\mathsf{bvand}(x, y)$, $\mathsf{bvor}(x, y)$ are unconstrained due to $\{x, y\}$. A comprehensive list of unconstrained terms can be found for example in Franzén's doctoral thesis [10].

On the contrary, multiplication by an even constant is not an unconstrained term. For example, the term $2 \cdot x$ over the theory of bit-vectors never evaluates to 3 as the number 3 does not have a multiplicative inverse in the ring of integers modulo 2^{32}. As a consequence, the term $x \cdot y$ is neither unconstrained due to $\{x\}$, nor unconstrained due to $\{y\}$.

Definition 2. *A subterm t of a formula φ is called* unconstrained in the formula φ *if it is unconstrained due to a set of variables that are unconstrained in φ.*

The following theorem states the correctness of simplification based on unconstrained terms.

Theorem 1 ([8,10]). *Let φ be a quantifier-free formula and t its subterm uncon-strained in φ. Then φ is equisatisfiable with the formula $\varphi[t \leftarrow v]$, where v is a fresh variable of the same sort as t.*

Note that our definition of unconstrained terms and the statement of The-orem 1 are slightly more general than the ones given by Brummayer and Bruttomesso, which consider unconstrained terms containing only a single unconstrained variable. The definition of unconstrained term used in this paper is due to Franzén [10].

The approach where subformulas are identified with terms of sort *Boolean* brings some additional benefits. In particular, a subformula can be an uncon-strained term even if it consists of terms that are not unconstrained. For example, let us consider the formula $\varphi \equiv (3x + 3y = 0 \ \wedge \ y > 0)$ over the theory of inte-gers. The term $3x + 3y$ is not unconstrained as its value is always a multiple of 3. However, term $3x + 3y = 0$ of sort *Boolean* is unconstrained due to $\{x\}$. As x is unconstrained in φ, we can simplify the formula to the equisatisfiable form $(v \ \wedge \ y > 0)$. Elimination of pure literals can then further reduce the formula to the form $(y > 0)$. As both $\top$ and $\bot$ can be obtained by suitable choices of the value of the variable y, the term $y > 0$ is unconstrained due to $\{y\}$, and thus the formula can be simplified to v', where v' is a *Boolean* variable.

Note on Models. The simplified formulas are in general equisatisfiable to the original ones, but not equivalent. For example, the formulas $(x + 3y = 0 \ \wedge \ y > 0)$ and $(v = 0 \ \wedge \ y > 0)$ mentioned above are both satisfiable, but they use different sets of variables and thus they have different models. In this case, a model of the original formula can be easily computed from the model μ of the simplified formula: it assigns to y the value $[\![y]\!]_\mu$ and to x the value $[\![-3y]\!]_\mu$. However, in some cases, the computation of a model for the original formula can be harder. For example, assume that we have replaced the unconstrained term $180423^{[32]} \cdot x$ over the bit-vector theory by a fresh variable y. To get the value of x such that the term $180423^{[32]} \cdot x$ evaluates to a given value of y then means to find the multiplicative inverse of 180423 in the ring of integers modulo 2^{32} and multiply it by the value of y. Although this inverse can be still computed using an extended Euclidean algorithm, it is computationally not trivial.

Note that algorithms for effective retrieval of models for the original formulas from models of the simplified formulas are beyond the scope of this paper.

4   Partially Constrained Terms in Quantifier-Free Formulas

The key property of the simplification presented in the previous section is that the possible values of an unconstrained term are precisely the same as the possi-ble values of a fresh variable of the same sort. This approach can be generalized even to terms that are *partially constrained*: a complex term can be replaced by a simpler one representing the same values. For example, the value of the term

$6 \cdot x$ over the bit-vector theory can be any number divisible by 2. Therefore, if $6 \cdot x$ is a subterm of a formula where x is unconstrained, then the subterm $6 \cdot x$ can be replaced by $2 \cdot y$ where y is a fresh variable of the same sort as x.

The following definition formalizes the notion that two terms represent the same set of possible values for any fixed valuation of variables in C. Intuitively, in applications of this definition, the set C will contain all constrained variables.

Definition 3. *Let C be a set of variables and t, s be terms of the same sort. Further, let $U = (vars(t) \cup vars(s)) \setminus C$. Terms t and s are called C-interchangeable, written $t \overset{C}{\rightleftharpoons} s$, if for every valuation μ of variables in C it holds that*

$$\{[\![t]\!]_{\mu \cup \nu} \mid \nu \text{ is a valuation of } U\} = \{[\![s]\!]_{\mu \cup \rho} \mid \rho \text{ is a valuation of } U\}.$$

Now we formulate the simplification principle for partially constrained terms and prove its correctness.

Theorem 2. *Let φ be a quantifier-free formula and C be the set of its constrained variables. For any subterm t of φ and any term s such that $t \overset{C}{\rightleftharpoons} s$ and $vars(s) \cap vars(\varphi) \subseteq C$, the formula φ is equisatisfiable with the formula $\varphi[t \leftarrow s]$.*

Proof. All variables of φ and $\varphi[t \leftarrow s]$ can be divided into three disjoint sets:

1. the set C of all constrained variables in φ,
2. the set $U = (vars(t) \cup vars(s)) \setminus C$ of all variables in t or s that are not in C,
3. the set U' containing all variables that are neither in C, nor in U.

The precondition $vars(s) \cap vars(\varphi) \subseteq C$ formulated in the theorem implies that every variable of U appears either only in t or only in s and not in any other part of the formula. Moreover, variables of U' appear neither in t, nor in s.

Suppose that φ is satisfiable. Hence, there exists a valuation μ of variables in C, a valuation ν of variables in U, and a valuation ν' of variables in U' such that $\mu \cup \nu \cup \nu'$ is a satisfying assignment of φ. As t and s are C-interchangeable and do not contain any variable from U', there exists a valuation ρ of variables in U such that $[\![t]\!]_{\mu \cup \nu} = [\![s]\!]_{\mu \cup \rho}$. As valuations ν and ρ concern only variables of U that do not appear outside t and s, we get that the assignment $\mu \cup \rho \cup \nu'$ satisfies $\varphi[t \leftarrow s]$.

It remains to show that satisfiability of $\varphi[t \leftarrow s]$ implies satisfiability of φ. However, the arguments are completely symmetric. □

Note that the definition of C-interchangeability generalizes Definition 1 in the sense that a term t is unconstrained due to U if and only if it is C-interchangeable with a fresh variable u of the same sort, where $C = vars(t) \setminus U$. Theorem 1 can then be seen as a corollary of Theorem 2.

Applications. Now we show some applications of the previous theorem. We start with an example from the theory of non-linear real arithmetic and then focus on terms over the bit-vector theory. In particular, we focus on simplification of partially constrained terms with multiplication as this operation is very expensive for some SMT solvers, especially these based on BDDs.

Example 2. Consider the term $t \cdot u$ in the theory of non-linear real arithmetic, where u is an unconstrained variable and t is an arbitrary term not containing the variable u. The term $t \cdot u$ can be replaced by $\texttt{ite}(t = 0, 0, v)$, where v is a fresh variable, as the terms $t \cdot u$ and $\texttt{ite}(t = 0, 0, v)$ are $vars(t)$-interchangeable. In general, this simplification can be performed in any theory in which addition and multiplication form a field.

Example 3. In the bit-vector theory, the term $4 \cdot u$ can be evaluated to any bit-vector where the two least significant bits are zeroes. The same holds for the term $12 \cdot u$. In general, the term $c^{[n]} \cdot u$ with a constant $c^{[n]}$ can represent any bit-vector ending with i zeroes, where i is the highest integer such that 2^i divides c. This follows from the fact that c can be expressed as $2^i \cdot m$ for some odd number m and every odd number has a multiplicative inverse m^{-1} in the bit-vector theory. All bit-vectors with i zeroes at the end can be also represented by the term $v \ll i$. Hence, the terms $c^{[n]} \cdot u$ and $v \ll i$ are $\emptyset$-interchangeable. Finally, Theorem 2 implies that a formula φ with an unconstrained variable u and a term $c^{[n]} \cdot u$ is equisatisfiable with the formula $\varphi[c^{[n]} \cdot u \leftarrow v \ll i]$ where v is a fresh variable and i is the constant described above. Note that the term $v \ll i$ is easier to compute and express as a circuit than the original multiplication by a potentially large constant $c^{[n]}$.

Example 4. More interestingly, we can simplify also the term $t \cdot u$ where u is unconstrained, even if t is a term with a non-constant value. As an example, consider the term $t \cdot u^{[3]}$ for a 3-bit variable u. We write $t[i]$ as a shortcut for the extraction of the i-th least significant bit of the term t where $0 \leq i \leq 2$, i.e. $t[i] \equiv \texttt{extract}_{2-i}^{1}(t)$. Then $t \cdot u^{[3]}$ is $vars(t)$-interchangeable with the term

$$\texttt{ite}\left(t[0] = 1^{[1]}, \, v_0, \, \texttt{ite}\left(t[1] = 1^{[1]}, \, v_1 \ll 1^{[3]}, \, \texttt{ite}\left(t[2] = 1^{[1]}, \, v_2 \ll 2^{[3]}, \, 0^{[3]}\right)\right)\right).$$

In general, the term $t \cdot u^{[k]}$ is $vars(t)$-interchangeable with the term defined as

$$\texttt{ite}\Big(t[0] = 1^{[1]}, \, v_0,$$
$$\texttt{ite}\big(t[1] = 1^{[1]}, \, v_1 \ll 1^{[k]},$$
$$\cdots$$
$$\texttt{ite}(t[k-1] = 1^{[1]}, v_{k-1} \ll (k-1)^{[k]}, 0^{[k]}) \dots\big)\Big).$$

Therefore, in a formula φ with an unconstrained variable $u^{[k]}$, a term $t \cdot u^{[k]}$ can be replaced by the term above with fresh variables $v_0, v_1, \dots, v_{k-1}$ and the resulting formula is equisatisfiable with φ.

In the previous two examples, the multiplications has been replaced by operations like bit-equality and bit-shift by a constant, which are very cheap for BDD-based SMT solvers.

Example 5. Now we discuss some simplification rules mentioned by Bruttomesso without a proof of correctness. For example, consider the simplification rule that rewrites the term $t >_u u$ containing an unconstrained bit-vector variable u by the term $b \wedge t \neq 0$, where b is a fresh *Boolean* variable. Intuitively, the rule is correct as $t >_u u$ can be evaluated to both $\top$ and $\bot$ unless t is evaluated to 0. If the value of t is 0, $t >_u u$ evaluates to $\bot$. Correctness of this rule follows directly from Theorem 2 and the fact that the term $t >_u u$ is $vars(t)$-interchangeable with the term $b \wedge t \neq 0$ assuming that $u, b \notin vars(t)$. Similar simplification rules can be derived from pairs of $vars(t)$-interchangeable terms presented in Table 1.

Table 1. Each line presents a pair of $vars(t)$-interchangeable terms, assuming that $b, u \notin vars(t)$. Terms on the right are considered simpler for SMT solvers than these on the left.

$t <_u u$	$b \wedge t \neq 2^k - 1$
$t <_s u$	$b \wedge t \neq 2^{k-1} - 1$
$u <_u t$	$b \wedge t \neq 0$
$u <_s t$	$b \wedge t \neq -2^{k-1}$
$t \leq_u u$	$b \vee t = 0$
$t \leq_s u$	$b \vee t = -2^{k-1}$
$u \leq_u t$	$b \vee t = 2^k - 1$
$u \leq_s t$	$b \vee t = 2^{k-1} - 1$

5   Unconstrained Terms in Quantified Formulas

In this section, we extend the treatment of unconstrained variables to formulas containing quantifiers. To simplify the presentation, we suppose that all formulas are in the prenex normal form and do not contain any free variables. That is, $\varphi = Q_1 B_1 Q_2 B_2 \dots Q_n B_n \psi$, where ψ is a quantifier-free formula, $Q_i \in \{\forall, \exists\}$ for all $1 \leq i \leq n$, and all B_i are pairwise disjoint sets of variables. Sequences $Q_i B_i$ are called *quantifier blocks*. Quantifier blocks are supposed to be maximal, that is $Q_i \neq Q_{i+1}$. A quantifier block $Q_i B_i$ is *existential* if $Q_i = \exists$ and *universal* otherwise. The *level of a variable* x is i such that $x \in B_i$. For a variable x, we denote as $level(x)$ its level and for a set of variables X we define $levels(X) = \{level(x) \mid x \in X\}$. If the set X contains only variables of the same level, we denote as $level(X)$ the level of all variables in that set. We say that an occurrence of a Boolean variable has the *positive polarity* if the occurrence is under an even number of negations and that it has the *negative polarity* otherwise. A variable is called *unconstrained* in the quantified formula φ if it is unconstrained in its quantifier-free part ψ.

It is easy to see that Theorem 1 can not be directly applied to quantified formulas. As an example, consider the formula $\varphi \equiv \exists x \forall y \, (x + y = 0)$. Although the variable x is unconstrained in the formula φ and the term $x + y$ is unconstrained due to $\{x\}$, the conclusion of Theorem 1 is not true regardless of the position of the quantifier for the fresh variable v: the formula φ is equisatisfiable neither with $\exists v \forall y \, (v = 0)$ nor with $\forall y \exists v \, (v = 0)$. The following modified definition of the unconstrained term solves this problem.

Definition 4. *Let φ be a quantified formula, t its subterm, and $U \subseteq vars(t)$ a set of variables such that $|levels(U)| = 1$. We say that the term t is unconstrained due to U if, for each valuation μ of variables in $(vars(t) \smallsetminus U)$ and every value b of the same sort as the term t, there exists a valuation ν of variables in U such that $[\![t]\!]_{\mu \cup \nu} = b$ and, furthermore,*

$$level(U) \geq \max\Big(levels\big(vars(t) \smallsetminus U\big)\Big).$$

For example, in the formula $\exists x \forall y\, (x + y = 0)$ mentioned above, the subterm $x + y$ is not unconstrained due to $\{x\}$, since $level(x) < level(y)$. On the other hand, it is unconstrained due to $\{y\}$.

The following theorem shows that a subterm that are unconstrained due to a set of unconstrained variables can be simplified even in quantified formulas.

Theorem 3. *Let φ be a formula, t be a term, U be a subset of $vars(t)$, and v be a variable not occurring in φ. If t is unconstrained due to the set of variables U and all variables in U are unconstrained in φ, then φ is equivalent to the formula φ in which the term t is replaced by v and the variables of U are replaced in their quantifier block by the variable v.*

Proof. The definition of unconstrained subterm implies that all variables in U have the same level. Let $k = level(U)$ and let $\varphi \equiv Q_1 B_1 \ldots Q_k B_k\, \psi$, where the formula ψ can contain quantifiers. We show that the formula $Q_k B_k\, \psi$ is equivalent to the formula $Q_k((B_k \smallsetminus U) \cup \{v\})\, (\psi[t \leftarrow v])$.

Let $V = \bigcup_{1 \leq i < k} B_i$. Observe that $U \subseteq B_k$ and the last line of the definition of an unconstrained term implies $(vars(t) \smallsetminus U) \subseteq V \cup (B_k \smallsetminus U)$. Let μ be an assignment of values to all variables in V. We distinguish two cases according to the quantifier Q_k.

- Suppose that $Q_k = \exists$. If $[\![\exists B_k \psi]\!]_\mu = \top$, then there is a valuation ν of variables in B_k such that $[\![\psi]\!]_{\mu \cup \nu} = \top$. Note that the function $\mu \cup \nu$ assigns values to a superset of $vars(t)$ and therefore can be used to evaluate the term t. Let b_v be the value $[\![t]\!]_{\mu \cup \nu}$. For this value, we have $[\![\psi[t \leftarrow v]]\!]_{\mu \cup \nu \cup \{v \mapsto b_v\}} = \top$ and therefore also $[\![\exists(B_k \cup \{v\})\psi[t \leftarrow v]]\!]_\mu = \top$. Since all variables in U are unconstrained, the formula $\psi[t \leftarrow v]$ does not contain any variable from U and therefore $[\![\exists((B_k \smallsetminus U) \cup \{v\})\, \psi[t \leftarrow v]]\!]_\mu = \top$.
  Conversely, if $[\![\exists((B_k \smallsetminus U) \cup \{v\})\, \psi[t \leftarrow v]]\!]_\mu = \top$, there is a valuation ν of variables in $(B_k \smallsetminus U) \cup \{v\}$ such that $[\![\psi[t \leftarrow v]]\!]_{\mu \cup \nu} = \top$. As t is unconstrained due to the set U, there is a valuation ν_U of variables in U such that $[\![t]\!]_{\mu \cup \nu \cup \nu_U} = \nu(v)$. Therefore $[\![\psi]\!]_{\mu \cup \nu \cup \nu_U} = \top$ and in turn $[\![\exists(B_k \cup \{v\})\, \psi]\!]_\mu = \top$, because $\nu \cup \nu_U$ is an assignment to variables from $B_k \cup \{v\}$. Finally, because the formula ψ does not contain the variable v, we know that $[\![\exists B_k\, \psi]\!]_\mu = \top$.
- If $Q_k = \forall$, the proof is dual to the $\exists$ case, but each existential quantifier is replaced by the universal quantifier and each $\top$ is replaced by $\bot$. $\square$

As an example, consider again the formula $\exists x \forall y \, (x + y = 0)$. According to the previous theorem, it is equivalent with $\exists x \forall y \, (v = 0)$, because the term $x + y$ is unconstrained due to $\{y\}$. Moreover, as the term $v = 0$ is unconstrained due to $\{v\}$, it is equisatisfiable with $\exists x \forall p \, p$, where p is a *Boolean* variable. This formula is trivially equivalent to $\perp$.

6 Partially Constrained Terms in Quantified Formulas

Both described extensions of unconstrained terms – i.e. partially constrained terms and unconstrained terms in quantified formulas – can be combined together in a fairly obvious way. The next theorem precisely describes this combination. The proof of this theorem is a straightforward extension of already presented proofs.

Theorem 4. *Let φ be a quantified formula and t be its subterm such that the set U of unconstrained variables appearing in t satisfies $|levels(U)| = 1$ and*

$$level(U) \; \geq \; \max \big(levels(C)\big),$$

where $C = vars(t) \smallsetminus U$. Further, let s be an arbitrary term such that $t \overset{C}{\rightleftharpoons} s$ and $vars(s) \cap vars(\varphi) \subseteq C$. Then the formula φ is equivalent with the formula φ where the term t is replaced by the term s and the variables of U are replaced in their quantifier block by the set of variables $vars(s) \smallsetminus vars(\varphi)$.

Note that due to this theorem, we can easily transfer simplification rules mentioned by Bruttomesso to quantified formulas, because they can be reformulated using the notion of interchangeable terms, as was described in Sect. 4.

Moreover, such simplifications can be combined with additional quantifier-specific simplification rules. The key observation is that the simplifications using unconstrained and partially constrained terms often introduce fresh – and therefore unconstrained – *Boolean* variables (see Table 1). For example, if b is an existentially quantified *Boolean* variable that is unconstrained in a formula φ, it can be replaced by $\top$ if it occurs in φ with the positive polarity and by $\perp$ if it occurs with the negative polarity and the resulting formula will be equivalent to the original one. Similarly, an unconstrained universally quantified *Boolean* variable can be replaced by $\perp$ if it has the positive polarity and by $\top$ if it has the negative polarity. Combining those simplifications with simplifications using unconstrained and partially constrained terms therefore yields more straightforward simplification rules, which are shown in Table 2. Although this table shows only rules for terms with positive polarity, the dual versions for terms with negative polarity are straightforward.

7 Experimental Evaluation

We have implemented all mentioned simplifications of quantified bit-vector formulas containing partially constrained terms – including all rules mentioned by

Table 2. Derived simplification rules for partially constrained terms (in the left column) with positive polarity in quantified formulas. We assume that u is an unconstrained variable and $level(u) \geq \max\left(levels(vars(t))\right)$.

Term	Quantifier type of u	
	Existential	Universal
$u = t$ or $t = u$	$\top$	$\bot$
$u \neq t$ or $t \neq u$	$\top$	$\bot$
$t <_u u$	$t \neq 2^k - 1$	$\bot$
$t <_s u$	$t \neq 2^{k-1} - 1$	$\bot$
$u <_u t$	$t \neq 0$	$\bot$
$u <_s t$	$t \neq -2^{k-1}$	$\bot$
$t \leq_u u$	$\top$	$t = 0$
$t \leq_s u$	$\top$	$t = -2^{k-1}$
$u \leq_u t$	$\top$	$t = 2^k - 1$
$u \leq_s t$	$\top$	$t = 2^{k-1} - 1$

Franzén and rules from Examples 3 and 4, and Table 2. The algorithm iteratively simplifies the input formula up to the fixed point. The implementation is written in C++, uses Z3 API to parse the input formula, and is freely available at https://gitlab.fi.muni.cz/xjonas/BVExprSimplifier. We have evaluated the effect of implemented simplifications on the performance of the solvers Z3 [9], Boolector [13], and the BDD-based solver Q3B [12] on two sets of benchmarks.

The first set of benchmarks contains all 191 formulas from the BV[1] category of the SMT-LIB benchmark repository [2]. The second set of benchmarks consists of 5 461 formulas generated by the model checker SYMDIVINE [4] when run on verification tasks from SV-COMP [5]. The generated quantified formulas arise from an equivalence check of two symbolic states, both of which are represented by path conditions. Since the input for the tool SYMDIVINE is an LLVM bit-code, the resulting formulas are expected to contain a large number of unconstrained variables.

All experiments were performed on a Debian machine with two six-core Intel Xeon E5-2620 2.00 GHz processors and 128 GB of RAM. Each benchmark run was limited to use 8 GB of RAM and 15 min of CPU time. All measured times are CPU times and include both the time of formula preprocessing and the time of the actual SMT solving, unless explicitly stated otherwise. For reliable benchmarking we employed BENCHEXEC [6], a tool that allocates specified resources for a program execution and measures their use precisely. We used the solver Z3 in the latest stable version 4.5.0, Boolector in the version attached to the paper [13], and the solver Q3B in the latest development version (commit **6830168** on GitHub).

---

[1] BV is a category of quantified bit-vector formulas without arrays and uninterpreted functions.

For all three SMT solvers, we compare the performance of two different configurations:

- In configurations Z3, Boolector, and Q3B, SMT solvers are run on the input formula, after performing cheap local simplifications done by Z3.
- In configuration Z3-s, Boolector-s, and Q3B-s, the input file is simplified using the implemented simplifications based on unconstrained variables and the same cheap local simplifications performed by Z3 up to the fixed point and the result is fed to SMT solvers.

Table 3 summarizes the obtained results. Thanks to the simplifications, the solver Z3 was able to decide 2 more benchmarks from the SMT-LIB benchmark set and 195 more benchmarks from the SYMDIVINE benchmark set. The solver Boolector can decide 8 more SMT-LIB benchmarks and 314 more SYMDIVINE benchmarks thanks to the simplifications. On the other hand, the number of benchmarks decided by Q3B does not change after performing simplifications. This is not surprising, as the remaining 122 formulas in the SYMDIVINE benchmark set, which Q3B did not solve, do not contain any unconstrained variables.

We have also examined the time needed to solve benchmarks. Even when the time of performing simplifications is counted, simplifications helped to reduce solving time of Z3 on SYMDIVINE benchmarks to 18 %, the solving time of Boolector on SMT-LIB benchmarks to 84 %, the solving time of Boolector on SYMDIVINE benchmarks to 7 %, the solving time of Q3B on SMT-LIB benchmarks to 75 %, and the solving time of Q3B on SYMDIVINE benchmarks to 35 % of the original CPU time. On the other hand, simplifications increased the solving time of Z3 on SMT-LIB benchmarks to 105 %, which is in part due to the time needed to perform simplifications and in part due to the fact that after the simplification, Z3's quantifier instantiation heuristics needed more instantiations to

Table 3. For each benchmark set, solver, and configuration, the table provides the numbers of formulas decided as satisfiable (*sat*), unsatisfiable (*unsat*), or undecided (*other*) because of an error, a timeout or a memory out. The table also shows the time necessary to decide benchmarks in the benchmark set in the form of *simplification time + solving time*, or just *solving time* if the simplification was not performed. Only benchmarks that were decided by both configurations of the given solver are counted into solving times.

	SMT-LIB				SYMDIVINE			
	Sat	Unsat	Other	Time (s)	Sat	Unsat	Other	Time (s)
Z3	70	92	29	426	1137	3999	325	3006
Z3-s	72	92	27	16 + 430	1137	4194	130	169 + 381
Boolector	78	95	18	1513	1137	3296	1028	20 269
Boolector-s	86	95	10	16 + 1257	1137	3610	714	169 + 1173
Q3B	94	94	3	2986	1137	4202	122	9046
Q3B-s	94	94	3	16 + 2233	1137	4202	122	169 + 3082

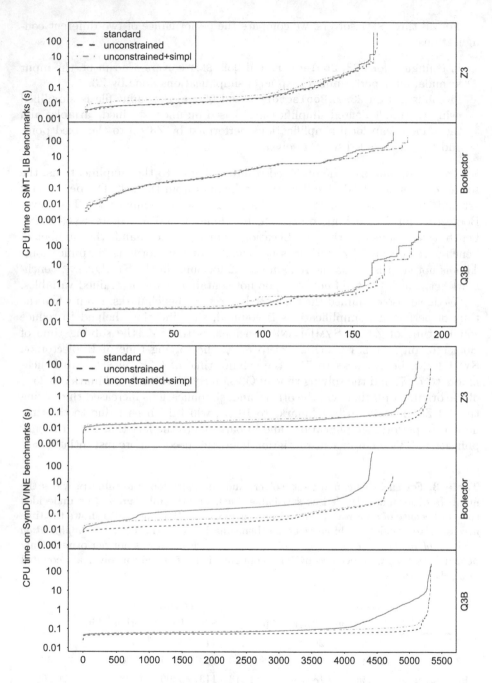

Fig. 1. The figure shows quantile plots for three configurations of all solvers run on both benchmark sets. The configuration *standard* runs on original formulas and the other two run on simplified formulas. The times of *unconstrained* do not include simplification time, the times of *unconstrained+simpl* do.

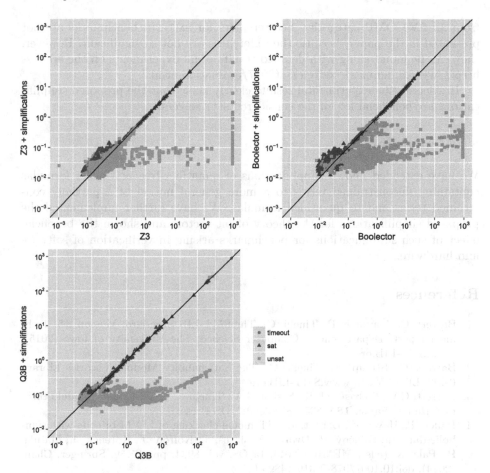

Fig. 2. The figure shows comparison of solving times with and without simplifications for all three SMT solvers on benchmarks from the SYMDIVINE benchmark set. The times are in seconds and include the time of performing simplifications.

solve some of the formulas. Overall, the time needed to perform simplifications is usually negligible when compared to the actual solving time – each formula was simplified in less than 1.3 s and the average simplification time is 0.03 s. Quantile plots on Fig. 1 show comparison of benchmarks solved within a given time for configurations with and without simplifications.

Additionally, scatter plots on Fig. 2 show comparison of time needed to solve benchmarks for formulas from the SYMDIVINE benchmark set. For this benchmark set, the proposed simplifications are clearly beneficial, as was expected. Interestingly, the benchmarks from SYMDIVINE set for which the simplifications improved the solving time are not the same for all three SMT solvers – the performance of Z3 improved mainly on benchmarks originating in the SV-COMP category ECA; the performance of Boolector and Q3B improved on benchmarks from all SV-COMP categories. Scatter plots for SMT-LIB benchmarks are not presented in the paper, because the differences are not so significant. However, they can be found along with detailed results of all experiments on the address http://www.fi.muni.cz/~xstrejc/sat2017/evaluation.html.

In general, the experimental results clearly show that the proposed simplifications are beneficial for all the considered SMT solvers.

8   Conclusion

We have extended known simplifications of quantifier-free first-order formulas containing unconstrained terms to a more general notion of a partially constrained terms and to quantified formulas. We have further implemented the proposed simplifications for the theory of bit-vectors and shown the beneficial effect of such simplifications for benchmarks arising in verification of software and hardware.

References

1. Barrett, C., Fontaine, P., Tinelli, C.: The SMT-LIB Standard: Version 2.5. Technical report, Department of Computer Science, The University of Iowa (2015). www.SMT-LIB.org
2. Barrett, C., Stump, A., Tinelli, C.: The Satisfiability Modulo Theories Library (SMT-LIB) (2010). www.SMT-LIB.org
3. Barrett, C.W., Sebastiani, R., Seshia, S.A., Tinelli, C.: Satisfiability modulo theories. Handb. Satisf. **185**, 825–885 (2009). IOS Press,
4. Bauch, P., Havel, V., Barnat, J.: LTL model checking of LLVM bitcode with symbolic data. In: Hliněný, P., Dvořák, Z., Jaroš, J., Kofroň, J., Kořenek, J., Matula, P., Pala, K. (eds.) MEMICS 2014. LNCS, vol. 8934, pp. 47–59. Springer, Cham (2014). doi:10.1007/978-3-319-14896-0_5
5. Beyer, D.: Software verification and verifiable witnesses. In: Baier, C., Tinelli, C. (eds.) TACAS 2015. LNCS, vol. 9035, pp. 401–416. Springer, Heidelberg (2015). doi:10.1007/978-3-662-46681-0_31
6. Beyer, D., Löwe, S., Wendler, P.: Benchmarking and resource measurement. In: Fischer, B., Geldenhuys, J. (eds.) SPIN 2015. LNCS, vol. 9232, pp. 160–178. Springer, Cham (2015). doi:10.1007/978-3-319-23404-5_12
7. Brummayer, R.: Efficient SMT solving for bit vectors and the extensional theory of arrays. Ph.D. thesis, Johannes Kepler University of Linz (2010)
8. Bruttomesso, R.: RTL Verification: From SAT to SMT(BV). Ph.D. thesis, University of Trento (2008)
9. de Moura, L., Bjørner, N.: Z3: an efficient SMT solver. In: Proceedings of 14th International Conference on Tools and Algorithms for the Construction and Analysis of Systems, TACAS 2008, Held as Part of the Joint European Conferences on Theory and Practice of Software, ETAPS 2008, Budapest, Hungary, 29 March - 6 April 2008, pages 337–340 (2008)
10. Franzén, A.: Efficient Solving of the Satisfiability Modulo Bit-Vectors Problem and Some Extensions to SMT. Ph.D. thesis, University of Trento (2010)
11. Hadarean, L.: An Efficient and Trustworthy Theory Solver for Bit-vectors in Satisfiability Modulo Theories. Ph.D. thesis, New York University (2015)

12. Jonáš, M., Strejček, J.: Solving quantified bit-vector formulas using binary decision diagrams. In: Creignou, N., Le Berre, D. (eds.) SAT 2016. LNCS, vol. 9710, pp. 267–283. Springer, Cham (2016). doi:10.1007/978-3-319-40970-2_17
13. Preiner, M., Niemetz, A., Biere, A.: Counterexample-guided model synthesis. In: Proceedings of 23rd International Conference on Tools and Algorithms for the Construction and Analysis of Systems, TACAS 2017, Held as Part of the European Joint Conferences on Theory and Practice of Software, ETAPS 2017, Part I, Uppsala, Sweden, 22–29 April 2017, pp. 264–280 (2017)

A Benders Decomposition Approach to Deciding Modular Linear Integer Arithmetic

Bishoksan Kafle[✉], Graeme Gange, Peter Schachte, Harald Søndergaard,
and Peter J. Stuckey

School of Computing and Information Systems, The University of Melbourne,
Melbourne, VIC 3010, Australia
{bishoksank,gkgange,schachte,harald,pstuckey}@unimelb.edu.au

Abstract. Verification tasks frequently require deciding systems of linear constraints over modular (machine) arithmetic. Existing approaches for reasoning over modular arithmetic use bit-vector solvers, or else approximate machine integers with mathematical integers and use arithmetic solvers. Neither is ideal; the first is sound but inefficient, and the second is efficient but unsound. We describe a linear encoding which correctly describes modular arithmetic semantics, yielding an optimistic but sound approach. Our method abstracts the problem with linear arithmetic, but progressively refines the abstraction when modular semantics is violated. This preserves soundness while exploiting the mostly integer nature of the constraint problem. We present a prototype implementation, which gives encouraging experimental results.

1 Introduction

Linear integer arithmetic (LIA) and its decision procedures have been studied for a long time. Here we consider the important variant in which the integers involved are in fact integers modulo m for some m. We refer to this variant as modular LIA, MLIA in short. Decision procedures for MLIA are needed for sound reasoning about "machine arithmetic" and related problems in software verification and analysis. In this paper we address the problem of deciding a Boolean combination of MLIA constraints, that is, whether a system of MLIA constraints is satisfiable, and if so, how.

Existing approaches for deciding MLIA either use bit-vector solvers, or else approximate fixed-size integers ($\mathbb{Z}_m$, the integers modulo m) with unbounded mathematical integers ($\mathbb{Z}$) and use LIA solvers. The theory of bit-vectors allows modeling the precise semantics of two's complement arithmetic and is a natural candidate for modelling MLIA constraints. However this approach is inefficient, especially for large bit-vector problems that are primarily arithmetic in nature, and for problems involving long bit-vectors [19,28,37]. Most bit-vector solvers are based on some form of SAT encoding which tends to obscure word-level information and consume excessive memory. On the other hand, the LIA approach is efficient but unsound because approximating fixed-size integers with mathematical integers ignores the "wrap around" nature of fixed-width integer

© Springer International Publishing AG 2017
S. Gaspers and T. Walsh (Eds.): SAT 2017, LNCS 10491, pp. 380–397, 2017.
DOI: 10.1007/978-3-319-66263-3_24

arithmetic. As a result, the (un)satisfiability of a problem over $\mathbb{Z}$ does not imply its (un)satisfiability over $\mathbb{Z}_m$, as we now show. To keep examples simple we shall usually assume unsigned 4-bit arithmetic (so that $m = 16$). (Our method handles signed or unsigned arithmetic equally well.) Constraints will often be given in the form $id : c$, so that constraint c can be referred to through the identifier id. We indicate a formula F interpreted in $\mathbb{Z}_m$ with the notation $[F]_m$.

Example 1. Consider $F_1 = c_1 : x \geq y \land c_2 : x + 1 = y$. Clearly $[F_1]_{16}$ allows solution $x = 15(1111)$ and $y = 0(0000)$, but F_1 is unsatisfiable in $\mathbb{Z}$ since c_1 is in conflict with c_2.   □

Example 1 is adapted from a benchmark in the pspace subset [18] of SMT-LIB QF_BV [11], translated to MLIA form. It is typical of software verification problems; it tests the overflow of integer addition. Its model consists of x with only 1-bits and y with only 0-bits regardless of the size of bit-vectors x and y. But it has proven to be a challenging example for sufficiently long bit-vectors for *bit-blasting* solvers, and the problem becomes increasingly intractable as the size of bit-vectors increases [37]. On the other hand, LIA solvers may be efficient but they produce an unsound result for Example 1 since F_1 is unsatisfiable over $\mathbb{Z}$ and satisfiable over $\mathbb{Z}_m$.

A formula may just as well be unsatisfiable over $\mathbb{Z}_m$ and satisfiable over $\mathbb{Z}$.

Example 2. Consider $F_2 = c_1 : y = x + 9 \land c_2 : z = y + 9 \land c_3 : x \leq y \land c_4 : y \leq z$. Here F_2 is satisfiable in $\mathbb{Z}$ with solutions like $x = 0$, $y = 9$, $z = 18$. But $[F_2]_{16}$ is unsatisfiable, as it requires $(x + 18) \equiv_{16} z$, so there must be a wrap around between x and z. And so at least one of the two inequalities fails not hold.   □

Hence we can trust neither "satisfiable" nor "unsatisfiable" verdicts from a LIA solver. In practice, neither conventional approach is satisfactory.

In this paper, we develop an optimistic but sound approach—abstracting the problem with linear arithmetic, but progressively refining the abstraction when modular semantics is violated. By transforming MLIA constraints to LIA constraints that correctly describe the modular arithmetic semantics, we can reuse existing approaches for LIA problems while maintaining soundness. In contrast to bit blasting, the method is independent of the number of bits used to represent the variables and it uniformly handles (large) moduli, not necessarily powers of 2, including large primes. It is rather easy to understand and applies beyond linear arithmetic. Moreover, assuming two's complement arithmetic, signed and unsigned integers can be treated uniformly.

In summary we make the following contributions:

1. We give a semantics preserving transformation of MLIA constraints to LIA constraints (Sect. 3).
2. We design novel algorithms for deciding MLIA constraints which use this transformation effectively (Sect. 4).
3. We present experimental results obtained with a prototype implementation (Sect. 5). As they are comparable to those of the best state-of-the-art solvers, we consider this a significant proof-of-concept.

2   Preliminaries

In this paper, we consider a logic of quantifier-free MLIA constraints defined as follows, where F ranges over formulas, C over atomic constraints, E over fixed-size linear arithmetic expressions, v over fixed-size integer variables and a over fixed-size integer constants:

$$F = C \mid \neg C \mid F \vee F \mid F \wedge F$$
$$C = E < E \mid E \leq E \mid E = E \qquad\qquad E = a \mid v \mid a \cdot E \mid E - E \mid E + E$$

The other linear constraints $\{>, \geq, \neq\}$ are rewritten as negated elements of C. For $e \in E$, let $vars(e)$ denote the set of variables appearing in e. We extend this to elements of F and C in the obvious way.

For an integer $k \in \mathbb{Z}$, let $[k]_m$ denote its value modulo m (the remainder on division by m). This is the unique value satisfying:

$$0 \leq [k]_m < m \wedge \exists q \in \mathbb{Z} \; . \; k = m \cdot q + [k]_m \tag{1}$$

The quotient q encodes the number of times k "wraps around" in a number circle before landing in the range $[0, m)$. Equation 1 can be equivalently written as Eq. 2, where $q' = -q$.

$$0 \leq [k]_m < m \wedge \exists q' \in \mathbb{Z} \; . \; [k]_m = m \cdot q' + k \tag{2}$$

We extend $[\;]_m$ onto integer-valued expressions such that all subexpressions are computed in $\mathbb{Z}_m$: $[E_1 + E_2]_m = [[E_1]_m + [E_2]_m]_m = (E_1 + E_2) \bmod m$, and similarly for subtraction and multiplication by a constant. Observe that $[\;]_m$ is preserved under translation by multiples of m. Thus for $e \in E$, $[e]_{2^b}$ reflects e computed with b bit fixed precision machine arithmetic. We also extend $[\;]_m$ to functions and maps pointwise.

Note that by our definition of $\mathbb{Z}_m$, negative E values are mapped to positive $[E]_m$, so it is necessary to distinguish signed two-complement comparisons $(<_s, \leq_s)$ from unsigned ones $(<_u, \leq_u)$. For $\lhd \in \{<, \leq, =\}$, the interpretation of $E_1 \lhd_u E_2$ under $\mathbb{Z}_m$ is $[E_1 \lhd_u E_2]_m \equiv [E_1]_m \lhd [E_2]_m$; for the signed case, $[E_1 \lhd_s E_2]_m \equiv \left[E_1 + \frac{m}{2}\right]_m \lhd \left[E_2 + \frac{m}{2}\right]_m$. In the following, we shall assume unsigned comparisons, but signed comparisons can be handled this way. Observe that we must be careful when manipulating modular linear constraints: rewriting $[A \leq B]_m$ as $[A - B \leq 0]_m$ does not preserve equivalence (for either signed or unsigned comparison).

An *assignment* μ to F is a mapping from $vars(F)$ onto $\mathbb{Z}_m$. We write $\mu(E)$ to denote the value of E under μ. μ is a *model* of F if F evaluates to TRUE under μ and we denote it by $\mu \models_{\mathbb{Z}_m} F$. Similarly if μ is a $\mathbb{Z}$-model of F, we simply write $\mu \models F$. Also observe that if μ is a model of F, and for any expression E appearing in an inequality $0 \leq \mu(E) < m$, then $[\mu]_m$ is a model of $[F]_m$ (as then $[\mu(E)]_m = \mu(E)$).

2.1 Benders Decomposition

Benders decomposition [4] is an approach for solving large integer linear programming problems, frequently applied where a problem consists of independent subproblems, connected via a small set of variables.

Consider the mixed integer programming problem (MIP) shown in Fig. 1(a). P is an optimisation problem, minimising $f(x)$ over variables x, y, consisting of constraints A over x, and additional constraints over x, y. Rather than solving P directly, we may instead fix x to some optimal value $\tilde{x}$, then check whether there exists a consistent extension $\tilde{y}$ satisfying $B(\tilde{x}, \tilde{y})$.

$$
\begin{array}{c|cc}
\begin{aligned}
P = \min_{x,y} \; & f(x) \\
\textbf{s.t. } & A(x) \geq b \, \wedge \\
& B(x, y) \geq c \, \wedge \\
& \ldots \\
& x \in D_x, \; y \in D_y
\end{aligned}
&
\begin{aligned}
P^{\sharp} = \min_{x} \; & f(x) \\
\textbf{s.t. } & A(x) \geq b \, \wedge \\
& B^{\sharp}(x) \geq c \, \wedge \\
& x \in D_x
\end{aligned}
&
\begin{aligned}
Q(\tilde{x}) = \min_{y} \; & 0 \\
\textbf{s.t. } & B(\tilde{x}, y) \geq c \, \wedge \\
& y \in D_y
\end{aligned}
\\
(a) & (b) & (c)
\end{array}
$$

Fig. 1. (a) A decomposable MIP, (b) initial relaxed master, (c) Benders subproblem

We thus construct a *relaxed master problem* $P^{\sharp}$, shown in Fig. 1(b), which is a relaxation of the projection $\exists y.\ P$. The *projection constraint* $B^{\sharp}(x) \geq c$ is a relaxation of $\exists y.\ B(x, y) \geq c$; this may be any constraint (or set of constraints) satisfying $B^{\sharp}(x) \geq \sup_{y \in D_y} B_i(x, y)$. As $B^{\sharp}(x)$ may permit invalid solutions, this is combined with a subproblem $Q(\tilde{x})$ which check feasibility. Solving $P^{\sharp}$ yields a candidate optimum $\tilde{x}$. If $Q(\tilde{x})$ is satisfiable, we have found an optimum. If not, we extract from Q a new constraint $a'x \geq b'$, excluding $\tilde{x}$, which is added to $P^{\sharp}$ (effectively strengthening $B^{\sharp}$). This procedure is repeated until either an assignment is found, or $P^{\sharp}$ is proven unsatisfiable.

The classical form of Benders decomposition requires the subproblems to be linear programs (i.e. over $\mathbb{Q}$). However, in *logic-based Benders decomposition* [25], the subproblem may be an arbitrary decision problem, and the feasibility cut is extracted from the unsatisfiability proof of Q.

3 From MLIA to Equivalent LIA Constraints

In this section, we describe a linear encoding over $\mathbb{Z}$ of the semantics of constraints over $\mathbb{Z}_m$. Recall the definition of $[k]_m$, being the unique value satisfying Eq. 2. By introducing fresh variables for the quotients and remainders, we can encode a modular linear constraint $[E_1 \lhd E_2]_m$ as a conjunction of linear constraints on $\mathbb{Z}$. We define a mapping Γ as:

$$\Gamma(E_1 \vartriangleleft E_2) = \exists q_1, q_2, e_1, e_2 \cdot \begin{pmatrix} 0 \leq e_1, e_2 < m \\ \wedge \ e_1 = E_1 + m \cdot q_1 \\ \wedge \ e_2 = E_2 + m \cdot q_2 \\ \wedge \qquad e_1 \vartriangleleft e_2 \end{pmatrix}$$

The first three conjuncts compute the interpretation of E_i under $\mathbb{Z}_m$, and the final constraint enforces the constraint of interest. Because $[[E_1]_m + [E_2]_m]_m = [E_1 + E_2]_m$, and similarly for all operations in a MLIA expression, it is sufficient to introduce one quotient variable q_i in each of E_1 and E_2; quotient variables are not needed for subexpressions. $\Gamma(E_1 \vartriangleleft E_2)$ can be simplified by eliminating the existential variables e_1 and e_2 leaving behind only the quotient variables.

$$\Gamma(E_1 \vartriangleleft E_2) = \exists q_1, q_2 \cdot \begin{pmatrix} 0 \leq E_1 + m \cdot q_1, E_2 + m \cdot q_2 < m \\ \wedge \qquad E_1 + m \cdot q_1 \vartriangleleft E_2 + m \cdot q_2 \end{pmatrix}$$

We extend Γ to Boolean combinations of constraints in the natural manner (eliminating $\neq$ through disjunction). The above encoding ($\Gamma(F)$) preserves equi-satisfiability as stated in Lemma 1. Note that variables are not bounded in $\Gamma(F)$.

Lemma 1 (Equi-satisfiability). *Let F be a formula and $\Gamma(F)$ be its LIA encoding. Then F and $\Gamma(F)$ are equi-satisfiable. Further if $\Gamma(F)$ is satisfiable, then there exists a model of $\Gamma(F)$ such that $\bigwedge_{v \in vars(F)} 0 \leq v < m$.* □

Let $\Gamma_1(F)$ be $\Gamma(F) \wedge \bigwedge_{v \in vars(F)} 0 \leq v < m$. These bounds restrict the search space of LIA solvers; possibly resulting in a faster convergence. $\Gamma_1(F)$ correctly encodes the semantics of modular arithmetic as stated in Proposition 1.

Proposition 1 (Soundness of the encoding). *Let F be a formula and $\Gamma_1(F)$ be its LIA encoding. Then F and $\Gamma_1(F)$ are logically equivalent.* □

Note that $\Gamma_1(F)$ is a quantified-formula though not F.

The formula $\Gamma_1(E_1 \vartriangleleft E_2)$ contains: two constraints expressing the lower and upper bounds of each expression $E_i (i = 1, 2)$, a constraint computing the interpretation of each expression and a constraint of interest totalling 5. In addition, there are $2n$ constraints representing the upper and lower bounds of the variables, where $n = \# \, vars(E_1 \vartriangleleft E_2)$. Then, the size of the transformed formula is given by the following proposition.

Proposition 2 (Bound on the size of the transformed formula). *Given a formula F with n variables and m atomic constraints, $\Gamma_1(F)$ contains at most $2n + 5m$ atomic constraints.* □

3.1   Encoding Simplification

While straightforward, the quotient variables have unbounded domains and large coefficients (the modulo integer m). This can have a dramatic impact on performance, as it substantially weakens the (real) linear relaxation underlying the LIA decision procedure. However, we can frequently infer reasonably tight bounds on feasible quotient values as we show next. For some expression E, let l_E and u_E be the minimum and maximum feasible values of E under $\mathbb{Z}$ (assuming *variables* are restricted to $[0, m)$). As we have $[E]_m = E + m \cdot q_E$, we may impose the constraint $-\lfloor \frac{u_E}{m} \rfloor \leq q_E \leq -\lfloor \frac{l_E}{m} \rfloor$ without changing satisfiability.

For example, given a modular expression $x + 2y$ over unsigned integer and its corresponding LIA expression $x + 2y + m.q$, we derive $-2 \leq q \leq 0$.

An expression bound can be extended to a constraint $E_1 \lhd E_2$ as follows.

Definition 1 (Quotient bound). *Given a MLIA constraint $E_1 \lhd E_2$, we have $E_1 + m \cdot q_{E_1} \lhd E_2 + m \cdot q_{E_2}$ as the corresponding LIA constraint. The following conjunction of constraints is called a* quotient bound *for $E_1 \lhd E_2$.*

$$-\left\lfloor \frac{u_{E_1}}{m} \right\rfloor \leq q_{E_1} \leq -\left\lfloor \frac{l_{E_1}}{m} \right\rfloor \qquad \wedge \qquad -\left\lfloor \frac{u_{E_2}}{m} \right\rfloor \leq q_{E_2} \leq -\left\lfloor \frac{l_{E_2}}{m} \right\rfloor.$$

If we infer a fixed value for q_E (as will be the case for variables and constant expressions), we may replace $m \cdot q_E$ with the appropriate constant. If bounds on q_E are narrow but do not fix a value for q_E, we may eliminate q_E by introducing a disjunction for the possible values of q_E—moving the wrapping decision from the LIA solver to the SAT solver. In case the bounds are not tight, we apply Benders decomposition, as discussed in Sect. 4.

Example 3. Consider the constraint $c_2 : [x + 1 = y]_m$ from Example 1. Computing bounds for the LHS and RHS, we find that y (unsurprisingly) cannot overflow, and $x + 1$ overflows at most once. This yields the encoding

$$x + 1 + m \cdot q_x = y \ \wedge \ 0 \leq x + 1 + m \cdot q_x \leq 15 \ \wedge \ -1 \leq q_x \leq 0$$
$$\wedge \ 0 \leq x \leq 15 \ \wedge \ 0 \leq y \leq 15$$

As the domain of q_x is small, we eliminate it by case-splitting:

$$x - 15 = y \wedge 0 \leq x - 15 \leq 15 \wedge 0 \leq x \leq 15 \wedge 0 \leq y \leq 15 \qquad \text{(where } q_x = -1)$$
$$\vee \ x + 1 = y \wedge 0 \leq x + 1 \leq 15 \wedge 0 \leq y \leq 15 \wedge 0 \leq x \leq 15 \qquad \text{(where } q_x = 0)$$

Similarly, the constraint $[x \geq y]_m$ yields $x \geq y \wedge 0 \leq x, y \leq 15$. $\qquad\square$

3.2 Challenges in Solving LIA Formula Directly

However, the resulting encodings turn out to be difficult to solve directly with current MIP solvers. This is, in part, due to weakening of the linear relaxation.

Example 4. Consider the pair of constraints:

$$P = [x = y + 3 \wedge y = x - 4]_m$$

These are inconsistent for any modulus $m \geq 2, m \neq 7$. The transformed constraints are:

$$\Gamma(P) = (x = m \cdot q_y + y + 3 \wedge y = m \cdot q_x + x - 4 \wedge x, y \in [0, m))$$

If integrality constraints on q_x, q_y are relaxed, $\Gamma(P)$ is easily satisfied—with $\{x = 0, y = 0, q_x = \frac{4}{m}, q_y = -\frac{3}{m}\}$. Indeed, any pair of (x, y) values admits a corresponding solution to the relaxation of the quotients. However, for any fixed integer assignment to (q_x, q_y), the residual subproblem can easily be shown to be inconsistent. □

A second pragmatic difficulty to solving these problems is due to numerical behaviour of MIP solvers. Commercial MIP solvers are well known to return non-solutions or claim optimality for non-optimal solutions due to rounding errors [30,31]. These problems are exacerbated in the presence of very large coefficients: not only is the solver forced to divide by large constants, intermediate computations and even integral solutions may have no exact floating-point representation. This becomes a significant problem when using moduli above 2^{32}.

We now consider how to solve these problems in practice.

4   Solving MLIA Constraints

For the reasons discussed above, the transformed formulae are difficult to solve directly. We can ameliorate this in several ways. First, we can try to avoid introducing quotient variables altogether; optimistically solving under integer semantics, and transforming only when necessary (Algorithm 1). This can accelerate solving, but in the worst case still requires transforming all constraints. Second, we can reduce the impact of quotient variables by adopting a Benders decomposition approach (Algorithm 2) to isolate quotient selection from the rest of the problem. We detail these approaches below.

4.1   Solving Algorithms

For simplicity of presentation we assume, without loss of generality, that the input is a conjunction of MLIA constraints (though our method applies more generally to a Boolean combination). Our algorithms assume the availability of the following methods.

LIA_DECIDE(H): An LIA solver (oracle) capable of generating either a model or an unsatisfiable core (a minimal set of unsatisfiable constraints). We assume that its output is a tuple of the form $\langle Result, Witness \rangle$, which can be either $\langle$SAT, $Model\rangle$ or $\langle$UNSAT, $Unsat\_Core\rangle$.

MLIA_MODEL(H, μ): Given an interpretation μ such that $\mu \models H$ (in this context, returned by LIA_DECIDE), the procedure checks whether $\mu \models_{\mathbb{Z}_m} H$ or not. It returns a tuple $\langle Result, Witness \rangle$, which can be either $\langle Yes, Model\rangle$ or $\langle No, Conflict\rangle$. We call c a conflict with respect to μ iff $\mu \not\models_{\mathbb{Z}_m} c$.

Putting all the pieces together, we present three algorithm for solving MLIA problems. The third is the final product; it combines Algorithms 1 and 2.

Algorithm 1. Solves MLIA constraints lazily

1: **function** MLIA_DECIDE_LAZILY(H)
2:     **Input:** A set of linear constraints of the form
3:         $E_1 \lhd E_2$, with $\lhd \in \{<, =, \leq\}$
4:     **Output:** $\langle$SAT, $Model\rangle$ or $\langle$UNSAT, $Unsat\_Core\rangle$
5:     $T \leftarrow \emptyset$ ▷ transformed constraints
6:     **while** TRUE **do**
7:         $\langle S, R\rangle \leftarrow$ LIA_DECIDE(H)
8:         **if** $S =$ SAT **then**
9:             $\langle IS, r\rangle \leftarrow$ MLIA_MODEL(H, R)
10:             **if** $IS = Yes$ **then**
11:                 **return** $\langle$SAT, $R\rangle$
12:             **else**
13:                 $H \leftarrow H \setminus \{r\} \cup \Gamma_1(r); T \leftarrow T \cup \{\Gamma_1(r)\}$
14:         **else**
15:             **if** $R \subseteq T$ **then return** $\langle$UNSAT, $R\rangle$
16:             **else**
17:                 **for each** $c \in R$ **do**
18:                       **if** $c \notin T$ **then**
19:                           $H \leftarrow H \setminus \{c\} \cup \Gamma_1(c); T \leftarrow T \cup \{\Gamma_1(c)\}$

---

1. Transforming Constraints Lazily. Algorithm 1 proceeds as follows. A set H of MLIA constraints is passed to an LIA solver (line 7). The solver returns one of the following:

- SAT(line 8): We have $R \models H$. Then we check whether $R \models_{\mathbb{Z}_m} H$ or not. If so, the procedure returns SAT and a model (line 11). Otherwise there exists a constraint c such that $R \not\models_{\mathbb{Z}_m} c$, which is replaced by $\Gamma_1(c)$ (line 13).
- UNSAT(line 14): In this case, there is an unsatisfiable core R over $\mathbb{Z}$. If all constraints in R are transformed then it is also the core over $\mathbb{Z}_m$ and the algorithm terminates (line 15). Otherwise, it replaces each $c \in R$ by $\Gamma_1(c)$ in the solver (line 17–19).

The algorithm runs until it exits from one of the above cases. During the run of the algorithm, we keep track of the set of the transformed constraints, which is initialized to empty set at the start of the algorithm (line 5).

Example 5. Let us run the algorithm on Example 1. Applied to the original set of constraints, the LIA solver returns UNSAT with $\{c_1, c_2\}$ as unsat core. Next we transform each MLIA constraint in the unsat core to LIA. Note that the resulting constraint system becomes equivalent to the system of constraints obtained by eager transformation as presented in Example 3. The resulting constraints are fed to the solver which finds them satisfiable, with $R = \{x = 15, y = 0\}$ as a model. One can easily verify that this is in fact a model of the original constraints over $\mathbb{Z}_m$. Then the algorithm terminates, returning $\langle$SAT, $R\rangle$.

Applied instead to the constraints from Example 2, the algorithm finds these satisfiable over $\mathbb{Z}$, returning the model $\{x = 0, y = 9, z = 18\}$. However, the

---

Algorithm 2. Solves MLIA constraints using Benders decomposition

---

1: **function** MLIA_DECIDE_BENDERS(H)
2:     **Input:** A set of linear constraints of the form
3:             $E_1 \lhd E_2$, with $\lhd \in \{<, =, \leq\}$
4:     **Output:** $\langle$SAT, $Model\rangle$ or $\langle$UNSAT, $Unsat\_Core\rangle$
5:     $Q \leftarrow true$ ▷ Quotient formula
6:     $H_r \leftarrow \emptyset$
7:     **for each** $c \in H$ **do**
8:         $H_r \leftarrow H_r \cup \{\Gamma_1(c)\}$
9:         $Q \leftarrow Q \wedge$ QUOTIENT_BOUND(c)   ▷ Definition 1
10:     $H_b \leftarrow H_r$
11:     **while** TRUE **do**
12:         **if** UNSAT(Q) **then**
13:             **return** $\langle$UNSAT, $H_b\rangle$
14:         Let μ_Q be a model of Q ($\mu_Q \models Q$)   ▷ Q is satisfiable
15:         $H_b \leftarrow \{E + m \cdot \mu_Q(q_i) \lhd F + m \cdot \mu_Q(r_i) \mid (E + m \cdot q_i \lhd F + m \cdot r_i) \in H_r\}$
16:         $\langle S, R\rangle \leftarrow$ LIA_DECIDE(H_b)
17:         **if** $S =$ SAT **then**
18:             **return** $\langle$SAT, $R\rangle$
19:         **else**   ▷ (R is the unsat core in this case)
20:             $C_R \leftarrow \bigvee\{\sigma(c) \mid c \in R\}$ ▷ (C_R: a cut generated from R, σ: Equation 3)
21:             $Q \leftarrow Q \wedge C_R$

---

constraint $c_4 : y \leq z$ is not satisfied under this model considering $\mathbb{Z}_{16}$ semantics, as $9 \nleq 2$. So we replace c_4 in the solver by $\Gamma_1(c_4)$. The resulting system of
constraints would still be satisfiable over $\mathbb{Z}$ but would still violate another constraint. We continue transforming MLIA constraints and finally we reach a state
where all the constraints are transformed and the LIA solver returns UNSAT,
thus proving the original system of constraints unsatisfiable over $\mathbb{Z}_m$. $\square$

2. Applying Benders Decomposition. Though we can infer tight bounds on
quotient variables, the presence of quotient variables with large coefficient (m) in
the constraints causes problems for LIA solvers. To avoid this, we adopt a logic-
based Benders decomposition strategy. The master problem Q assigns values to
the quotient variables, and the subproblem H_b tests whether the chosen quotients
can be extended to a model of H. If so, we terminate. Otherwise, we extract a
feasibility cut from the unsatisfiable core of H_b. See Algorithm 2.

Feasibility Cuts. Let $C_b = \{E_i + \tilde{q}_i m \leq F_i + \tilde{r}_i m \mid i = 1 \ldots n\}$ be the
unsatisfiable core of H_b, where $\tilde{q}_i$ and $\tilde{r}_i$ are the current assignments to quotients
q_i, r_i. To restore feasibility, at least one constraint in C_b must be relaxed—so
some q_i must decrease, or some r_i must increase. A valid cut, then, would be
$c = \bigvee_i q_i < \tilde{q}_i \vee r_i > \tilde{r}_i$. However, this is somewhat weak: the constraint is only
relaxed if the *difference* between q_i and r_i increases. Instead, then, we add the
more general cut $\bigvee_i r_i - q_i > \tilde{r}_i - \tilde{q}_i$.

Let σ be the mapping defined as follows:

$$\sigma(E_i + \tilde{q}_i m \lhd F_i + \tilde{r}_i m) = \begin{cases} r_i - q_i > \tilde{r}_i - \tilde{q}_i \text{ if } \lhd \in \{\leq, <\} \\ r_i - q_i \neq \tilde{r}_i - \tilde{q}_i \text{ if } \lhd \in \{=\} \end{cases} \tag{3}$$

Then for an unsat core $C_b = \{E_i + \tilde{q}_i m \lhd F_i + \tilde{r}_i m \mid i = 1 \ldots n\}$, the cut is given by the formula $\bigvee_{i=1}^n \sigma(E_i + \tilde{q}_i m \lhd F_i + \tilde{r}_i m)$.

Example 6. Consider again solving constraint $[c_1 : x \geq y \wedge c_2 : x + 1 = y]_{16}$, but this time using Algorithm 2. Encoding the constraints as LIA constraints, we obtain

$$c_1 : x \geq y \wedge 0 \leq y \leq 15 \wedge 0 \leq x \leq 15 \wedge$$
$$c_2 : x + 1 + m \cdot q_x = y \ \wedge \ 0 \leq x + 1 + m \cdot q_x \leq 15 \wedge$$
$$0 \leq x \leq 15 \wedge 0 \leq y \leq 15$$

The progress of the algorithm is outlined in Fig. 2. Computing bounds on q_x, we derive the master problem Q_1. Solving Q_1, we obtain a model $\mu_{Q_1} = \{q_x = 0\}$. Substituting μ_Q into H, we obtain the subproblem H_b. We find H_b is unsatisfiable, with an unsatisfiable core of $\{x + 1 + m \cdot 0_{q_x} = y, x \geq y\}$ (having noted occurrences of q_x). From this, we derive the feasibility cut $\{q_x > 0 \vee q_x < 0\}$ and derive a new master problem. Solving Q_2, we obtain $\mu_{Q_2} = \{q_x = -1\}$. We again substitute into H, obtaining H_{b_2}. Solving H_{b_2}, we obtain a model $\{x = 15, y = 0\}$, and terminate.   □

	ITERATION 1	ITERATION 2
	$Q_1 = [-1 \leq q_x \leq 0]$, $\mu_{Q_1} = \{q_x = 0\}$	$Q_2 = Q_1 \wedge [q_x \neq 0]$, $\mu_{Q_1} = \{q_x = -1\}$
	$c_1 : x \geq y \wedge 0 \leq y \leq 15 \wedge 0 \leq x \leq 15 \wedge$	$c_1 : x \geq y \wedge 0 \leq y \leq 15 \wedge 0 \leq x \leq 15 \wedge$
H_b	$c_2 : x + 1 + m \cdot 0 = y \ \wedge 0 \leq x + 1 + m \cdot 0 \leq 15 \wedge$	$c_2 : x - 15 = y \wedge 0 \leq x - 15 \leq 15 \wedge$
	$0 \leq x \leq 15 \wedge 0 \leq y \leq 15$	$15 \leq x \leq 15 \wedge 0 \leq y \leq 15$
R	$C = \{c_1, c_2\}$, $C_b = [q_x \neq 0]$	$\mu_{H_b} = \{x = 15, y = 0\}$

Fig. 2. Steps performed during Algorithm 2 solving $[x \geq y \wedge x + 1 = y]_{16}$.

Proposition 3 (Soundness of cut). *Given a system of MLIA constraints H, let H_b be an infeasible sub-problem of H for the quotient μ_Q and $C = \{E_i + \tilde{q}_i m \lhd F_i + \tilde{r}_i m \mid i = 1 \ldots n\}$ be any unsat core of H_b, and $C_b = \bigvee\{\sigma(c) \mid c \in C\}$. Then (1) C_b excludes μ_Q and (2) C_b does not exclude any model of H.*

Algorithm 3. Solves MLIA constraints using Benders decomposition lazily

---

1: **function** MLIA_DECIDE_MIX(H)
2:     **Input:** A set of linear constraints of the form
3:         $E_1 \lhd E_2$, with $\lhd \in \{<, =, \leq\}$
4:     **Output:** $\langle$SAT, $Model\rangle$ or $\langle$UNSAT, $Unsat\_Core\rangle$
5:     $T \leftarrow \emptyset$ ▷ transformed constraints
6:     $Q \leftarrow true$ ▷ Quotient formula
7:     $H_r \leftarrow H$
8:     **while** TRUE **do**
9:         **if** UNSAT(Q) **then**
10:             **return** $\langle$UNSAT, $H_r\rangle$
11:         Let μ_Q be a model of Q ($\mu_Q \models Q$)   ▷ Q is satisfiable
12:         $H_b \leftarrow \{E + m \cdot \mu_Q(q_i) \lhd F + m \cdot \mu_Q(r_i) \mid (E + m \cdot q_i \lhd F + m \cdot r_i) \in H_r\}$
13:         $\langle S, R\rangle \leftarrow$ LIA_DECIDE(H_b)
14:         **if** $S = $ SAT **then**
15:             $\langle IS, r\rangle \leftarrow$ MLIA_MODEL(H, R)
16:             **if** $IS = Yes$ **then**
17:                 **return** $\langle$SAT, $R\rangle$
18:             **else**
19:                 $H_r \leftarrow H_r \setminus \{r\} \cup \{\Gamma_1(r)\}; T \leftarrow T \cup \{\Gamma_1(r)\}$
20:                 $Q \leftarrow Q \wedge$ QUOTIENT_BOUND(r)   ▷ Definition 1
21:         **else**   ▷ R is unsat core in this case
22:             **if** $R \subseteq T$ **then**
23:                 $C_R \leftarrow \bigvee\{\sigma(c) \mid c \in R\}$ ▷ (C_R: a cut from R, σ: Equation 3)
24:                 $Q \leftarrow Q \wedge C_R$
25:             **else**
26:                 **for each** $s \in R$ **do**
27:                     **if** $s \notin T$ **then**
28:                         $H_r \leftarrow H_r \setminus \{s\} \cup \{\Gamma_1(s)\}; T \leftarrow T \cup \{\Gamma_1(s)\}$
29:                       $Q \leftarrow Q \wedge$ QUOTIENT_BOUND(s)   ▷ Definition 1

---

3. Applying Benders Decomposition Lazily. Algorithms 1 and 2 are complementary: the first algorithm transforms constraints lazily, whereas the second deals with large coefficients of quotient variables. We now present Algorithm 3 which in a sense combines them. The first two algorithms are presentation-purpose stepping stones for Algorithm 3 so the reader understands the two innovations separately. We do not evaluate and compare those algorithms.

All three algorithms are guaranteed to terminate. For the lazy approach (Algorithm 1), each iteration causes at least one (of the finitely many) initial constraints to be transformed. For the Benders decomposition, Q has finitely many models (as all variables are integral and bounded), and each iteration adds a cut eliminating at least one model. For the lazy Benders decomposition, each iteration either transforms at least one initial constraint, or eliminates some model of Q (without changing the number of transformed constraints), which again yields a finite descending chain.

Example 7. Recall again the problem of Example 1, extended with additional variables and constraints:

$$H = [c_1 : x \geq y \land c_2 : x + 1 = y \land c_3 : z + y \leq 7x \land c_4 : w - 2z \leq 3y + 2x]_{16}$$

Under Algorithm 3, the Q is initially trivial, and $H_b = H_r = H$ (as no constraints are initially transformed).

As before, this is unsatisfiable, with a core of $R = \{c_1, c_2\}$. R contains some constraint c_2 which is not yet transformed, so we replace c_2 with $\Gamma(c_2)$, introducing quotient variable q_x and appropriate bounds:

$$H' = (c_1 : x \geq y \land c'_2 : x + 1 + m \cdot q_x = y \land c_3 : z + y \leq 7x \land$$
$$c_4 : w - 2z \leq 3y + 2x \land c_5 : -1 \leq q_x \leq 0 \land c_6 : 0 \leq x \land y < m)$$

Resolving Q, we obtain $\mu_Q = \{q_x = 0\}$. As in Example 6, we find this to be unsatisfiable, having a core of $R = \{c_1, c'_2\}$, yielding a feasibility cut $q_x \neq 0$.

Solving again, we obtain a model $\mu_Q = \{q_x = -1\}$. Solving H_b now gives us a model: $\mu_{H_b} = \{x = 15, y = 0, z = 105, w = 240\}$. Evaluating this model under $\mathbb{Z}_{16}$, we obtain $\mu_H = \{x = 15, y = 0, z = 9, w = 0\}$. As μ_H is a valid model of H, we terminate.   □

Deciding Boolean Combinations of MLIA Constraints. Algorithms 2 and 3 can be extended to support Boolean combinations of MLIA constraints by adding a selection of 'active' constraints to the quotient problem, implicitly enumerating the Boolean skeleton.

5   Implementation and Experiments

Implementation. The algorithms are implemented in SoMoLIA (**So**lver for **Mo**dular **LIA**) which is written in Java. It uses Z3 [14] for input reading and pre-processing, and Gurobi [23] for solving LIA problems. The use of Gurobi is driven by unsatisfiable core extraction; in Z3 this incurs severe performance penalties (several orders of magnitude). Also, the Z3 LIA engine is ill-suited to our problems (optimized for incrementality, not for hard instances). To mitigate unsoundness, we fall back to Z3 if precision limits are exceeded as indicated by Gurobi (e.g., `pspace` instances) or non-integral solutions are obtained. The pre-processing we used includes simple word-level rewriting, constant propagation, Gaussian elimination and elimination of unconstrained variables [9], which are available from Z3 as tactics [14,15]. Such pre-processing, including the more advanced ones are common to all bit-blasting solvers (see [24]). Our tool supports input in SMT-LIB2 format [3] expressed over QF_BV or QF_LIA.

Benchmarks. Our main intent is to handle conjunctions of "pure" integer constraints efficiently. Accordingly, we chose a set of 1271 benchmarks from the CAV_2009, `dillig`, `check`, `pb2010`, `pidgeons`, `miplib2003` and `cut_lemmas` sub-categories of the QF_LIA category of SMT-COMP'16 [11] and interpreted them over $\mathbb{Z}_m$ (not over $\mathbb{Z}$). These are first translated into QF_BV logic so that we can reuse the pre-processing for bit-vector formulas as well as compare our results with the bit-vector solvers. Ideally, we would like to evaluate our approach on some challenging benchmarks from the QF_BV category, but unfortunately they contain problems with bit-wise operations (bvand, bvor, etc.), not purely word level operations (bvsub, bvmul, bvadd etc.) supported by SoMoLIA. However we selected the 41 problems from the pspace sub-category of the QF_BV category (containing only word level operations) and experimented with them. These contain very large bit-vectors (in the order of $\approx$ 23k bits).

Experimental Setting. We have conducted experiments on these benchmarks and compared the results (using Algorithm 3) with four state-of-the-art SMT(BV) solvers (the best of the currently available): Boolector [8] (winner SMT-COMP'16, QF_BVmain track), Yices2 [16] (winner SMT-COMP'16, QF_BV application track), Z3 [14] and CVC4 [2]. The pspace instances use very large moduli, making Gurobi unsound. Therefore, for these instances (and these instances only), SoMoLIA uses Z3 in place of Gurobi as the LIA solver. The experiments were carried out on a MacBook Pro with a 2.7 GHz Intel Core i5 processor and 16 GB memory running OS X 10.11.6. The timeout for each experiment is set to 10 min and the memory limit is 6 GB. The benchmarks and the tool itself are available at http://people.eng.unimelb.edu.au/gkgange/mod-arith/. The results are summarized in Table 1. The first column indicates the sub-category (and the number of problem instances in that sub-category). The columns that follow present the number of correctly solved instances (#correct) and average time taken (Time) to solve that many instances for the solvers Boolector [8] Yices2 [16], CVC4 [2], Z3 [14] and SoMoLIA respectively. The best result in each sub-category is bold-faced.

Discussion of the Results. The different tools are seen to have complementary success or timeout profiles. No solver consistently outperforms the others.

Table 1. Experimental results. Time (sec) is the average time over # solved instances of each category (timeout 10 min). † indicates SoMoLIA run using Z3 as LIA solver.

Category (#inst)	Boolector		Yices2		CVC4		Z3		SoMoLIA	
	#correct	Time	#correct	Time	#correct	Time	#correct	Time	#correct	Time
cut_lemmas (93)	**93**	14.33	**93**	7.45	91	0.81	**93**	**5.41**	34	10.14
dillig (233)	230	3.47	**233**	**0.08**	217	16.73	126	69.00	**233**	0.37
miplib2003 (16)	**16**	6.25	**16**	9.93	11	208.36	**16**	**3.81**	**16**	5.38
CAV_2009 (591)	**591**	3.62	**591**	0.45	544	3.71	343	27.51	**591**	**0.34**
check (5)	**5**	0.20	**5**	**0.01**	**5**	0.20	**5**	**0.01**	**5**	0.40
pb2010 (81)	**81**	2.62	**81**	**0.77**	53	161.90	**81**	1.49	**81**	1.71
pidgeons (19)	13	15.00	10	4.10	7	4.28	10	32.40	**19**	**0.78**
pspace (41)	**41**	155.70	25	92.04	**41**	**0.02**	0	0.00	**41**†	0.34

We note that SoMoLIA performed well on all sub-categories except `cut_lemmas`, owing to its ability to propagate "pure integer" level information, avoiding *bit-blasting* completely. We find that performance quite remarkable for a tool which is currently nothing more than a proof-of-concept. To us, it shows a great potential of a lazy form of Benders decomposition for this type of constraint problems.

Note that SoMoLIA leads in the `CAV-2009` and `pidgeons` cases, and it is the only solver which solves all the instances from `pidgeons`. The margin is significant in some cases. The problems from the `cut_lemmas` sub-category contain large coefficients, resulting in a large range of values for the quotient variables and requiring many iterations. The computation of unsat-cores also hindered performance. We note that 101 instances were solved by pre-processing alone. However we also note that it made a small number of problems timeout.

The problems from `pspace` sub-category are hard for all *bit-blasting* solvers (since they often make solvers run out of memory). Our approach, which is independent on the size of bit vectors, performed well on this subset and CVC4 performed extremely well. We believe CVC4's strong performance is due to its word level pre-processing of the formula. The approach taken by Zeljic *et al.* [37] also appears to perform well on SAT instances of this subset (albeit not on UNSAT instances as their experiment shows), but unfortunately we were not able to experiment with their tool and include it in the evaluation. A comparison of our approach against other solvers (with pre-processing turned off) allows us to measure the impact of pre-processing as well as the performance of purely bit-blasting approach against ours and we leave this task as future work.

6   Related Work

The ubiquity of two-complement semantics has raised considerable interest in the efficient solution of modular arithmetic constraints. Unfortunately, logics over modular arithmetic also resist efficient analysis and decision procedures.

Systems of linear arithmetic equations, as well as systems of linear congruences, can be solved in polynomial time [12]. However, the presence of inequalities hampers tractability; the general integer programming problem is NP-complete in the strong sense. For the special case of sets of integer difference constraints of the form $x - y \leq k$ and $x \leq k$ (as used in our examples), there are well-known efficient algorithms (for example, utilising the Bellman-Ford algorithm [12]), but the modular arithmetic version is already intractable for this special case [5, 21]. In the absence of efficient decision procedures for even restricted subclasses of constraints over modular arithmetic, it is typical to either use general bit-vector solvers, or to simply interpret the problem over $\mathbb{Z}$ (rather than $\mathbb{Z}_m$).

Bit-Vector Constraints. The theory of bit-vectors offers a natural encoding of modular arithmetic constraints (over modulus 2^w), yet is no more amenable to reasoning. The standard approach to solving bit-vector problems is *bit-blasting*: mapping bit-vector operations down to the corresponding Boolean circuits, and using a SAT solver to solve the resulting constraint system. They implement lazy/eager *bit-blasting* procedures; see Hadarean *et al.* [24] for comparison of these approaches. The most effective SMT(BV) solvers, such as Boolector [32], Yices2 [16], MathSAT [10], Z3 [14] CVC [2] and STP [20] also apply word-level simplification and rewriting, abstraction [32] and presolving over tractable sub-theories [24]. An alternate approach which avoids bit-blasting is to use a constraint programming (CP) approach—maintaining a compact abstraction of feasible assignments for bit-vector variables, and applying word-level filtering algorithms to prune inconsistent assignments. Such techniques have been integrated into pure CP [29] and CLP [1] frameworks. These approaches share the same domain abstraction, tracking which individual bits are fixed to particular values (and therefore have expressiveness equivalent to the bit-blasted representation). They offer compact representations and efficient local pruning, but cannot reason about *relationships* between variables. More recently, hybrid approaches which combine word-level filtering with SAT-style conflict reasoning have also arisen from both CP [36] and SMT [37] lineages.

Integer Arithmetic. A common alternative is to simply decide that mathematical integers are "near enough" to the desired semantics, and ignore wrapping effects altogether. Of course this is unsound, but it has the pragmatic advantage of allowing use of existing procedures for reasoning over $\mathbb{Z}$ and $\mathbb{R}$—particularly abstract domains (for static analysis), decision procedures (for constraint solving) and interpolation algorithms (for verification). As a result, interpretation over $\mathbb{Z}$ is a strategy adopted by many abstract interpretation-based static analysis [13,27] and program verification [22,26] tools.

LIA Encoding. Solving bit-vector constraints by translation into arithmetic constraints (linear and non-linear) is not new and has been studied for a while, though their efficient solving has been a challenge [1,6,7,17,34,35]. Bozzano et al. [6] discuss encoding of bit-vector formulae into LIA. Their focus, however, is on linearizing non-linear bit-vector constraints (e.g. bitwise expressions, non-constant multiplication) by decomposing a word v of width w into individual bits $b_{w-1}, \ldots, b_0$ with $w = \sum_i 2^i b_i$, then encoding bit-vector constraints using the introduced b_i variables. This approach effectively emulates bit-blasting with 0–1 variables; it allows handling of a more extensive range of operations, but suffers both the weak linear relaxation discussed in Sect. 3 and the encoding blow-up of bit-blasting.

7   Conclusion

We have presented a practical and efficient algorithm for solving MLIA constraints and evaluated it on a set of SMT-COMP'16 benchmarks. The main characteristic of the tool, SoMoLIA, is that it utilises Benders decomposition. Importantly, unlike bit-vector solvers, our approach uniformly handles (large) moduli, not necessarily powers of 2, including large primes, and the LIA encoding is bit-width independent. In spite of its increased scope, we find that our proof-of-concept implementation is competitive with state of the art bit-vector solvers, even when benchmarks are restricted to using moduli of form 2^w.

The experimental results are promising though there are many avenues for improvement. For example, additional simplification and pre-solving may lead to significant performance improvements (as has been seen in other solvers). Moreover, the current feasibility cuts are disjunctive, and somewhat weak; methods for deriving stronger cuts should greatly reduce the number of necessary iterations. Embedding this approach in a lazy DPLL(T) framework would provide several advantages: early detection of inconsistent quotient assignments, more efficient handling of Boolean combinations of constraints, and access to complementary theories, such as the theory of arrays.

Currently, we are also exploring ways to extend our work to non-linear bit-vectors problems. We can either encode non-linear bit-vector problems as non-linear integer arithmetic [6] and use non-linear integer solvers or combine linear integer and bit-blasted non-linear constraints and use a solver like IntSat [33], which is good at handling clausal linear constraints.

Acknowledgments. We are grateful for support from the Australian Research Council. The work has been supported by Discovery Project grant DP140102194, and Graeme Gange is supported through Discovery Early Career Researcher Award DE160100568.

References

1. Bardin, S., Herrmann, P., Perroud, F.: An alternative to SAT-based approaches for bit-vectors. In: Esparza, J., Majumdar, R. (eds.) TACAS 2010. LNCS, vol. 6015, pp. 84–98. Springer, Heidelberg (2010). doi:10.1007/978-3-642-12002-2_7
2. Barrett, C., Conway, C.L., Deters, M., Hadarean, L., Jovanović, D., King, T., Reynolds, A., Tinelli, C.: CVC4. In: Gopalakrishnan, G., Qadeer, S. (eds.) CAV 2011. LNCS, vol. 6806, pp. 171–177. Springer, Heidelberg (2011). doi:10.1007/978-3-642-22110-1_14
3. Barrett, C., Fontaine, P., Tinelli, C.: The SMT-LIB standard: version 2.5. Technical report, Department of Computer Science, The University of Iowa (2015). www.SMT-LIB.org
4. Benders, J.F.: Partitioning procedures for solving mixed-variables programming problems. Numer. Math. **4**(1), 238–252 (1962)
5. Bjørner, N., Blass, A., Gurevich, Y., Musuvathi, M.: Modular difference logic is hard, November 2008, Unpublished. arXiv:0811.0987v1
6. Bozzano, M., Bruttomesso, R., Cimatti, A., Franzén, A., Hanna, Z., Khasidashvili, Z., Palti, A., Sebastiani, R.: Encoding RTL constructs for MathSAT: a preliminary report. Electron. Notes Theor. Comput. Sci. **144**(2), 3–14 (2006)
7. Brinkmann, R., Drechsler, R.: RTL-datapath verification using integer linear programming. In: Proceedings of the ASPDAC/VLSI Design Conference 2002, pp. 741–746. IEEE Computer Society Press (2002)
8. Brummayer, R., Biere, A.: Boolector: an efficient SMT solver for bit-vectors and arrays. In: Kowalewski, S., Philippou, A. (eds.) TACAS 2009. LNCS, vol. 5505, pp. 174–177. Springer, Heidelberg (2009). doi:10.1007/978-3-642-00768-2_16

9. Bruttomesso, R., Cimatti, A., Franzén, A., Griggio, A., Hanna, Z., Nadel, A., Palti, A., Sebastiani, R.: A lazy and layered SMT($\mathcal{BV}$) solver for hard industrial verification problems. In: Damm, W., Hermanns, H. (eds.) CAV 2007. LNCS, vol. 4590, pp. 547–560. Springer, Heidelberg (2007). doi:10.1007/978-3-540-73368-3_54

10. Cimatti, A., Griggio, A., Schaafsma, B.J., Sebastiani, R.: The MathSAT5 SMT solver. In: Piterman, N., Smolka, S.A. (eds.) TACAS 2013. LNCS, vol. 7795, pp. 93–107. Springer, Heidelberg (2013). doi:10.1007/978-3-642-36742-7_7

11. Conchon, S., Déharbe, D., Heizmann, M., Weber, T.: SMT-COMP (2016). http://smtcomp.sourceforge.net/2016/

12. Cormen, T.H., Leiserson, C.E., Rivest, R.L., Stein, C.: Introduction to Algorithms. MIT Press, Cambridge (2009)

13. Cousot, P., Cousot, R., Feret, J., Mauborgne, L., Miné, A., Monniaux, D., Rival, X.: The ASTREÉ analyzer. In: Sagiv, M. (ed.) ESOP 2005. LNCS, vol. 3444, pp. 21–30. Springer, Heidelberg (2005). doi:10.1007/978-3-540-31987-0_3

14. de Moura, L., Bjørner, N.: Z3: an efficient SMT solver. In: Ramakrishnan, C.R., Rehof, J. (eds.) TACAS 2008. LNCS, vol. 4963, pp. 337–340. Springer, Heidelberg (2008). doi:10.1007/978-3-540-78800-3_24

15. de Moura, L., Passmore, G.O.: The strategy challenge in SMT solving. In: Bonacina, M.P., Stickel, M.E. (eds.) Automated Reasoning and Mathematics. LNCS, vol. 7788, pp. 15–44. Springer, Heidelberg (2013). doi:10.1007/978-3-642-36675-8_2

16. Dutertre, B.: Yices 2.2. In: Biere, A., Bloem, R. (eds.) CAV 2014. LNCS, vol. 8559, pp. 737–744. Springer, Cham (2014). doi:10.1007/978-3-319-08867-9_49

17. Ferrandi, F., Rendine, M., Sciuto, D.: Functional verification for SystemC descriptions using constraint solving. In: 2002 Design, Automation and Test in Europe Conference and Exposition (DATE 2002), pp. 744–751. IEEE Computer Society Press (2002)

18. Fröhlich, A., Kovásznai, G., Biere, A.: Efficiently solving bit-vector problems using model checkers. In: SMT Workshop (2013)

19. Fröhlich, A., Kovásznai, G., Biere, A.: More on the complexity of quantifier-free fixed-size bit-vector logics with binary encoding. In: Bulatov, A.A., Shur, A.M. (eds.) CSR 2013. LNCS, vol. 7913, pp. 378–390. Springer, Heidelberg (2013). doi:10.1007/978-3-642-38536-0_33

20. Ganesh, V., Dill, D.L.: A decision procedure for bit-vectors and arrays. In: Damm, W., Hermanns, H. (eds.) CAV 2007. LNCS, vol. 4590, pp. 519–531. Springer, Heidelberg (2007). doi:10.1007/978-3-540-73368-3_52

21. Gange, G., Søndergaard, H., Stuckey, P.J., Schachte, P.: Solving difference constraints over modular arithmetic. In: Bonacina, M.P. (ed.) CADE 2013. LNCS, vol. 7898, pp. 215–230. Springer, Heidelberg (2013). doi:10.1007/978-3-642-38574-2_15

22. Gurfinkel, A., Kahsai, T., Komuravelli, A., Navas, J.A.: The SeaHorn verification framework. In: Kroening, D., Păsăreanu, C.S. (eds.) CAV 2015. LNCS, vol. 9206, pp. 343–361. Springer, Cham (2015). doi:10.1007/978-3-319-21690-4_20

23. Gurobi Optimization, Inc.: Gurobi optimizer reference manual (2016). http://www.gurobi.com

24. Hadarean, L., Bansal, K., Jovanović, D., Barrett, C., Tinelli, C.: A tale of two solvers: eager and lazy approaches to bit-vectors. In: Biere, A., Bloem, R. (eds.) CAV 2014. LNCS, vol. 8559, pp. 680–695. Springer, Cham (2014). doi:10.1007/978-3-319-08867-9_45

25. Hooker, J.N., Ottosson, G.: Logic-based Benders decomposition. Math. Program. **96**(1), 33–60 (2003)

26. Jackson, D.: Software Abstractions: Logic, Language and Analysis. MIT Press, Cambridge (2006)
27. Jeannet, B., Miné, A.: APRON: a library of numerical abstract domains for static analysis. In: Bouajjani, A., Maler, O. (eds.) CAV 2009. LNCS, vol. 5643, pp. 661–667. Springer, Heidelberg (2009). doi:10.1007/978-3-642-02658-4_52
28. Kovásznai, G., Veith, H., Fröhlich, A., Biere, A.: On the complexity of symbolic verification and decision problems in bit-vector logic. In: Csuhaj-Varjú, E., Dietzfelbinger, M., Ésik, Z. (eds.) MFCS 2014. LNCS, vol. 8635, pp. 481–492. Springer, Heidelberg (2014). doi:10.1007/978-3-662-44465-8_41
29. Michel, L.D., Hentenryck, P.V.: Constraint satisfaction over bit-vectors. In: Milano, M. (ed.) CP 2012. LNCS, pp. 527–543. Springer, Heidelberg (2012). doi:10.1007/978-3-642-33558-7_39
30. Neumaier, A., Shcherbina, O.: Safe bounds in linear and mixed-integer linear programming. Math. Program. **99**(2), 283–296 (2004)
31. Neumaier, A., Shcherbina, O., Huyer, W., Vinkó, T.: A comparison of complete global optimization solvers. Math. Program. **103**(2), 335–356 (2005)
32. Niemetz, A., Preiner, M., Biere, A.: Boolector 2.0 system description. J. Satisf. Boolean Model. Comput. **9**, 53–58 (2014). (published 2015)
33. Nieuwenhuis, R.: The IntSat method for integer linear programming. In: O'Sullivan, B. (ed.) CP 2014. LNCS, vol. 8656, pp. 574–589. Springer, Cham (2014). doi:10.1007/978-3-319-10428-7_42
34. Parthasarathy, G., Iyer, M.K., Cheng, K., Wang, L.: An efficient finite-domain constraint solver for circuits. In: Malik, S., Fix, L., Kahng, A.B. (eds.) Proceedings of the 41th Design Automation Conference (DAC 2004), pp. 212–217. ACM Publ. (2004)
35. Vemuri, R., Kalyanaraman, R.: Generation of design verification tests from behavioral VHDL programs using path enumeration and constraint programming. IEEE Trans. VLSI Syst. **3**(2), 201–214 (1995)
36. Wang, W., Søndergaard, H., Stuckey, P.J.: A bit-vector solver with word-level propagation. In: Quimper, C.-G. (ed.) CPAIOR 2016. LNCS, vol. 9676, pp. 374–391. Springer, Cham (2016). doi:10.1007/978-3-319-33954-2_27
37. Zeljić, A., Wintersteiger, C.M., Rümmer, P.: Deciding bit-vector formulas with mcSAT. In: Creignou, N., Le Berre, D. (eds.) SAT 2016. LNCS, vol. 9710, pp. 249–266. Springer, Cham (2016). doi:10.1007/978-3-319-40970-2_16

SAT Encodings

SAT-Based Local Improvement for Finding Tree Decompositions of Small Width

Johannes K. Fichte[✉], Neha Lodha, and Stefan Szeider

TU Wien, Vienna, Austria
jfichte@dbai.tuwien.ac.at, {neha,sz}@ac.tuwien.ac.at

Abstract. Many hard problems can be solved efficiently for problem instances that can be decomposed by tree decompositions of small width. In particular for problems beyond NP, such as #P-complete counting problems, tree decomposition-based methods are particularly attractive. However, finding an optimal tree decomposition is itself an NP-hard problem. Existing methods for finding tree decompositions of small width either (a) yield optimal tree decompositions but are applicable only to small instances or (b) are based on greedy heuristics which often yield tree decompositions that are far from optimal. In this paper, we propose a new method that combines (a) and (b), where a heuristically obtained tree decomposition is improved locally by means of a SAT encoding. We provide an experimental evaluation of our new method.

1 Introduction

Treewidth is arguably the most prominent graph invariant with important application in discrete algorithms and optimization [5,8], constraint satisfaction [11,16], knowledge representation and reasoning [19], computational biology [30], and probabilistic networks and inference [10,24,26]. Treewidth was introduced by Robertson and Seymour in their Graph Minors Project and according to Google Scholar[1], the term is mentioned in over 17000 research articles.

Small treewidth of a graph indicates in a certain sense its tree-likeness and sparsity. Many otherwise NP-hard graph problems such as Hamiltonicity and 3-colorability, but also problems "beyond NP" such as the #P-complete problem of determining the number of perfect matchings in a graph are solvable in polynomial time for graphs of bounded treewidth [9]. Treewidth is based on certain decompositions of graphs, called tree decompositions, where sets of vertices of the input graph are arranged in bags at the nodes of a tree such that certain conditions are satisfied. The width of a tree decomposition is the size of a largest bag minus 1. A tree decomposition is optimal for a given graph if the graph has no tree decomposition of smaller width. The treewidth of a graph is the width of an optimal tree decomposition.

Research was supported by the Austrian Science Fund (FWF), Grants Y698, W1255-N23, and P-26696. The first author is also affiliated with the University of Potsdam, Germany.

[1] Retrieved on March 26, 2017.

S. Gaspers and T. Walsh (Eds.): SAT 2017, LNCS 10491, pp. 401–411, 2017.
DOI: 10.1007/978-3-319-66263-3_25

Algorithms that exploit the small treewidth of a graph usually proceed by dynamic programming along the tree decomposition where at each node of the tree, information is gathered in tables. The size of these tables is usually exponential or even double exponential in the size of the bag. Thus, it is important to obtain a tree decomposition of small width. However, since finding an optimal tree decomposition is an NP-hard task [2], the following two main approaches have been proposed in the literature:

(a) *Exact methods* that compute optimal tree decompositions. Optimal tree decompositions are found using specialized combinatorial algorithms based on graph separators [2], branch-and-bound algorithms [18], but also by means of *SAT encodings* [4,28]. These exact methods are limited to rather small graphs with about hundred vertices.

(b) *Heuristic methods* that compute sub-optimal tree decompositions. These algorithms are usually based on so-called elimination orderings which are found by a greedy approach [6,20]. The heuristic methods are quite fast and scale up to large graphs with thousands of vertices, but lead to tree decompositions that can be far from optimal.

In fact, because of the split into these two categories of algorithmic approaches, also the recent PACE challenge [13], where finding good tree decompositions was one of the main tasks, featured two respective categories: one asking for the exact treewidth of small graphs, and one asking for sub-optimal tree decompositions of large graphs.

SAT-Based Local Improvement. In this paper, we propose a new approach to finding tree decompositions, which combines exact methods with heuristics. The basic idea is to (i) start with a tree decomposition obtained with a heuristic method (the *global solver*) and (ii) subsequently select parts of the tree decomposition, trying to improve it with another method (the *local solver*). It turned out that SAT-based exact methods are particularly well-suited for providing the local solver.

Consider a given graph G and a tree decomposition T of G, obtained by the global solver. We select a small part S of T, which is a tree decomposition of the subgraph G_S of G, induced by all the vertices that appear in bags at nodes in S. Once the local solver finds a better tree decomposition of G_S, we would like to replace S in T with the new tree decomposition found by the local solver. This, however, does not work in general, as the new tree decomposition might not fit into the remaining parts of T. Fortunately we can make this approach work by using the following trick. We add to G_S certain cliques, which we call *marker cliques*, and which tell us how to replace the original local tree decomposition S with the new one. Due to a general property of tree decompositions, there is always a bag that contains all vertices of a clique. Hence, in particular, the new local decomposition will contain for each marker clique a bag that contains it, and this bag will be an anchor point for connecting the new decomposition to the parts of the old one. Details of this construction are explained in Sect. 3.

Related Work. A SAT-based local improvement approach was first proposed, implemented and evaluated by Lodha et al. [25] for finding *branch decompositions* of small width of graphs and hypergraphs. As the definitions of a branch decompositions and tree decompositions differ significantly, the methods for finding and replacing local decompositions are quite different. Also the SAT encoding of branchwidth and treewidth are different, as the former focuses on edges, while the latter focuses on vertices.

There are several approaches for improving treewidth heuristics based on elimination orderings. For instance, Kask et al. [21] use randomization to recompute the last few steps of the ordering computed so far, picking the best of the runs, whereas Gaspers et al. [17] use a different approach: as soon as a given width bound is exceeded during the computation of the ordering, the last c vertices of the ordering are recomputed with an exact method, trying to stay within the width bound.

2 Preliminaries

In this section we introduce some relevant graph theoretic notions.

All considered graphs are finite, simple, and undirected. Let G be a graph. $V(G)$ and $E(G)$ denote the vertex set and the edge set of G, respectively. We denote an edge between vertices u and v by uv (or equivalently by vu). The subgraph of G induced by a set $S \subseteq V(G)$ has as vertex set S and as edge set $\{ uv \in E(G) \mid u, v \in S \}$.

A *tree decomposition* of a graph G is a pair $\mathcal{T} = (T, \chi)$ where T is a tree and χ is a mapping that assigns to each node $t \in V(T)$ a set $\chi(t) \subseteq V(G)$, called a *bag*, such that the following conditions hold (we refer to the vertices of T as *nodes* to make the distinction between T and G clearer).

1. $V(G) = \bigcup_{t \in V(T)} \chi(t)$ and $E(G) \subseteq \bigcup_{t \in V(T)} \{ uv \mid u, v \in \chi(t) \}$.
2. The sets $\chi(t_1) \setminus \chi(t)$ and $\chi(t_2) \setminus \chi(t)$ are disjoint for any three nodes $t, t_1, t_2 \in V(T)$ such that t lies on a path from t_1 to t_2 in T.

The *width* of $\mathcal{T}$, denoted $\mathsf{w}(\mathcal{T})$, is $\max_{t \in V(T)} |\chi(t)| - 1$. The *treewidth* $\mathsf{tw}(G)$ of G is the minimum $\mathsf{w}(\mathcal{T})$ over all tree decompositions $\mathcal{T}$ of G.

We will make use of the following well-known fact.

Fact 1 ([23]). *Let (T, χ) be a tree decomposition of a graph G and K a clique in G. Then there exists at least one node $t \in V(T)$ such that $V(K) \subseteq \chi(t)$.*

3 Local Improvement of Tree Decompositions

3.1 Local Tree Decompositions

For the following considerations we fix a graph G and a tree decomposition $\mathcal{T} = (T, \chi)$ of G. We consider a subtree S of T.

We call $\mathcal{S} = (S, \chi_S)$ a *local tree decomposition of $\mathcal{T}$* (*induced by S*), where χ_S is the restriction of χ to the nodes of S. Let G_S denote the subgraph of G induced by all the vertices of G that appear in a bag of $\mathcal{S}$. The following observation is an immediate consequence of the definitions.

Observation 1. $\mathcal{S}$ *is a tree decomposition of G_S of width $\leq \mathsf{w}(\mathcal{T})$.*

Our goal is to replace $\mathcal{S}$ with an improved tree decomposition $\mathcal{S}'$ of G_S, i.e., one of smaller width, and to insert $\mathcal{S}'$ back into $\mathcal{T}$ so that we obtain a new tree decomposition $\mathcal{T}'$ of G of possibly smaller width. In order to make this work, we need to modify G_S such that any tree decomposition of the modified graph can be added back into $\mathcal{T}$.

Let us first introduce some auxiliary notions. For an edge st of $\mathcal{T}$ we define $\lambda_{\mathcal{T}}(st) = \chi(s) \cap \chi(t)$ to the *cut set* associated with st. We call an edge st of $\mathcal{T}$ to be a *boundary edge* (w.r.t. $\mathcal{S}$) if $s \in V(S)$ and $t \notin V(S)$.

Now we define the *augmented local graph* G_S^* by setting $V(G_S^*)$ to be the set of all vertices of G that appear in a bag of $\mathcal{S}$, and $E(G_S^*)$ to be the set of edges uv with $u, v \in V^*$ such that $uv \in E(G)$ or $u, v \in \lambda_{\mathcal{T}}(e)$ for a boundary edge e of $\mathcal{T}$. In other words, the augmented local graph G_S^* is obtained from G_S by forming cliques over cut sets associated with boundary edges. We will use these cliques as "markers" in order to connect a new tree decomposition of G_S^* to the parts of the tree decomposition $\mathcal{T}$ that we keep. Therefore we call these cliques *marker cliques*.

Observation 2. $\mathcal{S}$ *is a tree decomposition of G_S^* of width $\leq \mathsf{w}(\mathcal{T})$.*

Proof. In view of Observation 2 it remains to check that for each edge $uv \in E(G_S^*) \setminus E(G_S)$ there is a node s of S such that $u, v \in \chi(s)$. For such an edge uv there is a boundary edge e of $\mathcal{T}$ such that $u, v \in \lambda_{\mathcal{T}}(e)$. By definition of a boundary edge, exactly one end of e, say s, belongs to $V(S)$. Now $u, v \in \lambda_{\mathcal{T}}(e) \subseteq \chi(s)$. $\square$

Let $\mathcal{S}^* = (S^*, \chi^*)$ be another tree decomposition of G_S^* with $\mathsf{w}(\mathcal{S}^*) \leq \mathsf{w}(\mathcal{S})$. W.l.o.g., we assume that S^* and T do not share any vertices (if not, we can simply use a tree that is isomorphic to S^*). We define a new tree decomposition $\mathcal{T}' = (T', \chi')$ of G as follows.

Let $T_1, \ldots, T_r$ be the connected components of $T - S$ (each T_i is a tree). Each T_i gives raise to a local tree decomposition $\mathcal{T}_i = (T_i, \chi_i)$ where χ_i is the restriction of χ to the nodes of T_i.

For each T_i let t_i be the leaf of T_i that was incident with a boundary edge $e_i = t_i s_i$ in T. The boundary edge e_i is responsible for a marker clique $K(e_i)$ on the vertices in $\lambda_{\mathcal{T}}(e_i)$. By Fact 1, we can choose a node $s_i' \in V(S^*)$ such that $V(K(e_i)) = \lambda_{\mathcal{T}}(e_i) \subseteq \chi^*(s_i')$.

We define a new tree decomposition $\mathcal{T}' = (T', \chi')$ where T' is the tree defined by $V(T') = V(S^*) \cup \bigcup_{i=1}^{r} V(T_i) = V(S^*) \cup V(T) \setminus V(S)$ and $E(T') = E(S^*) \cup \bigcup_{i=1}^{r} E(T_i) \cup \{t_1 s_1', \ldots, t_r s_r'\}$. It remains to define the bags of the tree decomposition $\mathcal{T}'$. For $t \in V(T_i)$ we define $\chi'(t) = \chi(t)$ and for $s \in V(S^*)$ we define $\chi'(s) = \chi^*(s)$.

We denote T' as $T\left(\begin{smallmatrix}S\\S'\end{smallmatrix}\right)$ and say that T' *is obtained from T by replacing S with S'*.

Observation 3. $T\left(\begin{smallmatrix}S\\S'\end{smallmatrix}\right)$ *is a tree decomposition of G of width*

$$\max(\mathsf{w}(T_1), \ldots, \mathsf{w}(T_r), \mathsf{w}(S^*)) \leq \max(\mathsf{w}(T), \mathsf{w}(S^*)) \leq \max(\mathsf{w}(T), \mathsf{w}(S)) \leq \mathsf{w}(T).$$

Proof. Let $T\left(\begin{smallmatrix}S\\S'\end{smallmatrix}\right) = T' = (T'\chi')$. First we observe that T' is indeed a tree, as each tree T_i is connected to the central tree S^* with exactly one edge. Clearly T' satisfies the first of the two conditions in the definition of a tree decomposition. To see that it also satisfies the second condition, we observe that if a vertex v of G appears in bags at two different local tree decompositions T_i and T_j then v must also appear in the sets $\lambda_T(e_i)$ and $\lambda_T(e_j)$. Consequently, it appears in the bags of s'_i and s'_j (we use the notation from above). As S^* satisfies the second condition of a tree decomposition, v is contained in all the bags on the path between s'_i and s'_j in S^*. This shows that T' is indeed a tree decomposition of G. The claimed bound on its width follows directly from the construction. □

3.2 SAT Encodings for Tree Decompositions

A SAT encoding for tree decompositions was first proposed by Samer and Veith [28]. Given a graph G and an integer k, a CNF formula $\Phi(G, k)$ is produced which is satisfiable if and only if G has a tree decomposition of width $\leq k$. For the construction of $\Phi(G, k)$, an alternative characterization of tree decompositions in terms of *elimination orderings* is used. Here a linear ordering of the given graph G is guessed, and based on the ordering certain "fill-in edges" are added to the graph, providing a "triangulation" of G. The ordering is represented by Boolean variables, one for every pair of vertices, whose truth value indicates the relative ordering of the two vertices. Transitivity of the ordering is ensured by suitable clauses. Then, for each vertex v of G it is checked whether it has at most k neighbors that appear in the ordering right to v. This is checked via cardinality constraints [29]. The exact treewidth is then found by systematically calling a SAT solver for a heuristically computed upper bound u with $\Phi(G, k)$ for $k = u, u - 1, u - 2, \ldots$ and until $\Phi(G, k)$ is found unsatisfiable. From a satisfying assignment of $\Phi(G, k)$ one can obtain a tree decomposition of G of width k efficiently by a *decoding procedure*.

3.3 The Local Improvement Loop

We describe the overall algorithm. Let G be an input graph. First we obtain a tree decomposition $T = (T, \chi)$ of G using a standard heuristic method, which we refer to as the **global solver**.

The local improvement loop operates with the following parameters which are positive integers: the local budget **lb**, the local timeout **lt**, the global timeout **gt**, and the number of no-improvement rounds **ni**.

We select a node t from T with largest bag size, i.e., $|\chi(t)| = \mathsf{w}(T)$.

In T we perform a modified breadth-first-search (BFS) starting at t. We use an auxiliary set variable L which, at the beginning of the BFS is set to $\chi(t)$. For each node t' visited by the BFS, we add the new elements of $\chi(t')$ to L. If a node t' was visited via an edge e, a neighbor t'' of t' is only visited if $\lambda_T(t't'') < \lambda_T(e)$. The BFS terminates as soon as visiting another node would increase the size of L beyond the local budget **lb**. Now the visited nodes induce a subtree S of T, and in turn, this yields a local tree decomposition $\mathcal{S} = (S, \chi_S)$ of $\mathcal{T}$, as defined above. The set L contains the vertices of the local graph $G_{\mathcal{S}}$ (or equivalently, of the augmented local graph $G_{\mathcal{S}}^*$) which by construction can be at most **lb** many vertices.

Next we run the **local solver**, that is, we check satisfiability of the formula obtained by the SAT encoding, trying to get a tree decomposition $\mathcal{S}^*$ of $G_{\mathcal{S}}^*$ whose width is as small as possible. We start the SAT encoding with $k = \mathsf{w}(\mathcal{S}) - 1$ and upon success decrease k step by step. Each SAT-call has a timeout of **lt** seconds, and we stop if either we get an unsatisfiable instance or we hit the timeout. With the reached value of k, the treewidth of $G_{\mathcal{S}}^*$ is at most $k+1$. Since the SAT encoding with value $k+1$ is satisfiable, we can extract with a decoding procedure from the satisfying assignment a tree decomposition $\mathcal{S}^*$ of $G_{\mathcal{S}}^*$. Now we replace $\mathcal{S}$ in T by $\mathcal{S}^*$, and we repeat the local improvement loop with $\mathcal{T}\left(\begin{smallmatrix}\mathcal{S}\\\mathcal{S}_*\end{smallmatrix}\right)$. We note that a local replacement is done even if there was no local width improvement, i.e., if $\mathsf{w}(\mathcal{S}^*) = \mathsf{w}(\mathcal{S})$, as there is the possibility that the change triggers improvements in subsequent rounds of the local improvement loop.

We repeat the local improvement loop until either the global timeout **gt** is reached, or if the loop has been iterated **ni** times without any local width improvements.

4 Experimental Results

Solvers. As the global solver we used the greedy ordering heuristics-based algorithm from Abseher et al. [1, rev. 075019f] which we refer to as **heur**. It computes upper bounds for treewidth and outputs a certificate decomposition. The solver scored third in the heuristic track of the PACE 2016 challenge [13]. It is very space efficient and reports initial useful tree decompositions extremely fast compared to other solvers. It leaves almost the full time resource for the local improvement. We used the following three local solvers:

1. **sat**: a solver based on an improved version of Samer and Veith's [28] SAT encoding by Bannach et al. [3, rev. 25d6a98]. The solver employs Glucose as a SAT solver, PBLib for cardinality encodings, and progresses downwards from an upper bound. The solver scored third in the exact track of the PACE 2016 treewidth challenge and was there the best SAT-based solver.
2. **comb**: an implementation of Arnborg et al.'s combinatorial algorithm [2] by Tamaki [31, rev. d5ba92a], This solver won the exact track of the PACE 2016 treewidth challenge. It incrementally checks for the exact treewidth, it progresses upwards from 1.
3. **heur**: the same solver that we also use as global solver.

Our implementation is publicly available on GitHub [15]. Our experiments mainly focus on two questions: (i) can we improve with local improvement over traditional greedy heuristics and (ii) which solvers are favorable as local solver.

Instances. We considered an initial selection of overall 3168 graphs from various publicly available graph sets. Our sets consisted of the *TreewidthLIB* [7], networks from the *UAI competition* [12], publicly available transit graphs from *GTFS-transit feeds* [14], and graphs from the *PACE 2016 treewidth challenge* [13]. Since we aimed for larger graphs where exact methods cannot be used, we restricted ourselves to graphs that contain more than 100 vertices, resulting in 1946 graphs in total.

Experimental Setup. The experiments ran on a Scientific Linux cluster of 24 nodes (2x Xeon E5520 each) and overall 224 physical cores [22]. Due to the large number of instances, we started only from one initial decomposition (with random seed) and did not repeat the runs. In order to have reproducible results we used a benchmark cluster run generator and analysis tool[2]. All solvers have been compiled with gcc version 4.9.1, ran on Python 2.7.5, and Java 1.8.0_122 HotSpot 64-bit server VM, respectively. We executed solvers in single core mode. We limited available memory (RAM) to 8 GB, wall clock time of the global solver to 15 s, wall clock time of the overall search to 7800 s, and wall clock time of the local solver to 1800 s. For the SAT solver we imposed an additional restriction that the individual SAT call runs at most 900 s (**st**). Resource limits where enforced by *runsolver* [27].

For our experiments, we systematically tested the parameters **lb** $\in$ {75, 100,125,150}, **lt** $\in$ {90, 900, 1800}, **gt** = 7200, and **ni** = 10. For the parameter **ni** we also tried values 40 and 100 on a selected set of instances, but obtained no improvements. Individual results are publicly available [15].

Results. Table 1 summarizes the improvements we obtained with our experiments. Configurations in the legend are given in the form `solver-lb-lt(st)`. The best results in each column are highlighted in bold font. Table 2 shows some of the best and notable improvements we obtained with local improvements. The value "hash" provides the first four digits of sha-1 hash sum for the instance in DIMACS graph format. Column "htw" has the heuristically obtained treewidth, and "itw" has the treewidth after local improvement. The configuration with which we got these improvements are in the column "local solver." The best improvement we obtained is 20, for the instance or_chain_224.fg, from the graph set networks. Among further entries in the table are instance graph13pp with a width over 100, and instance Promedus_38 where we could reduce the width from 23 to 16, which makes this instance feasible for dynamic programming.

---

[2] The run and analysis tool is available online at https://github.com/daajoe/benchmark-tool. The file benchmark-tool/runscripts/treewidth/localimprovement. xml contains all solver flags to reproduce our benchmark runs.

Table 1. Summary of treewidth improvements.

# Improved	Improvements (sum)	Improvement (max)	Solver configuration
647	**2015**	13	sat-100-1800(900)
584	1984	16	sat-125-1800(900)
630	1805	15	comb-100-1800
493	1676	**20**	sat-150-1800(900)
631	1548	12	comb-075-1800
609	1460	12	comb-075-1800
447	1077	19	comb-125-1800
368	822	14	comb-150-1800
325	538	9	heur-150-1800
258	421	8	heur-100-1800

Table 2. Some of the best and notable improvements

Instance (hash)	$|V|$	$|E|$	Graphs	itw	htw	Local solver
or_chain_224.fg (a4cb)	1638	3255	networks	75	95	sat-150-1800-10
or_chain_54.fg (a6fc)	1404	2757	networks	65	84	comb-125-1800-10
or_chain_187.fg (826a)	1668	3197	networks	79	97	sat-150-1800-10
1bkr_graph (003a)	107	1340	twlib	44	56	comb-075-1800-10
dimacs_fpsol2.i.1-pp (69aa)	191	4418	pace2016	61	72	sat-150-1800-10
graph13pp (eb9d)	456	1874	twlib	115	125	comb-150-1800-10
Cell120 (b625)	600	1200	pace2016	94	104	comb-150-1800-10
bkv-zrt_20120422_0314 (fbca)	907	2209	transit	74	83	sat-150-1800-10
Promedus_38 (02d7)	668	1235	networks	16	23	sat-150-1800-10

Discussion. For our instance set, we can see that even a heuristic solver as local solver (**lb** = 150) improved the upper bounds. Both in terms of number of improved instances and when considering the cumulative sum of improvements, the SAT-based solver performed best. For both the combinatorial solver and the SAT-based solver, a local budget **lb** = 100 resulted in more solved instances. However, in terms of overall improvement the difference between the two local solvers is small. A local budget **lb** = 125 allowed us to increase the cumulative sum of improvements relatively early.

In consequence, we obtained the best results by using a SAT-based solver as local solver. Using a SAT-based solver, we can hope that an improved SAT encoding or new techniques in solvers immediately yield better upper bounds for treewidth using local improvement. We also computed the virtually best solver, which improved 200 instances more than the best SAT-based configuration. This indicates that we can very likely improve a much higher number of instances when applying a portfolio based solving approach.

5   Concluding Remarks

We have presented a new SAT-based approach to finding tree decompositions of small width based on a cross-over between standard heuristic methods and exact methods. Our work offers several directions for further research.

For instance, one could possibly improve the current setup by (a) upgrading the method for selecting the local tree decomposition, which is currently based on a relatively simple breadth-first-search, and (b) tuning and optimizing the SAT-based local solver specially to handle the type of instances that arise within the local improvement loop.

Another promising direction involves adding additional constraints to the SAT encoding, which yield local tree decompositions with special properties. For instance, when the local solver cannot improve the width of the current local tree decomposition, it could still replace it with one that increases the likelihood of success for further rounds of local improvements (for instance, by minimizing the number of large bags). Another application would be the computation of "customized tree decompositions" [1] which are designed to speed-up dynamic programming algorithms. Such additional constraints are relatively easy to build into a SAT-based local solver, but seem difficult to build into a local solver based on combinatorial methods.

Finally, due to the modularity of our approach (local solver, budget, time out, invoked SAT solver), it could benefit from automated algorithm configuration and parameter tuning, and it could provide the elements of a portfolio approach.

References

1. Abseher, M., Musliu, N., Woltran, S.: htd – a free, open-source framework for (customized) tree decompositions and beyond. In: Salvagnin, D., Lombardi, M. (eds.) CPAIOR 2017. LNCS, vol. 10335, pp. 376–386. Springer, Cham (2017). doi:10.1007/978-3-319-59776-8_30
2. Arnborg, S., Corneil, D.G., Proskurowski, A.: Complexity of finding embeddings in a k-tree. SIAM J. Algebraic Discrete Methods 8(2), 277–284 (1987)
3. Bannach, M., Berndt, S., Ehlers, T.: Jdrasil: a modular library for computing tree decompositions. Technical report, Lübeck University, Germany (2016)
4. Berg, J., Järvisalo, M.: SAT-based approaches to treewidth computation: an evaluation. In: Proceedings of the 26th IEEE International Conference on Tools with Artificial Intelligence, ICTAI 2014, pp. 328–335. IEEE Computer Society, Limassol, Cyprus, November 2014
5. Bodlaender, H.L., Koster, A.M.C.A.: Combinatorial optimization on graphs of bounded treewidth. Comput. J. 51(3), 255–269 (2008)
6. Bodlaender, H.L., Koster, A.M.C.A.: Treewidth computations I. Upper bounds. Inf. Comput. 208(3), 259–275 (2010)
7. van den Broek, J.W., Bodlaender, H.: TreewidthLIB - a benchmark for algorithms for treewidth and related graph problems. Technical report, Faculty of Science, Utrecht University (2010). http://www.staff.science.uu.nl/~bodla101/treewidthlib/

8. Chimani, M., Mutzel, P., Zey, B.: Improved Steiner tree algorithms for bounded treewidth. J. Discrete Algorithms **16**, 67–78 (2012)
9. Courcelle, B., Makowsky, J.A., Rotics, U.: On the fixed parameter complexity of graph enumeration problems definable in monadic second-order logic. Discr. Appl. Math. **108**(1–2), 23–52 (2001)
10. Darwiche, A.: A differential approach to inference in Bayesian networks. J. ACM **50**(3), 280–305 (2003)
11. Dechter, R.: Tractable structures for constraint satisfaction problems. In: Rossi, F., van Beek, P., Walsh, T. (eds.) Handbook of Constraint Programming, Chap. 7, vol. I, pp. 209–244. Elsevier, Amsterdam (2006)
12. Dechter, R.: Graphical model algorithms at UC Irvine. Technical report, UC Irvine (2013). The network instances consist of Bayesian and Markov network susedin UAI competition and protein folding/side-chain prediction problems. http://graphmod.ics.uci.edu/group
13. Dell, H., Rosamond, F.: The parameterized algorithms and computational experiments challenge (2016). https://pacechallenge.wordpress.com/
14. Fichte, J.K.: daajoe/gtfs2graphs - a GTFS transit feed to graph format converter (2016). https://github.com/daajoe/gtfs2graphs
15. Fichte, J.K., Lodha, N., Szeider, S.: Trellis: treewidth local improvement solver (2017). https://github.com/daajoe/trellis
16. Freuder, E.C.: A sufficient condition for backtrack-bounded search. J. ACM **32**(4), 755–761 (1985)
17. Gaspers, S., Gudmundsson, J., Jones, M., Mestre, J., Rümmele, S.: Turbocharging Treewidth Heuristics. In: Guo, J., Hermelin, D. (eds.) 11th International Symposium on Parameterized and Exact Computation (IPEC 2016). Leibniz International Proceedings in Informatics (LIPIcs), vol. 63, pp. 13:1–13:13. Schloss Dagstuhl-Leibniz-Zentrum fuer Informatik, Dagstuhl, Germany (2017)
18. Gogate, V., Dechter, R.: A complete anytime algorithm for treewidth. In: Proceedings of the Twentieth Conference Annual Conference on Uncertainty in Artificial Intelligence (UAI 2004), pp. 201–208. AUAI Press, Arlington (2004)
19. Gottlob, G., Pichler, R., Wei, F.: Bounded treewidth as a key to tractability of knowledge representation and reasoning. Artif. Intell. **174**(1), 105–132 (2010)
20. Hammerl, T., Musliu, N., Schafhauser, W.: Metaheuristic algorithms and tree decomposition. In: Kacprzyk, J., Pedrycz, W. (eds.) Springer Handbook of Computational Intelligence, pp. 1255–1270. Springer, Heidelberg (2015). doi:10.1007/978-3-662-43505-2_64
21. Kask, K., Gelfand, A., Otten, L., Dechter, R.: Pushing the power of stochastic greedy ordering schemes for inference in graphical models. In: Burgard, W., Roth, D. (eds.) Proceedings of the Twenty-Fifth AAAI Conference on Artificial Intelligence, AAAI 2011. AAAI Press (2011)
22. Kittan, K.: Zuse cluster (2017). http://www.cs.uni-potsdam.de/bs/research/labsZuse.html
23. Kloks, T.: Treewidth: Computations and Approximations. Springer, Heidelberg (1994)
24. Lauritzen, S.L., Spiegelhalter, D.J.: Local computations with probabilities on graphical structures and their application to expert systems. J. Roy. Statist. Soc. Ser. B **50**(2), 157–224 (1988)
25. Lodha, N., Ordyniak, S., Szeider, S.: A SAT approach to branchwidth. In: Creignou, N., Le Berre, D. (eds.) SAT 2016. LNCS, vol. 9710, pp. 179–195. Springer, Cham (2016). doi:10.1007/978-3-319-40970-2_12

26. Ordyniak, S., Szeider, S.: Parameterized complexity results for exact Bayesian network structure learning. J. Artif. Intell. Res. **46**, 263–302 (2013)
27. Roussel, O.: Controlling a solver execution with the runsolver tool. J. Satisfiability Boolean Model. Comput. **7**, 139–144 (2011)
28. Samer, M., Veith, H.: Encoding treewidth into SAT. In: Kullmann, O. (ed.) SAT 2009. LNCS, vol. 5584, pp. 45–50. Springer, Heidelberg (2009). doi:10.1007/978-3-642-02777-2_6
29. Sinz, C.: Towards an optimal CNF encoding of boolean cardinality constraints. In: Beek, P. (ed.) CP 2005. LNCS, vol. 3709, pp. 827–831. Springer, Heidelberg (2005). doi:10.1007/11564751_73
30. Song, Y., Liu, C., Malmberg, R.L., Pan, F., Cai, L.: Tree decomposition based fast search of RNA structures including pseudoknots in genomes. In: Proceedings of the 4th International IEEE Computer Society Computational Systems Bioinformatics Conference, CSB 2005, pp. 223–234. IEEE Computer Society (2005)
31. Tamaki, H.: TCS-Meiji (2016). https://github.com/TCS-Meiji/treewidth-exact

A Lower Bound on CNF Encodings
of the At-Most-One Constraint

Petr Kučera[1]([✉]), Petr Savický[2], and Vojtěch Vorel[1]

[1] Department of Theoretical Computer Science and Mathematical Logic,
Faculty of Mathematics and Physics, Charles University, Prague, Czech Republic
{kucerap,vorel}@ktiml.mff.cuni.cz
[2] Institute of Computer Science, The Czech Academy of Sciences,
Prague, Czech Republic
savicky@cs.cas.cz

Abstract. Constraint "at most one" is a basic cardinality constraint
which requires that at most one of its n boolean inputs is set to 1. This
constraint is widely used when translating a problem into a *conjunctive
normal form* (CNF) and we investigate its CNF encodings suitable for
this purpose. An encoding differs from a CNF representation of a function in that it can use auxiliary variables. We are especially interested
in propagation complete encodings which have the property that unit
propagation is strong enough to enforce consistency on input variables.
We show a lower bound on the number of clauses in any propagation
complete encoding of the "at most one" constraint. The lower bound
almost matches the size of the best known encodings. We also study
an important case of 2-CNF encodings where we show a slightly better
lower bound. The lower bound holds also for a related "exactly one"
constraint.

Keywords: Knowledge compilation · Cardinality constraint · At most
one constraint · Propagation complete encoding

1 Introduction

In this paper we study the properties of one of the most basic cardinality
constraints—the "at most one" constraint on n boolean variables which requires
that at most one input variable is set to 1. This constraint is widely used when
translating a problem into a propositional formula in *conjunctive normal form*
(CNF). Note that the "at most one" constraint is anti-monotone. This means
that if we increase the value of any input variable, the value of the constraint as a
boolean function does not increase. It follows that the "at most one" constraint
has a unique minimal prime CNF representation which requires $\binom{n}{2} = \Theta(n^2)$

P. Kučera—Supported by the Czech Science Foundation (grant GA15-15511S).
P. Savický—Supported by CE-ITI and GAČR under the grant GBP202/12/G061
and by the institutional research plan RVO:67985807.

S. Gaspers and T. Walsh (Eds.): SAT 2017, LNCS 10491, pp. 412–428, 2017.
DOI: 10.1007/978-3-319-66263-3_26

clauses, where n is the number of input variables. However, there are CNF encodings of size $O(n)$ which use additional auxiliary variables. Several encodings for this constraint were considered in literature. Let us mention *sequential encoding* [17] which addresses also more general cardinality constraints. The same encoding was also called *ladder encoding* by [13], and it forms the smallest variant of the *commander-variable encodings* [14]. After a minor simplification, it requires $3n - 6$ clauses and $n - 3$ auxiliary variables. Similar, but not smaller encodings can be also obtained as special cases of *totalizers* [5] and *cardinality networks* [1]. Currently the smallest known encoding is the *product encoding* introduced by Chen [9]. It consists of $2n + 4\sqrt{n} + O(\sqrt[4]{n})$ clauses and uses $O(\sqrt{n})$ auxiliary variables. Other encodings introduced in the literature for the "at most one" constraint use more clauses than either sequential or product encoding does. These include the *binary encoding* [6,12] and the *bimander encoding* [13].

All the encodings for the "at most one" constraint we have mentioned are in the form of a 2-CNF formula, which is a CNF formula where all clauses consist of at most two literals. This restricted structure guarantees that the encodings are propagation complete. The notion of propagation completeness was introduced by [8] as a generalization of unit refutation completeness introduced by [18]. We say that a formula φ is *propagation complete* if for any set of literals e_i, $i \in I$ the following property holds: either $\varphi \wedge \bigwedge_{i \in I} e_i$ is contradictory and this can be detected by unit propagation, or unit propagation started with $\varphi \wedge \bigwedge_{i \in I} e_i$ derives all literals f that logically follow from this formula. It was shown in [3] that a prime 2-CNF is always propagation complete. Since unit propagation is a standard tool which is used in state-of-the-art SAT solvers [7], this makes 2-CNFs as a part of a larger instance simple for them.

When encoding a constraint into a CNF formula, a weaker condition is often required. Namely, we require that unit propagation on the encoding is strong enough to enforce some kind of local consistency, for instance generalized arc consistency (GAC), see for example [4]. In this case we only care about propagation completeness with respect to input variables and not necessarily about behaviour on auxiliary variables. Later we formalize this notion as *propagation complete encoding* (PCE).

Chen [9] conjectures that the product encoding is the smallest possible. In this paper we provide support for the positive answer to this conjecture. Our lower bound almost matches the size of the product encoding. We show that any propagation complete encoding of the "at most one" constraint on n variables requires at least $2n + \sqrt{n} - O(1)$ clauses. The lower bound actually holds for a related constraint "exactly one" as well. We also consider the important special case of 2-CNF encodings for which we achieve a better lower bound—any 2-CNF encoding of "at most one" constraint on n variables requires at least $2n + 2\sqrt{n} - O(1)$ clauses.

We should note that having a smaller encoding is not necessarily an advantage when a SAT solver is about to be used. Adding auxiliary variables can be costly because a SAT solver has to deal with them and possibly use them for decisions. Encodings using auxiliary variables are mainly useful for constraints on a large number of input variables, when the full prime representation is too large. Moreover, the experimental results in [16] suggest that a SAT solver can be modified to minimize the disadvantage of introducing auxiliary variables. Another experimental evaluation of various cardinality constraints and their encodings appears in [11]. A propagation complete encoding can also be used as a part of a general purpose CSP solver where unit propagation can serve as a propagator of GAC, see [4]. As a conclusion, there are situations, when it is advantageous to consider encodings with auxiliary variables even for a simple constraint such as "at most one". If we allow using auxiliary variables, the number of clauses can decrease and it is natural to ask how many of them are necessary. Moreover, investigating lower bounds on the number of clauses needed in an encoding of a constraint can be helpful in search for better encodings of the constraint as well.

The paper is organized as follows. In Sect. 2 we give necessary definitions and recall the results we use in the rest of the paper. In Sect. 3 we describe reduction from a general propagation complete encoding to a regular form. Section 4 contains the proof of a lower bound on the size of any propagation complete encoding of the "at most one" constraint. In Sect. 5 we give a sketch of the proof of a stronger lower bound on the size of 2-CNF encodings. Due to space limitations, some of the proofs are omitted or replaced with sketches, this mostly affects Sect. 5. The full version of this paper can be found in [15].

2   Preliminaries

In this section we state various results which will be used throughout the paper.

2.1   Formulas in CNF

Given a finite vector $\mathbf{z}$ of *variables* with $|\mathbf{z}| = n$, a *boolean function* $f(\mathbf{z})$ is a mapping $f : \{0,1\}^n \to \{0,1\}$, which assigns a boolean value $f(\alpha)$ to each boolean *assignment* $\alpha \in \{0,1\}^n$. The value of assignment α on a variable x is denoted as $\alpha(x)$. For simplicity, we write $x \in \mathbf{z}$ for variable x which occurs in $\mathbf{z}$.

A *literal* is a variable $x \in \mathbf{z}$ or its negation $\neg x$. We use $\mathrm{var}\,(g)$ to denote the variable in literal g, i.e. $\mathrm{var}\,(g) = x$ if $g \in \{x, \neg x\}$. Given a set of literals C, $\mathrm{var}\,(C) = \{\mathrm{var}\,(g) \mid g \in C\}$. Given a variable x, we denote $\mathrm{lit}\,(x) = \{x, \neg x\}$. Given a set $\mathbf{z}$ of variables, we denote $\mathrm{lit}\,(\mathbf{z}) = \bigcup_{z \in \mathbf{z}} \mathrm{lit}\,(z)$.

A *clause* is a disjunction of a set of literals, which does not contain a complementary pair of literals. A formula φ is in *conjunctive normal form* (*CNF*) if it is a conjunction of clauses. We treat clauses as sets of literals and formulas in CNF as sets of clauses. Since in this paper we consider only formulas in conjunctive normal form, we often simply refer to *formulas*, by which we mean formulas in CNF. The empty clause (the contradiction) is denoted $\bot$ and the empty formula (the tautology) is denoted $\top$.

A *unit clause* consists of a single literal and we identify such a clause with its single literal. A *binary* clause consists of two literals. A formula φ in CNF where every clause has at most k literals is said to be in k-CNF.

A *partial assignment* ρ of variables $\mathbf{z}$ is a subset of lit $(\mathbf{z})$ such that $|\rho \cap$ lit $(x)| \leq 1$ for each $x \in \mathbf{z}$. By $\varphi(\rho)$ we denote the formula obtained by the application of a partial assignment ρ to a formula φ, i.e. $\varphi(\rho)$ originates from φ by substituting the values which satisfy literals in ρ.

2.2 Unit Resolution

We say that a clause C is an *implicate* of a formula φ if any satisfying assignment α of φ satisfies C as well. We denote this property with $\varphi \models C$. We say that implicate C of φ is a *prime implicate*, if no sub-clause C' of C is an implicate of φ (in other words clause C is a set-minimal implicate of φ). We say that CNF formula φ is *prime* if it consists only of prime implicates of φ.

Clauses C_1 and C_2 are *resolvable* if there is exactly one literal l such that $l \in C_1$ and $\neg l \in C_2$. Then clause $\mathcal{R}(C_1, C_2) = (C_1 \cup C_2) \backslash \{l, \neg l\}$ is called the *resolvent* of C_1 and C_2. It is a well known fact that if C_1 and C_2 are implicates of a CNF formula φ, then so is their resolvent. If one of C_1 and C_2 is a unit clause, we say that $\mathcal{R}(C_1, C_2)$ was derived from C_1 and C_2 by unit resolution. Repeated application of unit resolution on a given formula is also called unit propagation and it is an important derivation rule used in SAT solvers.

We use $\varphi \vdash_1 C$ to denote the fact that clause C can be derived from formula φ by a series of unit resolutions (or in other words by unit propagation). We will mostly consider the case when C is a unit clause or $\bot$. Given a CNF formula φ and literals $g_1, \ldots, g_k$ on variables in φ we denote $\mathcal{U}_\varphi(g_1, \ldots, g_k)$ the set of literals which can be derived by unit resolution from $\varphi \wedge g_1 \wedge \cdots \wedge g_k$ that is

$$\mathcal{U}_\varphi(g_1, \ldots, g_k) = \left\{ h \mid \varphi \wedge \bigwedge_{i=1}^{k} g_i \vdash_1 h \right\}$$

2.3 Encodings of Boolean Functions

We define an encoding of a boolean function as follows.

Definition 1. *Let $f(\mathbf{x})$ be a boolean function on variables $\mathbf{x} = (x_1, \ldots, x_n)$. Let $\varphi(\mathbf{x}, \mathbf{y})$ be a CNF formula on $n + \ell$ variables, where $\mathbf{y} = (y_1, \ldots, y_\ell)$.*

1. We call φ an encoding of f if for every $\alpha \in \{0,1\}^n$ we have that

$$f(\alpha) = 1 \iff (\exists \beta \in \{0,1\}^\ell) [\varphi(\alpha, \beta) = 1]. \tag{1}$$

2. We call φ a propagation complete encoding (PCE) of $f(\mathbf{x})$ if, moreover, for any $g_1, \ldots, g_p \in$ lit $(\mathbf{x})$, $p \geq 1$, it either holds that

$$\varphi \wedge \bigwedge_{i=1}^{p} g_i \vdash_1 \bot, \tag{2}$$

or

$$f(\mathbf{x}) \wedge \bigwedge_{i=1}^{p} g_i \models h \implies \varphi \wedge \bigwedge_{i=1}^{p} g_i \vdash_1 h \tag{3}$$

holds for each $h \in$ lit $(\mathbf{x})$. If φ is a prime CNF, we call it prime PCE.

If $\varphi(\mathbf{x}, \mathbf{y})$ is an encoding of $f(\mathbf{x})$, then the variables from $\mathbf{x}$ and $\mathbf{y}$ are called *input variables* and *auxiliary variables*, respectively. Note that the definition of a propagation complete encoding is less restrictive than requiring that φ is propagation complete as defined in [8]. The difference is that in a PCE we only consider literals on input variables as assumptions and consequences. The authors of [8] did not distinguish input and auxiliary variables and instead required condition (3) for all literals on all variables. The following propositions follow from definition of PCE.

Lemma 1. *A prime CNF obtained from a given PCE $\varphi(\mathbf{x}, \mathbf{y})$ of $f(\mathbf{x})$ by replacing every clause by a prime implicate contained in it, is also a PCE of $f(\mathbf{x})$.*

Lemma 2. *If $\varphi(\mathbf{x}, \mathbf{y})$ is a PCE of $f(\mathbf{x})$ of minimum size, then it does not contain a unit clause on an auxiliary variable.*

2.4   Identification of Variables in a Unit Resolution Proof

If $\varphi(\mathbf{z})$ is a formula and $g_1, g_2 \in \mathrm{lit}(\mathbf{z})$, we denote by $\varphi[g_1 \leftarrow g_2]$ the formula obtained from φ as follows. If the literal g_1 is positive, then the variable $\mathrm{var}(g_1)$ is substituted by the literal g_2. If g_1 is negative, then the variable $\mathrm{var}(g_1)$ is substituted by the literal $\neg g_2$. The following proposition easily follows from the properties of unit propagation. The proof can be found in [15].

Lemma 3. *Let $\varphi(\mathbf{z})$ be a formula, let $g_1, g_2, h_1, h_2 \in \mathrm{lit}(\mathbf{z})$, such that $\mathrm{var}(g_1) \notin \{\mathrm{var}(h_1), \mathrm{var}(h_2)\}$ and assume, $\varphi[g_1 \leftarrow g_2] \wedge h_1$ is satisfiable. Then*

$$\varphi \wedge h_1 \vdash_1 h_2 \implies \varphi[g_1 \leftarrow g_2] \wedge h_1 \vdash_1 h_2.$$

2.5   At-Most-One and Related Functions

In this paper we are interested in two special cases of cardinality constraints represented by "at most one" and "exactly one" functions. First we define the "at most one" function.

Definition 2. *The function $\mathrm{AMO}_n(x_1, \ldots, x_n)$ (at most one) is defined as follows: Given an assignment $\alpha \in \{0,1\}^n$, the value $\mathrm{AMO}_n(\alpha)$ is 1 if and only if there is at most one index $i \in \{1, \ldots, n\}$ for which $\alpha(x_i) = 1$.*

The "exactly one" function differs from AMO_n only on zero input.

Definition 3. *The function $\mathrm{EO}_n(x_1, \ldots, x_n)$ (exactly one) is defined as follows: Given an assignment $\alpha \in \{0,1\}^n$, the value $\mathrm{EO}_n(\alpha)$ is 1 if and only if there is exactly one index $i \in \{1, \ldots, n\}$ for which $\alpha(x_i) = 1$.*

One can easily verify the following lemma characterizing propagation complete encodings of AMO_n and EO_n.

Lemma 4. *Let $\varphi(\mathbf{x}, \mathbf{y})$ be a formula with $\mathbf{x} = (x_1, \ldots, x_n)$, $\mathbf{y} = (y_1, \ldots, y_\ell)$, $n \geq 1$, $\ell \geq 0$ and let us consider the following conditions on φ.*

(P1) $\varphi \wedge x_i$ *is satisfiable for each* $i \in \{1, \ldots, n\}$,
(P2) $\varphi \wedge x_i \vdash_1 \neg x_j$ *holds for each* $i, j \in \{1, \ldots, n\}$ *with* $i \neq j$,
(P3) $\varphi \wedge \bigwedge_{i=1}^{n} \neg x_i$ *is satisfiable*,
(P4) $\varphi \wedge \bigwedge_{j \in \{1, \ldots, n\} \setminus \{i\}} \neg x_j \vdash_1 x_i$ *holds for each* $i \in \{1, \ldots, n\}$.

Then the following equivalences hold:

(i) φ *is a PCE of* AMO_n *if and only if it satisfies (P1), (P2), and (P3).*
(ii) φ *is a PCE of* EO_n *if and only if it satisfies (P1), (P2), and (P4).*

The first two conditions (P1) and (P2) from Lemma 4 are frequently used in the rest of the paper. By Lemma 4 the propagation complete encodings of AMO_n and EO_n share these two properties. Although our main focus is on the function AMO_n, some of the induction arguments we use in proofs rely on the fact that we do not require that a formula satisfies condition (P3). It turns out that properties (P1) and (P2) are enough to show the lower bound which means that it holds for both constraints. In order to work with both functions AMO_n and EO_n in a unified way we introduce the following notation.

Definition 4. *Let* AMO_n^* *denote the set* $\{\mathrm{AMO}_n, \mathrm{EO}_n\}$.

The first two conditions of Lemma 4 allow us to characterize the notion of PCE of AMO_n^*.

Definition 5. *Let* $\varphi(\mathbf{x}, \mathbf{y})$ *be a CNF formula on* $n + \ell$ *variables, where* $\mathbf{x} = (x_1, \ldots, x_n)$ *and* $\mathbf{y} = (y_1, \ldots, y_l)$.

– *We say that the formula* φ *is an* encoding of AMO_n^*, *if it is an encoding of one of the functions in this set.*
– *We say that the formula* φ *is a* propagation complete encoding of AMO_n^* *(or PCE of* AMO_n^*), *if it moreover satisfies conditions (P1) and (P2).*

Let us point out that a PCE of AMO_n^*, which is an encoding of EO_n, may not be a PCE of EO_n. An example of such a formula is

$$\varphi' = (x_1 \vee \ldots \vee x_{n-2} \vee x_n \vee y) \wedge (x_{n-1} \vee x_n \vee \neg y) \wedge \varphi(\mathbf{x}),$$

where $\varphi(\mathbf{x})$ represents AMO_n. Note that $\varphi' \wedge \bigwedge_{i \in \{1, \ldots, n-1\}} \neg x_i \not\vdash_1 x_n$.

Definition 6. *The* size *of a formula is the number of its clauses. We denote the minimum size of a PCE of* AMO_n *with* $\mathcal{A}(n)$, *the minimum size of a PCE of* EO_n *with* $\mathcal{E}(n)$, *the minimum size of a PCE of* AMO_n^* *with* $\mathcal{S}(n)$, *and the minimum size of a 2-CNF encoding of* AMO_n *with* $\mathcal{A}_2(n)$.

2.6   Basic Size Estimates

The proof of the following lemma presents a variant of the *sequential encoding* [17], which addresses also more general cardinality constraints.

Lemma 5. *For every $n \geq 3$, we have $\mathcal{A}_2(n) \leq 3n - 6$.*

Proof. By induction using the formula $\varphi_3(x_1, x_2, x_3) = (\neg x_1 \vee \neg x_2) \wedge (\neg x_1 \vee \neg x_3) \wedge (\neg x_2 \vee \neg x_3)$ and the formula $\varphi_n(x_1, \ldots, x_n) = (\neg x_1 \vee \neg x_2) \wedge (\neg x_1 \vee y) \wedge (\neg x_2 \vee y) \wedge \varphi_{n-1}(y, x_3, \ldots, x_n)$ for each $n > 3$.

Let us now describe the *product encoding* φ_n^p of AMO_n introduced by Chen [9]. This encoding serves as an example of an encoding in regular form defined later. The base case for $n = 3$ is $\varphi_3^p(x_1, x_2, x_3) = (\neg x_1 \vee \neg x_2) \wedge (\neg x_1 \vee \neg x_3) \wedge (\neg x_2 \vee \neg x_3)$. For $n > 3$, denote $m = \lceil \sqrt{n} \rceil$ and arrange the input variables in n different cells of a square matrix of dimension $m \times m$. Let $r : \{1, \ldots, n\} \to \{1, \ldots, m\}$ and $c : \{1, \ldots, n\} \to \{1, \ldots, m\}$ be the functions, such that $r(i)$ is the row index and $c(i)$ the column index of the cell containing x_i. Let $y_j, j = 1, \ldots, m$ and $z_j, j = 1, \ldots, m$ be new auxiliary variables. Then we set

$$\varphi_n^p(\mathbf{x}) = \bigwedge_i (\neg x_i \vee y_{r(i)}) \wedge \bigwedge_i (\neg x_i \vee z_{c(i)}) \wedge \varphi_m^p(\mathbf{y}) \wedge \varphi_m^p(\mathbf{z}) \qquad (4)$$

Chen [9] shows that $|\varphi_n^p| = 2n + 4\sqrt{n} + O(\sqrt[4]{n})$. It turns out that $n = 25$ is the smallest value where the product encoding outperforms the sequential encoding (68 vs. 69 clauses). On the other hand, we show below that the sequential encoding is the smallest possible for $n \leq 6$. It is not clear whether this holds also for $7 \leq n \leq 24$.

We can observe the following basic relations between the sizes of encodings of AMO_n, AMO_n^*, and EO_n.

Lemma 6. *For each $n \geq 1$ we have that*

$$\mathcal{E}(n) \leq \mathcal{A}(n) + 1 \qquad (5)$$
$$\mathcal{S}(n) \leq \min(\mathcal{E}(n), \mathcal{A}(n)). \qquad (6)$$

Proof. The inequality (5) follows from the fact that $\mathrm{EO}_n(\mathbf{x}) \equiv \mathrm{AMO}_n(\mathbf{x}) \wedge (\bigvee_{i=1}^n x_i)$ and (6) follows from Lemma 4 and Definition 5, which imply that every propagation complete encoding of AMO_n or EO_n is also a propagation complete encoding of AMO_n^*.

We have the following sizes of minimum encodings for $n = 2, 3$.

Lemma 7. *We have $\mathcal{A}(2) = \mathcal{S}(2) = 1$ and $\mathcal{A}(3) = \mathcal{S}(3) = 3$.*

3   Reducing to Regular Form

Let us look at properties of propagation complete encodings of AMO_n^*, in particular, of encodings of minimum size. Using Lemma 2, we assume without loss of generality that a PCE of AMO_n^* does not contain unit clauses. The core of the proof of the lower bound lies in studying encodings in regular form defined in this section. In such an encoding, for every input variable x_i, there are exactly two clauses containing the negative literal $\neg x_i$. Moreover these two clauses are binary and the other literal in each of these clauses is an auxiliary variable or its negation. The aim of this section is to show that if n is large enough and φ is a minimum size PCE of AMO_n^* which is not in regular form, then there is a PCE φ' of AMO_{n-1}^* of size at most $|\varphi| - 3$. This allows to use induction on n for encodings that are not in regular form. Showing the lower bound then relies on analyzing encodings in regular form. This analysis differs for general encodings and for 2-CNF encodings. In the latter case a stronger lower bound can be shown. Analysis of the general case is presented in Sect. 4 and the analysis of the 2-CNF case is presented in Sect. 5.

We start with basic properties of PCE of AMO_n^*.

Lemma 8. *Let $\varphi(\mathbf{x}, \mathbf{y})$ be a formula with input variables $\mathbf{x} = (x_1, \ldots, x_n)$ and auxiliary variables $\mathbf{y} = (y_1, \ldots, x_\ell)$. Assume that φ is a propagation complete encoding of AMO_n^*. For each distinct $x_i, x_j \in \mathbf{x}$ it holds that*

(a) $\varphi \wedge x_i \wedge \neg x_j \not\vdash_1 \bot$,
(b) $\varphi \wedge x_i \not\vdash_1 x_j$,
(c) $\varphi \wedge \neg x_i \not\vdash_1 \neg x_j$,
(d) φ contains a binary clause containing the literal $\neg x_i$.

The following lemma shows that fixing any set of input variables to zero in a PCE of AMO_n^* gives us a PCE of AMO_m^* on the remaining m input variables.

Lemma 9. *Let $\varphi(\mathbf{x}, \mathbf{y})$ be a propagation complete encoding of $\text{AMO}_n^*(\mathbf{x})$. Let $I \subseteq \{1, \ldots, n\}$ be a set of indices and consider the partial assignment $\rho = \{\neg x_j \mid j \notin I\}$. Then $\varphi(\rho)$ is a propagation complete encoding of $\text{AMO}_{|I|}^*(x_I)$, where x_I denotes the vector of input variables x_i, $i \in I$.*

We now concentrate on clauses with negative literals on input variables.

Lemma 10. *Let $\varphi(\mathbf{x}, \mathbf{y})$ be a prime PCE of $\text{AMO}_n^*(\mathbf{x})$, $C \in \varphi$ and $\neg x_i \in C$. Then one of the following is satisfied*

(i) $C = (\neg x_i \vee A)$, where $\emptyset \neq A \subseteq \text{lit}(\mathbf{y})$,
(ii) $C = (\neg x_i \vee \neg x_j)$ for some $j \neq i$.

Proof. We have $C = (\neg x_i \vee A)$ for a non-empty set of literals A. If there is a literal $\neg x_j \in A$ for some $j \neq i$ then necessarily $C = (\neg x_i \vee \neg x_j)$ because this is a prime implicate of both functions in AMO_n^*. If $x_j \in A$ for some $j \neq i$ then $C' = \mathcal{R}(C, (\neg x_i \vee \neg x_j))$ is an implicate as well which is in contradiction with primality of φ. The proposition follows.

Lemma 11. *Let $n \geq 3$ and let $\varphi(\mathbf{x}, \mathbf{y})$ be a propagation complete encoding of $\text{AMO}_n^*(\mathbf{x})$. Let $x_i \in \mathbf{x}$. Suppose that $\neg x_i$ occurs only once in φ. Then there is a PCE φ' of AMO_n^* with $|\varphi| \geq |\varphi'| + 1$. Moreover, if φ is a 2-CNF formula, then so is φ'.*

Proof. Using Lemma 1, we can assume that φ is a prime formula. Lemma 8(d) provides a binary clause $C = (\neg x_i \vee e) \in \varphi$ with some $e \in \text{lit}(\mathbf{x} \cup \mathbf{y})$. Let us assume for a contradiction that var $(e) = x_j$ with $j \neq i$. By Lemma 10, $C = (\neg x_i \vee \neg x_j)$. Let $x_k \in \mathbf{x} \backslash \{x_i, x_j\}$. We have that $\varphi \wedge x_i \vdash_1 \neg x_k$. Since C is the only clause of φ containing $\neg x_i$, unit resolution uses x_i to derive $\neg x_j$ and does not use x_i in any of the later steps. Hence, we have $\varphi \wedge \neg x_j \vdash_1 \neg x_k$, which is a contradiction with Lemma 8. This implies $e \in \text{lit}(\mathbf{y})$.

Let $\varphi' = \varphi[e \leftarrow x_i]$. We can show that φ' satisfies the conditions (P1) and (P2). In particular (P2) follows by Lemma 3. The details can be found in [15]. Note that after substitution $\varphi' = \varphi[e \leftarrow x_i]$, clause C becomes $(\neg x_i \vee x_i)$ and is omitted in φ'. Hence φ' has size smaller than φ. This completes the proof.

Given a variable x_i, $i \in \{1, \dots, n\}$, unit propagation on formula $\varphi \wedge x_i$ starts with clauses which contain the negative literal $\neg x_i$. The structure of these clauses is important for the analysis of PCEs of minimum size. For each $i = 1, \dots, n$ let us denote

$$Q_{\varphi,i} = \{C \in \varphi \mid \neg x_i \in C\}. \tag{7}$$

Definition 7. *A propagation complete encoding $\varphi(\mathbf{x}, \mathbf{y})$ of AMO_n^* is in regular form if the following conditions hold for each $i \in \{1, \dots, n\}$:*

(R1) $|Q_{\varphi,i}| = 2$.
(R2) Clauses in $Q_{\varphi,i}$ contain no input variables other than x_i.
(R3) Clauses in $Q_{\varphi,i}$ are all binary.

Note that for $n > 3$ the product encoding φ_n^p of AMO_n given by Eq. (4) is in regular form. In particular for each x_i, $i = 1, \dots, n$ we have that $Q_{\varphi,i} = \{(\neg x_i \vee y_{r(i)}), (\neg x_i \vee z_{c(i)})\}$.

Proposition 1. *Let $\varphi(\mathbf{x}, \mathbf{y})$ be a propagation complete encoding of $\text{AMO}_n^*(\mathbf{x})$, $n \geq 3$, such that (R1) is not satisfied. Then, there is a formula φ', which satisfies one of the following*

(a) φ' is a PCE of AMO_n^* and $|\varphi| \geq |\varphi'| + 1$,
(b) φ' is a PCE of AMO_{n-1}^* and $|\varphi| \geq |\varphi'| + 3$.

Moreover, if φ is a 2-CNF formula, then so is φ'.

Proof. Assume that $|Q_{\varphi,i}| \neq 2$ for some $i \in \{1, \dots, n\}$. Lemma 8 implies that $|Q_{\varphi,i}| \geq 1$. Assume that $|Q_{\varphi,i}| = 1$. According to Lemma 11, there is a PCE φ' of $\text{AMO}_n^*(\mathbf{x})$ satisfying condition (a) of the conclusion. If $|Q_{\varphi,i}| \geq 3$, then setting $x_i = 0$ yields a formula φ' of size at most $|\varphi| - 3$. By Lemma 9, this formula is a PCE of AMO_{n-1}^*. Hence, condition (b) of the conclusion is satisfied.

Lemma 12. *Let φ be a PCE of AMO_n^*, $n \geq 4$, and let $i, j, k \in \{1, \dots, n\}$ be three different indices. Then $Q_{\varphi,i} \neq \{(\neg x_i \vee \neg x_j), (\neg x_i \vee \neg x_k)\}$.*

Proof. Let $\ell \in \{1, \dots, n\} \setminus \{i, j, k\}$. We have $\varphi \wedge x_i \vdash_1 \neg x_\ell$. Assume that $Q_{\varphi,i}$ consists only of two clauses $(\neg x_i \vee \neg x_j)$, $(\neg x_i \vee \neg x_k)$. Then, we have $\varphi \wedge \neg x_j \wedge \neg x_k \vdash_1 \neg x_\ell$. This implies that $(x_j \vee x_k \vee \neg x_\ell)$ is an implicate of φ which is in contradiction with the assumption that φ is an encoding of AMO_n^*.

Proposition 2. *Let $\varphi(\mathbf{x}, \mathbf{y})$ be a prime PCE of AMO_n^*, $n \geq 4$, such that (R1) is satisfied, but (R2) is not satisfied. Then there is a PCE φ' of AMO_{n-1}^*, such that $|\varphi| \geq |\varphi'| + 3$. If φ is a 2-CNF formula, then so is φ'.*

Proof. Since φ violates (R2), we get by Lemma 10 that there is an index i, such that $Q_{\varphi,i}$ contains the clause $(\neg x_i \vee \neg x_j)$ for some $j \neq i$. Without loss of generality, assume $i = 1$, $j = 2$, so φ contains clauses $(\neg x_1 \vee \neg x_2)$, $(\neg x_1 \vee B_1)$, $(\neg x_2 \vee B_2)$ for some sets of literals B_1, B_2. By lemmas 10 and 12, both B_1 and B_2 are sets of auxiliary literals. By Lemma 9 we have that $\psi = \varphi(\{\neg x_1\})$ is a propagation complete encoding of AMO_{n-1}^* on variables $x_2, \ldots, x_n$. Since $|Q_{\varphi,i}| = 2$, we have that $|\varphi| \geq |\psi| + 2$. Since the literal $\neg x_2$ occurs only once in ψ, Lemma 11 implies that there is a PCE φ' of AMO_{n-1}^* with $|\psi| \geq |\varphi'| + 1$. Together we get $|\varphi| \geq |\psi| + 2 \geq |\varphi'| + 3$ as required. $\qquad\square$

We are now ready to show the main result of this section.

Theorem 1. *If $\varphi(\mathbf{x}, \mathbf{y})$ is a prime PCE for AMO_n^*, $n \geq 4$, then at least one of the following holds:*

(a) There is a PCE φ' for AMO_n^*, such that $|\varphi| \geq |\varphi'| + 1$.
(b) There is a PCE φ' for AMO_{n-1}^*, such that $|\varphi| \geq |\varphi'| + 3$.
(c) Formula φ is in regular form.

Moreover if φ is a 2-CNF formula, then so is φ' in cases (a) and (b).

Proof. By Propositions 1 and 2, we have that either one of the conditions (a), (b) is satisfied, or φ satisfies (R1) and (R2). If φ is a 2-CNF formula, the condition (R3) is satisfied and we are done.

If φ is not a 2-CNF formula, assume, φ does not satisfy (R3) for some $i \in \{1, \ldots, n\}$. By Lemma 8 we have that one of the clauses in $Q_{\varphi,i}$ is a binary clause. Since $Q_{\varphi,i}$ does not satisfy (R3), the other clause consists of at least three literals (due to (R1) we have $|Q_{\varphi,i}| = 2$). Moreover, due to (R2) the only input variable which appears in some clause in $Q_{\varphi,i}$ is x_i. Thus we can write $Q_{\varphi,i} = \{C_1 = (\neg x_i \vee y), C_2 = (\neg x_i \vee z_1 \vee \ldots \vee z_\ell)\}$ for some literals $y, z_1, \ldots, z_\ell$ on auxiliary variables where $\ell > 1$. We claim that for every $j \in \{1, \ldots, \ell\}$ we have that

$$\varphi \wedge x_i \not\vdash_1 \neg z_j \tag{8}$$

and

$$\varphi \wedge y \not\vdash_1 \neg z_j. \tag{9}$$

Let us assume by contradiction that there is a $j \in \{1, \ldots, \ell\}$ satisfying negation of (8) or negation of (9). Using clause C_1, $\varphi \wedge y \vdash_1 \neg z_j$ implies $\varphi \wedge x_i \vdash_1 \neg z_j$, so we can assume $\varphi \wedge x_i \vdash_1 \neg z_j$. Then, $(\neg x_i \vee \neg z_j)$ is an implicate of φ. However resolvent $\mathcal{R}((\neg x_i \vee \neg z_j), C_2)$ is a strict subclause of C_2 which is in contradiction with primality of C_2.

Consider any input variable x_j, $j \neq i$. Since φ satisfies (P2) we have that $\varphi \wedge x_i \vdash_1 \neg x_j$. Since C_1 is the only clause in φ which becomes unit when resolved with x_i and considering (9) we get that necessarily

$$\mathcal{U}_\varphi(x_i) = \mathcal{U}_\varphi(y) \cup \{x_i\} \tag{10}$$

and in particular

$$\varphi \wedge y \vdash_1 \neg x_j. \tag{11}$$

Let $\psi = (\varphi \setminus \{C_2\}) \cup \{C_3\}$, where $C_3 = (\neg y \vee z_1 \vee \ldots \vee z_\ell)$. We shall prove that ψ is an encoding of AMO_n^*. Since $|\psi| = |\varphi|$ and $|\psi|$ contains only one occurrence of $\neg x_i$, we get by Lemma 11 that there is a formula φ' satisfying condition (a). According to Definition 5 it remains to show that ψ satisfies conditions (P1) and (P2).

(P1) Let x_j, $j \in \{1, \ldots, n\}$ be an arbitrary input variable and let us show that $\psi \wedge x_j$ is satisfiable.
- If $j = i$, we have $\varphi \wedge x_j \models z_1 \vee \cdots \vee z_\ell$, since C_2 is contained in φ and $x_j = x_i$. Consequently, $\varphi \wedge x_j \models C_3$.
- If $j \neq i$, we have $\varphi \models \neg y \vee \neg x_j$ by (11). Hence, $\varphi \wedge x_j \models \neg y$ and $\varphi \wedge x_j \models C_3$.

In both cases, since $\varphi \wedge x_j$ is satisfiable, so is $\psi \wedge x_j$ and ψ satisfies (P1).

(P2) Let $j, k \in \{1, \ldots, n\}$ be two different indices of input variables and let us show that $\psi \wedge x_j \vdash_1 \neg x_k$. Let us look at derivation of $\varphi \wedge x_j \vdash_1 \neg x_k$.
- If $j = i$, then clause C_2 is not used in the derivation $\varphi \wedge x_i \vdash_1 \neg x_k$. This follows by (8), because in order for C_2 to be used in a unit resolution derivation, at least one of $z_1, \ldots, z_\ell$ must be derived first. It follows that $\psi \wedge x_i \vdash_1 \neg x_k$ as well.
- Let us now suppose that $j \neq i$. If C_2 is not used in derivation of $\varphi \wedge x_j \vdash_1 \neg x_k$, then also $\psi \wedge x_j \vdash_1 \neg x_k$ and we are done. If C_2 were used to derive some z_k for $k \in \{1, \ldots, \ell\}$, then in order to do that we need $\varphi \wedge x_j \vdash_1 x_i$, which is not true. As the last case let us assume that C_2 is used to derive $\neg x_i$. Before that we have $\varphi \wedge x_j \vdash_1 \neg z_r$ for all $r \in \{1, \ldots, \ell\}$ and this is true in ψ as well. Hence, $\psi \wedge x_j \vdash_1 \neg z_r$ for all $r \in \{1, \ldots, \ell\}$. Moreover, we obtain $\psi \wedge x_j \vdash_1 \neg x_i$ because we can replace the step using C_2 in the original unit resolution derivation with two steps. The first uses C_3 to derive $\neg y$ and the second uses C_1 to derive $\neg x_1$. Together, we get that also in this case $\psi \wedge x_j \vdash_1 \neg x_k$.

This concludes the proof.

Let φ be a PCE of AMO_n in regular form. It follows that for every $i \in \{1, \ldots, n\}$ the two clauses in $Q_{\varphi,i}$ are binary and each consists of $\neg x_i$ together with an auxiliary variable or its negation. We will use the following sets which consist of these two auxiliary literals.

$$L_{\varphi,i} = \{e \mid (\neg x_i \vee e) \in Q_{\varphi,i}\}. \tag{12}$$

For example, in case of the product encoding φ_n^p of AMO_n given by equation (4), we have $L_{\varphi,i} = \{y_{r(i)}, z_{c(i)}\}$ for each $i = 1, \ldots, n$.

By Condition (R1) we have that $|L_{\varphi,i}| = |Q_{\varphi,i}| = 2$. Moreover, if $i, j \in \{1, \ldots, n\}$ are two different indices of input variables, then $L_{\varphi,i} \neq L_{\varphi,j}$. Indeed, assuming $L_{\varphi,i} = L_{\varphi,j}$, we get a contradiction with conditions (P1) and (P2) as follows. The formula $\varphi \wedge x_i$ is satisfiable and derives both literals in $L_{\varphi,i} = L_{\varphi,j}$. Hence, it is not possible to have $\varphi \wedge x_i \vdash_1 \neg x_j$.

We shall distinguish the following two types of clauses in φ.

- Clauses from $\bigcup_{i=1}^n Q_{\varphi,i}$ are of *type Q*.
- The remaining clauses in φ are of *type R*.

Note that since φ is in regular form, all clauses of type Q are binary and the number of clauses of type Q is $2n$. In the next two sections we aim to provide a lower bound on the number of clauses of type R in a general PCE in regular form (Sect. 4) and in a 2-CNF PCE in regular form (Sect. 5).

4   A Lower Bound for General Encodings

This section is devoted to the proof of a lower bound for PCE of AMO_n^* for general CNF formulas. The main part of the proof consists of showing a lower bound on the number of clauses of type R in a PCE in regular form. This is combined with an inductive argument based on Theorem 1.

Lemma 13. *Let* $\varphi(\mathbf{x}, \mathbf{y})$ *be a PCE of* AMO_n^* *in regular form. Let* i, j, k *be different indices with* $L_{\varphi,i} = \{g, h_1\}$, $L_{\varphi,j} = \{g, h_2\}$, *and* $L_{\varphi,k} = \{g, h_3\}$ *for* $g, h_1, h_2, h_3 \in \mathrm{lit}(\mathbf{y})$. *Then variables* $\mathrm{var}(h_1)$, $\mathrm{var}(h_2)$, *and* $\mathrm{var}(h_3)$ *are pairwise different.*

Proof. First, let us note that the literals h_1, h_2, h_3 have to be pairwise distinct, this follows from the arguments given at the end of Sect. 3. Let us show by contradiction that $\mathrm{var}(h_1) \neq \mathrm{var}(h_2)$. To this end, assume $\mathrm{var}(h_1) = \mathrm{var}(h_2)$. Since $L_{\varphi,i} \neq L_{\varphi,j}$ we have that $h_1 = \neg h_2$. By condition (P2) we have that $\varphi \wedge x_k \vdash_1 \neg x_i$ and $\varphi \wedge x_k \vdash_1 \neg x_j$. Since $\varphi \wedge x_k \vdash_1 g$, necessarily $\varphi \wedge x_k \vdash_1 \neg h_1$ and $\varphi \wedge x_k \vdash_1 \neg h_2$. However, then $\varphi \wedge x_k \vdash_1 \bot$ which is in contradiction with (P1).

The cases $\mathrm{var}(h_1) \neq \mathrm{var}(h_3)$ and $\mathrm{var}(h_2) \neq \mathrm{var}(h_3)$ are symmetrical.

Corollary 1. *Let* φ *be a PCE of* AMO_n^* *in regular form and let* h *be a literal which appears in* $L_{\varphi,i}$ *for some* $i \in \{1, \ldots, n\}$. *Let* $I_h = \{i \in \{1, \ldots, n\} \mid h \in L_{\varphi,i}\}$, *and* $L_h = \bigcup_{i \in I_h} L_{\varphi,i}$. *If* $|I_h| \geq 3$, *then* $|\mathrm{var}(L_h)| = |I_h| + 1$.

Proof. This is a simple corollary of Lemma 13. If we remove literal h from each $L_{\varphi,i}$, $i \in I_h$, then the remaining literals are on pairwise different variables different from $\mathrm{var}(h)$.

Lemma 14. *If* $\varphi(\mathbf{x}, \mathbf{y})$ *is a PCE of* AMO_n^* *in regular form and* $n \geq 5$, *then there exists* $i \in \{1, \ldots, n\}$ *such that* $|\mathcal{U}_\varphi(x_i) \cap \mathrm{lit}(\mathbf{y})| \geq \sqrt{n-1}$.

Proof. Let $L = \bigcup_{i=1}^n L_{\varphi,i}$ be the set of auxiliary literals in clauses of type Q. For each $h \in L$, let I_h and L_h be defined as in Corollary 1. Choose $g \in L$ that maximizes $|I_g|$ and fix some $i \in I_g$.

If $|I_g| \geq 3$ then according to Corollary 1, we get that $|\mathrm{var}(L_g)| = |I_g| + 1$. In order to derive all $\neg x_j$, $j \in I_g \setminus \{i\}$ from x_i, the literals in $L_g \setminus L_{\varphi,i}$ must be falsified by unit propagation. Moreover the literals in $L_{\varphi,i}$ are derived as well and thus we have

$$|M_i| \geq |I_g| + 1, \tag{13}$$

where $M_i = \mathcal{U}_\varphi(x_i) \cap \text{lit}\,(\mathbf{y})$.

On the other hand, each $\neg x_j$, $j \in \{1, \dots, n\} \setminus \{i\}$ must be derived from some $h \in M_i$ using a clause of type Q. As $|I_h| \leq |I_g|$ for each $h \in M_i$, any fixed h covers at most $|I_g|$ values of j. Thus,

$$|M_i| \cdot |I_g| \geq n - 1. \tag{14}$$

Finally, we get $|M_i| \geq \max\{|I_g|, (n-1)/|I_g|\}$ as follows:

- If $|I_g| \geq 3$, the claims (13) and (14) apply.
- If $|I_g| \leq 2$, we observe that by (14) we have $|M_i| \geq (n-1)/2$, which is at least 2 for each $n \geq 5$.

Clearly, the function $s \mapsto \max\{s, (n-1)/s\}$ on positive s achieves the smallest value when $s = (n-1)/s$ which is equivalent to $s = \sqrt{n-1}$.

Lemma 15. *If $n \geq 5$ and $\varphi(\mathbf{x}, \mathbf{y})$ is a PCE of AMO_n^* in regular form, then $|\varphi| \geq 2n + \sqrt{n-1} - 2$.*

Proof. Formula φ contains $2n$ clauses of type Q. Moreover, by Lemma 14, there is an index i, such that $\mathcal{U}_\varphi(x_i)$ contains at least $\sqrt{n-1}$ literals on auxiliary variables. As $\varphi \wedge x_i$ is satisfiable, these literals contain different variables. Only two of them are derived by clauses of type Q, while the others must be derived by clauses of type R, which implies the required estimate.

The following theorem is one of the main results of this paper.

Theorem 2. *For $n \geq 3$, the minimum size $\mathcal{S}(n)$ of a PCE of AMO_n^* satisfies*

1. *If $n \leq 6$, then $\mathcal{S}(n) = 3n - 6$.*
2. *If $n \geq 7$, then $\mathcal{S}(n) \geq 2n + \sqrt{n-1} - 2$.*

Proof. We treat the two claims separately:

1. It was shown in Lemma 5 that $\mathcal{S}(n) \leq 3n - 6$. To show that $\mathcal{S}(n) \geq 3n - 6$, we proceed by induction on n. The basis, i.e. $\mathcal{S}(3) = 3$, is given by Lemma 7. For $n > 3$, a minimum prime encoding φ of AMO_n^* must satisfy one of the conditions in Theorem 1. The condition (a) is excluded, since φ has minimum size. The condition (b) and the induction hypothesis imply

$$|\varphi| \geq \mathcal{S}(n-1) + 3 \geq 3(n-1) - 3 = 3n - 6,$$

   while (c) leads to $\mathcal{S}(n) = |\varphi| \geq 2n$, which is at least $3n - 6$ for $n \leq 6$.
2. Let φ be a minimum-size PCE of AMO_n^*, $n \geq 7$. It follows from Theorem 1 that either φ is regular and thus $|\varphi| \geq 2n + \sqrt{n-1} - 2$ due to Lemma 15, or $|\varphi| \geq \mathcal{S}(n-1) + 3$. In the latter case we observe that:
   - If $n = 7$, we obtain $\mathcal{S}(n-1) + 3 = 15 > 2n + \sqrt{n-1} - 2$ by the first claim of this theorem.
   - If $n > 7$, the induction hypothesis implies $\mathcal{S}(n-1) + 3 \geq 2(n-1) + \sqrt{n-2} + 1$, which exceeds $2n + \sqrt{n-1} - 2$, since $\sqrt{n-2} + 1 > \sqrt{n-1}$.

5   A Lower Bound for 2-CNF Encodings

In this section we present a lower bound $2n + 2\sqrt{n} - 5$ on the number of clauses for the special case of 2-CNF encodings of AMO_n. Due to space limitation we present only a sketch of the proof. For a detailed version see [15]. On the contrary to the general encodings, the result of this section implies a lower bound on the size of any 2-CNF encodings of AMO_n, not only for propagation complete encodings. This can be seen as follows. By Lemma 1, a size minimal 2-CNF encoding can be chosen as a prime formula and, moreover, a prime 2-CNF formula is propagation complete with respect to any literals, not just the input ones, see [3]. The importance of the special case of 2-CNF encodings of AMO_n stems from the fact that the smallest known encodings are in 2-CNF as well as all the other encodings suggested in the literature.

In order to prove a lower bound on the size of a 2-CNF encoding of AMO_n, we use Theorem 1 similarly as in Sect. 4 to handle encodings, which are not in regular form. However, the analysis of encodings in regular form is different and implies a stronger lower bound.

One of the differences in case of 2-CNF formulas is that a unit resolution derivation can be reversed in the following sense: Given a formula φ in 2-CNF and two literals g and h, each of which is consistent with the formula, we have that $\varphi \wedge g \vdash_1 \neg h$ if and only if $\varphi \wedge h \vdash_1 \neg g$. Due to this, it is useful to represent a 2-CNF formula with an *implication graph* introduced in [2], see also [10]— it is a directed graph on literals where each clause $(g \vee h)$ corresponds to two arcs $(\neg g, h)$ (represents the implication $(\neg g \rightarrow h)$) and $(\neg h, g)$ (represents the implication $(\neg h \rightarrow g)$).

We can exploit the properties of the implication graph to show stronger properties of sets $L_{\varphi,i}$ defined in (12) than in the case of general CNF encodings. In particular, we can show that the main part of the analysis can be reduced to the case where the sets $L'_{\varphi,i} = \mathrm{var}\,(L_{\varphi,i})$ are pairwise different for $i = 1, \ldots, n$. The following three lemmas show the important properties of sets $L'_{\varphi,i}$.

Lemma 16. *Let $\varphi(\mathbf{x}, \mathbf{y})$ be a minimum 2-CNF encoding of AMO_n. Let $r, s \in \{1, \ldots, n\}$ be different and let us suppose that $L'_{\varphi,r} = L'_{\varphi,s}$ and $L_{\varphi,r} = \{g, h\}$ for $g, h \in \mathrm{lit}\,(\mathbf{y})$. Then $L_{\varphi,s} = \{\neg g, \neg h\}$.*

Lemma 17. *Let $\varphi(\mathbf{x}, \mathbf{y})$ be a minimum size 2-CNF encoding of AMO_n. Let $r, s \in \{1, \ldots, n\}$ be two different indices and let us suppose that $L_{\varphi,r} = \{g, h\}$ and $L_{\varphi,s} = \{\neg g, \neg h\}$ for $g, h \in \mathrm{lit}\,(\mathbf{y})$. Then $|\varphi| \geq \mathcal{A}_2(n - 2) + 5$.*

Lemma 18. *Let $n \geq 4$ and let $\varphi(\mathbf{x}, \mathbf{y})$ be a minimum size 2-CNF encoding of AMO_n in regular form. If there are different indices $r, s, t \in \{1, \ldots, n\}$, s.t.*

1. *$L'_{\varphi,r}, L'_{\varphi,s}$, and $L'_{\varphi,t}$ are pairwise distinct,*
2. *$|L'_{\varphi,r} \cup L'_{\varphi,s} \cup L'_{\varphi,t}| = 3$,*

then $|\varphi| \geq \mathcal{A}_2(n - 2) + 5$.

The second main result of this paper is the following.

Theorem 3. *For* $n \geq 3$*, the minimum size* $\mathcal{A}_2(n)$ *of a 2-CNF PCE of* AMO_n *satisfies*

1. *If* $n \leq 6$*, then* $\mathcal{A}_2(n) = 3n - 6$.
2. *If* $n \geq 7$*, then* $\mathcal{A}_2(n) \geq 2n + 2\sqrt{n} - 5$.

Proof. (Sketch) Consider a minimum 2-CNF encoding $\varphi(\mathbf{x}, \mathbf{y})$ of AMO_n. We use induction on encodings which are not in regular form as in the proof of Theorem 2. Moreover, based on Lemmas 16 and 17, we can use induction also in case when the sets $L'_{\varphi,r}$, $i = 1, \ldots, n$ are not pairwise different. Let us now define graph $G = (\mathbf{y}, E)$ on auxiliary variables where two auxiliary variables y, y' form an edge if there is a binary clause in φ which contains both the variables y, y'. We can show that G consists of at most three connected components, it follows that there are at least $|\mathbf{y}| - 3$ binary clauses with both literals on auxiliary variables. We now use Lemma 18 to show that either we can use induction to show the required bound, or $L'_{\varphi,1}, \ldots, L'_{\varphi,n}$ are pairwise distinct and form edges of a triangle-free undirected graph G' on $\mathbf{y}$. Mantel's theorem implies that the number n of edges in such a graph is at most $\frac{1}{4}|\mathbf{y}|^2$. Thus, $|\mathbf{y}| \geq 2\sqrt{n}$ and there is at least $2\sqrt{n} - 3$ clauses containing only auxiliary variables. Together with $2n$ clauses of type Q we obtain $|\varphi| = 2n + |\psi| \geq 2n + 2\sqrt{n} - 3$ in this case.

6   Conclusion and Further Research

We have shown that any propagation complete encoding of the AMO_n or EO_n constraint for $n \geq 7$ contains at least $2n + \sqrt{n-1} - 2$ clauses. This shows that the best known upper bound of $2n + 4\sqrt{n} + O(\sqrt[4]{n})$ clauses achieved by product encoding introduced by [9] is essentially best possible. Let us point out that the product encoding is an encoding of AMO_n in regular form which is the notion playing central role in our proof.

The encodings of AMO_n which appear in the literature are 2-CNF formulas which have an advantage that they are always propagation complete. We have therefore considered this special case and were able to show that for $n \geq 7$ a stronger lower bound $2n + 2\sqrt{n} - 5$ holds.

A function can have a 2-CNF encoding only if it is expressible by a 2-CNF formula and function AMO_n can be represented by an anti-monotone 2-CNF formula (i.e. a 2-CNF formula containing only negative literals). It is quite natural to ask if there is a minimum PCE of AMO_n which is a 2-CNF formula or a CNF formula without positive occurrences of input variables. We conjecture that the answer to both questions is positive, but that remains open. We can only prove that a positive answer to the first question implies a positive answer to the second question.

Even in case of 2-CNF encodings there is still a gap between our lower bound and the upper bound given by the product encoding. We conjecture that the size of the optimal encoding matches the upper bound. The regular form used in our proof helps to understand the structure of the part of the encoding close to the input variables. A better understanding of the structure of the clauses of type can possibly lead to improving the lower bound.

References

1. Asín, R., Nieuwenhuis, R., Oliveras, A., Rodríguez-Carbonell, E.: Cardinality networks: a theoretical and empirical study. Constraints **16**(2), 195–221 (2011). doi:10.1007/s10601-010-9105-0
2. Aspvall, B., Plass, M.F., Tarjan, R.E.: A linear-time algorithm for testing the truth of certain quantified boolean formulas. Inf. Process. Lett. **8**(3), 121–123 (1979)
3. Babka, M., Balyo, T., Čepek, O., Gurský, Š., Kučera, P., Vlček, V.: Complexity issues related to propagation completeness. Artif. Intell. **203**, 19–34 (2013)
4. Bacchus, F.: GAC via unit propagation. In: Bessière, C. (ed.) CP 2007. LNCS, vol. 4741, pp. 133–147. Springer, Heidelberg (2007). doi:10.1007/978-3-540-74970-7_12
5. Bailleux, O., Boufkhad, Y.: Efficient CNF encoding of boolean cardinality constraints. In: Proceedings of 9th International Conference on Principles and Practice of Constraint Programming, CP 2003, Kinsale, Ireland, Berlin, Heidelberg, 29 September–3 October 2003, pp. 108–122 (2003). doi:10.1007/978-3-540-45193-8_8
6. Ben-Haim, Y., Ivrii, A., Margalit, O., Matsliah, A.: Perfect hashing and CNF encodings of cardinality constraints. In: Cimatti, A., Sebastiani, R. (eds.) SAT 2012. LNCS, vol. 7317, pp. 397–409. Springer, Heidelberg (2012). doi:10.1007/978-3-642-31612-8_30
7. Biere, A., Heule, M., van Maaren, H., Walsh, T.: Handbook of Satisfiability, Frontiers in Artificial Intelligence and Applications, vol. 185. IOS Press, Amsterdam, The Netherlands (2009)
8. Bordeaux, L., Marques-Silva, J.: Knowledge compilation with empowerment. In: Bieliková, M., Friedrich, G., Gottlob, G., Katzenbeisser, S., Turán, G. (eds.) SOFSEM 2012. LNCS, vol. 7147, pp. 612–624. Springer, Heidelberg (2012). doi:10.1007/978-3-642-27660-6_50
9. Chen, J.: A new SAT encoding of the at-most-one constraint. In: ModRef 2010 (2010)
10. Crama, Y., Hammer, P.: Boolean Functions: Theory, Algorithms, and Applications. Encyclopedia of Mathematics and Its Applications. Cambridge University Press, Cambridge (2011)
11. Frisch, A.M., Giannaros, P.A.: SAT encodings of the at-most-k constraint. some old, some new, some fast, some slow. In: Proceeding of the Tenth International Workshop of Constraint Modelling and Reformulation (2010)
12. Frisch, A.M., Peugniez, T.J., Doggett, A.J., Nightingale, P.W.: Solving nonboolean satisfiability problems with stochastic local search: A comparison of encodings. J. Autom. Reason. **35**(1), 143–179 (2005). doi:10.1007/s10817-005-9011-0
13. Hölldobler, S., Nguyen, V.H.: An efficient encoding of the at-most-one constraint. Technical Report MSU-CSE-00-2, Knowledge Representation and Reasoning Groupp. 2013–04, Technische Universitt Dresden, 01062 Dresden, Germany (2013). http://www.wv.inf.tu-dresden.de/Publications/2013/report-13-04.pdf
14. Klieber, W., Kwon, G.: Efficient CNF encoding for selecting 1 from n objects. In: Fourth Workshop on Constraints in Formal Verification (2007)
15. Kučera, P., Savický, P., Vorel, V.: A lower bound on CNF encodings of the at-most-one constraint (2017). https://arxiv.org/abs/1704.08934, arXiv:1704.08934 [cs.CC]
16. Marques-Silva, J., Lynce, I.: Towards robust CNF encodings of cardinality constraints. In: Bessière, C. (ed.) CP 2007. LNCS, vol. 4741, pp. 483–497. Springer, Heidelberg (2007). doi:10.1007/978-3-540-74970-7_35

17. Sinz, C.: Towards an optimal CNF encoding of boolean cardinality constraints. In: Proceedings of 11th International Conference on Principles and Practice of Constraint Programming, CP 2005, Sitges, Spain, 1–5 October 2005, pp. 827–831. Berlin, Heidelberg (2005). doi:10.1007/11564751_73

18. del Val, A.: Tractable databases: how to make propositional unit resolution complete through compilation. In: Knowledge Representation and Reasoning, pp. 551–561 (1994)

SAT-Encodings for Special Treewidth and Pathwidth

Neha Lodha, Sebastian Ordyniak$^{(\boxtimes)}$, and Stefan Szeider

Algorithms and Complexity Group, Tu Wien, Vienna, Austria
{neha,ordyniak,sz}@ac.tuwien.ac.at

Abstract. Decomposition width parameters such as treewidth provide a measurement on the complexity of a graph. Finding a decomposition of smallest width is itself NP-hard but lends itself to a SAT-based solution. Previous work on treewidth, branchwidth and clique-width indicates that identifying a suitable characterization of the considered decomposition method is key for a practically feasible SAT-encoding.

In this paper we study SAT-encodings for the decomposition width parameters special treewidth and pathwidth. In both cases we develop SAT-encodings based on two different characterizations. In particular, we develop two novel characterizations for special treewidth based on partitions and elimination orderings. We empirically obtained SAT-encodings.

1 Introduction

The decomposition of graphs is a central topic in combinatorics and combinatorial optimization where various decomposition methods have been developed. Decomposition methods gives rise to a so-called width parameters that indicates how well the graph is decomposable by the considered decomposition method. For instance tree decomposition, the most famous decomposition method, gives rise to the parameter *treewidth*, where the treewidth of a graph is the smallest width over all tree decompositions. Typically, finding an optimal decomposition (i.e., one of smallest width), is an NP-hard problem, for which various exponential-time algorithms have been suggested. Previous work indicates that SAT provides a valuable practical approach for finding optimal decompositions. This approach was pioneered by Samer and Veith for treewidth [16]; their methods was further improved [2] and achieved excellent results in a recent solver challenge [8]. Heule and Szeider [11] developed the first practically feasible approach for computing the decomposition parameter *clique-width* by means of a SAT encoding, which allowed for the first time to identify the clique-width of some well-known named graphs. A SAT-encoding for the decomposition parameter *branchwidth* was suggested by Lodha et al. [14], who also showed how the encoding can be used to improve heuristically obtained branch decompositions of large graphs.

Special Treewidth and Pathwidth. In this paper we consider new SAT encodings for the decomposition parameters *special treewidth* and *pathwidth*. Special treewidth, a decomposition parameter introduced by Courcelle [6,7], is

© Springer International Publishing AG 2017
S. Gaspers and T. Walsh (Eds.): SAT 2017, LNCS 10491, pp. 429–445, 2017.
DOI: 10.1007/978-3-319-66263-3_27

closely related to the well-known decompositional parameters pathwidth and treewidth [13]. A tree decomposition of a graph G is a tree T whose nodes are labeled with sets of vertices, called *bags*, such that for each edge of G there is a bag containing both ends of the edge, and for each vertex of G, the nodes of T labeled with bags containing this vertex form a non-empty connected subtree. The width of the tree decomposition is the size of a largest bag minus 1, and the treewidth of a graph is the smallest width over all its tree decompositions. Special treewidth is defined similar to treewidth, with the additional property that T is a rooted tree, and for each vertex of G there is some root-to-leaf path in T which contains all the nodes labeled with bags containing this vertex. Pathwidth is also defined similar to treewidth, where T itself is a path. It follows from these definitions that special treewidth is in-between treewidth and pathwidth, i.e., for every graph G we have

$$\text{treewidth}(G) \leq \text{special treewidth}(G) \leq \text{pathwidth}(G).$$

The motivation for special treewidth is that it allows for more efficient model-checking algorithms for variants of Monadic Second Order Logic than treewidth, but is often smaller than pathwidth. Special treewidth has been the subject of several theoretical investigations [4,5]. Pathwidth, on the other hand, was introduced by Robertson and Seymour in the first of their famous series of papers on graph minors [15] and has since then attracted a lot of attention. The computation of special treewidth and pathwidth are NP-hard problems. For the latter this has been known [1] for long, for the former we observe that it can be deduced from known results (Theorem 1).

Characterizations of Width Parameters. Previous work on SAT encodings for treewidth, branchwidth and clique-width indicates that identifying a suitable characterization of the considered decomposition method is key for a practically feasible SAT-encoding. In fact, the standard encoding for treewidth [16] is based on the characterization of treewidth in terms of *elimination orderings*, which are linear orderings of the vertices of the decomposed graph, where after adding certain "fill-in" edges, the largest number of neighbors of a vertex ordered higher than the vertex itself, gives the width of the decomposition. For clique-width, on the other hand, no characterization based on elimination ordering is known, and the known SAT-encoding [11] uses a *partition-based* characterization, where one considers a sequence of partitions of the vertex set. A similar partition-based characterization was used for the SAT encoding of branchwidth [14]. Recently, an encoding for pathwidth and similar decompositional parameters based on the interval model of a path-decomposition has been introduced by Biedl et al. [3].

In this paper we develop and compare SAT encodings based on two characterizations of special treewidth and two characterizations for pathwidth.

Results for Special Treewidth. For special treewidth we develop a new characterization based on elimination orderings (Theorem 3), as one could expect that a characterization that is similar to the characterization successfully used for a

SAT encoding of treewidth also works well for special treewidth. We also develop a partition-based characterization which is close to the original characterization by Courcelle [6]. Our experiments show that the partition-based encoding clearly outperforms the ordering-based encoding. For instance, the former could process square grids and complete graphs being almost twice as large as the square grids and complete graphs within the reach of the latter. The partition-based encoding also beats the ordering-based encoding on many of the well-known named graphs that we consider by an order of magnitude and is competitive in running-times to the currently leading encoding for treewidth.

Results for Pathwidth. For pathwidth, there exists a well known characterization in terms of linear orderings [12] which gives rise to a natural SAT encoding, similar in spirit to the Samer-Veith encoding for treewidth [16]. However, we also considered a partition-based encoding, similar in spirit to the Heule-Szeider encoding for clique-width [11]. Our experiments indicate that the ordering-based encoding has a slight advantage on average over the partition-based encoding. However, the partition-based encoding has an extraordinary advantage on dense graphs. This encourages the development of a portfolio-based approach for SAT-encodings for pathwidth.

2 Preliminaries

2.1 Satisfiability and SAT-Encodings

We consider propositional formulas in Conjunctive Normal Form (*CNF formulas,* for short), which are conjunctions of clauses, where a clause is a disjunction of literals, and a literal is a propositional variable or a negated propositional variable. A CNF formula is *satisfiable* if its variables can be assigned true or false, such that each clause contains either a variable set to true or a negated variable set to false. The satisfiability problem (SAT) asks whether a given formula is satisfiable.

We will now introduce a few general assumptions and notation that is shared among the encodings. Namely, for our encodings we will assume that we are given an undirected graph $G = (V, E)$ and an integer ω, which represents the width that we are going to test. Moreover, we will assume that the vertices of G are numbered from 1 to n and similarly the edges are numbered from 1 to m. Details on how we used the formulas to calculate the exact width of a graph are given in Sect. 6.1. For the counting part of all our encodings we will employ the *sequential counter* approach [16] since this approach has turned out to provide the best results in our setting. To illustrate the idea behind the sequential counter consider the case that one has a variable $S(v)$ for every vertex $v \in V(G)$ and one needs to restrict the number of vertices for which the variable $S(v)$ is set to true to be at most some integer ω. In this case one introduces a counting variable $\#(v, j)$ for every $v \in V(G)$ and j with $1 \leq j \leq \omega$, which is true whenever there are at least j variables $S(v)$ set to true in $\{ S(u) \mid 1 \leq u \leq v \}$. Then the following clauses ensure the semantics for the variable $\#(v, i)$ and ensure that at most ω of the variables S(v) are set to true. A clause $\neg S(v) \vee \#(v, 1)$ for every $v \in V(G)$, a clause $\neg \#(v - 1, j) \vee \#(v, j)$ for every $v \in V(G)$ and j with $v > 1$ and $1 \leq j \leq \omega$, a clause $\neg S(v) \vee \neg \#(v - 1, j - 1) \vee \#(v, j)$ for every $v \in V(G)$ and j with $v > 1$ and $1 < j \leq \omega$, and a clause $\neg S(v) \vee \neg \#(v - 1, \omega)$ for every $v \in V(G)$ with $v > 1$.

2.2   Graphs

We consider finite and simple undirected graphs. For basic terminology on graphs we refer to a standard text book [9]. For a graph G we denote by $V(G)$ the vertex set of G and by $E(G)$ the edge set of G. If $E \subseteq E(G)$, we denote by $G \setminus E$ the graph with vertices $V(G)$ and edges $E(G) \setminus E$.

We will often consider various forms of trees, i.e., connected acyclic graphs, as they form the backbone of tree decompositions. Let T be an undirected tree and $t \in V(T)$. We will often assume that T is rooted (in some arbitrary vertex r) and hence the parent and child relationships between its vertices are well-defined. We write T_t to denote the subtree of T rooted in t, i.e., the component of $T \setminus \{\{t, p\}\}$ containing t, where p is the parent of t in T. For a tree T, we denote by h(T), the *height* of T, i.e., the length of a longest path between the root and any leaf of T plus one.

2.3   Special Treewidth

To define special treewidth, it is convenient to first introduce treewidth and pathwidth and then show how to adapt the definition to obtain special treewidth.

A *tree decomposition* $\mathcal{T}$ of a graph $G = (V, E)$ is a pair (T, χ), where T is a tree and χ is a function that assigns each tree node t a set $\chi(t) \subseteq V$ of vertices such that the following conditions hold: **(T1)** for every vertex $u \in V$, there is a tree node t such that $u \in \chi(t)$, **(T2)** for every edge $\{u, v\} \in E$ there is a tree node t such that $\{u, v\} \subseteq \chi(t)$, and **(T3)** for every vertex $v \in V$, the set of tree nodes t with $v \in \chi(t)$ forms a subtree of T. The sets $\chi(t)$ for any $t \in V(T)$ are called *bags* of the decomposition $\mathcal{T}$ and $\chi(t)$ is the bag associated with the tree node t. The *width* of a tree decomposition (T, χ) is the size of a largest bag minus 1. A tree decomposition of minimum width is called *optimal*. The *treewidth* of a graph G is the width of an optimal tree decomposition of G. A *path decomposition* is a tree decomposition $\mathcal{T} = (T, \chi)$, where T is required to be a path and the *pathwidth* of a graph is the minimum width of any of its path decompositions.

A *special tree decomposition* $\mathcal{T} = (T, \chi)$ of a graph $G = (V, E)$ is a tree decomposition that is rooted at some node $r \in V(T)$ and additionally satisfies the following property [4,6]: **(ST)** for every vertex $v \in V$, the set of tree nodes t with $v \in \chi(t)$ forms a subpath of a path in T from r to a leaf. Note that **(ST)** subsumes **(T3)**, which implies that a special tree decomposition merely needs to satisfy **(T1)**, **(T2)**, and **(ST)**. The *width* of a special tree decomposition as well as the special treewidth of a graph G are defined analogously to the width of a tree decomposition and the treewidth, respectively. Figure 1 illustrates an (optimal) special tree decomposition of a graph.

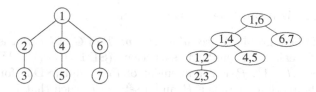

Fig. 1. A graph G (left) and an optimal (special) tree decomposition $\mathcal{T} = (T, \chi)$ of G (right).

As a prerequisite for the development of SAT-encodings for the problem, and since to the best of our knowledge this has never been explicitly stated previously, we first show that computing the special treewidth of a graph is NP-hard.

Theorem 1. *Given a graph G and an integer ω, then determining whether G has special treewidth at most ω is NP-complete.*

Proof. The problem is clearly in NP, because there is always an optimal (special) tree decomposition, where the number of nodes is at most the number of vertices in the graph. The NP-hardness follows from [10], where it was shown that pathwidth equals treewidth on the class of co-comparability graphs and moreover computing both width measures for co-comparability graphs is still NP-hard. Because special treewidth is in-between pathwidth and treewidth it equals both width measures on co-comparability graphs and its computation is therefore also NP-hard.

We remark that if ω is constant and not part of the input, then one can check in linear time whether a given graph has special treewidth at most ω (the running time depends exponentially on ω) [4]; similar results are well known to hold for treewidth and pathwidth.

2.4   Weak Partitions

A *weak partition* of a set S is a set P of nonempty subsets of S such that any two sets in P are disjoint. We denote by $\mathrm{U}(P)$ the union of all sets in P. If additionally $S = \mathrm{U}(P)$, then P is a *partition*. The elements of P are called *equivalence classes*. Let P, P' be weak partitions of S. Then P' is a *refinement* of P if $\mathrm{U}(P) \subseteq \mathrm{U}(P')$ and any two elements $x, y \in S$ that are in the same equivalence class of P' are not in distinct equivalence classes of P (this entails the case $P = P'$). Moreover, we say that P' is a *k-ary refinement* of P if additionally it holds that for every $p \in P$ there are $p_1, \ldots, p_k$ in P' such that $p \subseteq \bigcup_{1 \leq i \leq k} p_i$. Intuitively, if P' is a k-ary refinement of P, then P is obtained from P' by forgetting some elements and joining up to k equivalence classes.

3   Partition-Based Approach for Special Treewidth

In this section we introduce a novel characterization of special treewidth, in terms of special derivations. The characterization is inspired by the partition-based approaches employed for branchwidth and clique-width [11,14].

3.1   Characterization: Special Derivations

Let $G = (V, E)$ be a graph. A *special derivation* $\mathcal{P}$ of G of *length* l is a sequence $(P_1, \ldots, P_l)$ of weak partitions of V such that: **(SD1)** $U(P_1) = V$, **(SD2)** for every $i \in \{1, \ldots, l-1\}$, P_i is a refinement of P_{i+1}, and **(SD3)** for every edge $\{u, v\} \in E$ it holds that there is a P_i and a set $p \in P_i$ such that $\{u, v\} \subseteq p$. The *width* of $\mathcal{P}$ is the maximum size of any set in $P_1 \cup \ldots \cup P_l$ minus 1. We will refer to P_i as the i-th *level* of $\mathcal{P}$ and we will refer to elements in $\bigcup_{1 \leq i \leq l} P_i$ as *sets* of $\mathcal{P}$. We will show that any special tree decomposition can be transformed into a special derivation of the same width and vice verse. The following example illustrates the close connection between special tree decompositions and special derivations.

Example 1. Consider the special tree decomposition $\mathcal{T}$ given in Fig. 1. Then $\mathcal{T}$ can, e.g., be translated into the special derivation $\mathcal{P} = (P_1, \ldots, P_4)$ defined by setting $P_1 = \{\{1\}, \{2, 3\}, \{4, 5\}, \{6, 7\}\}$, $P_2 = \{\{1, 2\}, \{4, 5\}, \{6, 7\}\}$, $P_3 = \{\{1, 4\}, \{6, 7\}\}$, $P_4 = \{\{1, 6\}\}$. The width of $\mathcal{T}$ is equal to the width of $\mathcal{P}$.

The following theorem shows that special derivations provide an alternative characterization of special tree decompositions. The main observation behind the proof of the equivalence between the two characterizations is that after padding the special tree decomposition such that every leaf has the same distance from the root, it holds that the weak partition on a certain level of a special derivation is given by the set of bags that are at the same distance from a leaf in a special tree decomposition and vice versa.

Theorem 2. *A graph G has a special tree decomposition of width at most ω and height at most h if and only if G has a special derivation of width at most ω and length at most h. Moreover, there is a special derivation of optimal width with length at most $|V(G)| - \omega$.*

Note that the second statement of the above theorem allows us to restrict our search to special derivations of length at most $|V(G)| - \omega$. Its proof crucially uses the observation that a restricted form of tree decompositions, so called small tree decompositions, can be shown to have height at most $|V(G)| - \omega$.

3.2   SAT-Encoding of a Special Derivation

Here we will provide our encoding for special derivations. Namely, we will construct a CNF formula $F(G, \omega, l)$ that is satisfiable if and only if G has a special derivation of width at most ω and length at most l. Because of Theorem 2 (after setting l to the value specified in the theorem) it holds that $F(G, \omega, n - \omega)$ is satisfiable if and only if G has special treewidth at most ω. To achieve this aim we first construct a formula $F(G, l)$ that is satisfiable if and only if G has a special derivation of length at most l

The formula $F(G, l)$ uses a *set variable* $set(u, v, i)$, for every $u, v \in V(G)$ and i with $u \leq v$ and $1 \leq i \leq l$. Informally, $set(u, v, i)$ is true whenever either $u \neq v$ and u and v are contained in the same set at level i of the special derivation or

$u = v$ and u is contained in some set at level i. We now describe the clauses of the formula. The following clauses ensure transitive of the relation between two vertices $u, v \in V(G)$ defined by $\mathrm{set}(u, v, i)$ for every i with $1 \le i \le l$.

$$(\neg\mathrm{set}(u, v, i) \vee \neg\mathrm{set}(u, w, i) \vee \mathrm{set}(v, w, i))$$
$$\wedge(\neg\mathrm{set}(u, v, i) \vee \neg\mathrm{set}(v, w, i) \vee \mathrm{set}(u, w, i))$$
$$\wedge(\neg\mathrm{set}(u, w, i) \vee \neg\mathrm{set}(v, w, i) \vee \mathrm{set}(u, v, i))$$
$$\wedge(\neg\mathrm{set}(u, v, i) \vee \neg\mathrm{set}(u, u, i)) \qquad \text{for } u, v, w \in V(G), u < v < w, 1 \le i \le l.$$

To ensure Property **(SD1)**, we add the clause $\mathrm{set}(u, u, 1)$ for every $u \in V(G)$. The following clauses ensure **(SD2)**, i.e., P_i is a refinement of P_{i+1} for every $1 \le i < l$.

$$(\neg\mathrm{set}(u, u, i + 1) \vee \neg\mathrm{set}(v, v, i + 1) \vee \mathrm{set}(u, v, i + 1) \vee \neg\mathrm{set}(u, v, i))$$
$$wedge(\mathrm{set}(u, u, i) \vee \neg\mathrm{set}(u, u, i + 1)) \qquad \text{for } u, v \in V(G), u < v, 1 \le i < l$$

Towards presenting the clauses employed to ensure **(SD3)**, we will use the following property that is easily seen to be equivalent to **(SD3)**.
(SD3') For every edge $\{u, v\} \in E$, it holds that:

- if there is an i with $1 \le i < l$ such that $u, v \in \mathrm{U}(P_i)$ and $v \notin \mathrm{U}(P_{i+1})$, then $u, v \in p$ for some $p \in P_i$ and
- if $u, v \in \mathrm{U}(P_l)$, then $u, v \in p$ for some $p \in P_l$.

Note that **(SD3)** and **(SD3')** are equivalent because whenever there is a set $p \in P_i$ for some i with $1 \le i \le l$ containing two vertices u and v, then such a set also exists in every P_j for $j \ge i$ as long as $u, v \in \mathrm{U}(P_j)$. The following clauses now ensure **(SD3')** and thereby **(SD3)**.

$$((\neg\mathrm{set}(u, u, i) \vee \neg\mathrm{set}(v, v, i) \vee \mathrm{set}(u, u, i + 1)) \vee \mathrm{set}(u, v, i))$$
$$\wedge ((\neg\mathrm{set}(u, u, i) \vee \neg\mathrm{set}(v, v, i) \vee \mathrm{set}(v, v, i + 1)) \vee \mathrm{set}(u, v, i))$$
$$\wedge ((\neg\mathrm{set}(u, u, l) \vee \neg\mathrm{set}(v, v, l)) \vee \mathrm{set}(u, v, l))$$
$$\text{for } e \in E(G), u, v \in e, u < v, 1 \le i < l$$

We now ready to to extend $F(G, l)$ to the formula $F(G, \omega, l)$. We achieve this by restricting the sizes of all sets in P_i for every $1 \le i \le l$ to be at most $\omega + 1$, or in other words for every $v \in V(G)$ and i with $1 \le i \le l$, we need to restrict the number of variables $\mathrm{set}(v, u, i)$ set to true to be at most $\omega + 1$. We achieve this by using the sequential counter approach described in Subsect. 2.1. The obtained formula $F(G, l, \omega)$ contains $\mathcal{O}(n^3\omega)$ variables and $\mathcal{O}(n^4 + mn^3)$ clauses.

4 Ordering-Based Approach for Special Treewidth

In this section we introduce a second characterization of special treewidth, namely special elimination orderings, inspired by elimination orderings characterizing treewidth [13].

4.1 Characterization: Special Elimination Orderings

We start by introducing elimination orderings characterizing treewidth and then show how to adapt the notion in the context of special treewidth. Towards this aim we start with a slightly non-standard definition of elimination orderings for treewidth, from which it is particularly easy to obtain our adaptation for special treewidth.

Let G be a graph with n vertices and let $\leq_S$ be a total order $(v_1, \ldots, v_n)$ of the vertices of G. For two vertices u and v with $u \leq_S v$ we denote by $N_G^{\leq_S}(u, v)$ the set of all neighbors of u in G that are larger than v w.r.t. $\leq_S$. We extend this notation to sets $U \subseteq V(G)$, where $u \leq_S v$ for every $u \in U$, by setting $N_G^{\leq_S}(U, v)$ to be the set $\bigcup_{u \in U} N_G^{\leq_S}(u, v)$. We next define the sequence $G_0^{\leq_S}, \ldots, G_{n-1}^{\leq_S}$ of supergraphs of G inductively as follows: We set $G_0^{\leq_S} = G$ and for every i with $1 \leq i < n$ we let $G_i^{\leq_S}$ be the graph obtained from $G_{i-1}^{\leq_S}$ after adding all edges in the set $E_i^{\leq_S}$, which is defined as follows. Let $C_i^{\leq_S}$ be the set of all components of the graph $G_{i-1}[v_1, \ldots, v_{i-1}, v_i]$. Then $E_i^{\leq_S}$ is the set $\{ \{u, v\} \mid u, v \in N_{G_{i-1}}^{\leq_S}(C, v_i) \wedge C \in C_i^{\leq_S} \}$. We call $G_{\leq_S} = G_{n-1}^{\leq_S}$ the *fill-in graph* of G w.r.t. $\leq_S$ and $G_i^{\leq_S}$ the *i-th fill-in graph* of G w.r.t. $\leq_S$. Then any total ordering $\leq_S$ gives rise to an *elimination ordering* of G and the *width* of an elimination ordering $\leq_S$ is the maximum of $\max\{ |N_{G_{\leq_S}}^{\leq_S}(C, v_i)| \mid C \in C_i^{\leq_S} \}$ over all i with $1 \leq i < n$. Furthermore, the *elimination width* of a graph G is the minimum width of any elimination ordering of G. It is known that the elimination width of a graph is equal to the treewidth of a graph [13].

We are now ready to show how to adapt elimination orderings for special treewidth. Informally, the crucial observation here is that because of Property **(ST)** a special tree decomposition, in contrast to a normal tree decomposition, cannot have separate branches for components that have at least one common neighbor. This property directly translates to elimination orderings in the sense that whenever two components C and C' in $C_i^{\leq_S}$ share a neighbor that comes later in the ordering, they need to be handled together both for obtaining the fill-in edges as well as for determining the width of the ordering. To formalize this idea, we say that two components C and C' in $C_i^{\leq_S}$ *clash* if $N_{G_{i-1}}^{\leq_S}(C, v_i) \cap N_{G_{i-1}}^{\leq_S}(C', v_i) \neq \emptyset$. Moreover, let H be the graph with vertex-set $C_i^{\leq_S}$ having an edge between two vertices C and C' if and only if their associated components clash and let $P_i^{\leq_S}$ be the partition of $C_i^{\leq_S}$ that corresponds to the connected components of H. Then *special elimination orderings* are obtained from elimination orderings by using $P_i^{\leq_S}$ instead of $C_i^{\leq_S}$ to determine both the fill-in edges as well as the width of the ordering. Formally, for special elimination orderings the set $E_i^{\leq_S}$ becomes $\{ \{u, v\} \mid u, v \in N_{G_{i-1}}^{\leq_S}(P, v_i) \wedge P \in P_i^{\leq_S} \}$ and the width of $\leq_S$ becomes the maximum of $\max\{ |N_{G_{\leq_S}}^{\leq_S}(P, v_i)| \mid P \in P_i^{\leq_S} \}$ over all i with $1 \leq i < n$. We show next that special elimination orderings properly characterize special treewidth. The main ideas behind the proof of the theorem are similar to the proof showing the equivalence between eliminations orderings and treewidth [13], however, the proof is significantly more involved due to the properties of special treewidth.

Theorem 3. *A graph G has a special tree decomposition of width at most ω if and only if G has a special elimination ordering of width at most ω.*

4.2 SAT-Encoding for Special Elimination Orderings

Here we provide our encoding for special elimination orderings as introduced in the previous subsection. In particular, we will construct a CNF formula $F(G, \omega)$ that is satisfiable if and only if G has a special elimination ordering of width at most ω. Because of Theorem 3 it then holds that $F(G, \omega)$ is satisfiable if and only if G has special treewidth at most ω. Towards this aim we first construct the formula $F(G)$ that is satisfiable if and only if G has a special elimination ordering and building upon $F(G)$ we will then use cardinality constraints to obtain $F(G, \omega)$. For the definition of the formula we use the same notation as introduced in Sect. 4.1, i.e., we refer to the required elimination ordering by $\leq_S$, and use $\mathcal{C}_v^{\leq_S}$ and $\mathcal{P}_v^{\leq_S}$ to refer to the components and parts of the graph $G_{v-1}^{\leq_S}[1, \ldots, v]$ (recall that we assume that the vertices of G are numbered from 1 to n).

The formula $F(G)$ uses the following variables. An *order variable* $o(u, v)$ for all $u, v \in V(G)$ with $u < v$. The variable $o(u, v)$ will be true if and only if $u < v$ and $u \leq_S v$. The idea behind the variable $o(u, v)$ is that it can used to model the total ordering $\leq_S$ witnessing the elimination width of G by requiring that $u \leq_S v$ for arbitrary $u, v \in V(G)$ if and only if $u = v$ or $u < v$ and $u \leq_S v$ or $u > v$ and $\neg o(v, u)$. In order to be able to refer to $\leq_S$ in the clauses of $F(G)$, we define the "macro" $o^*(u, v)$ by setting $o^*(u, v) = \textbf{true}$ if $u = v$, $o^*(u, v) = o(u, v)$ if $u < v$ and $o^*(u, v) = \neg o(v, u)$ if $u > v$. Additionally, $F(G)$ contains an *arc variable* $a(u, v)$ for all $u, v \in V(G)$. The variable $a(u, v)$ is true if $u \leq_S v$ and $\{u, v\} \in E(G_{\leq_S})$ and moreover it is not true if $v <_S u$. Finally, $F(G)$ has a *part variable* $p(u, v)$ for all $u, v \in V(G)$. The variable $p(u, v)$ is true if and only if the vertices u and v belong to the same part in $\mathcal{P}_v^{\leq_S}$. Observe that whenever a vertex u belongs to the same part as a vertex v in $\mathcal{P}_v^{\leq_S}$, then u will also be in the same part as v in $\mathcal{P}_w^{\leq_S}$ for any w with $v \leq_S w$.

We will now provide the clauses for the formula $F(G)$. The following clauses ensure that $o^*(u, v)$ is a total ordering of $V(G)$ by ensuring that the relation between u and v defined by $o^*(u, v)$ is transitive:

$$(\neg o^*(u, v) \vee \neg o^*(v, w) \vee o^*(u, w))$$
$$\text{for } u, v, w \in V(G) \text{ where } u, v, \text{ and } w \text{ are pairwise distinct.}$$

We also introduce the clause $a(u, v) \vee a(v, u)$ for every $\{u, v\} \in E(G)$, which ensure that at least one of $a(u, v)$ or $a(v, u)$ is true for every edge $\{u, v\} \in E(G)$. Towards ensuring that the ordering $\leq_S$ represented by $o^*(u, v)$ is compatible with the direction of the edges given by $a(u, v)$, we introduce the clause $\neg a(u, v) \vee o^*(u, v)$ for every $u, v \in V(G)$. Moreover, to ensure that the relation given by

$p(u, v)$ is reflexive, i.e., every vertex belongs to its own part, we introduce the clause $p(v, v)$ for every $v \in V(G)$.

The following clauses ensure that if $p(u, v)$ is true, then also $p(w, v)$ is true for every w that is in the same component as u in $\mathcal{C}_{\bar{v}}^{\leq s}$. This is achieved by enforcing that whenever a vertex w with $w \leq_S v$ is connected via an edge in $G_{\leq_S}$ to some vertex u with $p(u, v)$ being true, then also $p(w, v)$ is true.

$$(\neg a(u, w) \lor \neg p(u, v) \lor \neg o^*(w, v) \lor p(w, v))$$
$$\land (\neg a(w, v) \lor \neg p(u, v) \lor \neg o^*(w, v) \lor p(w, v))$$
$$\text{for } u, w, v \in V(G) \text{ and } u \neq w \text{ and } w \neq v.$$

The following clauses complete the definition of $p(u, v)$ by enforcing that whenever there is a vertex u with $u \leq_S v$ that shares a neighbor x with some vertex w with $p(w, v)$ being true, then also $p(u, v)$ is true, as u must also be in this part.

$$\neg a(u, x) \lor \neg a(w, x) \lor \neg p(w, v) \lor \neg o^*(u, v) \lor p(u, v)$$
$$\text{for } u, w, x, v \in V(G) \text{ and } u \neq w \neq x.$$

The following clauses ensure that at least one of $a(u, v)$ or $a(v, u)$ is true for every "fill-in edge", i.e., for every edge in $E(G_{\leq_s}) \setminus E(G)$.

$$\neg p(u_1, v) \lor \neg p(u_2, v) \lor \neg a(u_1, w_1) \lor \neg a(u_2, w_2) \lor \neg o^*(v, w_1) \lor \neg o^*(v, w_2)$$
$$\lor a(w_1, w_2) \lor a(w_2, w_1) \quad \text{for } u_1, u_2, w_1, w_2, v \in V(G) \text{ with } w_1 \neq w_2.$$

This completes the construction of $F(G)$. Informally, the crucial parts to verify the correctness of the formula are that for any ordering of the vertices of G, which is defined by the setting of the ordering variables $o(u, v)$, the formula ensures that whenever $\{u, v\} \in G_{\leq_s}$ then either $a(u, v)$ or $a(v, u)$ is true. This way the formula ensures that all edges of $G_{\leq_s}$ are considered for the definition of the part variables $p(u, v)$, which in turn ensures the correctness of the formula.

We are now ready to construct the formula $F(G, \omega)$. To achieve this it only remains to restrict the sizes of the sets $N_{G_{\leq_s}}^{\leq_s}(P, v)$ to be at most ω for every $v \in V(G)$ and $P \in \mathcal{P}_v^{\leq_s}$. Indeed we need to restrict the number of vertices w satisfying the formula $a(u, w) \land p(u, v) \land o^*(v, w)$ for every $u, v \in V(G)$. We achieve this again by using the sequential cardinality counter described in Subsect. 2.1. This concludes the description of the formula $F(G, \omega)$, which contains $\mathcal{O}(n^2\omega)$ variables and $\mathcal{O}(n^5)$ clauses.

5   SAT-Encodings for Pathwidth

In this section we introduce our characterizations and encodings for pathwidth. Namely, we first introduce an encoding for pathwidth based on the well-known vertex separation number and then provide a second encoding based on path decompositions, which can be seen as a special case of the derivation-based encoding for special treewidth.

5.1   Partition-Based Encoding for Pathwidth

In this section we provide the partition-based encoding for pathwidth. Note that since a path decomposition has no branches, and therefore the partition on every level consists merely of a single set, the partition-based characterization of pathwidth becomes much simpler than its counterpart for special treewidth. In particular, the encoding is very closely based on the characterization of pathwidth in terms of a path decomposition, which can be equivalently stated as follows. A path decomposition can be seen as a sequence $(P_1, \ldots, P_\ell)$ of bags satisfying the following conditions: (**P1**) for every $v \in V(G)$ there is a bag P_i with $v \in P_i$, (**P2**) for every i with $1 \leq i < l$, if $v \in P_i$ and $v \notin P_{i+1}$, then $v \notin P_j$ for every $j > i$. We say that the vertex v has been forgotten at level $i + 1$. (**P3**) for every $u, v \in V(G)$ with $\{u, v\} \in E(G)$ and every i with $1 \leq i < \ell$, it holds that if u and v have not yet been forgotten at level i but u is forgotten at level $i + 1$, then u and v are contained in P_i. In the following we describe the CNF formula $F(G, \omega, \ell)$, which for a graph G and two integers ω and ℓ is satisfied if and only if G has a path decomposition of width at most ω with at most ℓ bags. Note that since path decompositions are a special case of special tree decompositions, we can bound the maximum number of bags in an optimal path decomposition by $n - \omega$ in accordance with Theorem 2. Therefore, the formula $F(G, \omega, n - \omega)$ is satisfied if and only if G has a path decomposition of width at most ω.

$F(G, \omega, \ell)$ contains the following variables for every $v \in V(G)$ and every i with $1 \leq i \leq \ell$: The *bag variable* $s(v, i)$, which is true if P_i contains the vertex v, and the *forgotten variable* $f(v, i)$, which is true if the vertex v has been forgotten at some step $j \leq i$. Moreover, $F(G, \omega, \ell)$ contains the following clauses:

- for every $v \in V(G)$, the clause $\neg f(v, 1)$ mirroring the property that no vertex is marked forgotten at (or before) the first bag of the path decomposition,
- for every $v \in V(G)$, the clause $f(u, \ell)$, which ensures Property (**P1**),
- for every $v \in V(G)$ and every i with $1 \leq i < \ell$, the clause $\neg s(v, i) \vee \neg s(v, i + 1) \vee f(v, i)$, which ensures that if a vertex does occur in the bag at level i but not in the bag at level $i + 1$, then it is marked as forgotten.
- for every $v \in V(G)$ and every i with $1 \leq i < \ell$, the clause $\neg f(v, i) \vee \neg s(v, i)$, which ensures that if a vertex has already been forgotten at level i, then it does not occur in the i-th bag of the path decomposition,
- for every $v \in V(G)$ and every i with $1 \leq i < \ell$, the clause $\neg f(v, i) \vee f(v, i + 1)$, which ensures that if a vertex is forgotten at level i then it remains forgotten at any level $j > i$ (note that these clauses together with the clauses defined in the previous item ensure Property (**P2**),
- for every $u, v \in V(G)$ with $\{u, v\} \in E(G)$ and every i with $1 \leq i < \ell$, the clauses $f(u, i) \vee f(v, i) \vee \neg f(u, i+1) \vee s(u, i)$ and $f(u, i) \vee f(v, i) \vee \neg f(u, i+1) \vee s(v, i)$, which together ensure Property (**P3**).

Finally, it remains to restrict the maximum size of the set $s(u, i)$ for any level i to be at most $\omega + 1$, i.e., for every level i with $1 \leq i \leq \ell$, we need to restrict the number of variables $s(u, i)$ set to true to be at most $\omega + 1$. We achieve this using the sequential cardinality counter described in Subsect. 2.1. This completes the construction of the formula $F(G, \omega, \ell)$, which including the counter variables and clauses contains $\mathcal{O}(n^2\omega)$ variables and $\mathcal{O}(n^3)$ clauses.

5.2   Ordering-Based Encoding for Pathwidth

Our second encoding for pathwidth is based on the characterization of pathwidth in terms of the vertex separation number, which is defined as follows. Given a graph G, an ordering $\leq_V$ of the vertices of G, and a vertex $v \in V(G)$, we denote by $S_{\leq_V}(v)$ the set of all vertices in G that are smaller or equal to v w.r.t. $\leq_V$. Moreover, for a subset S of the vertices of G, we denote by $\delta(S)$, the set of *guards* of S in G, i.e., the set of all vertices in S that have a neighbor in $V(G)\backslash S$. Then a graph G has *vertex separation number* at most ω if and only if there is an ordering $\leq_V$ of its vertices such that $|\delta(S_{\leq_V}(v))| \leq \omega$ for every $v \in V(G)$. It is well-known that G has vertex separation number at most ω if and only if G has pathwidth at most ω [12].

We will now show how to construct the formula $F(G, \omega)$ which is satisfiable if and only if G has vertex separation number (and hence pathwidth) at most ω. Apart from the variables needed for counting (which we will introduce later), the formula $F(G, \omega)$, has an *order variable* $o(u, v)$ for every $u, v \in V(G)$ with $u < v$. The variable $o(u, v)$ will be true if and only if $u < v$ and $u \leq_V v$. The idea behind the variable $o(u, v)$ is that it can used to model the total ordering $\leq_V$ witnessing the vertex separation number of G by requiring that $u \leq_V v$ for arbitrary $u, v \in V(G)$ if and only if $u = v$ or $u < v$ and $u \leq_V v$ or $u > v$ and $\neg o(v, u)$. In order to be able to refer to $\leq_V$ in the clauses, we define the "makro" $o^*(u, v)$ by setting $o^*(u, v) = \texttt{true}$ if $u = v$, $o^*(u, v) = o(u, v)$ if $u < v$ and $o^*(u, v) = \neg o(v, u)$ if $u > v$. Moreover, $F(G, \omega)$ has a *guard variable* $c(v, u)$ for every $u, v \in V(G)$, which is true if $u \leq_V v$ and vertex u has a neighbor vertex w such that $v \leq_V w$, i.e., vertex u contributes to the separation number for vertex v.

We will next provide the clauses for $F(G, \omega)$. Towards ensuring that $o^*(u, v)$ is a total ordering of $V(G)$, it is sufficient to ensure that the relation described by $o^*(u, v)$ is transitive, which is achieved by the following clauses:

$$\neg o^*(u, v) \vee \neg o^*(v, w) \vee o^*(u, w)$$
$$\text{for } u, v, w \in V(G) \text{ where } u, v, \text{ and } w \text{ are pairwise distinct.}$$

The next clauses provide the semantics for the variables $c(v, u)$. Namely, $c(v, u)$ is set to true if $u \leq_V v$ and there is an edge $\{u, w\} \in E(G)$ with $v \leq_V w$.

$$\neg o^*(u, v) \vee \neg o^*(v, w) \vee c(v, u) \qquad \text{for } v \in V(G), \{u, w\} \in E(G) \text{ and } v \neq w.$$

It remains to restrict the number of guards of each vertex set $S_{\leq_V}(v)$ given by the ordering $o^*(u, v)$. Using the variables $c(v, u)$ this is equivalent to restricting the number of variables $c(v, u)$ that are true to be at most ω for every $v \in V(G)$. Towards this aim, we again employ the sequential cardinality counter described in Subsect. 2.1. This completes the construction of the formula $F(G, \omega)$, which including the variables and clauses used for counting has $\mathcal{O}(n^2\omega)$ variables and $\mathcal{O}(n^3)$ clauses.

6 Experiments

We run the experiments on a 4-core Intel Xeon CPU E5649, 2.35 GHz, 72 GB RAM machine with Ubuntu 14.04 with each process having access to at most 8 GB RAM. For each individual SAT call we set a timeout of 1000 s and we do not impose an overall timeout for the whole process. The compilation of all encodings is implemented in C++ and we compared the performance of the encodings using the SAT solvers Minisat 2.2 (m), Glucose 4.0 (g), and MapleSAT (a). As benchmark instances we used the benchmark set of well-known named graphs from the literature [17] (previously also used in [11,14]) as well as uniformly generated instances like square grids and complete graphs. In the following we will refer to the two encodings introduced in Subsects. 3.2 and 5.1 as *partition-based encodings* (P) and to the encodings introduced in Subsects. 4.2 and 5.2 as *ordering-based encodings* (O). All our experimental results as well as the code for the compilation of our encodings can be found at https://github.com/nehal73/SATencoding.

6.1 Results

Our main experimental results are provided in Tables 1 and 3. Table 1 shows our results for the benchmark set of well-known named graphs from the literature. The benchmark set is a collection of well-known small to mid-sized graphs from the literature that has already been used in the comparison of encodings for other width measures such as clique-width [11] and branchwidth [14]. For each graph in the benchmark set we run our four encodings as well as, for comparison, the encoding for treewidth based on elimination orderings [16], using the three above mentioned SAT-solvers with the aim of computing the exact width of the graph. Namely, starting from width zero ($\omega = 0$) we increased ω by one as long as either the instance became satisfiable (in which case the current ω equals the width of the graph) or the SAT-call reached the timeout of 1000 s (in which case the current ω minus 1 is a lower bound for the width of the graph). If we reached a timeout, we further increased ω until the instance could be solved again within the timeout and returned satisfiable, thereby obtaining an upper bound for the width of the graph. In three cases (marked with an asterix in Table 1) we obtained the exact width using a longer timeout of 10000 s using the partition-based encoding for special treewidth. For each width parameter the obtained width of the graph (or an interval for the width giving the best possible lower bound and upper bound obtained by any encoding) is provided in the ω column of the table. Moreover, for special treewidth and pathwidth, the table contains the two columns (P) and (O), which show the best result obtained by any SAT-solver for the partition-based and ordering-based encodings, respectively. Namely, if the exact width of the graph could be determined, then the column shows the overall running-time in seconds (the sum of all SAT-calls) for the best SAT-solver, whose initial is given as a superscript. Otherwise the table shows the best possible interval that could be obtained within the timeout or "M.O." if every SAT-call resulted in a memory out.

Table 3 shows our results for square grids and complete graphs. The idea behind using square grids and complete graphs is that they represent two types of graphs with high treewidth, i.e., sparse and dense graphs. Moreover, square grids and complete graphs are also naturally well-suited to compare the encodings along the three considered width measures (i.e., pathwidth, special treewidth, and treewidth) as it is well-known that all three width measures coincide on square grids and complete graphs. Namely, the pathwidth, special treewidth, and treewidth of an $n \times n$-grid and a complete graph on n vertices is n and $n - 1$, respectively. For all our encodings, the table shows the largest size of square grids and complete graphs, whose width could be determined exactly within the timeout (using any of the three considered SAT-solvers). That is, starting from $n = 1$ we called each encoding for $\omega = n - 1$ and $\omega = n$ (in the case of the $n \times n$-grid) and for $\omega = n - 1$ and $\omega = n - 2$ (in the case of the complete graph on n vertices) and increased n as long as both calls completed within the timeout.

6.2   Discussion

In the case of special treewidth, our experiments indicate that the partition-based encoding is superior to the ordering-based encoding for all of the considered instances. Namely, the partition-based encoding can solve grids and complete graphs that are almost twice as large as the ones solvable using the ordering-based encoding (Table 3). The partition-based encoding almost always beats the ordering-based encoding by at least one order of magnitude on the well-known named graphs, and it also provides better lower bounds and upper bounds for the graphs that could not be solved exactly (Table 1). Overall, the partition-based encoding can be seen as the clear winner for special treewidth, which is somewhat unexpected for two reasons: (i) the ordering-based encoding is similar in spirit to the currently leading encoding for treewidth and (ii) asymptotically, the ordering-based encoding has fewer variables and almost the same number of clauses as the partition-based encoding (Table 2). It can be observed that the partition-based encoding for special treewidth is competitive with the leading encoding for treewidth, with both encodings showing advantages on different instances.

In the case of pathwidth the difference between the two encodings is far less pronounced. Whereas the ordering-based encoding has a clear advantage on the benchmark set of well-known named graph (Table 1), although far less significant than the advantage of the partition-based encoding for special treewidth, the partition-based encoding has an extraordinary advantage on complete graphs (Table 3). We note that both encodings have asymptotically the same numbers of clauses and variables (Table 2). It seems that in general the partition-based encoding has an advantage on dense graphs, whereas the ordering-based encoding

Table 1. Results for the benchmark set of the well-known named graphs.

Instance	$|V|$	$|E|$	Special treewidth			Pathwidth			Treewidth	
			ω	O	P	ω	O	P	ω	O
Petersen	10	15	5	5.03[m]	**0.48[m]**	5	**0.28[a]**	0.35[m]	4	0.15
Goldner-Harary	11	27	4	2.53[m]	**0.27[m]**	4	0.17[m]	0.17[m]	3	0.11
Grötzsch	11	20	5	4.27[m]	**0.43[m]**	5	**0.18[a]**	0.36[m]	5	0.28
Herschel	11	18	4	3.32[m]	**0.34[m]**	4	**0.17[a]**	0.20[m]	3	0.14
Chvátal	12	24	6	11.09[m]	**0.92[m]**	6	**0.44[g]**	0.69[a]	6	0.61
Dürer	12	18	4	7.28[a]	**0.63[m]**	4	**0.13[m]**	0.33[m]	4	0.25
Franklin	12	18	5	12.30[a]	**1.40[m]**	5	**0.30[m]**	0.60[m]	4	0.30
Frucht	12	18	4	7.56[g]	**0.71[m]**	4	**0.21[g]**	0.31[a]	3	0.12
Tietze	12	18	5	11.34[m]	**1.26[m]**	5	**0.27[m]**	0.53[m]	4	0.21
Paley13	13	39	8	22.13[m]	**1.16[m]**	8	**1.02[a]**	1.23[m]	8	2.60
Poussin	15	39	6	61.07[a]	**1.65[m]**	6	**0.39[m]**	0.65[m]	6	0.37
Clebsch	16	40	9	234.28[m]	**13.20[m]**	9	25.76[a]	**17.17[a]**	8	6.30
4 × 4-grid	16	24	4	97.98[m]	**1.13[m]**	4	**0.22[a]**	0.39[m]	4	0.28
Hoffman	16	32	7	204.73[a]	**20.22[m]**	7	**6.30[g]**	8.21[m]	6	2.39
Shrikhande	16	48	9	234.76[m]	**10.42[m]**	9	11.78[a]	**8.04[m]**	9	131.11
Sousselier	16	27	5	127.87[m]	**3.33[m]**	5	**0.24[m]**	0.62[m]	5	0.31
Errera	17	45	6	153.83[a]	**2.78[m]**	6	**0.40[m]**	0.76[m]	6	0.49
Paley17	17	68	12	504.54[a]	**15.76[m]**	12	106.99[a]	**27.52[a]**	11	35.23
Pappus	18	27	7	912.69[a]	**438.24[g]**	7	**16.47[g]**	54.62[g]	6	160.90
Robertson	19	38	8	1082.73[a]	**130.26[m]**	8	**11.84[g]**	36.02[g]	8	307.21
Desargues	20	30	6	1349.67[m]	**237.57[g]**	6	**0.84[m]**	10.16[m]	6	324.21
Dodecahedron	20	30	6	1564.23[a]	**337.20[g]**	6	**4[g]**	38.52[g]	4–6	4–6
FlowerSnark	20	30	6	1352.67[m]	**201.40[g]**	6	**1.04[m]**	10.99[m]	6	400.06
Folkman	20	40	7	1434.93[a]	**130.20[m]**	7	**2.84[g]**	23.15[m]	6	10.87
Brinkmann	21	42	8	2548.46[m]	**354.62[m]**	8	**14.85[g]**	63.71[g]	8	593.45
Kittell	23	63	7	160.33[g]	**24.70[m]**	7	**1.05[m]**	8.28[m]	7	4.38
McGee	24	36	8*	5–8	5–8	8	**62.47[a]**	524.21[g]	5–7	5–7
Nauru	24	36	8*	5–8	5–8	8	**181.73[a]**	6–8	6	457.92
Holt	27	54	10*	7–10	6–10	10	**386.16[a]**	8–10	7–9	7–9
Watsin	50	75	3–8	M.O.	3–8	7	**76.77[m]**	5–7	4–7	4–7
B10Cage	70	106	2–20	M.O.	2–20	8-16	8-16	6–16	4–17	4–17
Ellingham	78	117	3–9	M.O.	3–9	6	**22.88[m]**	5–7	4–6	4–6

Table 2. The number of variables and clauses for our four encodings in terms of the number n of vertices, the number m of edges, and the width ω

	sptw		pw	
	Vars	Cls	Vars	Cls
P	$\mathcal{O}(n^3\omega)$	$\mathcal{O}(n^4+mn^3)$	$\mathcal{O}(n^2\omega)$	$\mathcal{O}(n^3)$
O	$\mathcal{O}(n^2\omega)$	$\mathcal{O}(n^5)$	$\mathcal{O}(n^2\omega)$	$\mathcal{O}(n^3)$

Table 3. Experimental results for square-grids and complete graphs: number of vertices of largest graphs solved within the timeout

Graphs	sptw		pw	
	O	P	O	P
Square grids	16	36	81	64
Complete graphs	34	76	26	123

is better suited for sparse graphs. The results seem to indicate that different encodings should be employed for different classes of graphs. This underlines the importance of developing different encodings for the same width parameter and encourages the development of a portfolio-based approach for SAT-encodings.

Moreover, we would like to mention a few general observations concerning the performance of the three SAT-solvers. Generally the differences in the performance of the three SAT-solvers were quite minor over all encodings. In particular, the conclusions drawn about the comparison of the different encodings were the same for each of the three SAT-solvers. With respect to the special treewidth encodings, it can be inferred from Table 1 that MiniSAT has the best performance for more instances than Glucose or MapleSAT. However, we observed that Glucose was the most robust among the three solvers, since there are instances that could only be solved by Glucose and all instances that could be solved by any of the solvers could also be solved by Glucose. With respect to the pathwidth encodings, the differences between the solvers is less pronounced, each having advantages on about the same number of instances.

We also conducted initial experiments on random graphs, whose results (due to space limitations) can only be found in our github repository. Namely, we tested all our encodings on random graphs with 20, 40, and 60 vertices and edge probabilities $0.1, 0.2, \ldots, 0.9$. For each setting we generated 10 random graphs and reported the average running time for each of our encodings (we used a timeout of 2000s per SAT-call as well as an overall timeout of 6 h). Our results on random graphs strongly support our conclusions reported above concerning the relative performance of the various encodings.

7   Conclusion

We compared two SAT encodings for special tree width and pathwidth respectively. For the former we introduced two novel characterizations which might be of independent interest. Based on these characterizations for special treewidth and two related characterizations for pathwidth, we developed and empirically compared SAT-encodings for the computation of special treewidth and pathwidth. Our empirical results emphasize that the performance of SAT-encodings can strongly depend on the underlying characterization. Interestingly, for special treewidth, a partition-based encoding far outperforms an ordering-based encoding, although the latter encoding is closely related to the currently leading encoding for the prominent width parameter treewidth. It is only natural to ask whether a similar partition-based approach can be fruitful for treewidth. Moreover, for pathwidth, we obtained two SAT-encodings which both perform well, each of them having an advantage on different classes of instances; thus suggests a portfolio-based approach.

Acknowledgments. The authors kindly acknowledge the support by the Austrian Science Fund (FWF, projects W1255-N23 and P-26200).

References

1. Arnborg, S., Corneil, D.G., Proskurowski, A.: Complexity of finding embeddings in a k-tree. SIAM J. Algebr. Discret. Method. **8**(2), 277–284 (1987)
2. Berg, J., Järvisalo, M.: SAT-based approaches to treewidth computation: An evaluation. In: 26th IEEE International Conference on Tools with Artificial Intelligence, ICTAI 2014, Limassol, Cyprus, 10–12 November 2014, pp. 328–335. IEEE Computer Society (2014)
3. Biedl, T., Bläsius, T., Niedermann, B., Nöllenburg, M., Prutkin, R., Rutter, I.: Using ILP/SAT to determine pathwidth, visibility representations, and other grid-based graph drawings. In: Wismath, S., Wolff, A. (eds.) GD 2013. LNCS, vol. 8242, pp. 460–471. Springer, Cham (2013). doi:10.1007/978-3-319-03841-4_40
4. Bodlaender, H.L., Kratsch, S., Kreuzen, V.J.C.: Fixed-parameter tractability and characterizations of small special treewidth. In: Brandstädt, A., Jansen, K., Reischuk, R. (eds.) WG 2013. LNCS, vol. 8165, pp. 88–99. Springer, Heidelberg (2013). doi:10.1007/978-3-642-45043-3_9
5. Bodlaender, H.L., Kratsch, S., Kreuzen, V.J., Kwon, O.J., Ok, S.: Characterizing width two for variants of treewidth (part 1). Discr. Appl. Math. **216**, 29–46 (2017)
6. Courcelle, B.: Special tree-width and the verification of monadic second-order graph properties. In: Lodaya, K., Mahajan, M. (eds) 2010 IARCS Annual Conference on Foundations of Software Technology and Theoretical Computer Science, FSTTCS, LIPIcs, Chennai, India, vol. 8, pp. 13–29. Schloss Dagstuhl - Leibniz-Zentrum fuer Informatik, 15–18 Dec 2010
7. Courcelle, B.: On the model-checking of monadic second-order formulas with edge set quantifications. Discr. Appl. Math. **160**(6), 866–887 (2012)
8. Dell, H., Rosamond, F.: The 1st parameterized algorithms and computational experiments challenge–track A: Treewidth. Technical report (2016). https://pacechallenge.wordpress.com/2016/09/12/here-are-the-results-of-the-1st-pace-challenge/
9. Diestel, R.: Graph Theory: Graduate Texts in Mathematics, 2nd edn. Springer Verlag, New York (2000)
10. Habib, M., Möhring, R.H.: Treewidth of cocomparability graphs and a new order-theoretic parameter. Order **1**, 47–60 (1994)
11. Heule, M., Szeider, S.: A SAT approach to clique-width. ACM Trans. Comput. Log. **16**(3), 24 (2015)
12. Kinnersley, N.G.: The vertex separation number of a graph equals its path-width. Inf. Process. Lett. **42**(6), 345–350 (1992)
13. Kloks, T.: Treewidth: Computations and Approximations. Springer Verlag, Berlin (1994)
14. Lodha, N., Ordyniak, S., Szeider, S.: A SAT approach to branchwidth. In: Creignou, N., Le Berre, D. (eds.) SAT 2016. LNCS, vol. 9710, pp. 179–195. Springer, Cham (2016). doi:10.1007/978-3-319-40970-2_12
15. Robertson, N., Seymour, P.D.: Graph minors. I. excluding a forest. J. Combin. Theory Ser. B **35**(1), 39–61 (1983)
16. Samer, M., Veith, H.: Encoding treewidth into SAT. In: Kullmann, O. (ed.) SAT 2009. LNCS, vol. 5584, pp. 45–50. Springer, Heidelberg (2009). doi:10.1007/978-3-642-02777-2_6
17. Weisstein, E.: MathWorld online mathematics resource (2016). http://mathworld.wolfram.com

Tool Papers

MaxPre: An Extended MaxSAT Preprocessor

Tuukka Korhonen, Jeremias Berg, Paul Saikko, and Matti Järvisalo$^{(\boxtimes)}$

HIIT, Department of Computer Science, University of Helsinki, Helsinki, Finland
matti.jarvisalo@helsinki.fi

Abstract. We describe MaxPre, an open-source preprocessor for (weighted partial) maximum satisfiability (MaxSAT). MaxPre implements both SAT-based and MaxSAT-specific preprocessing techniques, and offers solution reconstruction, cardinality constraint encoding, and an API for tight integration into SAT-based MaxSAT solvers.

1 Introduction

We describe MaxPre, an open-source preprocessor for (weighted partial) maximum satisfiability (MaxSAT). MaxPre implements a range of well-known and recent SAT-based preprocessing techniques as well as MaxSAT-specific techniques that make use of weights of soft clauses. Furthermore, MaxPre offers solution reconstruction, cardinality constraint encoding, and an API for integration into SAT-based MaxSAT solvers without introducing unnecessary assumptions variables within the SAT solver. In this paper we overview the implemented techniques, implementation-level decisions, and usage of MaxPre, and give a brief overview of its practical potential. The system, implemented in C++, is available in open source under the MIT license via https://www.cs.helsinki.fi/group/coreo/maxpre/.

Due to space limitations, we will assume familiarity with conjunctive normal form (CNF) formulas and satisfiability. An instance of (weighted partial) maximum satisfiability (MaxSAT) consists of two CNF formulas, the hard clauses F_h and the soft clauses F_s, and a weight function w associating a non-negative weight with each soft clause. A truth assignment that satisfies the hard clauses is a solution, and is optimal if it minimizes cost, i.e., the weight of the soft clauses left unsatisfied, over all solutions.

The preprocessing flow of MaxPre is illustrated in Fig. 1. Given a MaxSAT instance in DIMACS format (extended with cardinality constraints, see Sect. 2.1), MaxPre starts by rewriting all cardinality constraints to clauses (step 1). MaxPre then enters the first preprocessing loop (step 2), using only techniques that are directly sound for MaxSAT (see Sect. 2.2). The sound use of SAT-based preprocessing more generally requires extending each (remaining) soft clause C with a fresh label variable l_C to form $C \vee l_C$ and then restricting

Work supported by Academy of Finland (grants 251170 COIN, 276412, and 284591) and DoCS Doctoral School in Computer Science and Research Funds of the University of Helsinki.

S. Gaspers and T. Walsh (Eds.): SAT 2017, LNCS 10491, pp. 449–456, 2017.
DOI: 10.1007/978-3-319-66263-3_28

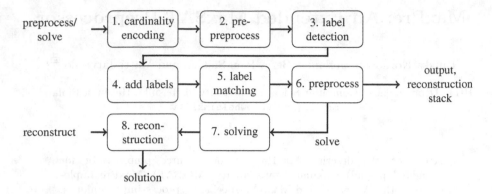

Fig. 1. MaxPre preprocessing flow

the preprocessor from resolving on the added labels [4,5]. Labelling of the soft clauses with assumption variables is done in steps 3–5. First (step 3) MaxPre applies group detection [7] to identify literals in the input that can be directly used as labels. All soft clauses that remain without a label are then given one (step 4); during the rest of the preprocessing, all clauses are treated as hard and the weight w_C of a soft clause C is associated with the label l_C attached to C. After each clause is instrumented with a label, a novel technique *label matching* (step 5, see Sect. 2.2) is used to identify labels which can be substituted with other labels in the instance. The main preprocessing loop (step 6) allows for applying all implemented techniques. Depending on how MaxPre was invoked (see Sect. 2.3), MaxPre will then either output the resulting preprocessed MaxSAT instance or invoke a given external MaxSAT solver binary on the instance. In the former case, MaxPre will provide a solution reconstruction stack in a separate file. In the latter, MaxPre will internally apply solution reconstruction on the solution output by the solver. We will give more details on the internals, usage, and API of MaxPre, as well as a brief overview of its performance in practice.

2 Supported Techniques

2.1 Cardinality Constraints

MaxPre offers cardinality network [2] based encodings of cardinality constraints, encoded as clauses (Step 1) before preprocessing. This allows the user to specify cardinality constraints over WCNF DIMACS literals as an extension of the standard WCNF MaxSAT input format. Cardinality constraints can be specified using lines of form CARD $l_1\ l_2\ \ldots\ l_n \circ K$, where each l_i is a literal of the formula, $\circ \in \{<,>,<=,>=,=,!=.\}$, and $K \in \mathbb{N}$. The constraints are encoded as hard clauses enforcing $\sum_{i=1}^n l_i \circ K$. MaxPre can also encode the truth value of $\sum_{i=1}^n l_i \circ K$ to a specific literal L. A line of form CARD $l_1\ l_2\ \ldots\ l_n \circ K$ OUT L is rewritten as $L \to \sum_{i=1}^n l_i \circ K$. Lines of form CARD $l_1\ l_2\ \ldots\ l_n \circ K$ OUT L IFF extend the implication to an equivalence, and are rewritten as $L \leftrightarrow \sum_{i=1}^{n} l_i \circ K$. Additionally, direct control on the output literals of the cardinality networks can be provided. A line of form CARD $l_1 \; l_2 \; \ldots \; l_n \; \square \; K$ OUT $o_1 \; \ldots \; o_K$ is encoded as a K-cardinality network [2] where $o_1, \ldots, o_K$ are the output literals of the network and $\square \in \{<:, >:, ::\}$. If $\square$ is $<:$, the network is encoded so that $\neg o_i \rightarrow \sum_{i=1}^{n} l_i < i$ for any i. If $\square$ is $>:$, the network is encoded so that $o_i \rightarrow \sum_{i=1}^{n} l_i \geq i$; and if $\square = ::$, the network is encoded so that $o_i \leftrightarrow \sum_{i=1}^{n} l_i \geq i$.

2.2 Preprocessing

SAT-Based Preprocessing. In addition to removing tautologies, soft clauses with 0 weight and duplicate clauses, MaxPre implements the following SAT-based techniques: unit propagation, bounded variable elimination (BVE) [11], subsumption elimination (SE), self-subsuming resolution (SSR) [11,13,17], blocked clause elimination (BCE) [15], Unhiding [14] (including equivalent literal substitution [1,9,18,24]), and bounded variable addition (BVA) [20]. In terms of implementation details, SE and SSR use three different approaches depending on the number of clauses they have to process. The asymptotically worst approach of checking all pairs of clauses can be improved by computing hashes of clauses, and by the asymptotically best approach using the AMS-Lex algorithm [3], implemented in MaxPre based on [21]. The average time complexity of AMS-Lex in finding subsumed clauses seems to be nearly linear (dominated by sorting). We have observed that in practice the BVA implementation can be in cases significantly faster than the implementation given in the original paper; this is achieved by using polynomial hashes of clauses. In contrast to using time stamping (directed spanning trees) as in the original work on Unhiding [14], MaxPre implements Unhiding using undirected spanning trees, which can be provably more effective in terms of the covered binary implications.

MaxSAT-Specific Techniques. MaxPre also includes the following (to the best of our knowledge unique) combination of MaxSAT-specific techniques that work directly on the label variables (recall Fig. 1). The first two are techniques meant to decrease the total number of fresh label variables that are introduced into the formula. Assume MaxPre is invoked on an instance $F = (F_h, F_s, w)$.

Group detection [7] (Step 3) Any literal l for which $(\neg l) \in F_s$, $l \notin C$ for any $C \in (F_s \setminus \{(\neg l)\})$ and $\neg l \notin C$ for any $C \in F_h \cup (F_s \setminus \{(\neg l)\})$ can be directly used as a label [7].

Label matching (Step 5) Label l_C is *matched*, i.e., substituted, with l_D if (i) $w_C = w_D$, (ii) l_C only appears in a single soft clause C, (iii) l_D only appears in a single soft clause D, and (iv) $C \vee D$ is a tautology. MaxPre implements label matching by computing a maximal matching using a standard greedy algorithm.

Group-subsumed label elimination (GSLE) Generalizing SLE [8], label l_D is subsumed by a group of labels $L = \{l_{C_1}, \ldots, l_{C_n}\}$ if for some $l_{C_i} \in L$, we have $l_{C_i} \in C$ whenever $l_D \in C$, and $\sum_{l_{C_i} \in L} w_{C_i} \leq w_D$. GSLE removes group-subsumed labels. SLE corresponds to GSLE with $n = 1$. Since GSLE corresponds to the NP-complete hitting set problem, MaxPre implements an approximate GSLE via a slightly modified version of a classical $ln(n)$–approximation algorithm for the hitting set problem [10,23].

Binary core removal (BCR) is the MaxSAT-equivalent of Gimpel's reduction rule for the binate covering problem [12]. Assume labels l_C, l_D with $w_C = w_D$ and the clause $(l_C \vee l_D)$. Let $F_{l_C} = \{C_i \mid l_C \in C_i\}$ and assume that (i) $F_{l_C} \cap F_{l_D} = \{(l_C \vee l_D)\}$, (ii) $|F_{l_C}| > 1$, and (iii) $|F_{l_D}| > 1$. BCR replaces the clauses in $F_{l_C} \cup F_{l_D}$ with the non-tautological clauses in $\{(C_i \vee D_j) \setminus \{l_C\} \mid C_i \in F_{l_C} \setminus (l_C \vee l_D), D_j \in F_{l_D} \setminus (l_C \vee l_D)\}$. MaxPre applies BCR whenever the total number of clauses in the formula does not increase. Notice that given l_C, l_D with $w_C = w_D$ and a clause $(l_C \vee l_D)$, assumptions (i)–(iii) for BCR follow by applying SE and SLE.

Structure-based labeling Given a label l and a clause C s.t. C is blocked (in terms of BCE) when assuming l to true, structure-based labelling replaces C by $C \vee l$. The correctness of structure-based labelling is based on the invariant that a clause C is redundant whenever l is true.

2.3   Options and Usage

MaxPre is called from the command line. Full details on command line options are available via `./maxpre -h`. One of the directives `preprocess`, `solve`, `reconstruct` needs to be specified after the input file. `preprocess` and `solve` assume a single input file containing a WCNF MaxSAT instance (possibly with cardinality constraints). Further, `solve` expects the solver binary and its command line arguments to be given via the `-solver` and `-solverflags` options. `reconstruct` expects as input a solution to a WCNF MaxSAT instance and the corresponding reconstruction stack file.

Solution Reconstruction. To map a solution of a preprocessed instance to a solution of the original instance, the reconstruction stack file produced by MaxPre needs to be specified via `-mapfile`. For example, to obtain an original solution from a solution `sol0.sol` of the preprocessed instances using the reconstruction stack in `input.map`, use `./maxpre sol0.sol reconstruct -mapfile=input.map`. To obtain the mapfile when preprocessing, `-mapfile` should be used in conjunction with `preprocess`.

Specifying Preprocessing Techniques. Following [19], MaxPre allows for specifying the order in which individual preprocessing techniques are applied via a technique string The default application order is specified by the string $[bu]\#[buvsrgc]$, i.e. to run BCE and UP in the first preprocessing loop and BCE, UP, BVE, SSR, GSLE and BCR in the second.

Enforcing Time Limits and Bounds. The running time of MaxPre is limited using the `-timelimit` option: e.g., `-timelimit=60` sets the time limit to 60 s. This limits the running time of preprocessing techniques somewhat independently of each other. Each technique gets allocated a proportion of the time to use, and if a technique leaves some of its time unused, it is dynamically real-located to other techniques. By default no time limits are imposed. Another way to prevent MaxPre from wasting efforts on the potentially more time-consuming techniques such as BVE via -skiptechnique: e.g., with -skiptechnique=100, MaxPre first tries to apply each of such preprocessing technique to 100 random variables/literals in a touched list. If no simplifications occur, MaxPre will subsequently skip the particular preprocessing technique. By default -skiptechnique is not enforced.

2.4 API

MaxPre is implemented in a modular fashion, and offers an API for tight integration with MaxSAT solvers. Via the API, the solver becomes aware of labels used for preprocessing which can be directly used as assumptions in SAT-based MaxSAT solving; without this, the solvers will add a completely redundant new layer of assumption variables to the soft clauses before search [6]. Unnecessary file I/O is also avoided.

To use MaxPre via its API, create a `PreprocessorInterface` object for the MaxSAT instance, call `preprocess` to preprocess, and then `getInstance` to obtain the preprocessed instance. After solving, an original solution is reconstructed by calling `reconstruct`. `PreprocessorInterface` encapsulates the preprocessing trace (mapfile) and maps/unmaps variables to/from internal preprocessor variable indexing. For more concreteness, `main.cpp` implements MaxPre using these API methods, and serves as an example of their use. The API handles literals as `int`-types and clauses as C++ standard library vectors of literals. The most central parts of the API are the following.

`PreprocessorInterface` constructs a `PreprocessorInterface` object from a vector of clauses, a vector of their weights and a top weight, which is used to identify hard clauses.

`preprocess` takes a technique string and preprocesses the instance using given techniques.

`getInstance` returns (by reference) the preprocessed instance as vectors of clauses, weights, and labels.

`reconstruct` takes a model for the preprocessed instance and returns a model for the original instance.

`setSkipTechnique` corresponds to -skiptechnique command line flag.

`print*` methods print solutions, instances, mapfiles, or logs to an output stream.

3   Experiments

While an extended empirical evaluation of different components of MaxPre is impossible within this system description, we shortly discuss the potential of MaxPre in practice. For the experiments, we used all of the 5425 partial and weighted partial MaxSAT instances collected and made available by the 2008–2016 MaxSAT Evaluations. Figure 2 show a comparison of MaxPre and the Coprocessor 2.0 [19] preprocessor in terms of the number of variables and clauses, and the sums of weights of the soft clauses, in the output instance. Here we used Coprocessor by first adding a label to each soft clauses, and used the "white listing" option of Coprocessor on the labels to maintain soundness of preprocessing for MaxSAT. The SAT-based techniques used with Coprocessor and MaxPre are the same; the preprocessing loop used in coprocessor consisted of BVE, pure literal elimination, UP, SE, SSR and BCE. MaxPre additionally applied the MaxSAT-specific techniques shortly described earlier in this paper; the first preprocessing loop used BCE and UP and the second BCE (which also removes pure literals), UP, BVE, SE, SSR, GSLE and BCR. MaxPre provides noticeably more simplifications in terms of these parameters, while the

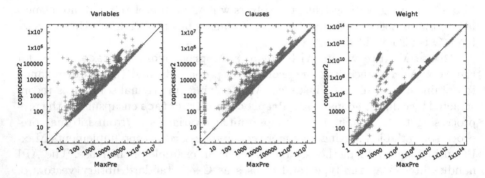

Fig. 2. Comparison of MaxPre and Coprocessor in terms of preprocessing effects: number of clauses and variables, and the sum of the weights of soft clauses, in the preprocessed instances.

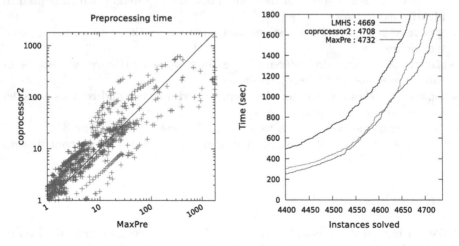

Fig. 3. Left: Comparison of preprocessing times: MaxPre vs Coprocessor. Right: Performance impact on LMHS.

preprocessing times as comparable (see Fig. 3 left); in fact, MaxPre performs faster on a majority of the instances. The few timeouts observed for MaxPre are on very large instances (>10M clauses) which are in fact not solved by current state-of-the-art MaxSAT solvers. Coprocessor contains fixed constants which switch off preprocessing on very large instances; while we did not enforce any limits on MaxPre for this experiment, the `-timelimit` and `-skiptechniques` options would enable faster preprocessing on such very large intances. In terms of potential impact on solver performance, our LMHS SAT-IP MaxSAT solver [22] previously integrated Coprocessor 2.0 as its internal preprocessor. Figure 3 right shows that, although the effects are mild, replacing Coprocessor with MaxPre in LMHS improves its performance.

4 Conclusions

We introduced MaxPre, an open-source MaxSAT preprocessor with extended capabilities, including cardinality constraint encodings and MaxSAT-specific simplification techniques not implemented in the current (Max)SAT preprocessors. The API of MaxPre allows for tight integration with MaxSAT solvers, e.g., avoiding unnecessary introduction of assumption variables after preprocessing, and potentially opening up further avenues for inprocessing MaxSAT solving [16]. Empirical results suggest that MaxPre is a viable option for integrating preprocessing into MaxSAT solvers.

References

1. Aho, A., Garey, M., Ullman, J.: The transitive reduction of a directed graph. SIAM J. Comput. **1**(2), 131–137 (1972)
2. Asín, R., Nieuwenhuis, R., Oliveras, A., Rodríguez-Carbonell, E.: Cardinality networks: a theoretical and empirical study. Constraints **16**(2), 195–221 (2011)
3. Bayardo, R.J., Panda, B.: Fast algorithms for finding extremal sets. In: Proceedings of SDM, pp. 25–34. SIAM/Omnipress (2011)
4. Belov, A., Järvisalo, M., Marques-Silva, J.: Formula preprocessing in MUS extraction. In: Piterman, N., Smolka, S.A. (eds.) TACAS 2013. LNCS, vol. 7795, pp. 108–123. Springer, Heidelberg (2013). doi:10.1007/978-3-642-36742-7_8
5. Belov, A., Morgado, A., Marques-Silva, J.: SAT-based preprocessing for MaxSAT. In: McMillan, K., Middeldorp, A., Voronkov, A. (eds.) LPAR 2013. LNCS, vol. 8312, pp. 96–111. Springer, Heidelberg (2013). doi:10.1007/978-3-642-45221-5_7
6. Berg, J., Saikko, P., Järvisalo, M.: Improving the effectiveness of SAT-based preprocessing for MaxSAT. In: Proceedings of IJCAI, pp. 239–245. AAAI Press (2015)
7. Berg, J., Saikko, P., Järvisalo, M.: Re-using auxiliary variables for MaxSAT preprocessing. In: Proceedings of ICTAI, pp. 813–820. IEEE Computer Society (2015)
8. Berg, J., Saikko, P., Järvisalo, M.: Subsumed label elimination for maximum satisfiability. In: Proceedings of ECAI. Frontiers in Artificial Intelligence and Applications, vol. 285, pp. 630–638. IOS Press (2016)
9. Brafman, R.: A simplifier for propositional formulas with many binary clauses. IEEE Trans. Syst. Man Cybern. Part B **34**(1), 52–59 (2004)

10. Chvatal, V.: A greedy heuristic for the set-covering problem. Math. Oper. Res. **4**(3), 233–235 (1979)
11. Eén, N., Biere, A.: Effective preprocessing in SAT through variable and clause elimination. In: Bacchus, F., Walsh, T. (eds.) SAT 2005. LNCS, vol. 3569, pp. 61–75. Springer, Heidelberg (2005). doi:10.1007/11499107_5
12. Gimpel, J.F.: A reduction technique for prime implicant tables. In: Proceedings of SWCT, pp. 183–191. IEEE Computer Society (1964)
13. Groote, J., Warners, J.: The propositional formula checker HeerHugo. J. Autom. Reasoning **24**(1/2), 101–125 (2000)
14. Heule, M.J.H., Järvisalo, M., Biere, A.: Efficient CNF simplification based on binary implication graphs. In: Sakallah, K.A., Simon, L. (eds.) SAT 2011. LNCS, vol. 6695, pp. 201–215. Springer, Heidelberg (2011). doi:10.1007/978-3-642-21581-0_17
15. Järvisalo, M., Biere, A., Heule, M.: Blocked clause elimination. In: Esparza, J., Majumdar, R. (eds.) TACAS 2010. LNCS, vol. 6015, pp. 129–144. Springer, Heidelberg (2010). doi:10.1007/978-3-642-12002-2_10
16. Järvisalo, M., Heule, M.J.H., Biere, A.: Inprocessing rules. In: Gramlich, B., Miller, D., Sattler, U. (eds.) IJCAR 2012. LNCS (LNAI), vol. 7364, pp. 355–370. Springer, Heidelberg (2012). doi:10.1007/978-3-642-31365-3_28
17. Korovin, K.: iProver – an instantiation-based theorem prover for first-order logic (system description). In: Armando, A., Baumgartner, P., Dowek, G. (eds.) IJCAR 2008. LNCS, vol. 5195, pp. 292–298. Springer, Heidelberg (2008). doi:10.1007/978-3-540-71070-7_24
18. Li, C.: Integrating equivalency reasoning into Davis-Putnam procedure. In: Proceedings of AAAI, pp. 291–296 (2000)
19. Manthey, N.: Coprocessor 2.0 – a flexible CNF simplifier. In: Cimatti, A., Sebastiani, R. (eds.) SAT 2012. LNCS, vol. 7317, pp. 436–441. Springer, Heidelberg (2012). doi:10.1007/978-3-642-31612-8_34
20. Manthey, N., Heule, M.J.H., Biere, A.: Automated reencoding of boolean formulas. In: Biere, A., Nahir, A., Vos, T. (eds.) HVC 2012. LNCS, vol. 7857, pp. 102–117. Springer, Heidelberg (2013). doi:10.1007/978-3-642-39611-3_14
21. Marinov, M., Nash, N., Gregg, D.: Practical algorithms for finding extremal sets. J. Exp. Algorithmics **21**, Article 1.9 (2016)
22. Saikko, P., Berg, J., Järvisalo, M.: LMHS: a SAT-IP hybrid MaxSAT solver. In: Creignou, N., Le Berre, D. (eds.) SAT 2016. LNCS, vol. 9710, pp. 539–546. Springer, Cham (2016). doi:10.1007/978-3-319-40970-2_34
23. Slavík, P.: A tight analysis of the greedy algorithm for set cover. In: Proceedings of STOC, pp. 435–441. ACM (1996)
24. Van Gelder, A.: Toward leaner binary-clause reasoning in a satisfiability solver. Ann. Math. Artif. Intell. **43**(1), 239–253 (2005)

The GRAT Tool Chain

Efficient (UN)SAT Certificate Checking with Formal Correctness Guarantees

Peter Lammich[✉]

Technische Universität München, Munich, Germany
`lammich@in.tum.de`

Abstract. We present the GRAT tool chain, which provides an efficient and formally verified SAT and UNSAT certificate checker. It utilizes a two phase approach: The highly optimized gratgen tool converts a DRAT certificate to a GRAT certificate, which is then checked by the formally verified gratchk tool.

On a realistic benchmark suite drawn from the 2016 SAT competition, our approach is faster than the unverified standard tool drat-trim, and significantly faster than the formally verified LRAT tool. An optional multithreaded mode allows for even faster checking of a single certificate.

1 Introduction

The complexity and high optimization level of modern SAT solvers makes them prone to bugs, and at the same time hard to (formally) verify. A common approach in such situations is certification, i. e. to make the SAT solver produce a certificate for its output, which can then be checked independently by a simpler algorithm. While SAT certificates describe a satisfying assignment and are straightforward to check, UNSAT certificates are more complex. The de facto standard are DRAT certificates [15] checked by drat-trim [3]. However, efficiently checking a DRAT certificate still requires a quite complex and highly optimized implementation.[1] A crucial idea [2] is to split certificate checking into two phases: The first phase produces an enriched certificate, which is then checked by the second phase. This effectively shifts the computationally intensive and algorithmically complex part of checking to the first phase, while the second phase is both computationally cheap and algorithmically simple, making it amenable to formal verification.

Cruz-Filipe et al. [2] originally implemented this approach for the weaker DRUP certificates [14], and later extended it to DRAT certificates [1,6], obtaining the LRAT tool chain. Independently, the author also extended the approach to DRAT certificates [9]. While Cruz-Filipe et al. use an extended version of drat-trim to enrich the certificates, the author implemented the specialized gratgen tool for that purpose. Compared to drat-trim, gratgen's distinguishing feature is

---

[1] We found several bugs in drat-trim. Most of them are already fixed [4,9].

© Springer International Publishing AG 2017
S. Gaspers and T. Walsh (Eds.): SAT 2017, LNCS 10491, pp. 457–463, 2017.
DOI: 10.1007/978-3-319-66263-3_29

its support for multithreading, allowing it to generate enriched certificates several times faster at the cost of using more CPU time and memory. Moreover, we have implemented some novel optimizations, making gratgen faster than drattrim even in single-threaded mode. While [9] focuses on the formal verification of the certificate checker (gratchk), this paper focuses on gratgen. The GRAT tools and raw benchmark data are available at http://www21.in.tum.de/~lammich/grat/.

2   The GRAT Toolchain

To obtain a formally verified solution for a CNF formula, it is first given to a SAT solver. If the formula is satisfiable, the SAT-solver outputs a valuation of the variables, which is then used by gratchk to verify that the formula is actually satisfiable. If the formula is unsatisfiable, the SAT-solver outputs a DRAT certificate. This is processed by gratgen to produce a GRAT certificate, which, in turn, is used by gratchk to verify that the formula is actually unsatisfiable. We have formally proved that gratchk only accepts satisfiable/unsatisfiable formulas.

2.1   DRAT Certificates

A DRAT certificate [15] is a list of clause addition and deletion items. Clause addition items are called *lemmas*. The following pseudocode illustrates the *forward checking* algorithm for DRAT certificates:

```
F := F₀   // F₀ is CNF formula to be certified as UNSAT
F := unitprop(F); if F == conflict then exit "s UNSAT"

for item in certificate do
  case item of
    delete C => F := remove_clause(F,C)
  | add C =>
      if not hasRAT(C,F) then exit "s ERROR Lemma doesn't have RAT"
      F := F ∧ C
      F := unitprop(F); if F == conflict then exit "s UNSAT"

exit "s ERROR Certificate did not yield a conflict"
```

The algorithm maintains the invariant that F is satisfiable if the initial CNF formula F_0 is satisfiable. Deleting a clause and unit propagation obviously preserve this invariant. When adding a clause, the invariant is ensured by the clause having the RAT property. The algorithm only reports UNSAT if F has clearly become unsatisfiable, which, by the invariant, implies unsatisfiability of F_0. A clause C has the *RAT property* wrt. the formula F iff there is a *pivot literal* $l \in C$, such that for all *RAT candidates* $D \in F$ with $\neg l \in D$, we have $(F \wedge \neg(C \cup D \setminus \{\neg l\}))^{\mathrm{u}} = \{\emptyset\}$. Here, F^{u} denotes the unique result of unit propagation, where we define $F^{\mathrm{u}} = \{\emptyset\}$ if unit propagation yields a conflict. Exploiting that $(F \wedge \neg(C \cup D))^{\mathrm{u}}$ is equivalent to $((F \wedge \neg C)^{\mathrm{u}} \wedge \neg D)^{\mathrm{u}}$, the candidates do not have to be checked if the first unit propagation $(F \wedge \neg C)^{\mathrm{u}}$ already yields a conflict. In this case, the lemma has the *RUP property*. This optimization is essential, as most lemmas typically have RUP, and gathering the list of RAT candidates is expensive.

2.2 GRAT Certificates

The most complex and expensive operation in DRAT certificate checking is unit propagation,[2] and highly optimized implementations like two watched literals [11] are required for practically efficient checkers. The main idea of enriched certificates [2] is to make unit propagation output a sequence of the identified unit and conflict clauses. The enriched certificate checker simulates the forward checking algorithm, verifying that the clauses from the certificate are actually unit/conflict, which is both cheaper and simpler than performing fully fledged unit propagation. For RAT lemmas, the checker also has to verify that all RAT candidates have been checked.

A GRAT certificate consists of a lemma and a proof part. The lemma part contains a list of lemmas to be verified by the forward checking algorithm, and the proof part contains the unit and conflict clauses, deletion information, and counters how often each literal is used as a pivot in a RAT proof.

The lemma part is stored as a text file roughly following DIMACS CNF format, and the proof part is a binary file in a proprietary format. The gratchk tool completely reads the lemmas into memory, and then streams over the proof during simulating the forward checking algorithm. We introduced the splitting of lemmas and proof after gratchk ran out of memory for some very large certificates.[3]

2.3 Generating GRAT Certificates

Our gratgen tool reads a DIMACS CNF formula and a DRAT certificate, and produces a GRAT certificate. Instead of the simple forward checking algorithm, it uses a multithreaded backwards checking algorithm, which is outlined below:

```
fun forward_phase:
  F := unitprop(F); if F == conflict then exit "s UNSAT"

  for item in certificate do
    case item of
      delete C => F := remove_clause(F,C)
    | add C =>
        F := F ∧ C
        F := unitprop(F);
        if F == conflict then truncate certificate; return

  exit "s ERROR Certificate did not yield a conflict"
```

---

[2] We found that more than 90% of the execution time is spent on unit propagation.

[3] LRAT [6] uses a similar streaming optimization, called *incremental mode*.

```
fun backward_phase(F):
  for item in reverse(certificate) do
    case item of
      delete C => F := F ∧ C
    | add C =>
        remove_clause(F,C); undo_unitprop(F,C)
        if is_marked(C) && acquire(C) then
          if not hasRAT(C,F) then exit "s ERROR Lemma doesn't have RAT"

fun main:
  F := F₀  // F₀ is formula to be certified as UNSAT
  forward_phase
  for parallel 1..N do
    backward_phase(copy(F))
  collect and write out certificate
```

The forward phase is similar to forward checking, but does not verify the lemmas. The backward phase iterates over the certificate in reverse order, undoes the effects of the items, and verifies the lemmas. However, only *marked* lemmas are actually verified. Lemmas are marked by unit propagation, if they are required to produce a conflict. This way, lemmas not required for any conflict need not be verified nor included into the enriched certificate, which can speed up certificate generation and reduce the certificate size. Moreover, we implement core-first unit propagation, which prefers marked lemmas over unmarked ones, aiming at reducing the number of newly marked lemmas. While backwards checking and core-first unit propagation are already used in drat-trim, the distinguishing feature of gratgen is its parallel backward phase: Verification of the lemmas is distributed over multiple threads. Each thread has its own copy of the clause database and watch lists. The threads only synchronize to ensure that no lemma is processed twice (each lemma has an atomic flag, and only the thread that manages to acquire it will process the lemma), and to periodically exchange information on newly marked lemmas (using a spinlock protected global data structure).

We implemented gratgen in about 3k lines of heavily documented C++ code.

2.4   Checking GRAT Certificates

We have formalized GRAT certificate checking in Isabelle/HOL [12], and used program refinement techniques [8,10] to obtain an efficient verified implementation in Standard ML, for which we proved:

theorem verify_unsat_impl_correct:
  <DBi ↦ₐ DB>
    verify_unsat_impl DBi prf_next F_end it prf
  <λresult. DBi ↦ₐ DB * ↑(¬isl result ⟹ formula_unsat_spec DB F_end)>

This Hoare triple states that if DBi points to an integer array holding the elements DB, and we run verify_unsat_impl, the arwill be unchanged, and if the return value is no exception, the formula represented by the range 1...F_end in the array is unsatisfiable. For a detailed discussion of this correctness statement, we refer the reader to [7,9]. Similarly, we also defined and proved correct a `verify_sat_impl` function.

The gratchk tool contains the `verify_unsat_impl` and `verify_sat_impl` functions, a parser to read formulas into an array, and the logic to stream the proof file. As the correctness statement does not depend on the parameters `prf_next`, `prf`, and `it`, which are used for streaming and iterating over the lemmas, the parser is the only additional component that has to be trusted.

The formalization is about 12k lines of Isabelle/HOL text, and gratchk is 4k lines of Standard ML, of which 3.5k lines are generated from the formalization by Isabelle/HOL.

2.5 Novel Optimizations

Apart from multithreading, gratgen includes two key optimizations that make it faster than drat-trim, even in single-threaded mode: First, we implement core-first unit propagation by using separate watch lists for marked and unmarked clauses. Compared to the single watch list of drat-trim, our approach reduces the number of iterations over the watch lists in the inner loop of unit propagation, while requiring some more time for marking a lemma.

Second, if we encounter a run of RAT lemmas with the same pivot literal, we only collect the candidate clauses once for the whole run. As RAT lemmas tend to occur in runs, this approach saves a significant amount of time compared to drat-trim's recollection of candidates for each RAT lemma.

3   Benchmarks

We have benchmarked GRAT with one and eight threads against drat-trim and (incremental) LRAT [6] on the 110 problems from the 2016 SAT competition main track that CryptoMiniSat could prove unsatisfiable, and on the 128 problems that silver medalist Riss6 proved unsatisfiable. Although not among the Top 3 solvers, we included CryptoMiniSat because it seems to be the only prover that produces a significant amount of RAT lemmas.

Using a timeout of 20,000 s (the default for the 2016 SAT competition), single-threaded GRAT verified all certificates, while drat-trim and LRAT timed out on two certificate, and segfaulted on a third one. For fairness, we exclude these from the following figures: GRAT required 44 h, while drat-trim required 72 h and LRAT required 93 h. With 8 threads, GRAT ran out of memory for one certificate. For the remaining 234 certificates, the wall-clock times sum up to only 21 h.

The certificates from CryptoMiniSAT contain many RAT lemmas, and thanks to our RAT run optimization, we are more than two times faster than drat-trim, and three times faster than LRAT. (17 h/42 h/51 h) The certificates from Riss6 contain no RAT lemmas at all, and we are only slightly faster. (26 h/30 h/42 h) The scatter plot in Fig. 1 compares the wall-clock times for

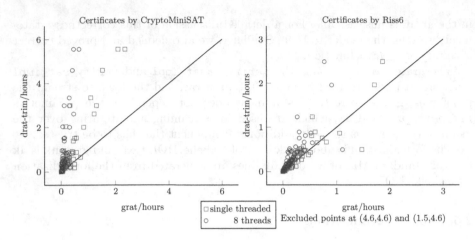

Fig. 1. Comparison of drat-trim and GRAT, ran on a server board with a 22-core XEON Broadwell CPU @ 2.2 GHz and 128 GiB RAM.

drat-trim against GRAT, differentiated by the SAT solver used to generate the certificates.

We also compare the memory consumption: In single threaded mode, gratgen needs roughly twice as much memory as drat-trim, with 8 threads, this figure increases to roughly 7 times more memory. Due to the garbage collection in Standard ML, we could not measure meaningful memory consumptions for gratchk. The extra memory in single-threaded mode is mostly due to the proof being stored in memory, the extra memory in multithreaded mode is due to the duplication of data for each thread.

The certificates for the 64 satisfiable problems that CryptoMiniSat solved at the 2016 SAT competition main track [13] have a size of 229 MiB and could be verified in 40 s.

4   Conclusion

We have presented a formally verified (un)satisfiability certificate checker, which is faster than the unverified state-of-the-art tool. An optional multithreaded mode makes it even faster, at the cost of using more memory. Our tool utilizes a two-phase approach: The highly optimized unverified gratgen tool produces an enriched certificate, which is then checked by the verified gratchk tool.

Future Work. We plan to reduce memory consumption by writing out the proof on the fly, and by sharing the clause database between threads. While the former optimization is straightforward, the latter has shown a significant decrease in performance in our initial experiments: Reordering of the literals in the clauses by moving watched literals to the front seems to have a positive effect on unit propagation, which we have not fully understood. However, when using a shared clause database, we cannot implement such a reordering, and the algorithm performs significantly more unit propagations before finding a conflict. Note that parallelization at the level of unit propagation is conjectured to be hard [5].

Acknowledgement. We thank Simon Wimmer for proofreading, and the anonymous reviewers for their useful comments.

References

1. Cruz-Filipe, L., Heule, M., Hunt, W., Kaufmann, M., Schneider-Kamp, P.: Efficient certified RAT verification. In: de Moura, L. (ed.) CADE 2017. LNCS, pp. 220–236. Springer, Cham (2017). doi:10.1007/978-3-319-63046-5_14
2. Cruz-Filipe, L., Marques-Silva, J., Schneider-Kamp, P.: Efficient certified resolution proof checking. In: Legay, A., Margaria, T. (eds.) TACAS 2017. LNCS, vol. 10205, pp. 118–135. Springer, Heidelberg (2017). doi:10.1007/978-3-662-54577-5_7
3. DRAT-trim homepage. https://www.cs.utexas.edu/marijn/drat-trim/
4. DRAT-trim issue tracker. https://github.com/marijnheule/drat-trim/issues
5. Hamadi, Y., Wintersteiger, C.M.: Seven challenges in parallel SAT solving. AI Mag. **34**(2), 99–106 (2013)
6. Heule, M., Hunt, W., Kaufmann, M., Wetzler, N.: Efficient, verified checking of propositional proofs. In: Proceeding of ITP. Springer (2017, To appear)
7. Lammich, P.: Gratchk proof outline. http://www21.in.tum.de/~lammich/grat/outline.pdf
8. Lammich, P.: Refinement to imperative/HOL. In: Urban, C., Zhang, X. (eds.) ITP 2015. LNCS, vol. 9236, pp. 253–269. Springer, Cham (2015). doi:10.1007/978-3-319-22102-1_17
9. Lammich, P.: Efficient verified (UN)SAT certificate checking. In Proceeding of CADE. Springer (2017, To appear)
10. Lammich, P., Tuerk, T.: Applying data refinement for monadic programs to hopcroft's algorithm. In: Beringer, L., Felty, A. (eds.) ITP 2012. LNCS, vol. 7406, pp. 166–182. Springer, Heidelberg (2012). doi:10.1007/978-3-642-32347-8_12
11. Moskewicz, M.W., Madigan, C.F., Zhao, Y., Zhang, L., Malik, S.: Chaff: Engineering an efficient sat solver. In Proceeding of DAC, pp. 530–535. ACM (2001)
12. Nipkow, T., Wenzel, M., Paulson, L.C. (eds.): Isabelle/HOL — A Proof Assistant for Higher-Order Logic. LNCS, vol. 2283. Springer, Heidelberg (2002)
13. SAT competition (2016). http://baldur.iti.kit.edu/sat-competition-2016/
14. Wetzler, N., Heule, M.J.H., Hunt, W.A.: Mechanical verification of SAT refutations with extended resolution. In: Blazy, S., Paulin-Mohring, C., Pichardie, D. (eds.) ITP 2013. LNCS, vol. 7998, pp. 229–244. Springer, Heidelberg (2013). doi:10.1007/978-3-642-39634-2_18
15. Wetzler, N., Heule, M.J.H., Hunt, W.A.: DRAT-trim: Efficient checking and trimming using expressive clausal proofs. In: Sinz, C., Egly, U. (eds.) SAT 2014. LNCS, vol. 8561, pp. 422–429. Springer, Cham (2014). doi:10.1007/978-3-319-09284-3_31

CNFgen: A Generator of Crafted Benchmarks

Massimo Lauria[1]([⊠]), Jan Elffers[2], Jakob Nordström[2], and Marc Vinyals[2]

[1] Università degli studi di Roma "La Sapienza", Rome, Italy
massimo.lauria@uniroma1.it
[2] KTH Royal Institute of Technology, 100 44 Stockholm, Sweden

Abstract. We present CNFgen, a generator of combinatorial benchmarks in DIMACS and OPB format. The proof complexity literature is a rich source not only of hard instances but also of instances that are theoretically easy but "extremal" in different ways, and therefore of potential interest in the context of SAT solving. Since most of these formulas appear not to be very well known in the SAT community, however, we propose CNFgen as a resource to make them readily available for solver development and evaluation. Many formulas studied in proof complexity are based on graphs, and CNFgen is also able to generate, parse and do basic manipulation of such objects. Furthermore, it includes a library cnfformula giving access to the functionality of CNFgen to Python programs.

1 Introduction

The Boolean satisfiability problem (SAT) is a foundational problem in computational complexity theory. It was the first problem proven NP-complete [21], and is widely believed to be completely infeasible to solve in the worst case—indeed, a popular starting point for many other impossibility results in computational complexity theory is the *Exponential Time Hypothesis (ETH)* [33] postulating that there are no subexponential-time algorithms for SAT.

From an applied perspective SAT looks very different, however. In the last 15–20 years there has been a dramatic increase in the performance of satisfiability algorithms, or *SAT solvers*, and so-called *conflict-driven clause learning (CDCL) solvers* [5,37,41] are now routinely used to solve real-world instances with hundreds of thousands or even millions of variables.

Surprisingly, although the performance of current state-of-the-art SAT solvers is very impressive indeed, our understanding of *why* they work so well (at least most of the time) leaves much to be desired. Essentially the only known rigorous method for analysing SAT solvers is to use tools from *proof complexity* [22] to study the potential and limitations of the methods of reasoning they use.

The basic CDCL algorithm searches for *resolution proofs* [12]. Some solvers such as *PolyBoRi* [14,15] use algebraic *Gröbner basis computations*, but it seems hard to make them competitive with resolution-based solvers. A compromise is to have *Gaussian elimination* inside a resolution-based solver as in [30,48].

Webpage: https://massimolauria.github.io/cnfgen/.

© Springer International Publishing AG 2017
S. Gaspers and T. Walsh (Eds.): SAT 2017, LNCS 10491, pp. 464–473, 2017.
DOI: 10.1007/978-3-319-66263-3_30

The power of these algebraic methods is captured by the *polynomial calculus (PC)* proof system [1, 20]. There are also *pseudo-Boolean solvers* such as [18, 24, 35, 47] exploring the geometric proof system *cutting planes (CP)* [23], although again it seems like a tough challenge to make these solvers as efficient as CDCL. We refer to the survey [42] and references therein for a more detailed discussion about the connections between proof complexity and SAT solving.

It seems fair to say that research in proof complexity into the proof systems mentioned above has not yielded too much by way of interesting insights for applied SAT solving so far. This is natural, since this research is driven mainly by theoretical concerns in computational complexity theory. However, what this body of work has produced is a wide selection of combinatorial formulas with interesting properties, and these we believe could be fruitfully mined for insights by SAT practitioners. As the SAT community starts to focus not only on producing blisteringly fast SAT solvers, but also on understanding better why these SAT solvers work the way they do, we expect that a study of combinatorial benchmarks could be particularly useful.

This immediately raises a question, however: Why do we need more crafted SAT problems? Is there really a need for more combinatorial benchmarks on top of what is already available in the standard SAT competition benchmarks?

We believe the answer is an emphatic "yes." In fact, it is our feeling that the SAT community has made quite limited use of crafted benchmarks so far. Most of these benchmarks are known to be dead hard for the resolution proof system, and will hence quickly grow out of reach of any CDCL solver (except if these solvers have dedicated preprocessing techniques to deal with such formulas, such as cardinality detection or Gaussian reasoning, but even then further minor tweaks to the benchmarks can easily make them infeasible).

This does not seem to be very informative—these benchmarks are hard simply because the method of reasoning employed by CDCL solvers cannot solve them efficiently in principle. A more interesting question is how well SAT solvers perform *when there are short proofs to be found*, and the solvers therefore have the potential to run fast. Studying solvers performance on such benchmarks can shed light on the quality of proof search, and indicate potential for improvement.

As a case in point, for the first time (to the best of our knowledge) many of the crafted benchmarks used in the *SAT Competition 2016* [4] (and generated by CNFgen) had the property that they possess extremely short resolution proofs and that SAT solvers can even be guided to find these proofs by, e.g., simply following a good fixed variable decision order. Yet the competition results showed that many of these benchmarks were beyond reach of even the best solvers.

It would seem that such formulas that are easy in theory for resolution but hard in practice for CDCL would merit further study if we want to understand what makes CDCL solvers fast and how they can be improved further, and CNFgen is a convenient tool for providing such formulas. An obvious downside is that such benchmarks can appear to be somewhat artificial in that one would not really run into them while solving applied problems. We readily concede this point. However, these formulas have the very attractive property that they can be scaled freely to yield instances of different sizes—as opposed to applied benchmarks, that typically exist for a fixed size—and running the solvers on instances from the same family while varying the instance size makes it possible to tease out the true asymptotic behaviour.

By judiciously choosing formulas with different theoretical properties one can "stress-test" CDCL solvers on memory management (using formulas with size-space trade-off properties), restart policy (for formulas that are hard for strict subsystems of resolution), decision heuristic (for formulas that are easy with a good fixed variable order), et cetera, as done, e.g., in [26,34].

Furthermore, even theoretically hard crafted benchmarks can yield interesting insights in that they can be used to compare SAT solvers based on different methods of reasoning, for instance by benchmarking CDCL against algebraic solvers on formulas that are hard for resolution but easy for algebraic methods of reasoning, or against pseudo-Boolean solvers on formulas easy for cutting planes. CNFgen has been heavily used in work on analysing pseudo-Boolean solvers [25,52], which has so far generated quite intriguing and counter-intuitive results. (In particular, state-of-the-art pseudo-Boolean solvers sometimes struggle hopelessly with instances that are dead *easy* for the cutting planes method which they use to search for proofs, as also confirmed by benchmarks submitted to the *Pseudo-Boolean Competition 2016* [43].)

The CNFgen tool generates all of the CNF formulas discussed above in the standard DIMACS and OPB formats, thus making these benchmarks accessible to the applied SAT community. The included Python library allows formulas construction and manipulation, useful when encoding problems in SAT.

In Sect. 2 we present a small selection of the benchmarks in CNFgen and in Sect. 3 we illustrate some of its features. Concluding remarks are in Sect. 4.

2   Some Formula Families in CNFgen

A formula generator is a Python function that outputs a CNF, given parameters. A CNF is represented in our cnfformula library as a sequence of constrains (e.g., clauses, linear constraints, ...) defined over a set of named variables. CNFgen command line tool is essentially a wrapper around the available generators and the others CNF manipulation and SAT solving utilities in cnfformula.

Let us now describe briefly some examples of formulas available in CNFgen. Due to space constraints we are very far from giving a full list, and since new features are continuously being added such a list would soon be incomplete anyway. Typing cnfgen --help shows the full list of available formulas. The command cnfgen <name> <params> generates a formula from the family <name>, where the descriptions of the parameters needed is shown by cnfgen <name> --help.

Pigeonhole principle formulas (php) claim that m pigeons can be placed in n separate holes, where the variable $x_{i,j}$ encodes that pigeon i flies to hole j and the indices range over all $i \in [m]$ and $j \in [n]$ below. *Pigeon clauses* $\bigvee_{j=1}^{n} x_{i,j}$ enforce that every pigeon goes to a hole, and *hole clauses* $\overline{x}_{i,j} \vee \overline{x}_{i',j}$ for $i < i'$

forbid collisions. One can optionally include *functionality clauses* $\overline{x}_{i,j} \vee \overline{x}_{i,j'}$ for $j < j'$ and/or *onto clauses* $\bigvee_{i=1}^{m} x_{i,j}$ specifying that the mapping is one-to-one and onto, respectively. PHP formulas are unsatisfiable if and only if $m > n$ and if so require exponentially long proofs for all variants in resolution [29,44]. Functional onto-PHP formulas are easy for polynomial calculus (PC) but the other versions are hard (at least for a linear number of pigeons $m = O(n)$) [40]. All versions are easy for cutting planes (CP).

Tseitin formulas (`tseitin`) encode linear equation systems over $GF(2)$ generated from connected graphs $G = (V, E)$ with *charge function* $\chi : V \to \{0, 1\}$. Edges $e \in E$ are identified with variables x_e, and for every vertex $v \in V$ we have the equation $\sum_{e \ni v} x_e \equiv \chi(v) \pmod 2$ encoded in CNF, yielding an unsatisfiable formula if and only if $\sum_{v \in V} \chi(v) \not\equiv 0 \pmod 2$. When G has bounded degree and is well-connected, the formula is hard for resolution [50] and for PC over fields of characteristic distinct from 2 [16], but is obviously easy if one can do Gaussian elimination (as in PC over $GF(2)$). Such Tseitin formulas are also believed to be hard for CP, but this is a major open problem in proof complexity. For long, narrow grid graphs, Tseitin formulas exhibit strong time-space trade-offs for resolution and PC [6,7].

Ordering principle formulas (`op`) assert that there is a partial ordering $\preceq$ of the finite set $\{e_1, \ldots, e_n\}$ so that no element is minimal, where variables $x_{i,j}$, $i \neq j \in [n]$, encode $e_i \preceq e_j$. Clauses $\overline{x}_{i,j} \vee \overline{x}_{j,i}$ and $\overline{x}_{i,j} \vee \overline{x}_{j,k} \vee x_{i,k}$ for distinct $i, j, k \in [n]$ enforce asymmetry and transitivity, and the non-minimality claim is encoded as clauses $\bigvee_{i \in [n] \setminus \{j\}} x_{i,j}$ for every $j \in [n]$. The **total ordering principle** also includes clauses $x_{i,j} \vee x_{j,i}$ specifying that the order is total. The **graph ordering principle (`gop`)** is a "sparse version" where the non-minimality of e_j must be witnessed by a neighbour e_i in a given graph (which for the standard version is the complete graph). For well-connected graphs these formulas are hard for DPLL but easy for resolution [13,49]. If the well-connected graphs are sparse, so that all initial clauses have bounded size, the formulas have the interesting property that any resolution or PC proof must still contain clauses/polynomials of large size/degree [13,27].

Random k-CNF formulas (`randkcnf`) with m clauses over n variables are generated by randomly picking m out of the $2^k \binom{n}{k}$ possible k-literal clauses without replacement. These formulas are unsatisfiable with high probability for $m = \Delta_k \cdot n$ with Δ_k a large enough constant depending on k, where $\Delta_2 = 1$ (provably) and $\Delta_3 \approx 4.26$ (empirically). Random k-CNFs for $k \geq 3$ are hard for resolution and PC [3,19] and most likely also for CP, although this is again a longstanding open problem.

Pebbling formulas (`peb`) are defined in terms of directed acyclic graphs (DAGs) $G = (V, E)$, with vertices $v \in V$ identified with variables x_v, and contain clauses saying that (a) source vertices s are true (a unit clause x_s) and

(b) truth propagates through the DAG (clauses $\bigvee_{i=1}^{\ell} \overline{x}_{u_i} \vee x_v$ for each non-source v with predecessors $u_1, \ldots, u_\ell$) but (c) sinks z are false (a unit clause $\overline{x}_z$). Pebbling formulas are trivially refuted by unit propagation, but combined with transformations as described in Sect. 3 they have been used to prove time-space trade-offs for resolution, PC, and CP [7–9,32] and have also been investigated from an empirical point of view in [34].

Stone formulas (`stone`) are similar to pebbling formulas, but here each vertex of the DAG contains a stone, where (a) stones on sources are red and (b) a non-source with all predecessors red also has a red stone, but (c) sinks have blue stones. This unsatisfiable formula has been used to separate general resolution from so-called *regular* resolution [2] and has also been investigated when comparing the power of resolution and CDCL without restarts [17].

k-clique formulas (`kclique`) declare that a given graph $G = (V, E)$ has a k-clique. Variables $x_{i,v}$, $i \in [k]$, $v \in V$, constrained by $\sum_{v \in V} x_{i,v} = 1$ identify k vertices, and clauses $\overline{x}_{i,u} \vee \overline{x}_{j,v}$ for every non-edge $\{u, v\} \notin E$ and $i \neq j \in [k]$ enforce that these vertices form a clique. For k constant it seems plausible that their proof length should scale roughly like $|V|^k$ in the worst case but this remains wide open even for resolution and only partial results are known [10,11].

Subset cardinality formulas (`subsetcard`). For a $0/1$ $n \times n$ matrix $A = (a_{i,j})$, identify positions where $a_{i,j} = 1$ with variables $x_{i,j}$. Letting $R_i = \{j \mid a_{i,j} = 1\}$ and $C_j = \{i \mid a_{i,j} = 1\}$ record the positions of 1s/variables in row i and column j, the formula encodes the cardinality constraints $\sum_{j \in R_i} x_{i,j} \geq |R_i|/2$ and $\sum_{i \in C_j} x_{i,j} \leq |C_i|/2$ for all $i, j \in [n]$. In the case when all rows and columns have $2k$ variables, except for one row and column that have $2k + 1$ variables, the formula is unsatisfiable but is hard for resolution and polynomial calculus if the positions of the variables are "scattered enough" (such as when M is the bipartite adjacency matrix of an expander graph) [39,51]. Cutting planes, however, can just add up all constraints to derive a contradiction immediately.

Even colouring formulas (`ec`) are defined on connected graphs $G = (V, E)$ with all vertices having bounded, even degree. Edges $e \in E$ correspond to variables x_e, and for all vertices $v \in V$ constraints $\sum_{e \ni v} x_e = \deg(v)/2$ assert that there is a $0/1$-colouring such that each vertex has an equal number of incident 0- and 1-edges. The formula is satisfiable if and only if the total number of edges is even. For suitably chosen graphs these formulas are empirically hard for CDCL [36], but we do not know of any formal resolution lower bounds. Despite being easy for CP, they still seem hard for pseudo-Boolean solvers.

3   Further Tools for CNF Generation and Manipulation

Formula transformations. A common trick to obtain hard proof complexity benchmarks is to take a CNF formula and replace each variable x by a Boolean function $g(x_1, \ldots, x_\ell)$ of arity ℓ over new variables. As an example, XOR substitution $y \leftarrow y_1 \oplus y_2$, $z \leftarrow z_1 \oplus z_2$ applied to the clause $y \vee \overline{z}$ yields

$$(y_1 \vee y_2 \vee z_1 \vee \overline{z}_2) \wedge (y_1 \vee y_2 \vee \overline{z}_1 \vee z_2) \wedge (\overline{y}_1 \vee \overline{y}_2 \vee z_1 \vee \overline{z}_2) \wedge (\overline{y}_1 \vee \overline{y}_2 \vee \overline{z}_1 \vee z_2) .$$

Note that such transformations can dramatically increase formula size, and so they work best when the size of the initial clauses and the arity ℓ is small. Similar substitutions, and also other transformations such as *lifting*, *shuffling*, and *variable compression* from [45], can be applied either in CNFgen during formula generation (using command line options -T), or alternatively to a DIMACS file using the included cnftransform program. Multiple occurrences of -T <params> results in a chain of transformations as in, e.g., this 2-xorified pebbling formula over the pyramid graph of height 10, with random shuffling.

```
$ cnfgen peb --pyramid 10   -T xor 2 -T shuffle
```

Formulas based on graphs. Many formulas in CNFgen are generated from graphs, which can be either read from a file or produced internally by the tool. In the next example we build a Tseitin formula over the graph in the file G.gml and then a graph ordering principle on a random 3-regular graph with 10 vertices.

```
$ cnfgen tseitin -i G.gml --charge randomodd | minisat
UNSATISFIABLE
$ cnfgen gop --gnd 10 3 | minisat
UNSATISFIABLE
```

The CNFgen command line provides some basic graph constructions and also accepts graphs in different formats such as, e.g., Dot [46], DIMACS [38], and GML [31]. Inside Python there is more flexibility since any NetworkX [28] graph object can be used, as sketched in the next example.

```
from cnfformula import GraphColoringFormula
G= ...   # build the graph
GraphColoringFormula(G,4).dimacs() # Is G is 4-colourable?
```

As already discussed in Sect. 2, the hardness of many formulas generated from graphs are governed by (different but related notions of) *graph expansion*. Going into details is beyond the scope of this paper, but in many cases a randomly sampled regular graph of bounded vertex degree almost surely has the expansion required to yield hard instances.

OPB output format. CNFgen supports the OPB format used by pseudo-Boolean solvers, which use techniques based on cutting planes. CNFgen can produce formulas that are easy for cutting planes but seem quite hard for pseudo-Boolean solvers (e.g., subset cardinality formulas, even colouring formulas, some kinds of k-colouring instances).

4   Concluding Remarks

We propose CNFgen as a convenient tool for generating crafted benchmarks in DIMACS or OPB. CNFgen makes available a rich selection of formulas appearing in the proof complexity literature, and new formulas can easily be added by using the cnfformula library. It is our hope that this tool can serve as something of a one-stop shop for, e.g., SAT practitioners wanting to benchmark their solvers on tricky combinatorial formulas, competition organizers looking for crafted instances, proof complexity researchers wanting to test theoretical predictions against actual experimental results, and mathematicians performing theoretical research by reducing to SAT.

Acknowledgments. The first author performed most of this work while at KTH Royal Institute of Technology. The authors were funded by the European Research Council under the European Union's Seventh Framework Programme (FP7/2007–2013) / ERC grant agreement no. 279611 as well as by Swedish Research Council grant 621-2012-5645. The first author was also supported by the European Research Council under the European Union's Horizon 2020 Research and Innovation Programme / ERC grant agreement no. 648276 AUTAR.

References

1. Alekhnovich, M., Ben-Sasson, E., Alexander, A., Razborov, A.A., Wigderson, A.: Space complexity in propositional calculus. SIAM J. Comput. **31**(4), 1184–1211 (2002). Preliminary version in STOC '00
2. Alekhnovich, M., Johannsen, J., Pitassi, T., Urquhart, A.: An exponential separation between regular and general resolution. Theor. Comput. **3**(5), 81–102 (2007). Preliminary version in STOC '02
3. Alekhnovich, M., Alexander, A. Razborov, A.A.: Lower bounds for polynomial calculus: Non-binomial case. In: Proceedings of the Steklov Institute of Mathematics, 242, 18–35 (2003). http://people.cs.uchicago.edu/~razborov/files/misha.pdf. Preliminary version in FOCS '01
4. Balyo, T., Marijn, J., Heule, H., Järvisalo, M.: Proceedings of SAT competition 2016: Solver and benchmark descriptions. Technical report B-2016-1, University of Helsinki (2016). http://hdl.handle.net/10138/164630
5. Roberto, J., Bayardo, Jr., Schrag, R.: Using CSP look-back techniques to solve real-world SAT instances. In Proceedings of the 14th National Conference on Artificial Intelligence (AAAI 1997), pp. 203–208 (1997)
6. Beame, P., Beck, C., Impagliazzo, R.: Time-space tradeoffs in resolution: Super-polynomial lower bounds for superlinear space. In: Proceedings of the 44th Annual ACM Symposium on Theory of Computing (STOC 2012), pp. 213–232 (2012)
7. Beck, C., Nordström, J., Tang, B.: Some trade-off results for polynomial calculus. In: Proceedings of the 45th Annual ACM Symposium on Theory of Computing (STOC 2013), pp. 813–822 (2013)
8. Ben-Sasson, E., Nordström, J.: Short proofs may be spacious: An optimal separation of space and length in resolution. In: Proceedings of the 49th Annual IEEE Symposium on Foundations of Computer Science (FOCS 2008), pp. 709–718 (2008)

9. Ben-Sasson, E., Nordström, J.: Understanding space in proof complexity: Separations and trade-offs via substitutions. In: Proceedings of the 2nd Symposium on Innovations in Computer Science (ICS 2011), pp. 401–416 (2011)

10. Beyersdorff, O., Galesi, N., Lauria, M.: Parameterized complexity of DPLL search procedures. ACM Trans. Comput. Logic **14**(3), 20:1–20:21 (2013). Preliminary version in SAT '11

11. Beyersdorff, O., Galesi, N., Lauria, M., Razborov, A.A.: Parameterized bounded-depth Frege is not optimal. ACM Trans. Comput. Theor. **4**, 7:1–7:16 (2012). Preliminary version in ICALP '11

12. Blake, A.: Canonical expressions in boolean algebra. PhD thesis, University of Chicago (1937)

13. Bonet, M.L., Galesi, N.: Optimality of size-width tradeoffs for resolution. Comput. Complex. **10**(4), 261–276 (2001). Preliminary version in FOCS '99

14. Brickenstein, M., Dreyer, A.: PolyBoRi: A framework for Gröbner-basis computations with Boolean polynomials. J. Symb. Comput. **44**(9), 1326–1345 (2009)

15. Brickenstein, M., Dreyer, A., Greuel, G.-M., Wedler, M., Wienand, O.: New developments in the theory of Gröbner bases and applications to formal verification. J. Pure Appl. Algebr. **213**(8), 1612–1635 (2009)

16. Buss, S.R., Grigoriev, D., Impagliazzo, R., Pitassi, T.: Linear gaps between degrees for the polynomial calculus modulo distinct primes. J. Comput. Syst. Sci. **62**(2), 267–289 (2001). Preliminary version in CCC '99

17. Buss, S.R., Kołodziejczyk, L.: Small stone in pool. Logic. Method. Comput. Sci. **10**, 10:16–16:22 (2014)

18. Chai, D., Kuehlmann, A.: A fast pseudo-Boolean constraint solver. IEEE Trans. Comput. Aided Des. Integr. Circuits Syst. **24**(3), 305–317 (2005). Preliminary version in DAC '03

19. Chvátal, V., Szemerédi, E.: Many hard examples for resolution. J. ACM **35**(4), 759–768 (1988)

20. Clegg, M., Edmonds, J., Impagliazzo, R.: Using the Groebner basis algorithm to find proofs of unsatisfiability. In: Proceedings of the 28th Annual ACM Symposium on Theory of Computing (STOC 1996), pp. 174–183 (1996)

21. Cook, S.A.: The complexity of theorem-proving procedures. In: Proceedings of the 3rd Annual ACM Symposium on Theory of Computing (STOC 1971), pp. 151–158 (1971)

22. Cook, S.A., Reckhow, R.: The relative efficiency of propositional proof systems. J. Symbol. Logic **44**(1), 36–50 (1979)

23. Cook, W.: Collette Rene Coullard, and György Turán. On the complexity of cutting-plane proofs. Discr. Appl. Math. **18**(1), 25–38 (1987)

24. Dixon, H.E., Ginsberg, M.L., Hofer, D.K., Luks, E.M., Parkes, A.J.: Generalizing Boolean satisfiability III: Implementation. J. Artif. Intell. Res. **23**, 441–531 (2005)

25. Elffers, J., Giráldez-Crú, J., Nordström, J., Vinyals, M.: Using combinatorial benchmarks to probe the reasoning power of pseudo-Boolean solvers (2017, Submitted)

26. Elffers, J., Nordström, J., Simon, L., Sakallah, K.A.: Seeking practical CDCL insights from theoretical SAT benchmarks. In: Presentation at the Pragmatics of SAT 2016 workshop (2016). http://www.csc.kth.se/~jakobn/research/TalkPoS16.pdf

27. Galesi, N., Lauria, M.: Optimality of size-degree trade-offs for polynomial calculus. ACM Trans. Comput. Logic **12**, 4:1–4:22 (2010)

28. Hagberg, A.A., Schult, D., Swart, P.S.: Exploring network structure, dynamics, and function using NetworkX. In: Proceedings of the 7th Python in Science Conference (SciPy2008), Pasadena, CA USA, pp. 11–15 (2008)
29. Haken, A.: The intractability of resolution. Theor. Comput. Sci. **39**(2–3), 297–308 (1985)
30. Heule, M., van Maaren, H.: Aligning CNF- and equivalence-reasoning. In: Hoos, H.H., Mitchell, D.G. (eds.) SAT 2004. LNCS, vol. 3542, pp. 145–156. Springer, Heidelberg (2005). doi:10.1007/11527695_12
31. Himsolt, M.: GML: A portable graph file format. Technical report, Universität of Passau (1996)
32. Huynh, T., Nordström, J.: On the virtue of succinct proofs: Amplifying communication complexity hardness to time-space trade-offs in proof complexity (Extended abstract). In: Proceedings of the 44th Annual ACM Symposium on Theory of Computing (STOC 2012), pp. 233–248 (2012)
33. Impagliazzo, R., Paturi, R.: On the complexity of k-SAT. J. Comput. Syst. Sci. **62**(2), 367–375 (2001). Preliminary version in CCC '99
34. Järvisalo, M., Matsliah, A., Nordström, J., Živný, S.: Relating proof complexity measures and practical hardness of SAT. In: Milano, M. (ed.) CP 2012. LNCS, pp. 316–331. Springer, Heidelberg (2012). doi:10.1007/978-3-642-33558-7_25
35. Le Berre, D., Parrain, A.: The Sat4j library, release 2.2. J. Satisf. Boolean Model. Comput. **7**, 59–64 (2010)
36. Markström, K.: Locality and hard SAT-instances. J. Satisf. Boolean Model. Comput. **2**(1–4), 221–227 (2006)
37. Marques-Silva, J.P., Sakallah, K.A.: GRASP: A search algorithm for propositional satisfiability. IEEE Trans. Comput. **48**(5), 506–521 (1999). Preliminary version in ICCAD '96
38. Massey, B.: DIMACS graph format (2001). http://prolland.free.fr/works/research/dsat/dimacs.html. Accessed 11 Feb 2016
39. Mikša, M., Nordström, J.: Long proofs of (seemingly) simple formulas. In: Sinz, C., Egly, U. (eds.) SAT 2014. LNCS, vol. 8561, pp. 121–137. Springer, Cham (2014). doi:10.1007/978-3-319-09284-3_10
40. Mikša, M., Nordström, J.: A generalized method for proving polynomial calculus degree lower bounds. In: Proceedings of the 30th Annual Computational Complexity Conference (CCC 2015). Leibniz International Proceedings in Informatics (LIPIcs), vol. 33, pp. 467–487 (2015)
41. Moskewicz, M.W., Madigan, C.F., Zhao, Y., Zhang, L., Chaff, M.S.: Engineering an efficient SAT solver. In: Proceedings of the 38th Design Automation Conference (DAC 2001), pp. 530–535 (2001)
42. Nordström, J.: On the interplay between proof complexity and SAT solving. ACM SIGLOG News **2**(3), 19–44 (2015)
43. Pseudo-Boolean competition (2016). http://www.cril.univ-artois.fr/PB16/
44. Razborov, A.A.: Resolution lower bounds for perfect matching principles. J. Comput. Syst. Sci. **69**(1), 3–27 (2004). Preliminary version in CCC '02
45. Razborov, A.A.: A new kind of tradeoffs in propositional proof complexity. J. ACM **63**, 16:1–16:14 (2016)
46. AT and T Research: Dot Language. http://www.graphviz.org/content/dot-language. Accessed 11 Feb 2016
47. Sheini, H.M., Sakallah, K.A.: Pueblo: a hybrid pseudo-Boolean SAT solver. J. Satisf. Boolean Model. Comput. **2**(1–4), 165–189 (2006). Preliminary version in DATE '05

48. Soos, M., Nohl, K., Castelluccia, C.: Extending SAT solvers to cryptographic problems. In: Kullmann, O. (ed.) SAT 2009. LNCS, vol. 5584, pp. 244–257. Springer, Heidelberg (2009). doi:10.1007/978-3-642-02777-2_24
49. Stålmarck, G.: Short resolution proofs for a sequence of tricky formulas. Acta Inform. **33**(3), 277–280 (1996)
50. Urquhart, A.: Hard examples for resolution. J. ACM **34**(1), 209–219 (1987)
51. Gelder, A., Spence, I.: Zero-one designs produce small hard SAT instances. In: Strichman, O., Szeider, S. (eds.) SAT 2010. LNCS, vol. 6175, pp. 388–397. Springer, Heidelberg (2010). doi:10.1007/978-3-642-14186-7_37
52. Vinyals, M., Elffers, J., Giráldez-Crú, J., Gocht, S., Nordström, J.: In between resolution and cutting planes: A study of proof systems for pseudo-Boolean SAT solving (2017, Submitted)

Author Index

Printed in the United States
By Bookmasters